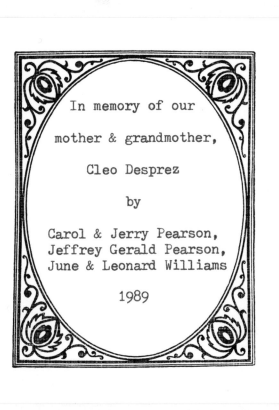

In memory of our

mother & grandmother,

Cleo Desprez

by

Carol & Jerry Pearson,
Jeffrey Gerald Pearson,
June & Leonard Williams

1989

DICTIONARY
OF THE
ENVIRONMENT

DICTIONARY OF THE ENVIRONMENT
THIRD EDITION

Michael Allaby

NEW YORK UNIVERSITY PRESS
Washington Square, New York

Copyright © 1977, 1983, 1989 by Michael Allaby
All rights reserved
Printed in Hong Kong

Published in the U.S.A. in 1989 by
NEW YORK UNIVERSITY PRESS
Washington Square, New York, NY 10003

LIBRARY OF CONGRESS
Library of Congress Cataloging-in-Publication Data

Allaby, Michael
  Dictionary of the environment/Michael Allaby — 3rd ed.
  p.   cm.
  ISBN 0-8147-0591-X
  1. Ecology—Dictionaries. 2. Natural history—Dictionaries.
  I. Title.
QH540.4.A44 1989
574.5'03'21—dc19   88-19184   CIP

PREFACE TO THE THIRD EDITION

It is now five years since the Second Edition of the *Dictionary of the Environment* appeared, and more than a decade since the publication of the First Edition. The format has been changed for the present edition but, a mark of the times perhaps, the entire text has been transferred to computer disks. This has meant the entire content of the book had to be retyped. The retyping has made possible a fresh and very thorough revision.

Every entry has been examined critically and every cross-reference has been checked. Where necessary entries have been updated and some have been reworded in order to simplify them.

When the First Edition was being prepared it was decided not to include internal cross-references. As entries accumulated terms used in them that I felt required explanation were explained either within the entries where they were used, or as the subject of entries in their own right, but q.v. was used as an internal referencing system on the index cards which preceded computerization. Eventually, the internal references became so numerous as to occupy a significant amount of space and the proliferation of q.v. threatened to interrupt the flow of entries, making them difficult to read. So at the final typing it was decided to abandon cross-references within entries. Now they have been restored, but more neatly.

Many new entries have been added, and one or two which appeared in the First Edition, but were dropped from the Second, have been returned. These concern topics which were current in the 1970s, seemed to be less so in the early 1980s, but which are now appearing once again in the newspapers.

In recent years, as environmental concerns have come to attract more urgent attention from governments and international institutions, the number of relevant agencies and regulations has increased. Students and supporters of voluntary conservation and environmental organizations may need such rudimentary information about these as dictionary entries can provide, and I have tried at least to list the more important ones and explain their policy or purpose. I have also added the names and objects of a number of non-governmental organizations.

Continuing environmentalist worries about nuclear power have led me to expand the number of entries on that broad topic, and I have included definitions of the SI units which are coming into use, relating them to the more familiar units they replace. New causes of concern, such as acid rain, or ones whose importance has increased in recent years, such as the greenhouse effect, are covered in new or expanded entries.

Environmental disasters continue to occur, and this edition of the Dictionary includes those that have happened since the last revision. However, by listing them separately, each under its own name, inevitably they are scattered and I felt it might be useful to group them together, the better to compare them. So, in addition to the entry allotted to each incident there is a table, under the heading ENVIRONMENTAL DISASTERS, listing all of them very briefly. For this revision I have not approached those who contributed entries to the earlier editions. The revisions are my own, and I must bear responsibility for any errors I may have introduced. Nevertheless, many of the definitions included here are essentially those contributed and verified by Ailsa Allaby, Dr G. Browning, Dr M.D. Hooper, John Macadam, Margaret Palmer, Professor F. Roberts, Professor R.S. Scorer, and Professor E.K. Walton, to whom I remain deeply grateful.

Michael Allaby
Wadebridge, Cornwall
November 1987

A

A. *See* AMPERE.

a. *See* ATTO-.

aa. (1) In many parts of Europe, a small river, the word (derived originally from the Latin *aqua* meaning water) often forming part or all of the name. (2) In volcanology, a Hawaiian term describing a BASALTIC LAVA with a rough, blocky surface, often covered with clinker and formed by rivers of lava that may overflow. *See also* PAHOEHOE.

abacá (Manila hemp, *Musa textilis*). A plant grown mainly in the Philippines; the toughest of all natural fibres. The leaf stalks are used to make ropes. Abacá is resistant to salt water. *See also* MUSA.

abaxial. The surface of a leaf that faces away from the stem (i.e. the dorsal surface). *Compare* ADAXIAL.

abiocoen. All non-living components of the environment. *Compare* BIOCOEN.

abioseston. The non-living matter floating in water. *Compare* BIOSESTON. *See also* SESTON.

abiotic. Non-biological. *Compare* BIOTIC.

ablation. The removal of a surface layer. The term is applied especially to the melting and evaporation of the surface of ice and to the removal of loose surface material by the wind (i.e. deflation).

abscisin (abscisic acid, dormin). An AUXIN that induces leaf fall and dormancy in seeds and buds, probably by inhibiting the synthesis of nucleic acid (*see* DNA, RNA) and protein.

absolute age. The age of rock, mineral or fossil in years, determined as a RADIOMETRIC AGE or by counting VARVES. Radiometric dating involves experimental errors, so such dates are usually quoted with a plus or minus error.

absolute humidity (humidity mixing rate). The amount of water present in a unit mass of air, usually expressed as grams of water per kilogram of air.

absorbate. *See* ABSORPTION.

absorbent. *See* ABSORPTION.

absorbing duct. The tube used in a ventilator to attenuate sound waves while offering low resistance to a continuous flow of air.

absorption. (1) A process in which one material (the absorbent) takes up and retains another (the absorbate) to form an homogeneous solution. (2) The process by which substances become attached to a solid surface by physical forces (*see* ADSORPTION) (e.g., absorption of sulphur dioxide by stone, vegetation, particulate AEROSOLS, etc.). (3) The transfer of energy from radiation passing through the atmosphere to a substance such as aerosols or to a gaseous atmospheric component (e.g., of ULTRAVIOLET RADIATION by ozone or of infrared radiation by carbon dioxide or water vapour). Absorption also occurs in the ocean. The absorption of gases by plants depends on the state of the vegetation, humidity, temperature and various physical laws. The absorption of light by water may be expressed as the path length in which the intensity is reduced by the factor *e* (approximately 2.73) or by the reduction of intensity per unit of path length.

absorption coefficient. In acoustics, if a surface is exposed to a field of sound, the ratio of the sound energy absorbed by the surface to the total sound energy that strikes it. An absorption coefficient of 1 means that all of the sound energy is absorbed. *See also* ANECHOIC.

absorption tower. A structure, most commonly found in chemical factories, in which a liquid is made to absorb a gas (e.g., in the production of sulphuric acid from sulphur dioxide/sulphur trioxide and water).

absorptive capacity (assimilative capacity). A measure of the amount of waste that can be deposited in a particular environment without causing adverse ecological or aesthetic change.

abstractive use. Of water, a use which removes it so that it is lost temporarily as a resource (e.g., in a COOLING TOWER). *Compare* NON-ABSTRACTIVE USE.

abyssal. Very deep; applied to the sea bed at water depths greater than about 2000 metres. The term may also be applied to the zone in lakes below the depth of effective (i.e. for photosynthesis) penetration of light. *Compare* BATHYAL. *See also* ABYSSOPELAGIC.

abyssal gap. A gap in a SILL or RIDGE or rise that separates two ABYSSAL PLAINS and through which the sea floor slopes from one plain to the other.

abyssal hill. A relatively small topographic feature of the deep ocean floor, ranging up to 1000 metres high and a few kilometres wide.

abyssal plain. A large, relatively flat area of the deep sea floor lying seaward of the CONTINENTAL SLOPE and CONTINENTAL RISE, where gradients become less than 1:1000.

abyssobenthos. An ocean floor at great depths. *See also* BENTHOS.

abyssopelagic. Applied to PELAGIC organisms living at water depths greater than about 3000 metres. *Compare* BATHYPELAGIC, EPIPELAGIC, MESOPELAGIC.

Acacia (wattles). A genus of leguminous (*see* LEGUMINOSAE) trees of the tropics and subtropics, especially Australasia, that may form the dominant vegetation in arid areas. Dyes, perfumes, timber and many other commercial products are derived from acacias, and some (e.g., *ACACIA ALBIDA*) might be exploited more as a species of great value to people living in arid regions.

Acacia albida. A leguminous (*see* LEGUMINOSAE) tree, native to semiarid Mediterranean regions, that bears its leaves during the dry season and is leafless throughout the rainy season. The leaves and pods (whose nutritional value is not reduced by drying) are palatable to livestock and the seeds (containing up to 27 percent crude protein) are also palatable to humans, usually mixed with meal.

Acacia harpophylla. *See* BRIGALOW FOREST.

Acanthaster planci. *See* CROWN OF THORNS STARFISH.

acanthite (Ag_2S). A major ore mineral of silver. It occurs in HYDROTHERMAL deposits, characteristically with lead, zinc and copper minerals, which also contain silver by atomic substitution. Acanthite also occurs in SUPERGENE deposits. Nearly all the silver produced is a by-product from mining for lead, zinc and copper. *Compare* ARGENTITE.

Acanthocephala (spiny-headed worms). A phylum of about 600 species of parasitic worms (*see* PARASITISM), with affinities to the NEMATODA. The common name refers to the spiny proboscis by means of which they attach themselves to their hosts. The larvae live in insects or CRUSTACEA, whereas the adults live in the gut of vertebrates, including mammals, where they can cause serious illness and sometimes death. An example is *Echynorhynchus proteus*, the adult of which lives in ducks and the larva in freshwater shrimps.

Acanthodii. An extinct group of small fishes, originating in the SILURIAN Period, that were the first vertebrates known to possess jaws. Their fins were supported by long spines.

acaricides. Chemicals (e.g., DERRIS and some ORGANOPHOSPHORUS PESTICIDES, DINITRO PESTICIDES and ORGANOCHLORINES) that are used to kill ticks and mites (i.e. ACARINA).

Acarina (Acarida; mites, ticks). An order of small ARACHNIDA with rounded bodies. Mites are very abundant in the soil, feeding on plant material and invertebrate animals. Some parasitic (*see* PARASITISM) mites (e.g., red spider) damage crops and can be serious pests. Others cause diseases in animals (e.g., mange). Ticks are blood-suckers, some being VECTORS of diseases such as Rocky Mountain spotted fever in humans, relapsing fever in humans and fowls, and louping ill in cattle and sheep.

acceleration. Rate of change of velocity with time. According to Newton's laws, acceleration x mass = force = rate of change of momentum. Momentum is a VECTOR, and motion in a curved path therefore requires the application of a force. In the atmosphere or ocean, vertical motion always requires horizontal acceleration, which results from BUOYANCY forces.

access agreement. As defined by the COUNTRYSIDE COMMISSION, in UK planning, an agreement allowing the public access to privately owned land, being OPEN COUNTRY suitable for open-air recreation.

access order. As defined by the COUNTRYSIDE COMMISSION, in UK planning, an order allowing the public access to privately owned land, being OPEN COUNTRY where an ACCESS AGREEMENT is impracticable or is not adequately securing public access to the land for open-air recreation.

accessory mineral. A mineral occurring in small amounts in a rock and disregarded in the classification of that rock. Accessory minerals can yield evidence about the origin of the rock. For example, the presence of metamorphic (*see* METAMORPHISM) minerals in a sandstone suggests a provenance, at least in part, from a metamorphic belt. *Compare* ESSENTIAL MINERAL.

accessory species. A species that occurs in one-fourth to one-half of a STAND. *Compare* ACCIDENTAL SPECIES.

accidental species. A species that occurs in less than one-fourth of a STAND. *Compare* ACCESSORY SPECIES.

accident scenario. A simulation of an imagined disaster (e.g., the catastrophic failure of a nuclear reactor, a major fire releasing toxic fumes, etc.) to test the response of emergency services and to help to estimate the extent of damage and injury. Such scenarios may be planned at any organizational level, but those involving major incidents with international implications commonly involve extensive international collaboration and the sharing of resulting data.

Accipitridae. *See* FALCONIFORMES.

acclimatization. The process of adapting to ABIOTIC environmental conditions, by phenotypic (*see* PHENOTYPE) rather than genetic variation.

accretion. The attachment of airborne material to fixed, falling or flying objects. Ice accretion occurs on wires, hailstones or aircraft wings when the air contains supercooled (*see* SUPERCOOLING) cloud droplets, DRIZZLE or rain, and is particularly dangerous on the rigging of ships in polar regions or on TV masts on hills in winter. Pollution accretion is exemplified by smoke deposition on window frames, ventilation intakes, etc., where the air flow is swift and curved.

Acer (maples). A genus of trees and shrubs found in temperate regions. They yield charcoal and timber, and *A. saccharum* is the source of maple sugar.

acetaldehyde (ethanal, CH_3CHO). A direct

oxidation product of ETHANOL (C_2H_5OH), made industrially from ethene (C_2H_4). It can be further oxidized to ACETIC ACID. It is an important raw material for certain organic compounds, has medical uses and, being very volatile, is used to make metaldehyde, which is used as solid pellets to fuel cooking stoves. It has the flavour of apple and is used as a food additive.

acetate film. Non-flammable cinema film based on cellulose acetate.

acetic acid (ethanoic acid, CH_3COOH). The acid in vinegar and an important industrial raw material, obtained by FERMENTATION from ethanol.

acetone (propanone, CH_3COCH_3). An important laboratory and industrial solvent, and raw material for making PLASTICS. It is miscible with water.

acetylcholine (ACh). A substance released in minute amounts at many nerve endings when impulses arrive, so transmitting the impulses to other nerve cells or effectors (e.g., muscles). Its effects disappear rapidly after secretion because it is destroyed by the enzyme cholinesterase.

acetylene (ethyne, C_2H_2). A colourless, poisonous, gaseous HYDROCARBON that can be prepared by the action of water on calcium carbide (CaC_2), although other methods are also used industrially. It is used for welding, the synthesis of ACETIC ACID and as a starting material for many chemicals (e.g., PVC, polyvinyl chloride).

ACh. See ACETYLCHOLINE.

achene. A dry, one-seeded fruit that does not split open (e.g., the fruit of the buttercup). Dispersal may be aided by wings (e.g., sycamore), plumes (e.g., old man's beard) or hooks (e.g., wood avens).

achira (Queensland arrowroot). The starchy root of *Canna edulis*, first domesticated in Peru before 2200 BC and still cultivated for human consumption. The tops are sometimes fed to cattle.

achondrite. A stony meteorite, without chondrules. *Compare* CHONDRITE.

acicular. Needle-shaped; applied especially to elongated crystals.

aciculilignosa. Needle-leafed forest and bush comprising evergreen coniferous vegetation.

acid. (1) In geology (*see* ACIDIC). (2) In chemistry (*see* pH). (3) *See* LYSERGIC ACID DIETHYLAMIDE.

acid dipping. The immersion of a metal object into a tank of suitable acid or acids to remove scale and clean the surface. The process often produces hazardous fumes and acid mists.

acid droplets. Minute liquid particles emitted by certain industrial processes that act as condensation nuclei (*see* CONDENSATION NUCLEUS). *See also* ACID RAIN, SULPHURIC ACID.

acidic. (1) Applied to IGNEOUS rocks containing more than a certain percentage (commonly set at 65 percent) of SILICA in their chemical composition. Most of the silica is in the form of silicate minerals (e.g., FELDSPAR, MICA, AMPHIBOLE), but the excess silica manifests itself in the presence of 10 percent or more free QUARTZ. GRANITE, RHYOLITE and OBSIDIAN are all acidic rocks. In petrology, acidic is contrasted with INTERMEDIATE, BASIC and ULTRABASIC, but not with ALKALINE. (2) In chemistry, describing an acid (*see* pH). *Compare* ALKALINE.

acidophile. *See* CALCIFUGE.

acid rain. Generally, PRECIPITATION in any form, or dry deposition, with a pH lower than would be expected from natural causes (most rain is slightly acid, with a pH of about 5). Acid rain was first reported in 1852, downwind from Manchester, England, but it emerged as a serious problem in the early 1970s first in Scandinavia, where poorly buffered (*see* BUFFER) lakes were affected, and later in central Europe, where forests were damaged. The term 'rain' is somewhat misleading, since mist and dry deposition are more injurious to vegetation than rain. The cause of acid rain is uncertain, but in some areas it is probably due to NITROGEN OXIDES, mainly from vehicle exhausts, leading to photochemical reactions yielding OZONE, in other areas to SULPHUR DIOXIDE, mainly from coal-fired power generation. Natural causes (e.g., prolonged dry weather, disease) may produce symptoms similar to those of acid rain damage, and high emissions of dimethyl sulphide from marine phytoplankton may also contribute sulphur, especially in southern Scandinavia.

acid refractory. Materials composed mainly of SILICA and designed to resist acid SLAGS. They are used to line furnaces. *See also* BESSEMER PROCESS.

acid soot (acid smut). Particles of carbon held together by water made acidic due to combination with SULPHUR TRIOXIDE. The carbon particles are emitted during combustion, the soot particles being roughly 1–3 millimetres in diameter. Where oil-burning installations have metal chimneys, acid soot can acquire iron sulphate, which produces brown stains on materials and damages paintwork. Acid soot emissions can be reduced by using low-sulphur fuels, by reducing the air flow to minimize sulphur trioxide formation, by making flues airtight, by insulating chimneys, by raising the temperature, etc.

acoustic. Applied to properties or characteristics connected with sound (e.g., the acoustic qualities of an auditorium). It is not used to refer to people, in which case the term is acoustical (e.g., acoustical engineer).

acoustical. *See* ACOUSTIC.

acoustic reflex. The mechanisms by which the mammalian EAR protects itself against sounds that are too loud, by adjusting the connecting muscles that regulate the relative positions of the auditory ossicles.

acquired character. A variation in an organism that appears as a response to environmental influence. *See also* LAMARCK, JEAN BAPTISTE DE.

acquired immune deficiency syndrome (AIDS). A condition in humans in which the immune system suffers a progressive failure, leaving the victim susceptible to opportunistic infections. It is caused by the human immunodeficiency virus (HIV), a slow-acting RETROVIRUS that invades and kills T_4 helper cells. These are integral to the immune system. AIDS is believed to have occurred first in the late 1950s and was identified as a distinct medical condition in the early 1980s. It is believed to have originated in Africa, probably by several mutations of a virus transmitted from green monkeys to humans in an area where green monkeys are eaten; within a few years further mutations produced a number of distinct viral strains. Estimates of the number of infected persons who will develop the full range of symptoms varies widely, but in the absence of an effective antiviral drug the great majority of those who develop symptoms will die. Some epidemiologists fear that by the end of this century the death toll in Africa will number tens of millions, in which case there is reason to fear severe disruption of development programmes, increasing poverty and consequent social and political unrest.

Acrania (Cephalochordata). A small subphylum of the CHORDATA comprising the living lancelets (AMPHIOXUS) and the extinct *Jaymoytius* which lived in the SILURIAN. Lancelets are small, fish-like, ciliary (*see* CILIA) feeders with poorly developed heads, no brain, bone or cartilage, and nephridia (*see* NEPHRIDIUM) as excretory organs. They may be similar to the ancestors of fish.

Acraniata. *See* INVERTEBRATA.

acre-foot. The volume of any substance that will cover 1 acre of a level surface to a depth of 1 foot (i.e. 43 560 cubic feet; in SI units, 1232.75 cubic metres).

Acrididae (short-horned grasshoppers). A

family of grasshoppers (order: ORTHOPTERA) whose antennae are shorter than their bodies. Some (e.g., the locust, *Locusta migratoria*), although commonly solitary, under certain conditions develop a gregarious and migratory form which causes incalculable harm to crops.

acrodont. Having teeth fused to the bones, a condition found, for example, in most bony fishes. *Compare* PLEURODONT, THECODONT.

acrosome. The projection on the head of a SPERMATOZOON that contains enzymes that play a part in the fusion of egg and sperm.

acrylic resins. A group of synthetic resins, obtained by the polymerization (*see* POLYMER) of monomers derived from acrylic acid (propenoic acid, $CH_2CHCOOH$). They are transparent, resistant to light, weak acids, alkalis and alcohols, but are attacked by oxidizing acids, ORGANOCHLORINES, KETONES and ESTERS. They are used widely. Acrilan and perspex are acrylic resins.

actinides (actinoids). The elements ranging in the periodic table from actinium ($A_r = 89$) to lawrencium ($A_r = 103$), and including: actinium (89), thorium (90), proactinium (91), uranium (92), neptunium (93), plutonium (94), americium (95), curium (96), berkelium (97), californium (98), einsteinium (99), fermium (100), mendelevium (101), nobelium (102), lawrencium (103). Actinium, thorium, proactinium and uranium occur naturally; the remainder may be produced in nuclear reactions. All are radioactive.

actinium. *See* ACTINIDES.

actinoids. *See* ACTINIDES.

actinomorphic (radially symmetrical). Applied to animals (e.g., CNIDARIA, ECHINODERMATA) and flowers (e.g., buttercup) that have more than one plane of symmetry. SESSILE animals are commonly actinomorphic. *Compare* BILATERALLY SYMMETRICAL.

Actinomycetales. An order of bacteria, mostly Gram-positive, that form fine filaments (*see* MYCELIUM). They are important constituents of soils, where they assist in the decomposition of organic matter, although a few are pathogens in mammals. *Streptomyces griseus* produces the antibiotic STREPTOMYCIN.

actinomycin. An antibiotic that blocks the synthesis of RNA by combining with DNA. It is produced by some species of the ACTINOMYCETALES.

Actinopterygii. A large subclass of the Osteichthyes that contains most of the modern bony fishes and many fossil forms, characterized by having the paired fins supported by horny fin rays, with no skeletal axis. *See also* CHOANICHTHYES, TELEOSTI.

Actinozoa (Anthozoa). A class of marine CNIDARIA that includes the sea anemones, stony corals, sea pens and 'dead men's fingers' (*Alcyonium digitatum*). Some species are solitary, some colonial. The medusa stage typical of other cnidarians is absent. *See also* HYDROZOA, SCYPHOZOA.

activated alumina. A granular, porous form of aluminium trioxide (alumina) capable of absorbing (*see* ABSORPTION) water, oil vapour or certain other substances from gases or liquids. It is used in pollution control, chromatographic analysis (*see* CHROMATOGRAPHY) and as a CATALYST.

activated carbon (activated charcoal). A form of carbon with a high adsorptive (*see* ADSORPTION) capacity for gases, vapours and colloidal solids (*see* COLLOID). It is made by heating carbon to 900°C with steam or carbon dioxide, which gives it a porous, particulate structure. It is used for odour, fume and other pollution control, and in gas masks.

activated carbon process. A Japanese process for removing SULPHUR DIOXIDE from flue gases. There are three versions: (a) water washing, in which the gas is absorbed on dry ACTIVATED CARBON and the carbon washed with water to give dilute sulphuric acid or GYPSUM; (b) gas DESORPTION in which the gas is absorbed dry and then desorbed to give sulphur dioxide; and (c) steam desorption in which the gas is absorbed dry, then desorbed to give sulphur dioxide.

activated charcoal. *See* ACTIVATED CARBON.

activated manganese oxide process. A Japanese process for removing SULPHUR DIOXIDE from flue gases by dry ABSORPTION to produce ammonium sulphate.

activated sludge. The active material, consisting largely of PROTOZOA and BACTERIA, used to purify sewage. When mixed with aerated sewage the organisms break down the organic matter that is present, using it as food, and multiply, so producing more activated sludge.

active factors. The factors that supply energy and nutrient for the active operation of natural processes in plants.

active transport. The passage, accompanied by the expenditure of energy, of a substance from a region of low concentration to one of high concentration (i.e. against the concentration gradient). This usually occurs across cell membranes.

activity. In ecology, the total flow of energy through a system in a unit of time.

actual vegetation. The vegetation that actually exists at the time of observation, regardless of the character, condition and stability of its constituent species.

Aculeata (ants, bees, wasps). A division of the HYMENOPTERA, most of whose members are parasitic. Gall wasps induce the formation of GALLS on oak and other plants. Many are PARASITOIDS (e.g., ichneumons), laying their eggs in the eggs, larvae or pupae of other insects, and thus play an important role in controlling pests, especially LEPIDOPTERA.

adamantine. *See* LUSTRE.

adaptation. (1) The fitness of a structure,

function or entire organism for life in a particular environment (e.g., the webbed feed of water birds); the process, brought about by natural selection, of becoming so fitted. (2) The modification of an organism in response to environmental conditions (e.g., an increase in specific enzyme production by bacteria in response to certain substances).   (3) A reduction in the excitability of a sense organ that is continuously stimulated. *See also* ADAPTIVE RADIATION.

adaptive radiation. The evolution from primitive stock of divergent forms, each adapted to survive under different conditions (e.g., on the Galapagos Islands, the 14 species of Darwin's finches, each with a different mode of life, must all have evolved from an ancestral species that colonized the islands from the mainland).

adaxial. The surface of a leaf that faces towards the stem (i.e. the upper side). *Compare* ABAXIAL.

additives (food additives). Substances that have no nutritive value in themselves (or are not being used as nutrients) which are added to food to preserve, colour or flavour it. They may be added during domestic cooking (e.g., salt, pepper, cochineal) or industrially. In most countries the use of additives by the food industry is confined to lists of permitted substances that have been in use for a long time without evidence of harmful effects on consumers or have been tested for safety, about half of which are synthesized chemically. In EEC countries approved additives in commercial use are given 'E' numbers. Some individuals are sensitive to particular additives. *See also* DELANEY CLAUSE.

adenine. One of the nitrogenous bases in DNA and RNA.

adenosine diphosphate (ADP). *See* ATP.

adenosine triphosphate. *See* ATP.

adiabatic. Occurring without a gain or loss of heat by the system involved.

adiabatic lapse rate (ALR). The rate of decrease of temperature with height of a parcel of air rising without exchange of heat (by mixing or conduction) with surrounding air, but taking at each height the ambient pressure. It is deduced from the equations of state, hydrostatic equilibrium and ADIABATIC change. If, in the adiabatic ascent, the parcel has the same temperature as the surroundings, no BUOYANCY force will result from vertical displacement, and the lapse rate in the surroundings is adiabatic and neutral. A larger or superadiabatic lapse rate is unstable, and a smaller one stable. The adiabatic lapse rate for unsaturated air, usually called the dry adiabatic lapse rate, and denoted by Υ has the value $(\Upsilon - 1)\, g/\Upsilon R \simeq 9.86$ °C/km, where Υ is the ratio of the specific heats of air ($\simeq 1.4$), g is gravity and R is the gas constant for air. *See also* WET ADIABATIC LAPSE RATE.

adipose tissue. Tissue composed of large cells containing stored fat, often found in CONNECTIVE TISSUE.

adit. In mining, a horizontal or nearly horizontal opening from the surface to the ore that is used for access or to drain the mine.

adjustment. The behavioural response of organisms to a change in environmental conditions.

adobe. In North America, fine ROCK FLOUR deposits produced by ice abrasion during recent glaciation and transported by wind to the site of its deposition. It is used for brick making, so the term has come to mean a sundried brick. *See also* RAMMED EARTH.

ADP. *See* ATP.

adrenaline. *See* AUTONOMIC NERVOUS SYSTEM.

adsere. That part of a SERE that precedes its development into another at any time before the CLIMAX is reached.

adsorption. The physical or chemical bonding of molecules of gas, liquid or a dissolved substance to the external surface of a solid or the internal surface if the material is porous in a very thin layer. *Compare* ABSORPTION.

advanced gas-cooled reactor (AGR). A NUCLEAR REACTOR in which enriched uranium dioxide (*see* URANIUM ENRICHMENT) is used as fuel, gaseous carbon dioxide as coolant and graphite as moderator. Operating temperatures (about 675 °C) are higher than those in the earlier MAGNOX REACTOR.

advanced waste treatment. Any process for the treatment of waste that follows other physical, chemical or biological treatments and aims to improve the quality of effluent prior to reuse or discharge. The term often refers to the removal of NITRATE and PHOSPHATE plant nutrients. *See also* EUTROPHICATION, PRIMARY TREATMENT, SECONDARY TREATMENT.

advection. Transport by motion of the air, water or other fluid. Advection has the same general meaning as CONVECTION, but is used particularly to refer to the horizontal transport by wind of something carried by the air (e.g., pollutants, heated air, fog, etc.).

advection fog. *See* FOG.

adventitious. Applied to parts of plants that arise in unusual positions (e.g., roots that grow from stems, buds produced elsewhere than in the axils of leaves).

adventive plant (casual plant). An introduced or alien plant that grows without human assistance, but is not permanently established.

adversarial approach. The resolution of controversial issues by argument between adversaries, as in a court of law, applied widely in the USA, and in the UK public inquiry system. Legally it has advantages, but because it requires each side to select data in support of its case it tends to obscure the overall assessment needed to make a sound scientific judgement.

AEC. *See* ATOMIC ENERGY COMMISSION.

Aegypiinae. A subfamily comprising the Old World VULTURES.

aeolian deposit (eolian deposit). A sediment deposited after having been carried by the wind.

aeon. *See* EON.

aeration. Any process whereby a substance becomes permeated with air or another gas. The term is usually applied to aqueous liquids that are brought into intimate contact with air by spraying, bubbling or agitating the liquid, and it is especially important with reference to the oxygen required by fishes and other AEROBIC aquatic organisms and in soil aeration.

aerator. A device for introducing air into a liquid.

aerial plankton. Spores, bacteria and other microorganisms floating in the air. *See also* PLANKTON.

aeroallergen. Pollen or organic dust that causes hay fever and other allergic conditions in those susceptible to it.

aerobic. Living or active only in the presence of oxygen. *Compare* ANAEROBIC.

aerobic respiration. Process whereby organisms, using gaseous or dissolved oxygen, release energy by the chemical breakdown of food substances. *Compare* ANAEROBIC RESPIRATION. *See also* RESPIRATION.

aerobiosis. Biological processes that require oxygen.

aerodynamic drag. The force required to move a solid body through air. At low speeds, when flow is laminar (*see* LAMINAR MOTION), the drag is viscous and proportional to the speed. When the body leaves a turbulent wake (*see* TURBULENCE), the drag is due to the release of kinetic energy and is proportional to the square of the speed. At speeds greater than sound, the energy of shock and sound waves is the predominant form of energy released, and the drag is proportional to a higher power of the speed.

aerodynamic roughness. The capacity of a surface to retard the flow of air over it, its extent depending on the size of the roughness elements causing the retardation and the height to which the retardation is felt corresponding to the height of the wakes of the roughness. Below that height the LOGARITHMIC WIND PROFILE does not apply.

aerogenerator. A device for exploiting the movement of air to generate electrical power. It consists of rotor blades, attached by a horizontal, and in a few designs vertical, axis to a tower. *Compare* WINDMILL. *See also* DARRIEUS GENERATOR, GRANDPA'S KNOB GENERATOR, SAVONIUS ROTOR, WIND FARM.

aerosol. A solid or liquid particle suspended in a gaseous medium and so small that its fall speed is small compared with the vertical component of air motion. Haze and cloud are the commonest atmospheric aerosols, with fall speeds much less than 10 millimetres per second. Aerosols in the TROPOSPHERE generally fall to the surface in a matter of hours or days; those in the STRATOSPHERE may remain there for months or years. Volcanoes are the major source of atmospheric aerosols, but human activities (e.g., cultivating dry soils, quarrying, industrial manufacturing, etc.) contribute about 30 percent of tropospheric aerosols. Tropospheric aerosols may act as condensation nuclei (*see* CONDENSATION NUCLEUS); some stratospheric aerosols, especially sulphate particles, have a climatic effect by increasing the Earth's ALBEDO. *See also* AEROSOL SPRAY, RAYLEIGH SCATTERING.

aerosol spray. A container in which a propellant (e.g., ammonia, butane or CHLOROFLUOROCARBONS) is mixed with a substance (e.g., paint, perfume, hair lacquer, polish or wound dressing) and held under pressure. When the pressure is released the substance is propelled through a nozzle as a mist of

AEROSOLS. A more primitive version of the aerosol spray is the liquid atomizer, in which the propulsion pressure is produced by a small hand-operated pump.

aestidurilignosa. Mixed evergreen and deciduous HARDWOOD forest.

aestilignosa. Broadleaf deciduous forest and brush, in which the trees are leafless in winter.

aestivation. (1) The DORMANCY of certain animals (e.g., lungfish) during the dry season, the summer or a prolonged drought. (2) The folding of parts of a flower bud. *Compare* HIBERNATION.

aetiology. The science of the cause or origin of disease.

afforestation. (1) The planting of trees in an area, or the management of an area to allow trees to regenerate or colonize naturally, in order to produce a forest. (2) In the UK, historically, the designation of an area, which might or might not be wooded, within which forest laws would henceforth apply.

afterblow. In the BESSEMER PROCESS (now rapidly becoming obsolete) the removal of phosphorus by continuing to blow air after the carbon has been consumed. It can be a cause of pollution from steel mills.

afterburner. In incinerators, a burner located so that the combustion gases are made to pass through its flames to remove smoke and smells.

after-ripening. The period of chemical and physical change undergone by the EMBRYOS of certain seeds (e.g., hawthorn) after they are apparently fully developed, without which they will not germinate. A similar phenomenon may occur in BULBS and TUBERS.

aftershocks. A series of smaller shocks following an earthquake of large magnitude and occurring fairly close to the FOCUS of the main shock.

Ag. *See* SILVER.

agar-agar. A gelatinous substance, derived from certain seaweeds, used as a bacteriological culture medium, as a thickening agent in foodstuffs and in pharmaceutical products.

Agaricaceae (agarics, gill fungi). A family of BASIDIOMYCETES that are characterized by the production of SPORES on GILLS. Examples include common mushrooms and toadstools.

agate. *See* CHALCEDONY.

Agave. A genus of American plants. *A. americana* produces large quantities of sap which, when fermented, gives *pulque*, the Mexican national drink from which *mescal* is distilled. Other species are cultivated for their fibre.

Agency for International Development. *See* US AID.

ageostrophic. A form of motion in air in which the horizontal pressure gradient force is not in balance with the deviating force (CORIOLIS FORCE) due to the wind velocity. Ageostrophic motion cannot be deduced from the pressure field and may be caused by friction, acceleration (either linear or due to curvature), or changing pressure field. It is necessarily associated with vertical motion, cloud formation and weather.

agglomerate. A rock composed of fragments of volcanic material with a preponderance of pieces greater than 20 millimetres in diameter. Some agglomerates are bedded and were sedimented out after ejection of the fragments from a vent. Others (vent agglomerates) are vent fillings.

agglomeration. The gathering together of particles (e.g., smoke particles in air under the influence of ultrasonic radiation). *Compare* FLOCCULATION.

agglutinin. *See* BLOOD GROUP.

agglutinogen. *See* BLOOD GROUP.

aggradation. The raising of the level of the land surface by the accumulation of deposited material.

aggregate. Construction material made from particles, most commonly of crushed stone. For road construction (*see* PAVEMENT) UK regulations require particles to have a diameter of 0.075–38 millimetres, depending on the use to which they are put, and to be well graded (*see* SORTING). For surfacing layers the aggregates must be bonded well with BITUMEN.

aglycon. *See* GLYCOSIDES.

Agnatha. The lampreys, hagfishes and their extinct relatives, forming a class (or subphylum in some classifications) of primitive, jawless, fish-like vertebrates. *Compare* GNATHOSTOMATA. *See also* CYCLOSTOMATA.

agonistic behaviour. Threatening or otherwise aggressive behaviour displayed by an animal towards another of the same species.

agora. In ancient Greek cities, a public open space at the centre of the city, of no special shape, used as a market place and meeting place. It developed into a place for sporting activities, and eventually into the plaza, campo, piazza, etc. which survive in many modern Latin towns.

AGR. *See* ADVANCED GAS-COOLED REACTOR.

Agreement on Conservation of Polar Bears. An international agreement, drafted by the INTERNATIONAL UNION FOR CONSERVATION OF NATURE AND NATURAL RESOURCES, and signed by the USSR, Norway, Canada, Denmark and the USA, which came into force in May 1976. It forbids the killing or taking of polar bears, except for scientific purposes or by local Eskimo people, and requires governments to protect polar bear habitats.

agric horizon. The depositional SOIL HORIZON, formed by cultivation, that contains clay and humus.

agricultural revolution. (1) The introduction of agriculture to supplement hunting and gathering in Neolithic times (around 10 000 – 8000 BC) which made food much more readily available and meant less land was needed to support a given population, thus permitting a great increase in population size. (2) The introduction of mechanization, enclosure of open land, rotation of crops and other agricultural improvements that began in northern Europe in the 18th century.

agroecosystem. A community of microorganisms, plants and animals, together with their ABIOTIC environment, that occurs on farmed land, and including the crop species.

agroforestry. The interplanting of farm crops and trees, especially leguminous species (*see* LEGUMINOSAE). In semiarid regions and on denuded hillsides, agroforestry helps control erosion and restores soil fertility, as well as supplying valuable food and commodities at the same time.

agronomy. The study of rural economy and husbandry.

Agung. *See* MOUNT AGUNG.

A horizon. *See* SOIL HORIZONS.

AID. (1) Artificial insemination by donor; used widely in livestock breeding. (2) The Agency for International Development, the major US organization concerned with AID projects in developing countries.

aid. Assistance given by developed to developing countries, in financial or technical form. *See also* ECONOMIC DEVELOPMENT.

AIDS. *See* ACQUIRED IMMUNE DEFICIENCY SYNDROME.

air conditioning. The process of bringing atmospheric or other air to conditions of cleanliness, temperature and humidity for use in buildings. In industrial countries this is often a prodigiously energy-intensive process. In poor countries it is more usually achieved by draughts, humidifiers and wind traps.

aircraft wake. Two parallel, contrarotating vortices (*see* VORTEX) trailing approximately from the wing tips, which drive one another downwards together with a body of air surrounding them that contains the engine exhaust. The intensity of the vortices corresponds to the wing downwash, which supports the aircraft and is reduced with increased speed. It is greatest behind a slow-moving, heavy aircraft, and on the approach to landing it must be avoided by other aircraft, particularly smaller ones which might be overturned in one of the vortices.

air lock. A device to provide access from one region or enclosed space to another without direct contact of the fluids in those regions. Air locks are used in control of radioactive contamination and for air/water and air/space transitions.

air mass. A mass of air extending over a large distance (e.g., 1000 kilometres or more) and possessing distinct characteristics of temperature, humidity and LAPSE RATE.

air parcel. A theoretical concept, useful in atmospheric physics, of a body of air whose properties and transformations may be considered by application of the laws of physics and mechanics.

air pollution. Gaseous or AEROSOL material in the air that is considered to be not a normal constituent of the air, or a normally minor constituent (e.g., sulphur dioxide, carbon monoxide, nitrogen oxides, etc.) present in excess. The definition is confused by the occurrence of natural pollutants, (e.g., methane, hydrogen sulphide in boggy areas, dust in deserts, volcanic debris, etc.).

air quality. The degree to which air is polluted, such that air quality is deemed to be high when pollution levels are low. This may be judged aesthetically or by reference to

medical, botanical or material damage. Standards of quality are not absolute and have cultural variations. They are also confused by possible synergistic effects (see SYNERGISM) or effects of humidity, wind speed, time of year, duration of the incidence of the pollutant, etc., so that a single value for the concentration of a pollutant is not a complete specification of air quality in respect of that pollutant.

Air Quality Act, 1967. US Federal law that empowers the Department of Health, Education and Welfare to designate areas within which air quality is to be controlled, to set AMBIENT AIR STANDARDS, to specify technologies to be used in pollution control and to prosecute offenders if local agencies fail to do so. *See also* CLEAN AIR ACT, 1970.

airshed. A concept, invented by analogy with watershed, used in crude computations of the air pollution concentrations to be expected from a given catalogue of pollution sources, expected air movement, and chemical rates of change.

Aitken counter. A device that causes the rapid condensation of droplets in air by the rapid ADIABATIC expansion of it. It makes possible the counting of the number of condensation nuclei (*see* CONDENSATION NUCLEUS) present in a unit volume by means of a microscope. *See also* AITKEN NUCLEI.

Aitken nuclei. Particles that produce a droplet during the expansion of air in an AITKEN COUNTER, corresponding to those naturally available for cloud formation in rising air or fog. In a slow expansion only some of the Aitken nuclei form droplets. They are named after the physicist John Aitken.

aitogenic. *See* PARATONIC.

Al. *See* ALUMINIUM.

AL63. *See* DERRIS.

alabaster. Massive, fine-grained GYPSUM used for carvings and vessels.

Alamogordo cattle. A herd of 32 Hereford cattle that were exposed to FALLOUT, about 16–24 kilometres (10–15 miles) north-east of ground zero, from the first atomic bomb test at Alamogordo, New Mexico, in July 1945. Subsequently the animals were moved to Oak Ridge, Tennessee, where the effects of radiation were studied. The animals developed patches of grey hair on their backs and sides, horn-like growths apparently no deeper than the DERMIS, but which took two to three times longer than ordinary tissue to heal when wounded, and three cows developed skin cancer. Their general health and reproductive efficiency were otherwise no different from those of 74 control cattle.

alanine. An AMINO ACID with the formula $CH_3CH(NH_2)COOH$ and a molecular weight of 89.1.

Alaska Lands Bill, 1980. The US law under which more than 42 million hectares (104 million acres) of land in Alaska were designated as national parks, wildlife refuges and wilderness areas; oil exploration, mining and logging were permitted in other specified areas.

albatrosses. *See* DIOMEDEIDAE.

albedo. The whiteness of an object or material, or the degree of reflection of incident light from it Smooth-surfaced materials often have a high albedo, reflecting most of the incident light. Snow and cloud surfaces have a high albedo, near to unity because most of the energy of the visible solar spectrum is reflected or scattered. Vegetation and sea have a low albedo, nearer to zero, because they absorb a large fraction of the energy. Cloud is the chief cause of variations in the Earth's average albedo, which has a value of around one-half.

Albian. A stage of the CRETACEOUS System.

albic. Applied to SOIL HORIZONS from which CLAY and iron oxides have been removed.

albinism. The lack of normal skin pigments

caused by the absence of the enzyme tyrosinase, in mammals often due to an autosomal RECESSIVE gene. In a pure albino the eye appears pinkish. Most bird albinos are partial, with white patches on the plumage.

albite. *See* FELDSPAR.

Alcedinidae (kingfishers). A family of brightly coloured, hole-nesting birds that dive for fish and hence are highly susceptible to any environmental change that reduces the quality of water and so reduces fish populations. They have large heads, short tails and long, sharp beaks. Along with the hoopoe and bee-eaters, they belong to the mainly tropical order Coraciiformes.

Alcidae (auks, razorbills, guillemots, puffins, etc.). A family of marine, fish-eating birds with webbed feet, short wings and heavy bodies. They come ashore only to breed and swim under water using their wings. They often suffer heavy casualties after oil spillages, and in coastal areas popular with tourists, disturbance by humans has reduced the number of breeding sites acceptable to them.

alcoholic fermentation. The fermentation of ETHANOL from sugar (*see* CARBOHYDRATES) by YEASTS.

alcynonarian coral. *See* CORAL.

aldehydes. A class of chemical compounds, mostly colourless, volatile and with suffocating vapours, that are intermediate between CARBOXYLIC ACIDS and ALCOHOLS. *See also* ACETALDEHYDE.

aldrin. An ORGANOCHLORINE; a persistent insecticide similar to DIELDRIN, whose use in the UK, the USA and most industrial countries is now very restricted. It is effective against many insects, but is poisonous to vertebrates. *See also* CYCLODIENE INSECTICIDES.

ALECSO. *See* ARAB LEAGUE EDUCATIONAL, CULTURAL AND SCIENTIFIC ORGANIZATION.

aleurone grains. Large granules of protein found in the food-storage organs of plants, notably in seeds.

Aleutian low. A low-pressure area in the neighbourhood of the Aleutian Islands (north Pacific) which is evident on long-term average-pressure charts on account of the high frequency with which CYCLONES occur in that area.

Alfisols. *See* SOIL CLASSIFICATION.

algae. Informal (non-taxonomic) name for a large group of plants that contain CHLORO-PHYLL and have a relatively simple body construction, varying from a single cell to a multicellular, ribbon-like THALLUS. They are either aquatic (e.g., diatoms; *see* BACILLARIO-PHYTA; seaweed) or live in damp places (e.g., the green stain on damp walls, etc.). Plank-tonic (*see* PLANKTON) algae form the bases of marine and freshwater FOOD CHAINS. The bacterial culture medium AGAR-AGAR is ex-tracted from seaweeds. *See also* BACTERIA, CHLOROPHYTA, CHRYSOPHYTA, CHAROPHYTA, CRYPTOPHYTA, CYANOPHYTA, EUGLENOPHYTA, PHAEOPHYTA, PYRROPHYTA, RHODOPHYTA, XANTHOPHYTA.

algal bloom. *See* BLOOM.

aliphatic. Applied to organic compounds containing open chains of carbon atoms, and comprising paraffins, olefins and acety-lenes, their derivatives and substitution products. *Compare* AROMATIC.

alkali. *See* BASE, pH.

alkali–aggregate reaction (concrete can-cer). A chemical reaction that weakens and may cause the collapse of structures built from ALUMINOUS CEMENT, in which SILICA reacts with other constituents of the cement to produce compounds that absorb water, swell and crack the concrete. This allows water to enter, and repeated wetting and drying, and possibly freezing, widen the cracks and may corrode the reinforcing steel.

alkalic. *See* ALKALINE.

Alkali Inspectorate. In the UK, the official body, formed in 1880, that was most concerned with the control of AIR POLLUTION from industrial sources, including the nuclear industry. The Health and Safety at Work Act, 1974, brought the Inspectorate within the HEALTH AND SAFETY EXECUTIVE, where its official title became H.M. Alkali and Clean Air Inspectorate, with responsibility for England and Wales. Its Scottish equivalent is H.M. Industrial Pollution Inspectorate for Scotland, which acts as an agent for the Health and Safety Commission and Executive.

alkaline (alkalic). (1) *See* pH. (2) Applied to IGNEOUS rocks containing an abundance of the alkaline metals, indicated by the pres-ence of such minerals as sodium-rich PYROX-ENES and AMPHIBOLES, and sodium- and/or potassium-rich FELDSPARS. In petrology. the opposite of alkaline is CALC-ALKALINE, not acidic. Alkaline is also used to refer to igne-ous rocks rich in FELDSPATHOIDS, and to min-erals rich in sodium and/or potassium (e.g., alkali feldspar).

alkaloids. Complex, BASIC organic com-pounds containing nitrogen, found in certain families of flowering plants (e.g., RANUNCULACEAE, SCROPHULARIACEAE, SOLAN-ACEAE, UMBELLIFERAE). They are important for their poisonous and medicinal qualities (e.g., morphine is derived from the fruit of the opium poppy, nicotine from tobacco leaves, atropine from deadly nightshade, quinine from the bark of *Cinchona* species, strychnine from the seeds of *Strychnos nux-vomica*). *See also* STRYCHNOS.

alkylmercury. A highly toxic mercurial compound used as a fungicide and seed dressing. It is the cause of many accidental deaths and permanent injuries resulting from the consumption of dressed grain.

alkyl sulphonates. SURFACE-ACTIVE AGENTS used in synthetic DETERGENTS. Those with a linear molecular structure are degraded fairly readily by microorganisms; those with a branched structure (e.g., alkyl benzene sulphonate) are stable and resist biodegradation, causing foaming in water into which they are discharged. In the UK and most industrial countries commercial detergents sold for domestic use, but not necessarily those for industrial use, are biodegradable.

allantois. One of the embryonic membranes of AMNIOTA. It is a large outgrowth from the hind gut of the EMBRYO and is covered in a network of blood vessels. In reptiles and birds excretory products are stored in the allantoic cavity, and the allantoic blood vessels are responsible for gaseous exchange as they lie close to the shell. In mammals these blood vessels supply the PLACENTA.

allele. *See* ALLELOMORPH.

allelochemistry. The chemistry of substances released by one population that affect another population.

allelomorph (allele, allelomorphic gene). The alternative form of a gene. Allelomorphs lie at the same locus (relative position) on HOMOLOGOUS chromosomes and pair during MEIOSIS. They behave in a mendelian manner (*see* MENDEL), producing different effects on the same developmental process.

allelomorphic gene. *See* ALLELOMORPH.

allelopathy. The influence that one living plant exerts upon another by means of chemical exudates.

Allen's rule. As the mean temperature of the environment decreases (e.g., with increasing latitude) the relative sizes of the appendages (e.g., ears, tails, limbs) of warm-blooded animals tends to decrease. This is correlated with the increased need to conserve heat.

alligators. *See* CROCODILIA.

allochthonous. (1) Applied to material that originated elsewhere (e.g., drifted plant debris on the bottom of a lake). (2) Applied to rocks whose major constituents have not been formed *in situ*. *Compare* AUTOCH-THONOUS.

allogeneic (allogenic).   Having different genes. *Compare* ISOGENEIC.

allogenic. (1) Applied to those constituents of a rock that were formed previously and separately from the rock itself (e.g., the pebbles in a CONGLOMERATE). *Compare* AU-THIGENIC. (2) *See* ALLOGENEIC.

allogenic succession.   A SUCCESSION that takes place because of changes in the environment not produced by the plants themselves, but by external factors (e.g., alteration in drainage or climate). *Compare* AUTO-GENIC SUCCESSION.

allopatric. Applied to different species or SUBSPECIES whose areas of distribution do not overlap. *Compare* SYMPATRIC.

allopolyploid. An organism with more than two haploid sets of CHROMOSOMES derived from two or more species. Cultivated wheats are probably allopolyploids.

allotetraploid (amphidiploid).   The plant species produced when a diploid hybrid doubles its CHROMOSOME number. Unlike most diploid hybrids, allotetraploids are fertile. Artificially induced allotetraploidy is an important means of producing new varieties in agriculture. *See also* ENDOMITOSIS.

allotropy. The existence of an element in two or more physical forms (e.g., carbon as diamond and graphite, red and white phosphorus).

alloy. A metallic substance produced by adding other metals to a major metallic ingredient to obtain certain required properties (e.g., bronze, brass, solder). The term is often used metaphorically for intimate non-metallic mixtures.

alloy steel. Any special type of STEEL made by adding other elements (e.g., chromium, nickel, cobalt) to ordinary carbon steel. Many steels with properties of corrosion resistance, great hardness, specific magnetic characteristics, etc. have been produced in this way.

alluvium. Sediment transported and deposited by flowing water. *See also* FLOOD PLAIN.

almanac. A catalogue of predictions. Some almanacs contain reliably computed astronomical predictions and tidal tables, others contain guesses with a variety of bases in experience, yet others are founded in fancy.

alpha bronze. A solid solution of tin in copper, the former usually forming 4–5 percent. Unlike phosphor bronze and GUNMETAL, it is workable and is used for springs, turbine blades, coins, etc.

alpha diversity (niche diversification). Diversity resulting from competition between species that reduces the variation within particular species as they become more precisely adapted to the NICHES they occupy.

alpha-mesosaprobic. *See* SAPROBIC CLASSIFICATION.

alpha-particles. Heavy particles (i.e. HELIUM nuclei) produced by radioactive decay (e.g., decay of PLUTONIUM) with little penetrative power, but damaging when in contact with living tissue such as occurs following inhalation or ingestion. *See also* IONIZING RADIATION.

alpine. (1) Applied to the parts of a mountain above the tree-line, but below permanent snow. (2) Applied to processes or products associated with the ALPINE OROGENY.

Alpine orogeny. The orogeny (mountain building) that culminated in the MIOCENE Period, forming the Alpine–Himalayan belt.

ALR. *See* ADIABATIC LAPSE RATE.

alternation of generations. The alternation of sexual and asexual reproduction in the life cycle of an organism. True alternation of generations occurs in plants that have a haploid, GAMETE-producing generation (gametophyte) alternating with a diploid, SPORE-producing generation (sporophyte). The gametophyte is dominant in mosses, liverworts and some ALGAE, but in ferns it is an insignificant THALLUS, the familiar fern being the sporophyte. In flowering plants, the gametophyte is microscopic, forming part of the pollen grain or ovule. What has been called an alternation of generations occurs in some animals (e.g., tapeworms, jellyfish), but in this case both generations are diploid. *See also* CHROMOSOMES.

alternative energy. Energy derived from sources other than the burning of COAL, PETROLEUM or NATURAL GAS, or from NUCLEAR FUSION or NUCLEAR FISSION; usually derived on a domestic or small-community scale. Examples of alternative energy installations include those on small-scale based on BIOGAS, SOLAR POWER, HYDROELECTRIC POWER or WIND POWER. *See also* SOFT-ENERGY PATHS.

alternative technology. Technology that aims to use resources sparingly and to do the minimum damage to the environment or to species inhabiting it, while permitting the greatest possible degree of personal control over the technology on the part of its user. *Compare* APPROPRIATE TECHNOLOGY. *See also* ALTERNATIVE ENERGY.

altimeter. An aneroid BAROMETER calibrated to read height above some reference level whose barometric pressure (QFE) is set on the altimeter dial.

altitude. (1) Height above some reference level (e.g., ordnance datum, mean sea level) that is equipotential in the Earth's gravitational field. (2) In astronomy, a measure of the angular elevation above the horizon.

altocumulus. CUMULUS clouds generated by upcurrents not having their origin at the ground or sea, but at some higher level. The cloud elements thus appear smaller from the ground than cumulus originating at the ground. Altocumulus is typically dappled.

altostratus. A high-level layer of cloud, lacking dappled or fibrous features, but of uniform appearance over a large part of the sky. Typically altostratus appears before and in the rain of a WARM FRONT of a CYCLONE.

altricial. Applied to birds that hatch blind and naked, or nearly so, and at first are unable to run about or look after themselves (e.g., pigeon). *Compare* PRECOCIAL.

altruistic. Applied to behaviour that contributes to the survival of the group rather than of the individual and that may be performed at some cost to the individual. *Compare* EGOCENTRIC.

alumina. Aluminium trioxide. (Al_2O_3). *See also* ACTIVATED ALUMINA.

aluminium (aluminum, Al). A very light, ductile metallic element used in many industries. It is produced from BAUXITE by an energy-intensive electrolytic process, more energy being required to produce aluminium from its ore than to produce steel from iron ores. The process is also highly polluting, unless measures are applied to control plant emissions. Aluminium is very resistant to corrosion due to the monomolecular layer of the trioxide (alumina, Al_2O_3) formed on exposure to air and water. $A_r = 26.9815$; $Z = 13$; SG 2.7; mp 659.7 °C.

aluminous cement (high-alumina cement). PORTLAND CEMENT containing 30–50 percent ALUMINA, 30–50 percent lime (calcium hydroxide) and not more than 30–50 percent iron oxide, SILICA and other ingredients. It hardens rapidly and resists seawater. After extensive use in the UK in the 1960s, it was found to deteriorate with time due to the ALKALI–AGGREGATE REACTION. This caused the collapse of a few large buildings and the weakening of bridges and other structures, and led to the closing of certain buildings pending further investigation.

aluminum. *See* ALUMINIUM.

alveoli. The innumerable small sacs in the lung through whose walls oxygen enters the bloodstream and carbon dioxide leaves. *See also* BRONCHIAL DISEASES.

amalgam. A solution of a metal in MERCURY.

Amaranthus. A genus of cereal-like plants,

three species of which (*A. caudatus, A. cruentus, A. hypochondriacus*) are fast-growing and produce seed heads resembling *SORGHUM* containing seeds rich in protein (about 15 percent protein and 63 percent starch), with a high LYSINE content. The leaves are edible and also rich in protein. Amaranths are grown in Latin America, where they yield crops containing much more protein than MAIZE, the principal staple crop in the region. Other species are grown as pot herbs.

ambari hemp. *See* KENAF.

amber. Fossil resin, known from the CRETACEOUS onwards, that often contains excellently preserved insects with only their soft inner tissues missing.

ambergris. A secretion of the sperm whale. At one time it was used extensively in perfumery, but now is scarce because of the decline in whaling and has largely been discarded by the cosmetic industry.

ambient. Surrounding a phenomenon, or present before a phenomenon.

ambient air standard. A quality standard for air in a particular place defined in terms of pollutants. Industries discharging pollutants are required to limit emissions to levels that will not reduce the quality of air below the standard. The principle is used widely in the USA, but not generally in the UK.

ambient noise. The background or prevailing noise.

ambient quality standards. *See* ENVIRONMENTAL QUALITY STANDARDS.

ambient temperature. The temperature of the surrounding air.

ambient turbulence. The eddy motion present that is not caused by the phenomenon being observed. It is a cause of BUMPINESS and the dispersion of air pollutants.

amenity. An abstract concept expressing those natural or man-made qualities of the environment from which people derive pleasure, enrichment and satisfaction.

amensalism. The opposite of COMMENSALISM.

americium. *See* ACTINIDES.

Ametabola. *See* APTERYGOTA.

amethyst. QUARTZ made purple by slight impurities.

ametoecious parasite. *See* PARASITISM.

amines. Compounds formed by replacing hydrogen atoms in ammonia (NH_3) by organic RADICALS (R). Amines are classified according to the number of hydrogen atoms so replaced as primary (NH_2R), secondary (NHR_2) or tertiary (NR_3). The release of nitrites into the environment (either directly as nitrites or as NITRATES reduced to nitrites naturally by bacteria) may lead to random reactions with secondary amines to form NITROSAMINES.

amino acids. Organic compounds containing a carboxyl group (–COOH) and an amino group ($-NH_2$) (e.g., glycine (CH_2NH_2COOH)). About 30 amino acids are known. They are fundamental constituents of living matter because protein molecules are made up of many amino acid molecules combined together. Amino acids are synthesized by green plants and some bacteria, but some (e.g., PHENYLALANINE) cannot be synthesized by animals and therefore are essential constituents of their diets. Proteins from specific plants may lack certain amino acids, so a vegetarian diet must include a wide range of plant products.

ammonia (NH_3). A chemical compound used in liquid or gaseous form directly as a FERTILIZER or as the basis of ammonium (NH_4^+) fertilizer compounds. Ammonia has been used industrially in many processes and was formerly employed as a refrigerant, in which use it has largely been replaced by the cheaper, less reactive and more efficient CHLOROFLUOROCARBONS. Ammonia is produced naturally by the decomposition of animal urine, but its industrial production is energy-intensive (*see* COAL GAS, HABER PROCESS). Ammonia is very toxic, but its irritant vapours are immediately recognizable.

ammonification. The biochemical process that produces nitrogen, as AMMONIA, from organic compounds. *See also* NITRIFICATION.

ammonifying bacteria. Those nitrifying bacteria (e.g., *Pseudomonas*) that break down the protein in the tissues of dead plants and animals and release AMMONIA. *See also* NITRIFICATION, NITROGEN CYCLE.

ammonites. A group of extinct CEPHALOPODA, similar to the present-day pearly nautilus, that were marine and lived in the JURASSIC and CRETACEOUS. Their shells were curved mostly in one plane and divided by internal septa. *See also* INDEX FOSSIL.

ammonium (NH_4^+). The cation produced by the reaction of AMMONIA with hydrogen.

ammonium nitrate (NH_4NO_3). A compound used as a FERTILIZER and as a component of explosives, HERBICIDES and INSECTICIDES. It is the main source of the nitrate RUNOFF from farm land that results from excessive fertilizer use. In water it can be a pollutant and may cause EUTROPHICATION.

Ammophila. A genus of grasses that includes *A. arenaria* (marram grass), which is often used to stabilize blowing SAND DUNES because of its ability to colonize open sand rapidly, spreading partly by seed and partly by producing extensive RHIZOMES with large, tough leaves that appear above the surface. Its roots also bind sand particles. When other grasses become established on the site, the marram is suppressed and eventually disappears.

amnion. A fluid-filled sac that is an outgrowth of the EMBRYO of a reptile, bird or mammal. The amnion envelops the embryo and provides it with the fluid environment necessary for its protection and development.

Amniota (reptiles, birds mammals). The land-living vertebrate classes whose embryos develop an AMNION.

Amoco Cadiz. A Liberian-registered tanker owned by Amoco International Oil that ran aground and broke in two off Portsall, Brittany, on 17 March 1978. The tanker released the whole of its cargo of 1.6 million barrels of oil, mostly light crude, in the worst oil pollution incident up to that time. Almost all the oil drifted on to the Brittany coast. *See also* PANHONLIB GROUP.

Amoeba. A genus of minute, acellular animals (PROTOZOA) that move and ingest food by means of pseudopoda (*see* PSEUDOPODIUM) and thus are continually changing shape. Members of the genus *Amoeba* are freeliving, but some closely related genera are commensal (*see* COMMENSALISM) or parasitic (*see* PARASITISM). An example is *Entamoeba histolytica*, which causes amoebic dystentry.

amoebic dysentry. *See* AMOEBA.

amoebocytes. Cells that move by means of pseudopoda (*see* PSEUDOPODIUM) and are found in the tissues and body fluids of vertebrates and invertebrates. Many are phagocytic (*see* PHAGOCYTE) (e.g., some types of white blood cell in vertebrates).

amorphous. (1) Having no distinct shape. (2) Of a mineral, non-crystalline.

ampere (A). The SI unit of electric current, defined in 1948 as the intensity of a constant current which, if maintained in two parallel, rectilinear conductors of infinite length, negligible circular section and positioned 1 metre apart in a vacuum, will produce between the conductors a force equal to 2 x 10^{-7} newtons per metre of length.

Amphibia (newts, salamanders, frogs, toads). A class of vertebrates which, during their evolution, have succeeded only partly in conquering the difficulties of life on land. The adults are aquatic or live in damp situations and have moist smooth skins, often used in addition to lungs for respiratory purposes. Apart from a few VIVIPAROUS species, most have to return to the water to breed, because the eggs have no shell to prevent desiccation. The larvae have gills. The earliest (CARBONIFEROUS) amphibians resembled OSTEOLEPID fishes, from which they differed in having PENTADACTYL limbs instead of fins, and a middle ear apparatus. *See also* ANURA, APODA, LABYRINTHODONTIA, RESPIRATION, URODELA.

amphibole. A group of rock-forming silicate minerals with a very wide range of compositions. The general formula is $X_{2-3}Y_5Z_8O_{22}(OH)_2$, where X is usually calcium, sodium or potassium, Y is usually magnesium, iron or aluminium and Z is silicon or aluminium (with a maximum of 2Al). Amphiboles occur in IGNEOUS rocks, being characteristic of those of INTERMEDIATE character, and in metamorphic rocks (*see* METAMORPHISM). *See also* HORNBLENDE.

amphidiploid. *See* ALLOTETRAPLOID.

amphimixis. True sexual reproduction, involving FERTILIZATION. *Compare* APOMIXIS.

Amphineura. A class of marine MOLLUSCA whose most familiar members (chitons, coat-of-mail shells) are limpet-like, bearing eight calcareous plates and live on seaweed in the intertidal zone. *See also* SHORE ZONATION.

amphioxus (lancelets). Small, rather eel-like animals, of the genera *Branchiostoma* and *Asymmetron*, which live in clean sand in shallow water where there is a good flow of current. They exhibit primitive chordate (*see* CHORDATA) characteristics but, although their general structure is related to that of vertebrates, they lack a true head and backbone. Thus, any theory concerning the origin of the VERTEBRATA must rest heavily on the true nature of amphioxus. The term amphioxus was formerly confined to the genus *Branchiostoma*, but it is now used in respect of both genera. *See also* ACRANIA.

amphiphyte. An amphibious plant (e.g., amphibious bistort, *Polygonum amphibium*). Aquatic and terrestrial forms of plants often differ considerably in vegetative features.

Amphipoda. *See* MALACOSTRACA.

amplitude. The peak or maximum value in a wave-type transmission.

amygdale. *See* VESICLE.

amylases (diastases). ENZYMES, widely distributed in animals and plants, that break down starch or glycogen to dextrin, maltose or glucose (e.g., ptyalin in human saliva).

Anabaena. A genus of blue–green algae (CYANOPHYTA) that are able to fix atmospheric nitrogen (*see* NITROGEN FIXATION). They form symbiotic (*see* SYMBIOSIS) relationships with some aquatic ferns (e.g., *AZOLLA*).

anabatic wind. An upslope wind generated when a slope is warmed (e.g., by sunshine). It is local and unstable. *Compare* KATABATIC WIND.

anabiont. A perennial plant that fruits many times.

anabiosis. Resuscitation after apparent death (e.g., after desiccation).

anabolism. The synthesis of complex molecules from simpler ones. *Compare* CATABOLISM.

Anacardiaceae. A family of mainly tropical plants, with some outliers (e.g., *RHUS*), many of which are grown for ornament, useful products or fruit (e.g., *Anacardium occidentale*, cashew; *Mangifera indica*, mango).

anadromy. The migration of some fish (e.g., salmon) from the sea into fresh water for breeding. *Compare* CATADROMY.

anaerobic. Applied to organisms or processes that live or occur only in the absence of gaseous or dissolved oxygen. Anaerobic bacteria can obtain oxygen from NITRATES releasing nitrogen, nitrous oxide or ammonium, or from sulphates releasing hydrogen sulphide, provided that oxidizable material (e.g., vegetable waste, METHANOL or other carbon compounds) is available. *Compare* AEROBIC.

anaerobic biological treatment. Any treatment that utilizes ANAEROBIC organisms to reduce organic matter in wastes.

anaerobic decomposition. Organic breakdown in the absence of air. The term is commonly applied to the digestion of sewage, with the production of METHANE.

anaerobic respiration. The process by which organisms release energy by the chemical breakdown of food substances without consuming gaseous or dissolved oxygen. Examples include the breakdown of glycogen to lactic acid in the muscles of vertebrates and the breakdown of sugar to ETHANOL and carbon dioxide by yeast (*see* FERMENTATION). Anaerobic respiration is less efficient than AEROBIC RESPIRATION. Many organisms are able to respire anaerobically when the oxygen supply is insufficient for aerobic respiration, but some (e.g., a few bacteria) never use free oxygen. *See* RESPIRATION.

ana-front. A FRONT at which the air on one side is ascending. The term is most commonly applied to fronts at which the cold air is ascending, because this is unusual. *Compare* KATA-FRONT.

analog. *See* ANALOGUE.

analogous. In zoology, an organ of one species is said to be analogous to that of another species when both organs have the same function, but did not originate in the same way and therefore do not indicate a close evolutionary relationship (e.g., the wings of a bat and of an insect). *Compare* HOMOLOGOUS.

analogue (analog). (1) A physical MODEL in which the properties of the original are replaced by the consistent, but physically different, properties of the model (e.g., an electric spark travelling along a length of gunpowder fuse as an analogue of an impulse travelling along a nerve fibre). *See also* ANALOGUE COMPUTER. (2) In engineering, a model of a mechanical or electrical system whose main properties are governed by the same kind of mathematical equation. Thus waves in a channel are analogous to sound waves in a pipe, but are easier to study visually.

analogue computer. A device used in modelling (*see* MODEL) that represents time-varying properties of the original by a set of time-varying voltages, the correspondence between original and model being dictated by a mathematical representation common to both. *Compare* DIGITAL COMPUTER.

analogue methods. In meteorology, procedures whereby predictions are made on the basis of the study of other (theoretical or actual) systems having analogous properties. In weather forecasting, the analogue might be a similar situation taken from historical records. *See also* ANALOGUE.

analysis of variance. *See* ANOVAR.

anaphase. *See* MITOSIS.

Anatidae. A family (order: Anseriformes) of waterfowl, which includes swans, geese and ducks, aquatic birds with webbed feet, long necks and broad, flat beaks edged with horny layers. The most representative of the geese is the greylag goose (*Anser anser*) from which the domestic goose is derived. The genus *Anser* comprises the grey geese (e.g., bean goose, *A. fabatis*; pink-footed goose, *A. f. brachyrhynchus*; white-fronted goose, *A. albifrons*). Other members of the family are black geese (e.g., barnacle goose,

Branta leucopsis; Canada goose, *B. canadensis*). All are dependent on WETLAND habitats, and the survival of some is threatened. The Hawaiian goose or nene (*B. sandivencis*) became almost extinct, but was saved by the breeding and rearing of captive birds, notably at the WILDFOWL TRUST.

Ancient Monument. In the UK, an official designation given in the public interest by the Secretary of State for the Environment to any structure of historic, architectural, traditional, artistic or archaeological value, other than an ecclesiastical building or a building that for the time being is used for ecclesiastical purposes. Once designated, the structure is afforded official protection and may be maintained by the Property Services Agency of the Department of the Environment.

ancient soil. An older, weathered (*see* WEATHERING) zone, the product of past climates and processes, that lies beneath present surface deposits.

Andean Floral Region. *See* PACIFIC SOUTH AMERICAN FLORAL REGION.

andesite. A fine-grained, INTERMEDIATE volcanic rock associated with continental crust. Andesites are the volcanic equivalent of DIORITE.

androecium. *See* FLOWER.

androgens. Male sex HORMONES (e.g., testosterone, produced by the interstitial cells of the testis). Natural androgens are steroids, responsible for the initiation and maintenance of many male sexual characteristics.

anechoic. Applied to surfaces that are almost totally sound-absorbent (*see* ABSORPTION COEFFICIENT). An anechoic chamber gives almost free field conditions.

anemochore. A plant whose seeds, spores or other reproductive structures are dispersed by the wind (e.g., thistles).

anemograph. *See* ANEMOMETER.

anemometer. An instrument for measuring wind speed, usually employing a PITOT TUBE directed by a vane, a rotor carrying three or four hemispherical cups, or a pressure plate deflected against a spring or against gravity. An anemograph makes a record of the wind speeds detected by the anemometer to which it is connected. *See also* PANEMONE.

anemophily. Wind pollination.

aneroid barometer. *See* BAROMETER.

aneuploid. Applied to an organism that is genetically unbalanced because its CHROMOSOME number is not a multiple of the basic haploid number. *Compare* EUPLOID, NULLISOMIC, POLYPLOID, TRISOMIC.

aneurine. *See* THIAMINE.

Angiospermae. The class (or division or subdivision in some classifications) of seed plants (SPERMATOPHYTA) that includes all the flowering plants, characterized by the possession of flowers, although these may be relatively inconspicuous structures (e.g., in grasses). The OVULES, which become seeds after FERTILIZATION, are enclosed in ovaries. The XYLEM contains true vessels. The Angiospermae are divided into two subclasses (or classes): MONOCOTYLEDONEAE and DICOTYLEDONEAE. *Compare* GYMNOSPERMAE.

angle of indraught. The angle of inclination of a steady wind to the ISOBARS, and called positive when the direction is towards low pressure. The angle of indraught is a measure of the departure of the surface wind from the GEOSTROPHIC WIND and is usually towards low pressure, and at a greater angle over land than over sea because of greater friction. The angle over sea may be zero or negative when there is a strong THERMAL WIND.

angle of rest. The steepest angle, measured from the horizontal, at which a fragmental material remains stable. *See also* TALUS.

ångström (A). A unit of measurement equal to one ten-millionth of a millimetre (10^{-7} mm)

or one-tenth of a NANOMETRE.

angular frequency. In acoustics, the frequency of sound expressed in radians per second and obtained by multiplying the frequency in HERTZ by 2π.

angular unconformity. An UNCONFORMITY marked by angular divergence between the older and younger rocks.

anhydrite. An EVAPORITE mineral, calcium sulphate ($CaSO_4$), found in SEDIMENTARY ROCKS.

anhydrous. Generally, containing no water; applied specifically to oxides, salts and other compounds to indicate the absence of water of crystallization or combined water.

animal. A living organism characterized chiefly by the holozoic mode of feeding (*see* HOLOZOIC NUTRITION). Most animals possess powers of locomotion (unlike most plants), their cells lack CHLOROPHYLL and cell walls, but the distinction between the plant and animal kingdoms becomes blurred in some simple organisms (e.g., PROTOZOA, MYXOMYCOPHYTA, unicellular ALGAE) which are now often regarded as a distinct kingdom, the PROTISTA.

animal unit. A unit of livestock, related to food needs and used in calculating stocking densities for management purposes, based on the equivalent of a mature cow of approximately 1000 pounds (454 kilogrammes) live weight. One cow is considered equal to one horse, one mule, five sheep, five swine or six goats.

anion. A negatively charged ION. In an electrolytic process, a potential gradient causes anions to migrate towards the anode, the (positive) electrode at which oxidation occurs. The other (negative) electrode is the cathode. *Compare* CATION.

Anisian. *see* TRIASSIC.

anisogamy. *See* HETEROGAMY.

anisotropic. Having different optical or other physical properties in different directions (e.g., QUARTZ, GRAPHITE).

ankerite. A carbonate mineral with the formula $Ca(Mg,Fe)(CO_3)_2$.

Annelida (Annulata). A phylum of animals comprising the ringed, or metamerically segmented worms. They possess a COELOM, well-developed blood and nervous systems, nephridia (*see* NEPHRIDIUM) and, typically, chaetae (bristles). The two main classes are the CHAETOPODA, which includes earthworms and ragworms, and the leeches (HIRUDINEA). In addition, there is a small class of marine worms, the Archiannelida, whose members usually lack bristles, also two other classes of marine worms (Echiuroidea and Sipunculoidea), which are often included in the phylum, although their members show few signs of segmentation.

annual. (1) Applied to events that occur once each year. (2) Applied to a plant that completes its life cycle from seed to seed in a single year. *Compare* BIENNIAL, EPHEMERAL, PERENNIAL.

annual cycle. One of the two cycles dominating the terrestrial environment (the other being the DIURNAL) and determined by obvious celestial mechanisms, in this case emphasized as a result of the inclination of the Earth's axis to the normal to the ECLIPTIC.

annual rings (growth rings). Concentric rings of secondary wood evident in a cross-section of the stem of a woody plant. The marked difference between the dense, small-celled late wood of one season and the wide-celled early wood of the next enables the age of a trunk to be estimated.

Annulata. *See* ANNELIDA.

anode. *See* ANION.

anodizing. The formation of a thin film of ALUMINA on the surface of ALUMINIUM by making the treated object the anode (*see* ANION) in an appropriate acid electrolyte.

This improves resistance to CORROSION.

anomaly. A departure from some defined value (e.g., AMBIENT or normal). Atmospheric anomalies include the temperature anomaly of rising warm air and anomalous visibility due to refraction (SUPERIOR IMAGE).

Anoplura (Siphunculata, sucking lice). Wingless, parasitic insects (division: EXOPTERYGOTA) with flattened bodies that live on mammalian blood and may transmit diseases to their hosts (e.g., the human body louse, *Pediculus humanus*, is a VECTOR of typhus). *See also* MALLOPHAGA, PARASITISM.

anorthite. *See* FELDSPAR.

anorthosite. A coarse-grained, IGNEOUS rock, usually PRECAMBRIAN, composed almost entirely of plagioclase FELDSPAR.

ANOVAR (analysis of variance). In statistics, a method of attributing amounts of variation to different causes.

anoxia (hypoxia). A deficiency of oxygen in tissues, blood or a body of water.

Anseriformes. *See* ANATIDAE.

antagonism. (1) Opposing effects produced by drugs, HORMONES, etc. on living systems (e.g., adrenaline accelerates heart beat, whereas ACETYLCHOLINE slows it). (2) The adverse effect of one organism on the growth of another. The interference may be due to the production of antibiotic substances, competition for food, etc. (3) The action of opposing pairs or sets of muscles (e.g., biceps and triceps) one of which must contract, whereas the other relaxes.

Antarctic. Applied to articles, substances, organisms, qualities, etc. on or near the continent (Antarctica) or seas surrounding the South Pole. *Compare* ARCTIC.

Antarctic Treaty. The international agreement, which came into force in 1961, stating that Antarctica shall be used for peaceful purposes only and containing provisions for the protection of Antarctic plants and animals. These provisions were further strengthened when the Treaty signatories, plus West Germany and East Germany agreed the Convention for the Conservation of Antarctic Seals (1972) and the Convention on the Conservation of Antarctic Marine Living Resources (CCAMLR, 1980). The Treaty has been signed by all 16 nations participating in Antarctic research: Argentina, Australia, Belgium, Chile, Czechoslovakia, Denmark, France, Japan, the Netherlands, New Zealand, Norway, Poland, South Africa, UK, USA and USSR.

antecedent drainage. A drainage system that maintains its original direction across a line of localized uplift that occurred after the system was established.

Antedon. *See* CRINOIDEA.

anthelion. A bright spot of white light at the same altitude as the Sun, but at opposite azimuth, produced by reflection and refraction in hexagonal ice prisms in the air whose axes are vertical.

anthelmintics. Drugs used against parasitic worms (HELMINTHS).

anther. *See* FLOWER.

antheridium. The male sex organ of algae, Fungi, Bryophyta and Pteridophyta, which produces SPERMATOZOIDS.

antherozoid. *See* SPERMATOZOID.

Anthocerotae. *See* BRYOPHYTA.

anthocyanins. GLYCOSIDE pigments, dissolved in cell sap, which impart colour to petals, autumn leaves, fruits, etc. Their colour depends on the pH of the cell sap, being red in acidic conditions and blue or violet in alkaline conditions.

Anthozoa. *See* ACTINOZOA.

anthracene. An AROMATIC HYDROCARBON obtained during coal distillation and used as an intermediate in the production of dyes.

anthracite. A coal of high RANK, with a vitreous lustre, hard enough to leave no mark on the finger when rubbed; it can even be made into jewelry. It contains a minimum of 86 percent carbon and a maximum of 14 per cent volatile matter. It burns slowly and (compared with other coals) cleanly.

anthrax (splenic fever, malignant pustule, woolsorters' disease). An acute, specific, infectious, febrile disease of mammals (including humans) caused by *Bacillus anthracis*, whose spores may persist in virulent form for many years in soil or other contaminated material. It occurs throughout the world and has caused major epidemics. It is transmitted to humans most commonly from infected wool or carcasses of diseased herbivores, and may begin as a localized skin infection which can lead to fatal septicaemia if it becomes generalized, or as a pulmonary infection (woolsorters' disease) caused by the inhalation of spores. The disease proceeds rapidly and is fatal. It may also occur as an intestinal infection following the ingestion of contaminated food. The spores can be transmitted aerially under certain conditions. The disease is one that has been considered as a weapon of war.

anthropic. Applied to a surface SOIL HORIZON, similar to a MOLLIC horizon, which is enriched in phosphate derived from cultivation over a long period. *See also* AGRIC HORIZON, PLAGGEN.

anthropic zone. The area of the Earth's surface that has been entirely transformed, or could be transformed, by humans for their own purposes.

anthropocentrism. The attitude that humans are separate from and, to some extent, in competition with all other species and that human activities need take no account of non-humans. It is a version of SPECIESISM.

anthropochore. A plant introduced into an area by humans.

anthropogenic. Produced as a result of human activities.

Anthropoidea. The suborder of PRIMATES that includes monkeys, apes (gibbons, orang-utan, gorilla, chimpanzee) and humans, all characterized by having large brains, forward-facing eyes and a high degree of manual dexterity.

anthropomorphism. The projection on to non-humans, especially mammals, of motives, emotions or thoughts that a human might experience in circumstances apparently similar to those affecting the individual animal being observed. It is a common attitude for a person to hold when confronting a non-human, but it has no scientific validity.

antibiosis. A special case of ANTAGONISM in which an organism produces a substance (an ANTIBIOTIC) injurious to other organisms.

antibiotic. A substance given out by one species that has an antagonistic (*see* ANTAGONISM) effect on organisms of other species. Some antibiotics produced by microorganisms are highly effective against other microorganisms (usually microfungi or bacteria) and are thus of outstanding medical importance. Examples include penicillin, produced by the mould *Pencillium notatum*, which inhibits the growth of many PATHOGENIC bacteria, and streptomycin, produced by the bacterium *Streptomyces griseus*, which is effective against the tuberculosis bacillus, which is not affected by penicillin.

antibody. A protein produced by an animal in response to the presence in its tissues of a foreign body (antigen), usually a protein or carbohydrate. Thus the presence of a parasite or its toxins stimulate the formation of specific antibodies which can kill or damage the parasite or neutralize its toxins. Natural immunity to diseases occurs because antibodies remain in the blood, in some cases throughout life, after the parasite has disappeared.

anticline. A FOLD that is convex upwards,

with the oldest rocks occupying the inner core. *Compare* SYNCLINE.

anticlinorium. A complex ANTICLINE in which the limbs are folded. The term is applied to large-scale structures.

anticyclone. A centre of high atmospheric sea-level pressure, where the air circulates in the direction, relative to the Earth, opposite to the vertical component of the Earth's rotation (anticyclonically, clockwise in the northern hemisphere, anticlockwise in the southern hemisphere). The circulation is CYCLONIC in space, but slower than the Earth's rotation. In a cold anticyclone, the high pressure is due to cold air in the lowest 2–5 kilometres, but in a warm anticyclone it is due to cold air at greater altitudes, usually above the TROPOPAUSE.

antigen. A substance that stimulates the production of an ANTIBODY.

antigibberellin. An organic compound that causes stunted growth in plants or has opposite effects to those of the GIBBERELLIN auxins. MALEIC HYDRAZIDE is an example.

antiknock additive. A compound added to petrol (gasoline) to prevent preignition (knocking or pinking) in internal combustion engines. The most widely used is tetraethyllead, the main source of lead in air, and hence a pollutant, for which reason 'leaded petrol' is being phased out in many countries.

antimetabolite. A substance that inhibits the action of a METABOLITE. Vitamins, amino acids, some proteins and other substances necessary to living organisms can be prevented from performing their metabolic functions by an interaction between an antimetabolite and an enzyme. Antimetabolites are often related structurally to the nutrients or metabolites they inhibit.

antimony (Sb). An element whose compounds are used in metal used in letterpress printing, dyestuffs and lead–acid batteries. Prolonged exposure can lead to heart disease and dermatitis. A_r=121.75; Z=51; SG 6.69; mp 630°C.

antinatalist. Applied to views or policies, and persons holding or supporting them, designed to reduce BIRTH RATES. The opposing pronatalist views or policies aim to increase the birth rate and so achieve an increase in human numbers.

antinode. In acoustics, the point, line or surface where the amplitude of sound is at its maximum. It is the opposite of node, where sound amplitude is zero.

antisolar point. The point opposite the Sun, in the direction of the observer's shadow.

antistatic agents. Compounds that give sufficient conductivity to ordinarily non-conducting surfaces to prevent the accumulation of charges of STATIC ELECTRICITY.

antitriptic wind. A wind blowing against significant drag (*see* AERODYNAMIC DRAG) at the surface, the drag being comparable to the force due to the pressure gradient and acting in the opposite direction to the wind. Examples include sea breezes early in the day, winds in canyons blowing towards the lower-pressure end and KATABATIC WINDS, especially in ravines.

antlers. Horns of deer, which are equivalent to the horns of other members of the RUMINANTIA. In most species antlers are grown only by the males, but in reindeer they are grown by both sexes. They grow as bony processes from the frontal bones of the skull, and reach their full size quickly. They are shed and grown anew each year, usually with more branches being added at each regrowth. Antlers are employed in ritualistic mating contests between rival stags, but not for combat with other species, when the forehoofs are used. Reindeer use the widened part of their antlers to shovel snow away from the plants on which they feed.

ants. *See* HYMENOPTERA.

Anura (Salientia; frogs, toads). An order of AMPHIBIA characterized by the absence of a tail and the possession of long hind legs used for jumping.

anvil cloud. *See* CUMULONIMBUS.

AONB. *See* AREA OF OUTSTANDING NATURAL BEAUTY.

apatetic coloration. Protective coloration which misleads an animal's predators (e.g., by masking the body outline — disruptive coloration — or by producing the appearance of eyes where there are none).

apatite ($Ca_5(OH,F,C1)$ $(PO_4)_3$). A group of minerals found in IGNEOUS rocks and metamorphosed (*see* METAMORPHISM) LIMESTONES. It is the main source of PHOSPHATE. Bones and teeth are formed from varieties of apatite.

apetalous. Of flowers, having no true petals.

aphagia. Failure to eat, which can lead to death and may necessitate forced feeding in animals. *Compare* HYPERPHAGIA.

Aphaniptera (Siphonaptera; fleas). An order of wingless, blood-sucking insects (division: ENDOPTERYGOTA) that are parasitic on mammals and birds. The adults have laterally compressed bodies and strong legs for jumping, and the larvae are grub-like. Most are specific to one or a few hosts. Some rat fleas transmit the plague bacillus from rats to humans, and the rabbit flea is the vector of MYXOMATOSIS. *See also* PARASITISM.

aphanitic. Applied to an IGNEOUS rock in which the individual crystals of the ground mass cannot be distinguished by the naked eye.

Aphidae. *See* APHIDIDAE.

Aphididae (in North American usage, Aphidae, aphids). Members of the insect order HEMIPTERA, whose mouthparts are adapted for piercing and sucking plants.

They are notable for their widespread distribution and prolific reproduction, and are of great economic importance as VECTORS of virus diseases of crops (e.g., sugar-beet yellows). The annual life cycle of *Aphis rumicis*, which is typical, involves the production of many generations of wingless, parthenogenetic (*see* PARTHENOGENESIS) females that produce VIVIPAROUS young, the host plants being the spindle tree (*Euonymus europaeus*) and crops such as beans. In the autumn, OVIPAROUS females are produced, which mate with winged males and lay eggs that overwinter in the spindle tree. New adults are distributed by air motion, becoming airborne on warm days. They exemplify the principle of survival by dispersion and often travel many kilometres when carried to heights of several kilometres by convection currents.

aphids. *See* APHIDIDAE.

aphotic zone. The deeper parts of seas and lakes, where light does not penetrate.

aphytic zone. The part of a lake floor that, by reason of its depth, lacks plants.

apical meristem. *See* MERISTEM.

API scale. An arbitrary gravity scale adopted by the American Petroleum Institute (API) for use with crude oil. API gravity values, expressed in degrees API or, more usually, just degrees, are related to the formula: degrees API = (141.5-131.5)/specific gravity at 60°F. Hence a heavy crude has a numerically low API gravity. Most Middle Eastern crudes fall in the range 27.0–43.0°API.

Apocrita. *See* HYMENOPTERA.

Apoda (Gymnophiona, caecilians).   An order of limbless, worm-like, burrowing AMPHIBIA from tropical countries. Unlike other present-day amphibians, some caecilians have small scales embedded in the skin.

apodeme. *See* ENDOSKELETON.

Apodidae (Micropodidae, swifts). A family (order: Apodiformes or Micropodiformes, which also includes the humming birds) of swallow-like birds. They are accomplished fliers and highly adapted for life in the air. Swifts fly for very long periods, catching insects in their wide mouths and are important in controlling insect numbers. They can roost on the wing and have difficulty taking off if they are grounded.

apomixis. Reproduction by seeds formed without fertilization. This is the usual form of reproduction in some plants (e.g., dandelion) and can be taken to include PARTHENOGENESIS. *Compare* AMPHIMIXIS.

aposematic coloration (sematic coloration).   Warning coloration consisting of conspicuous markings on an animal that is poisonous or distasteful to others (e.g., the black and yellow markings on wasps and on the poisonous amphibian *Salamandra maculosa*).

apparent mortality. *See* SUCCESSIVE PERCENTAGE MORTALITY.

appetitive behaviour. Behaviour that increases the chance of an animal satisfying a need (e.g., when hungry, moving around until food is found). As defined by W.H. Thorpe, it is the variable introductory phase of an instinctive pattern or sequence.

apple capsid. *See* CAPSIDAE.

apple scale. *See* SCALE INSECTS.

Appleton layer. The atmospheric region around 200 kilometres high where IONIZATION by solar radiation causes the refraction and reflection of radio waves. It is named after Sir Edward Appleton (1892–1965), the inventor of radar.

application factor. The ratio between that concentration of a substance which produces a particular chronic response in an organism and that which causes death in 50 percent of the population within a given period of time (the LD_{50}). The application factor is used to determine the maximum safe concentration of substances for the organisms under consideration.

appressed. Pressed together, but not united; applied most commonly to plant organs.

appropriate technology. A technology that is suitable for a particular group of people at their particular stage of economic development. A technology is considered appropriate if its use can be mastered fully by those it is intended to benefit (i.e. it does not require operators from elsewhere), it does not depend on fuel or other resources that are difficult to obtain locally, and its machinery and equipment can be maintained by the communities using it.

Apterygota (Ametabola). A subclass of primitive, wingless insects whose young closely resemble the adults, including the Collembola (springtails) and Thysanura (bristletails, such as silverfish). Some Thysanura are pests of stored food. *Compare* PTERYGOTA.

Aptian. A stage of the CRETACEOUS System.

aquaculture (pisciculture, fish farming, fish ranching). The breeding and rearing of freshwater or marine fish in captivity.

aquafalfa. Ground where the WATER TABLE is high.

aquamarine. *See* BERYL.

aqua regia. *See* GOLD.

aquiclude (in North American usage, aquitard). A geological rock or soil formation that is sufficiently porous to absorb water slowly, but does allow water to pass quickly enough to furnish a supply for a well or spring. CLAYS and SHALES are commonly regarded as aquicludes.

aquifer. A geological formation through which water can percolate, sometimes very slowly, for long distances. Springs and wells are charged from aquifers and the contamination of an aquifer may lead to the contamination of wells and springs over a wide area. Aquifers are described as confined when they are overlain by AQUICLUDES and unconfined when their upper surface is at ground level. *See also* ARTESIAN BASIN, GROUND WATER, PERMEABILITY, POROSITY, WATER TABLE.

aquiherbosa. Submerged aquatic vegetation (marsh and fen vegetation) and SPHAGNIHERBOSA.

aquiprata. Plant communities in wet meadows.

aquitard. *See* AQUICLUDE.

Ar. *See* ARGON.

A_r. *See* ATOMIC WEIGHT.

Arab League Educational, Cultural and Scientific Organization (ALECSO). The intergovernmental body, based in Tunis, that is responsible for monitoring pollution in the Red Sea and Gulf of Aden. It works closely with UNEP, UNESCO, IUCN and IMCO.

arachidonic acid. A constituent of VITAMIN F.

Arachnida (spiders, mites, ticks, harvestmen, scorpions, king crabs). A class of ARTHROPODA, all of whose members have four pairs of walking legs, prehensile first appendages instead of antennae and simple eyes. *See also* ACARINA, ARANEAE, PENTASTOMIDA, SCORPIONIDEA, TARDIGRADA, XIPHOSURA.

aragonite. *See* CARBONATE MINERALS.

Araneae (spiders). ARACHNIDA whose bodies are divided into two parts and which have abdominal silk glands and poison glands at the bases of their first appendages. All are carnivores and are important in regulating populations of small invertebrates, some chasing their prey, others using webs or trapdoors. Food is digested externally with saliva, then sucked up as a fluid.

arboreal. Pertaining to trees.

arboretum. An area (often parkland) dominated by trees.

arboriculture. The cultivation of trees.

arborist. A person who studies trees.

arc furnace. Any furnace heated by an electric arc, rather than by a fuel or external heat.

Archaean. Refers to the oldest known PRECAMBRIAN rocks.

arch clouds. WAVE CLOUDS that extend a considerable distance along a mountain range and remain stationary, with a wind blowing through them, having well-defined upwind and downwind edges, and giving the appearance of an arch in the sky (e.g., the Chinook Arch in the Rocky Mountains, the Southern Arch in the Southern Alps, New Zealand).

Archegoniatae. A group of plants comprising the BRYOPHYTA and PTERIDOPHYTA, members of the CRYPTOGAMIA in which the female sex organs are archegonia (*see* ARCHEGONIUM).

archegonium. The female sex organ of BRYOPHYTA, PTERIDOPHYTA, and most GYMNOSPERMAE, comprising a hollow, multicellular, flask-shaped organ containing an egg cell.

Archiannelida. *See* ANNELIDA.

Archimedes' principle. The buoyancy force on an object immersed in a fluid is equal to the weight of the fluid the object displaces. Archimedes, a Greek engineer of the Alexandrian school (c. 250 BC), first appreciated that by means of this principle the density of the material of an object of complex shape could be measured accurately as a multiple of the density of the fluid.

Arctic. The region around the North Pole, within the Arctic Circle (i.e. all latitudes higher than 60°30′N). The Arctic Circle defines the region north of which there is at least one period of 24 hours in summer during which the Sun does not set, and one 24-hour period in winter during which it does not rise. The word is derived from the Greek *arktos* meaning bear, referring to the constellation of The Bear. The climate is characterized by extreme fluctuations between summer and winter temperatures, permanent snow and ice in the uplands, and PERMAFROST, which may melt superficially in summer, in the lowlands.

arctic. Applied to high-latitude regions, which may or may not lie inside the geographical ARCTIC Circle, from which tree growth is usually absent because of unfavourable conditions (e.g., shortness of the growing season). Also applied to the plants and animals living there. *Compare* ANTARCTIC.

arctic brown earth. *See* TUNDRA SOIL.

Arctic Circle. *See* ARCTIC.

Arctic Floral Region. The region of the HOL-ARCTIC REALM that comprises all the northern and arctic lands as far south as central Alaska, Labrador, central Scandinavia and northern Siberia.

arctic smoke. Steaming FOG, produced when very cold air passes over warm water. It is common off the coast of northern Norway and, on cold mornings, on rivers warmed by cooling water from power stations.

Arctogea. The ZOOGEOGRAPHICAL REGION that includes most of the northern hemisphere as far south as Mexico and the Himalayas, and all of Africa, distinguished by the similarity of the fauna found throughout it. Arctogea is subdivided into the PALAEARCTIC REGION, NEARCTIC REGION, ETHIOPIAN REGION and ORIENTAL REGION.

Area of Outstanding Natural Beauty (AONB). As defined by the COUNTRYSIDE

COMMISSION, an area in England or Wales, not being a NATIONAL PARK, designated for planning purposes under the National Parks and Access to the Countryside Act, 1949, in order to preserve and enhance its natural beauty. AONB designation enables local planning authorities to operate more stringent policies of development control.

Arecaceae. *See* PALMAE.

arenaceous. *See* ARENITE.

Arenigian. The oldest series of the ORDOVICIAN System in the UK. In continental Europe, the Tremadocian (*see* TREMADOC SERIES) is considered to be Ordovician rather than CAMBRIAN.

arenite. A SEDIMENTARY rock made up of grains of sand 0.06–2 millimetres in diameter. Sediments formed mainly from arenite grains are termed arenaceous. *Compare* ARGILLITE, RUDITE.

arête. A sharp-edged ridge left between two CIRQUES as the ice hollows them out.

Argentinian Floral Region. The part of the AUSTRAL REALM that includes Argentina, Paraguay, the southern half of Chile and their offshore islands, including the Falklands (Malvinas).

argentite. A mineral; a form of silver sulphide (Ag_2S) stable above 180°C. ACANTHITE is the form stable below 180°. These two minerals are the main source of silver. Argentite is found in HYDROTHERMAL deposits, characteristically with lead, zinc and copper minerals.

argillaceous (argillic). Rich in CLAY minerals. *Compare* ARGILLITE.

argillic. *See* ARGILLACEOUS.

argillite (lutite). SEDIMENTARY ROCK with grains less than 0.06 millimetres in diameter. It is slightly metamorphosed (*see* METAMORPHISM) and does not split easily. *Compare* ARENITE, RUDITE.

arginine. An AMINO ACID with the formula $NH_2C(:NH)NH(CH_2)_3(NH_2)COOH$ and with a molecular weight of 174.2.

argon (Ar). An element ($A_r = 39.948$); an inert gas that comprises nearly 1 percent of the Earth's atmosphere by weight. It has a radioactive ISOTOPE, ^{14}Ar, with a HALF-LIFE of 30 days, which is formed in the air-cooling systems of some nuclear reactors. $Z = 18$.

Aridisols. *See* SOIL CLASSIFICATION.

arid zone. A zone of latitude (15–30° in both hemispheres) of very low rainfall, which contains most of the world's deserts. Rain does occur there, but either evaporates or sinks underground, so it is of little use to vegetation. An oasis occurs where the WATER TABLE is close to the surface. Even in arid zones, there are sometimes years in which cultivation or grazing is possible, but there is a danger of overexploitation at such times. High populations cannot be supported in a long sequence of dry years, and at such times overgrazing or tillage can cause EROSION and the destruction of TOPSOIL. *See also* DROUGHT, SAHEL.

aril. An additional covering to the seed in some plants (e.g. yew, *see* TAXACEAE).

Aristotle's lantern. *See* ECHINOIDEA.

arithmetic growth. *See* LINEAR GROWTH. *Compare* EXPONENTIAL GROWTH.

arithmetic mean. In statistics, the sum of a set of VARIABLES divided by the number of variables in the set.

arkose. A FELDSPAR-rich SANDSTONE, strictly containing 25 percent or more feldspar.

armadillos. *See* EDENTATA.

Armorican. The processes and products of mountain building in Europe in late PALAEOZOIC times. Armorican was originally restricted geographically to Brittany and the

Massif Central, but most current usage is broader; Armorican and HERCYNIAN are used interchangeably.

aromatic. Applied to organic compounds derived from BENZENE and having closed rings of carbon atoms. *Compare* ALIPHATIC. *See also* BENZENE RING.

arrow-worms. *See* CHAETOGNATHA.

arroyo (wadi). A steep-sided, flat-floored stream channel; a landform of arid regions.

arsenic (As). An element that exists in three allotropic (*see* ALLOTROPY) forms: ordinary grey crystalline arsenic (SG 5.727); black, metalloid arsenic (SG 4.5); and yellow arsenic (SG 2.0). It occurs in elemental form, combined with sulphur as realgar (As_2S_2) or orpiment (As_2S_3), with oxygen as white arsenic (As_2O_3), and with some metals. Its compounds are very poisonous and are used medicinally. In the elemental form arsenic is used in alloys for sporting ammunition, cable sheathing and transistors. It was formerly used widely in insecticides. A_r=74.9216; Z=33.

arsenopyrite (FeAsS). A mineral found in HYDROTHERMAL deposits. Arsenopyrite is a major ore of ARSENIC and was formerly mined extensively to provide insecticides against such pests as cotton-boll weevil. It is a common GANGUE mineral of hypothermal deposits of gold and tin, and of tungsten ores. It was formerly known as mispickel.

Artemisia. A genus of COMPOSITAE common in arid areas, some species of which are cultivated elsewhere. *A. tridentata* is American sagebrush; *A. absinthum* is wormwood and is used to flavour absinthe.

artesian basin. A SYNCLINE in which the level of GROUND WATER in the OUTCROPS of an AQUIFER is sufficient to produce natural flow from wells tapping the confined part of the aquifer. *See also* ARTESIAN WELL.

artesian well. A well bored into a confined

AQUIFER which overflows the well head (i.e. the PIEZOMETRIC LEVEL of the aquifer is above the level of the well head).

Arthropoda. The largest phylum of animals, containing more than 700 000 species, and the only major invertebrate phylum with members adapted for life on truly dry land. Arthropods are characterized by the possession of a jointed EXOSKELETON containing CHITIN, paired jointed limbs, a well-developed head and a haemocoelic (blood-containing) body cavity. *See also* ARACHNIDA, CRUSTACEA, INSECTA, MYRIAPODA, ONYCHOPHORA, TARDIGRADA, TRILOBITA .

artificial fertilizer. *See* NPK.

artificial formation. A vegetation pattern caused by the continuous activity of humans.

artificial insemination (AI). The practice of introducing semen artificially into female animals for breeding purposes. The semen of a male with desirable hereditary qualities can be preserved under refrigeration and used to inseminate numerous females when optimum conditions occur.

artificial insemination by donor. *See* AID.

artificial rain. Rain caused by an artificial stimulus. Examples include common salt, which forms large droplets because it is HYGROSCOPIC, dry ice (solid carbon dioxide) pellets which sublime (*see* SUBLIMATION) at −72°C and so leave a trail of air cooled to near the temperature at which supercooled droplets freeze, or silver iodide (AgI) which has a crystal structure close enough to that of ice to initiate crystal growth in droplets supercooled to below about -4°C. In the latter cases the crystals formed grow by the BERGERON-FINDEISEN MECHANISM; in all cases the larger particles grow by ACCRETION of smaller, slower-falling cloud droplets. Although FALLSTREAK HOLES have been made in layers of supercooled clouds and some CUMULUS appears to have been made to grow into showers more quickly, the technique is unreliable and of limited value.

artificial recharge. The introduction of surface water into an underground AQUIFER, through recharge wells.

artificial reef. A REEF-like structure made in the sea from solid waste materials to provide cover for marine organisms. As a method of waste disposal it is effective, although limited by the availability of suitable sites. Its ecological advantage is more doubtful, and its effects on commercial fisheries are negligible.

Artiodactyla (Paraxonia). The even-toed ungulates (hoofed animals), including cattle, sheep, goats, pigs, deer, giraffes, hippopotamus and camels. They are large, herbivorous, cloven-hoofed animals, some of which chew the cud. *Compare* PERISSODACTYLA. *See also* RUMINANTIA.

Arusha Declaration. The declaration issued following an International Union for Conservation of Nature and Natural Resources conference at Arusha, Tanzania, in 1961, which recognized the importance of the survival of wildlife in Africa 'as a source of wonder and inspiration (and as) an integral part of our national resources and of our future livelihood and well-being'.

As. *See* ARSENIC.

asbestos. The fibrous form of several SILICATE MINERALS, at one time widely used for electrical and thermal insulation and fire protection, and as a strong, durable and cheap construction material (e.g., in tiles and roofing sheets). Chrysotile, a fibrous serpentine mineral, accounted for about 90 percent of the asbestiform minerals mined; blue asbestos (crocidolite) is natural sodium iron silicate. Asbestos fibres are non-flammable and stable in many corrosive environments, and they can be woven or bound by inert media. Prolonged exposure to asbestos fibres can cause ASBESTOSIS and cancers (e.g., MESOTHELIOMA), and the use of all forms of asbestos is now either banned or strictly controlled in many countries, although it is still encountered (e.g., in demolition work).

asbestosis. A dust disease caused by the inhalation of ASBESTOS fibres. Each fibre entering the lungs may scar the oxygen-absorbing lung tissue, reducing lung function and leading to chronic shortness of breath.

Aschelminthes. A phylum used by some taxonomists that includes the ROTIFERA, NEMATODA, NEMATOMORPHA, and some of the Gephyrea.

ascidians. Sea squirts. *See* UROCHORDATA.

Ascomycetes (powdery mildews). A group of fungi characterized by the production of cylindrical or spherical spore sacs (asci) containing eight SPORES. Some species cause diseases of trees, others impart flavour to cheese and some are edible (e.g., truffles, morels). A few are used to produce important drugs (e.g., *Penicillium* gives penicillin; *Claviceps* causes ERGOT poisoning, but provides ergotamine). The *Saccharomyces* yeasts used in brewing and baking also belong to this group.

ascorbic acid. *See* VITAMIN C.

asexual reproduction. Reproduction that does not involve FERTILIZATION. The term includes SPORE formation and propagation by means of BULBS, cuttings, etc. in plants and reproduction by FISSION or budding (*see* GEMMATION) in animals. *Compare* SEXUAL REPRODUCTION.

ash. (1) The mineral content of a substance that remains after complete combustion. (2) PYROCLASTIC fragments, less than 4 millimetres in their long dimension, ejected by a volcano. Volcanic ash is a major atmospheric pollutant. Injected into the atmosphere it may cause haze and reduce surface temperatures, and where it settles out in large amounts it may coat leaves, block STOMA and so retard plant growth. (3) *See FRAXINUS EXCELSIOR.*

Ashgillian. The youngest series of the ORDOVICIAN System in Europe.

asparagine. An AMINO ACID with the formula $NH_2COCH_2CH(NH_2)COOH$ and with a molecular weight of 132.1.

aspartic acid. An AMINO ACID with the formula $COOHCH_2CH(NH_2)COOH$ and with a molecular weight of 133.1.

asphalt. A brown to black, solid or semi-solid bituminous substance (*see* BITUMEN) that occurs naturally but is also obtained as a residue during the refining of certain petroleums. Asphalt melts between 65 and 100°C. It is used in the construction of roads, as a waterproofing component in roofing, in FUNGICIDES and in paints.

assembly. In ecology, the smallest COMMUNITY unit of plants or animals (e.g., a colony of blackfly (*Aphis fabae*) on a broad bean plant).

assimilation. (1) The incorporation of food substances into the cells, tissues or organs of a living organism. (2) The natural process by which a water body purifies itself of organic pollution, largely by the action of micro-organisms.

assimilative capacity. *See* ABSORPTIVE CAPACITY.

association. (1) In ecology, a stable plant COMMUNITY of definite floristic composition, dominated by particular species (*see* DOMINANT) and growing under uniform habitat conditions. It is the largest natural vegetation group recognized (e.g., oak–ash woodland) in some systems, but the term is also applied to smaller units in Europe. (2) In statistics, the degree of dependence or independence between two or more variables, especially in cases of simple two-fold classification into A or not A, B or not B. *Compare* CORRELATION, REGRESSION.

assortative mating. Mating that is non-random and involves the tendency of females and males with particular characteristics to breed together.

astatine. *See* HALOGENS.

Asteroidea (starfishes). A class of predaceous ECHINODERMATA whose members have flattened, star-shaped bodies with arms not sharply distinct from the central region of the body. They move by means of tube feet, feed mainly on molluscs, and some may cause losses on commercial oyster beds. *See also* OPHIUROIDEA .

asthenosphere. (1) The LOW-VELOCITY ZONE in which seismic (earthquake) waves travel at much reduced speeds. The development of PLATE TECTONICS has made this much the most common use of the term. (2) The less rigid part of the Earth's interior (i.e. the outer CORE). (3) The MANTLE and core. (4) The lower part of the CRUST, where adjustment for ISOSTASY is believed to occur.

asulam. A TRANSLOCATED HERBICIDE; a CARBAMATE used to control docks in grassland and orchards, and bracken in grassland, forest plantations and upland pastures.

asymmetrical fold. A FOLD in which one limb dips (*see* DIP) away from the axis more steeply than the other. Such a fold will have an inclined AXIAL PLANE.

Athens Treaty on Land-Based Sources of Pollution. An international treaty drawn up in 1980, which came into force in 1983 when six of its 16 signatory governments ratified it. It aims to reduce pollution in the Mediterranean. *See also* MEDITERRANEAN ACTION PLAN, REGIONAL SEAS PROGRAMME.

Atlantic North American Floral Region. The region of the HOLARCTIC REALM that comprises North America south of the Great Lakes, west of the Rocky Mountains and south to the coast of the Gulf of Mexico.

atmosphere. (1) The gaseous envelope surrounding a planet. The Earth's atmosphere consists by volume of NITROGEN (79.1 percent), OXYGEN (20.9 percent), CARBON DIOXIDE (about 0.03 percent) and traces of the noble gases (i.e. ARGON, krypton, xenon, HELIUM) plus water vapour, traces of AMMONIA, organic matter, OZONE, various salts and suspended solid particles. *See also* EVOLUTION OF

THE ATMOSPHERE. (2) A unit of pressure, defined as that pressure which will support a column of mercury 760 millimetres high at 0°C, at sea level, at a latitude of 45°; 1 atmosphere is equal to 101.325 newtons per square metre.

atmospheric dispersion. The mechanism of dilution of gaseous or smoke pollution whereby the concentration is progressively decreased. The dilution rate is very variable, and pollution incidents occur when the atmospheric dispersion is insufficiently effective. Atmospheric dispersion is a most important worldwide mechanism for the distribution of salts by rain and is relied upon for the removal of the products of combustion. Because of its complexity and variability it is subject to only very approximate mathematical treatment.

atmospheric tides. Tides in the atmosphere, like ocean tides produced by the gravitational influence of the Sun and Moon, but of negligible magnitude. A solar semidiurnal tide is generated by the daily variations in atmospheric heating and consequent expansion. The wave travels around the world in 24 hours, with two crests 12 hours apart. It is of the order of 2 millibars at the equator, with crests at 10 h and 22 h, solar time. The gravitational field of the Sun exerts a couple on this tide, accelerating the Earth's rotation.

atoll. *See* CORAL REEF.

atomic energy. Energy, as heat and radiation, derived in a controlled fashion from NUCLEAR FISSION or NUCLEAR FUSION in a NUCLEAR REACTOR.

Atomic Energy Authority. *See* UNITED KINGDOM ATOMIC ENERGY AUTHORITY.

Atomic Energy Commission (AEC). The US federal agency responsible for all aspects of ATOMIC ENERGY in the USA.

atomic number (Z). The number of electrons surrounding the nucleus of the neutral atom of an element, or the number of

protons in the nucleus.

atomic spectrum. The spectrum emitted by an excited atom (*see* EXCITATION) due to changes within the atom. It is often used to identify trace metals in the environment.

atomic weight (A_r). The relative atomic mass, being the ratio of the average mass per atom of a specified isotopic composition (*see* ISOTOPE) of an element to 1/12 of the mass of an atom of $^{12}_{6}C$.

atomization. The dividing of a liquid into extremely small particles, so increasing its surface area. *See also* AEROSOL, AEROSOL SPRAY.

atomizer. *See* AEROSOL SPRAY.

ATP (adenosine triphosphate). A COENZYME that provides a source of energy for activities such as muscle contraction and the synthesis of organic compounds in plants and animals. By removing a phosphate group from ATP, thus forming ADP (adenosine diphosphate), and transferring it to other substances, ENZYMES are able to make this energy available. ADP is subsequently rebuilt to ATP, using energy from CATABOLISM or from light during PHOTOSYNTHESIS.

ATPase. An ENZYME that breaks down ATP to ADP.

attenuation. The loss of virulence in bacteria or other PATHOGENIC microorganisms.

Atterberg limits. A set of three limits, based on tests, beyond which a brittle soil which is drying changes from a viscous liquid to a plastic solid to a brittle solid. Below the shrinkage limit the water content is such that no change in volume occurs on drying. The liquid limit represents the minimum water content at which soil will flow under a specified applied force. The plastic limit represents the minimum water at which plastic deformation can occur. The limits are of importance in SOIL CREEP, SOLIFLUCTION, EROSION, etc. and in civil engineering.

atto- (a). Prefix used in conjunction with SI units to denote the value of the unit x 10^{-18}.

attribute. A qualitative characteristic of an individual (e.g., male or female). *See also* VARIABLE.

audio frequency. A sound frequency within the audible range, about 20–20 000 Hz.

audiogram. A graph, usually plotted by an AUDIOMETER, that describes the response of a subject (listener) to sound as a function of sound frequency, usually measuring each ear separately.

audiometer. An instrument that produces an AUDIOGRAM. It feeds calibrated pure sound tones into an earphone and records signals made by the subject (listener) to indicate whether or not the sounds have been heard.

auditory ossicles. *See* EAR.

Audubon, John James (1785–1851). An early American conservationist, famous for his paintings of birds, who initiated the ringing of birds in order to study their migration. Modern conservation groups, most notably the NATIONAL AUDUBON SOCIETY, are named after him.

Audubon Society. *See* NATIONAL AUDUBON SOCIETY.

aufeis. Layer upon layer of frozen water, discharged sequentially, as a spring.

auger. A tool for drilling and collecting cores of soil or surface deposits.

augite. A calcium magnesium aluminosilicate mineral that is one of the PYROXENES.

auks. *See* ALCIDAE.

aureole. The bright orange glow sometimes seen around the Sun in HAZE or thin cloud. *See also* CORONA.

aurochs (*Bos primigenius*). Literally, the original ox; the species of wild cattle from which modern domestic cattle are descended, and which became extinct in the 17th century. *See also* BOVIDAE.

aurora borealis (northern lights). The curtain-like flashing formations of white and coloured lights seen in the sky in high latitudes that occur in the ionized layers (*see* IONIZATION) about 400 kilometres above the ground. The phenomenon is caused by particles emitted by sunspots, which travel down the lines of force of the Earth's magnetic field. The lights appear to emanate from near the magnetic pole.

Austausch coefficient. The coefficient originally proposed by the German physicist Ludwig Prandtl (1875–1953) to represent the diffusive effect of eddies in a fluid according to the mixing length theory. The German word *Austausch* means exchange.

Australasian Region. *See* NOTOGEA.

Australian Floral Region. The region of the AUSTRAL REALM that includes Australia and Tasmania.

Australian Region. *See* NOTOGEA.

Austral Realm (Southern Realm). The FLORAL REALM that includes the southern part of South America, Australasia, the southern tip of South Africa and the southern oceanic islands. It is divided into five regions: SOUTH AFRICAN FLORAL REGION; ARGENTINIAN FLORAL REGION; AUSTRALIAN FLORAL REGION; NEW ZEALAND FLORAL REGION; SOUTH OCEANIC FLORAL REGION.

autecology. The study of the relationships between a single species and its environment. *Compare* ECOLOGY, SYNECOLOGY.

authigenic. Applied to MINERALS that grew *in situ*, during or after deposition. *Compare* ALLOGENIC.

autochthonous. (1) Applied to soil organisms whose activity is not affected by the addition of organic material to the soil. *Compare* ZYMOGENOUS. (2) Applied to rocks or organic material in which the major constituents have been formed *in situ. Compare* ALLOCHTHONOUS.

autoclave. A strong pressure vessel, working on the same principle as a domestic pressure cooker, used to sterilize instruments and equipment and to carry out reactions at high pressures and sometimes at high temperatures.

autocoprophagy. *See* REFECTION.

autoecious parasite. *See* PARASITISM.

autogamy. (1) In plants, self-fertilization. (2) (paedogamy) A process occurring in some diatoms (*see* BACILLARIOPHYTA) and PROTOZOA in which the cell divides to form two GAMETES, which then reunite.

autogenic. *See* AUTONOMIC.

autogenic succession. A SUCCESSION that results from changes created by the organisms themselves. *Compare* ALLOGENIC SUCCESSION.

autoimmunity. The condition in which the immune responses (*see* IMMUNITY) of an animal are directed against its own tissues.

autolysis. The breakdown of animal and plant tissues that occurs after the death of the cells and is carried out by enzymes within those cells.

autonomic (autogenic). Applied to spontaneous movements (e.g., NUTATION) that occur in response to external stimuli. *Compare* PARATONIC.

autonomic nervous system. The sympathetic and parasympathetic nerve supply to the involuntary muscle (e.g., heart, blood vessels, gut) and glands of vertebrates, together with the sensory nerve fibres which pass impulses from internal sense organs to the central nervous system (brain and spinal

cord). Some of the cell bodies of the autonomic nervous system are outside the central nervous system, situated in ganglia (e.g., the solar plexus). Where an organ is supplied by both sympathetic and parasympathetic systems, the two act antagonistically (*see* ANTAGONISM). For example, the heart beat is accelerated by the sympathetic supply and slowed by the parasympathetic, whereas peristaltic movements of the gut are stimulated by the parasympathetic and inhibited by the sympathetic supply. ACETYLCHOLINE is released at endings of parasympathetic nerves, whereas adrenaline is released at many sympathetic nerve endings.

autopolyploid. An organism with more than two haploid sets of CHROMOSOMES all derived from the same species. *See* ENDOMITOSIS.

autosome. A CHROMOSOME other than a sex chromosome.

autotomy. The automatic self-amputation of part of an animal, usually followed by its regeneration. For example, by intense muscular contraction a lobster will break off a limb when seized by a predator, and some lizards can lose their tails.

autotrophic. Applied to organisms that produce their own organic constituents from inorganic compounds, utilizing energy from sunlight or oxidation processes. CHLOROPHYLL-containing plants are autotrophs which elaborate organic compounds from simple substances (i.e. carbon dioxide, water) during PHOTOSYNTHESIS. A few bacteria are autotrophic. Those containing BACTERIOCHLOROPHYLL are PHOTOTROPHIC; others (e.g., those that obtain energy by oxidizing hydrogen sulphide) are CHEMOTROPHIC. *Compare* HETEROTROPHIC.

auxins. Plant HORMONES (e.g., INDOLE -3- ACETIC ACID) that are made in the growing tips of stems and roots and move to other parts of the plant, where they regulate growth. They may initiate, for example, cell division in cambium (*see* MERISTEM), increase the rate of cell enlargement causing a part of the plant to curve (as in twining), initiate root formation in cuttings, inhibit bud development and fruit drop, or initiate flower formation. *See also* ABSCISIN, CYTOKININS, GEOTROPISM, GIBBERELLINS, HORMONE WEEDKILLERS, PHOTOTROPISM.

auxin-type growth regulators.   *See* HORMONE WEEDKILLERS.

auxotroph.   A strain of microorganism which, usually because of a MUTATION, has lost the ability to synthesize a particular nutrient and thus has nutritional requirements additional to those of the organism in its natural state. *Compare* PROTOTROPH.

avalanche.   A catastrophic slide of rock debris and/or ice and snow; the most rapid form of MASS WASTING.

average.   A value that summarizes a set of values (e.g., geometric, or more usually, ARITHMETIC MEAN). It is not always obvious which averages are significant statistically. Usually they are defined for reference purposes and then called normal. Atmospherically, average wind speed, for example, represents the run of the wind, but the average of the cube of the speed of the wind represents the power obtainable aerodynamically from an AEROGENERATOR or WINDMILL, if it can make use of strong winds. The average day temperature is often defined as being midway between the maximum and minimum, but this has many disadvantages if the aim is to discern trends.

Aves (birds).   The class of warm-blooded vertebrates that have feathers and forelimbs modified as wings. Birds have marked affinities to reptiles and probably evolved from reptiles related to primitive dinosaurs. Some fossil forms (*Archaeopteryx* and *Hesperornis*) had teeth on the jaws. Birds lay large-yolked, hard-shelled eggs and exhibit a high degree of parental care.

avocets.   *See* CHARADRIIDAE.

awn.   A bristle-like projection from the tip of

the back of the GLUME in some GRAMINEAE.

axenic culture.   A pure culture that contains only a single kind of organism.

axerophthol.   *See* VITAMIN A.

axial plane.   The imaginary plane that divides a FOLD as symmetrically as possible and that passes through the AXIS.

axial velocity.   The component of velocity along an AXIS, usually the axis of symmetry of a moving system. Thus there is a velocity along the axes of the vortices behind an aircraft, with a maximum towards the aircraft in the VORTEX cores. Swirling flow in a pipe or a tornado has a velocity component along the axis which varies with the distance from it.

axillary.   Botanically, the angle between the leaf stalk of a plant and the stem.

axis.   (1) The median line between the limbs of a FOLD, along the trough of a SYNCLINE, or the crest of an ANTICLINE. (2) The direction of flow at the centre of a JETSTREAM. (3) *See* FLOWER.

axon.   The process of a NEURON, usually an especially long one, that conducts impulses away from the nerve cell body.

Azolla.   A genus of aquatic (floating) ferns that form a symbiotic relationship (*see* SYMBIOSIS) with the blue–green algae *Anabaena* (*see* CYANOPHYTA), which can fix nitrogen (*see* NITROGEN FIXATION) and thus may be of significance in rice culture.

Azores high.   *See* SUBTROPICAL HIGH.

Azotobacter.   A genus comprising the most important of the nitrogen-fixing (*see* NITROGEN FIXATION) bacteria. It is AEROBIC and obtains its energy by breaking down carbohydrates in the soil. Along with other nitrogen-fixing bacteria, it is responsible for the gradual increase in the nitrogen content of unmanured grassland. *See also* CLOSTRIDIUM PASTEURIANUM, NITROGEN CYCLE.

B

B. (1) Billion (i.e. one thousand million; 10^9) (2) *See* BORON.

Ba. *See* BARIUM.

Bacillariophyta (Bacillariophyceae, diatoms). Unicellular ALGAE, some of which are colonial, green or brownish in colour (but all contain CHLOROPHYLL) and with siliceous and often highly sculptured cell walls. Diatoms make up much of the producer level in marine and freshwater FOOD CHAINS, and they have contributed to the formation of oil reserves. Deposits of DIATOMACEOUS EARTHS were formed by the accumulation of diatom cell walls.

Bacillus. A genus of bacteria, including *Bacillus tetani*, responsible for lockjaw (tetanus) and *Bacillus subtilis.*

bacillus. Any rod-shaped bacterium (*see* BACTERIA).

Bacillus anthracis. *See* ANTHRAX.

Bacillus subtilis. A species of bacteria that produces a pectin-splitting ENZYME that is very stable, enabling it to derive nutrients from wood. It synthesizes vitamin B_{12} (*see* COBALAMINE) and is a symbiotic inhabitant (*see* SYMBIOSIS) of the gut of some animals.

backcross. The offspring of a cross-mating between a HYBRID (HETEROZYGOTE) with one of its parents or with an organism genetically similar to one of its parents.

backing wind. A wind showing a turn of direction. As a FRONT approaches from the west (in the northern hemisphere) the wind backs from west to southwest or south. A wind that changes its direction in the opposite sense is a veering wind.

backscatter. The dispersion of radiation in an upbeam direction. The backscattering of incoming solar radiation in the atmosphere tends to increase ALBEDO. Solar radiation is scattered in a forward (downbeam) direction by atmospheric AEROSOLS. *See also* MIE SCATTERING, RAYLEIGH SCATTERING.

backyardism. Local opposition to a proposed development based on the fear that it would reduce the quality of living for residents and lower the value of property, rather than to any fundamental opposition in principle to the type of development. For example, communities may oppose the building of a nuclear power station or waste depository, although continuing to use nuclear-generated electricity and hospital facilities that generate radioactively contaminated waste.

bacteria (Bacteriophyta, Schizomycophyta). A group of ubiquitous, microscopic organisms, mostly unicellular, classified as plants or PROTISTA. They may be rod-shaped (bacilli), spherical (cocci), roughly spiral (spirelli) or filamentous (actinomycetes) and may possess flagella (*see* FLAGELLUM) which enable them to move. They reproduce by simple FISSION, SPORE formation and, sometimes, by sexual means. A few are AUTOTROPHIC, obtaining energy from oxidation processes or by PHOTOSYNTHESIS if they contain BAC-TERIOCHLOROPHYLL. Many are SAPROPHYTES and play a vital role in the decomposition of organic matter (*see* CARBON CYCLE, NITROGEN CYCLE). Many are parasites (*see* PARASITISM), causing such animal diseases as bubonic plague, tuberculosis, cholera and tetanus (lockjaw), and a few plant diseases (e.g. soft rot in carrots). Some are symbiotic (*see* SYMBIOSIS) (e.g., *BACILLUS SUBTILIS*). *See also* MYCOPLASMAS, MYXOBACTERIA.

bacteriochlorophylls. Bacterial pigments that trap light energy, thus enabling the bacteria that possess them to be AUTOTROPHIC. *See also* PHOTOSYNTHESIS.

bacteriophages (phages). VIRUSES that are parasitic (*see* PARASITISM) on bacteria. A bacteriophage consists largely of nucleic acid (usually DNA) which takes over the nucleic acid-producing mechanisms of the bacterial cell, forcing it to make virus nucleic acid instead of its own. *See also* LYSOGENIC BACTERIUM.

Bacteriophyta. *See* BACTERIA.

bacteriostat. A substance that retards or inhibits the growth of bacteria.

baffle. A plate, grating or refractory wall used especially to block, hinder or divert a flow, or to redirect the passage of a substance.

bagasse. Crushed sugar-cane. It is left as a fibrous residue after the sugar has been extracted and can be treated for use as a fertilizer or for protein extraction. *See also* PROTEIN FOODS.

bag filter. A device based on a filter made from woven or felted fabric, often tube-shaped, and used to remove particulate matter from industrial waste gases. Bag filters may be up to 10 metres long and 1 metre wide. They are closed at one end, the other end being attached to a gas inlet. Provided the velocity of the gas is low, more than 99 percent of all particles may be trapped using a bag filter. The material from which the filter is made is limited by operating temperatures. Natural fibres cannot be used above 90°C, nylon above 200°C, glass fibres, sometimes impregnated with GRAPHITE and siliconized for greater durability, above 260°C.

Baikal. One of the largest freshwater lakes in the world, with a surface area of 31 500 square kilometres (12 162 square miles), in the Buryat Autonomous SSR in eastern Siberia. Its long isolation has allowed species to survive or evolve, many of which are found nowhere else (e.g., the Baikal seal; *Phoca sibirica*). Factories, mainly pulp and cellulose mills, on the shores of the lake are a source of serious pollution, and protecting the lake is the subject of a major Soviet conservation programme.

Baikal seal. *See* BAIKAL.

Bajocian. A stage of the TRIASSIC System.

Balaenoidea (Mysticeti). Toothless whales (CETACEA) that have large heads and baleen ('whalebone') hanging from their palates, used for straining shrimps and other small animals from the water. The group includes the rorquals, right whales and the blue whale.

balance of nature. *See* ECOLOGICAL BALANCE.

ball lightning. A rare and elusive form of electric discharge, usually taking the form of an incandescent ball moving at a rate comparable with the air motion.

balloons. Lighter-than-air devices that are used widely for experimental and monitoring purposes. Pilot balloons are tracked with THEODOLITES to measure wind direction and speed; sounding balloons carry instruments that record and transmit data (radiosonde) or are tracked by radar. Typically, a sounding balloon carries a few kilograms to heights of up to 20 kilometres, and occasionally 30. Constant level balloons are made by enclosing the balloon envelope in a silk net or sheath to maintain a constant volume; without the sheath a rubber balloon will expand and maintain a fairly constant rate of rise in the order of 5–10 metres per second. *See also* CONSTANT-LEVEL BALLOON.

bamboo. *See* BAMBUSACEAE.

Bambusaceae (bamboos). A group of woody plants of GRAMINEAE that often grow to a great height. They form a very important part of the ECOSYSTEM in tropical and subtropical areas.

banana. *See MUSA.*

banana republic. *See ONE-CROP ECONOMY.*

band. *See WAVEBAND.*

banded ironstone. A PRECAMBRIAN deposit of alternating layers, usually less than 1 millimetre thick, of CHERT and HAEMATITE. The cherts often contain fossils of microscopic plants. Banded ironstones have been dated at 1700–3200 million years old, at which time, it is believed, the atmosphere was far poorer in oxygen and richer in carbon dioxide than it is now. Unleached banded ironstone is known as taconite and contains 15–40 percent iron, but surface LEACHING of siliceous and CARBONATE MINERALS has produced a more desirable ore grading 55 percent or more iron. After BENEFICIATION the taconite is pelleted and used for smelting. *See also* LAKE SUPERIOR-TYPE IRON ORE.

banding. *See* BIRD RINGING.

banner cloud. A WAVE CLOUD that is fixed over a mountain, extending downwind of the peak.

banyan. *See FICUS.*

bar. (1) The meteorological unit of pressure, equal to 10^5 newtons per square metre (10^6 dynes per square centimetre), or 10^3 millibars, which is roughly the sea-level atmospheric pressure (usually taken to be 1013.2 millibars). (2) A barrier that is wholly or partially submerged beneath the sea, and traditionally a navigation hazard, or (POINT BAR) formed in a river by a meander.

barban. A TRANSLOCATED HERBICIDE; a CARBAMATE used to control wild oat and other seeding grasses in cereal crops, beans and sugar-beet. It is harmful to fish.

barberry. *See BERBERIS VULGARIS.*

Barcelona Convention. A convention drawn up in 1976 by nations bordering the Mediterranean, forbidding the discharge of a wide range of substances into that sea, including organosilicon compounds, all petroleum hydrocarbons, all radioactive wastes and acids and alkalis. The sources of these substances were not specified.

barchan (barchane, barkan, barkham, barkhan). A sand dune, crescent-shaped in plan, with the horns of the crescent pointing downwind and the steeper slope on the leeward side.

barium (Ba). A silvery–white, soft metallic element that tarnishes readily in air. It occurs as BARYTE and as barium carbonate. Its compounds resembles those of calcium, but are poisonous and are used in the manufacture of paints, glass and fireworks. $A_r = 137.34$; $Z = 56$; SG 3.5; mp 710° C.

bark. The protective layer of dead cells outside the cork cambium (phellogen, *see* MERISTEM) in the older stem or root of a woody plant. Bark may consist of cork alone or alternating layers of cork and dead cortex or PHLOEM tissues. The term is used more loosely to mean all the tissue outside the XYLEM.

barkham. *See* BARCHAN.

barklice. *See* PSOCOPTERA.

barley. *See* GRAMINEAE.

barnacles. *See* CIRRIPEDIA.

baroclinic. Having a different distribution of pressure at different heights owing to horizontal variations in density. This is the normal state of the atmosphere.

barograph. A recording BAROMETER, usually employing a pen activated by an aneroid barometer, that writes on a rotating cylinder, completing one rotation each week.

barometer. An instrument for measuring atmospheric pressure, the most accurate being the mercury-type, which measures the height of a column of mercury supported under a vacuum in a glass tube. Different designs use various methods for correcting for temperature and the strength of the gravitational field (which varies with height and latitude, and also locally due to anomalies). The aneroid (without fluid) type, which measures the extent to which an evacuated metal cylinder is compressed by varying pressures, is small and portable, and so used in altimeters and barographs, but it is not capable of the 0.1 millibar accuracy required for meteorological observations.

barotropic. Having the same pressure field at all heights and no horizontal density gradients. A barotropic fluid conserves the vertical component of vorticity (*see* VORTEX). A barotropic model of the atmosphere was used in early computed weather forecasts, but it had serious limitations.

barrage. A DAM constructed across the flow of tidal water to hold back water in a basin for subsequent release, usually in order to generate electrical power by the flow of water in one or both directions. *See also* TIDAL POWER.

barrel. (1) A container for liquids, used for storage or transport. (2) In brewing, a cask containing 36 imperial gallons (164 litres). (3) A volumetric unit of measurement used for petroleum. The most commonly used unit is equivalent to 42 US gallons (or 5.61 cubic feet or 0.159 cubic metres). Depending on the specific gravity (or °API, *see* API SCALE) of the particular crude oil, between 6.5 and 7.5 barrels weigh 1 tonne. Barrel is usually abbreviated to brl, bbl and bl, and may be prefixed by M for million (10^6), B for billion (10^9) or T for trillion (10^{12}).

Barremian. *See* NEOCOMIAN.

Berriasian. *See* NEOCOMIAN.

barrier forest. In mountainous regions, a forest that holds back snow and so prevents AVALANCHES.

barrier island. Rows of parallel ridges of sand, caused by repeated deposition and erosion, forming an elongated island lying close to the shore and parallel to it. Depending on wind direction, during storms they may lose or acquire sand from dunes on the beach, and the combination of dunes and islands protects the coast against storm damage. Barrier islands often support vegetation, which traps more sand, and they may also trap fresh water, leading to the formation of WETLANDS rich in wildlife, but they are extremely fragile and rarely survive onshore developments that alter the beach dune system.

barrier reef. *See* BARRIER REEF MARINE PARK, CORAL REEF, GREAT BARRIER REEF.

Barrier Reef Marine Park. An area on the GREAT BARRIER REEF designated in 1979 by the Australian federal government to protect the Reef from damage and conserve its wildlife.

barrier spit. A BEACH BARRIER connected to the mainland at one end and terminating in open water. As it builds downdrift of the LONGSHORE CURRENTS it may develop a hook at the open end as currents are refracted shorewards. *See also* BAR, BAYHEAD BARRIER, BAYMOUTH BARRIER.

baryte ($BaSO_4$). A BARIUM mineral chiefly found as GANGUE with GALENA in HYDROTHERMAL veins. Because of its high specific gravity (4.5), baryte is used to make drilling mud more dense. It is also used in paints and as an opaque material in radiography.

basal metabolism. *See* METABOLISM.

basalt. A fine-grained, BASIC IGNEOUS rock, consisting essentially of a calcium-rich plagioclase FELDSPAR and a PYROXENE, with or without OLIVINE.

basaltic lava. LAVA with the composition of BASALT. Basaltic lava is the least viscous of the lavas and the most common; 90 percent of lavas are basaltic.

base. A substance that reacts with an acid to give a salt and water. If the base gives hydroxide ions (OH⁻) when dissolved in water it is called an alkali.

base course. *See* PAVEMENT.

base exchange. The transfer of CATIONS between an aqueous solution and a mineral. The principle has many important applications (e.g., in the softening of water when Ca^{2+} in the water are replaced by Na^+ in the softening agent). It is also used to improve soil fertility by the addition of solutions rich in a mineral that is deficient in the soil (e.g., Ca^{2+}, Mg^{2+}, Na^+, K^+, to replace another cation, say H^+, adsorbed on to the soil particle). The efficiency with which cations can replace other cations depends on such variables as the number of charges on the ion, the HYDRATION property of the ion, its size and the relative concentrations of the ions present. The replacement series in soil is usually: $Al^{3+}>$ (will replace) $Ca^{2+}> Mg^{2+}> K^+> Na^+$.

base level. The lowest level to which a stream can erode its channel. This may be sea level, lake level or the level of the main stream into which the tributary stream flows.

baselines. The tolerance levels of organisms to particular concentrations of substances. These vary from species to species.

base-pairing. *See* DNA, RNA.

base saturation. The condition in which the cation exchange (*see* BASE EXCHANGE) capacity of a soil is saturated with the exchangeable bases Ca^{2+}, Mg^{2+}, K^+ or Na^+. It is expressed as a percentage of the total cation exchange capacity.

base subsistence density. The human population density above which the continued survival of the population is impossible.

basic. (1) Applied to an IGNEOUS rock that contains a relatively low amount of SILICA (commonly set in the range 45–52 percent). The silica is in the form of silicate minerals (e.g., plagioclase FELDSPAR, PYROXENE and OLIVINE). QUARTZ is rarely present in basic rocks. BASALT, DOLERITE and GABBRO are basic rocks. In petrology, basic is contrasted with ACIDIC, INTERMEDIATE and ULTRABASIC. *See also* MAFIC. (2) In chemistry, applied to a substance that reacts with an acid to form a salt. *See* PH.

basic refractory. A heat-resistant material containing a large proportion of metallic oxides that is used as a furnace lining, (e.g., DOLOMITE, MAGNETITE).

basidia. *See* BASIDIOMYCETES.

basidiomycetes. A group of fungi in which SPORES are borne externally, usually in fours, on mother cells (basidia). The group includes mushrooms, puffballs, bracket fungus, most of the fungi of MYCORRHIZAS, AGARICS, SMUT FUNGI, RUST FUNGI and the dry rot fungus *Serpula (Merulius) lacrymans*.

basket-of-eggs topography. *See* DRUMLIN.

batesian mimicry (pseudaposematic coloration). The resemblance of a harmless animal to a poisonous, dangerous or distasteful one, which is often conspicuously marked (*see* APOSEMATIC COLORATION). This affords protection as predators tend to avoid both (e.g., bee hawk moths resemble bees and wasps). *Compare* MÜLLERIAN MIMICRY.

batholith. A large IGNEOUS INTRUSION with steeply dipping contacts and no apparent floor. Exposed batholiths may cover hundreds of thousands of square kilometres and are associated with OROGENIC BELTS. Most batholiths are granitic (*see* GRANITE).

Bathonian. A stage of the JURASSIC System.

bath plug vortex. The most homely model of a tornado or hurricane, demonstrating the conservation of angular momentum with spin increasing as fluid approaches the centre of the VORTEX. The shape of the funnel being a constant, the pressure surface is like that of a tornado cloud and resembles roughly the shape of the TROPOPAUSE as it is sucked down into the eye of a hurricane.

bathyal. Applied to the sea bed and sediments deposited between the edge of the CONTINENTAL SHELF and the start of the ABYSSAL zone at a water depth of 2000 metres. *See also* BATHYPELAGIC.

bathypelagic. Applied to marine organisms that live at depths between 1000 and 3000 metres. *Compare* ABYSSOPELAGIC, EPIPELAGIC, MESOPELAGIC.

bats. *See* CHIROPTERA.

Battersea gas-washing process. A method of wet scrubbing (*see* WET SCRUBBER) the flue gases to remove SULPHUR DIOXIDE. It was invented in 1930, first applied at Battersea Power Station, London (which is now closed) and introduced later at other power stations. The gases are washed with river water and chalk, using about 157 litres to each tonne of coal burned and producing an effluent that is an almost saturated solution of calcium sulphate (i.e. GYPSUM). The disadvantages of the process are that the cooling of the PLUME causes a local FALLOUT of residual sulphur dioxide, the river is polluted, the recovered sulphur cannot be used, the process causes corrosion within the plant and costs are increased.

battery. (1) A device for storing electrical energy by a reversible chemical reaction, such that the application of electrical power causes a reaction which can be reversed later to release electrical power. (2) An energy-intensive, semi-industrialized system of poultry husbandry in which birds are confined in cages which commonly rise in banks, and are fed and watered by partly automated systems. Their eggs are collected by rolling along channels beneath each tier of cages. It is the most common source of hens' eggs in most industrial countries, and one unit may house many thousands of birds. Battery units are also sources of polluting effluent whose

recycling or disposal is often difficult. *Compare* BROILER.

bauxite. The most common ore of ALUMINIUM, mainly its hydroxide, but also used directly as a filler in plastics and rubber, in CATALYSTS and as an abrasive.

bayhead barrier. A barrier that encloses a lagoon near the inland head of a bay. *See also* BAR, BARRIER SPIT, BAYMOUTH BARRIER, BEACH BARRIER.

baymouth barrier. A barrier across the mouth of a bay. *See also* BAR, BARRIER SPIT, BAYHEAD BARRIER, BEACH BARRIER.

bbl. *See* BARREL.

Be. *See* BERYLLIUM.

beach barrier. A wide ridge of sand, protruding above normal high-tide level, that lies parallel to the coast and is separated from it by a lagoon or marsh. There are two proposed modes of origin: (a) that barriers are formed by LONGSHORE DRIFT; (b) that barriers are produced by wave action transporting sand shoreward and turbulence, due to surf, piling sand on to the barrier site, the process being augmented by longshore drift. *See also* BAR, BARRIER ISLAND, BARRIER SPIT, BAYHEAD BARRIER, BAYMOUTH BARRIER.

beach drift. The zig-zag movement of material along a beach platform when waves swash obliquely on to the shore. *See also* LITTORAL CURRENTS, LITTORAL DRIFT, LONGSHORE CURRENTS, LONGSHORE DRIFT.

bear animalcules. *See* TARDIGRADA.

beat. In acoustics, a periodic increase and decrease in amplitude caused by the superposition of two TONES of different FREQUENCIES, the beat frequency being the difference between the two frequencies.

Beaufort scale. A scale of wind force devised by Admiral Sir Francis Beaufort (1774–1857), originally related to the state of the sea, but adapted for use on land. It has a scale of 0 (calm) to 12 (hurricane), although forces above 8 are rare over land. The wind force was described in terms of its effect on the behaviour of smoke, trees, waves, umbrellas, ease of walking, etc. and facilitated an internationally agreed system for reporting wind speed. It is still used, especially by shipping; it has been largely superseded since 1950 at land-based weather stations by direct reports of speeds based on instrument readings.

Beckmann thermometer. A type of mercury thermometer with a large bulb, which gives it high sensitivity over a limited range. Mercury can be added to or removed from the column for the measurement of different temperature ranges.

Beckman process. A process, designed in the 1930s, for producing a livestock feed additive by treating cereal straw with sodium hydroxide (caustic soda), followed by washing. Ruminants can eat straw, but the Beckman process increases the amount of nourishment they derive from it. The process was not developed for use on farms, mainly because of the dangerous chemicals it uses, but from time to time it is reconsidered for use off farms, or on farms in a modified form.

becquerel (Bq). The SI unit for the radioactivity of a substance that is decaying spontaneously at the rate of 1 disintegration per second. The becquerel replaces the curie (Ci): $1 \text{ Bq} = 2.7 \times 10^{-11} \text{ Ci}$.

bed. A layer of sediment more than 1 centimetre thick and distinguished from adjacent layers by its composition, structure or texture.

bedding plane. The surface separating successive layers (BEDS) of stratified rock. Such surfaces form planes of weakness.

bed load (bottom load). The sand and gravel particles that are rolled, slid or bounced along the bed of a stream by the movement of water.

bedrock. The consolidated rock that lies below loose, superficial material such as SOIL, GLACIAL DRIFT or ALLUVIUM.

beech. See FAGUS SYLVATICA.

bee flies. See BRACHYCERA.

bees. See ACULEATA.

beetles. See COLEOPTERA.

beetroot. See BETA VULGARIS.

behaviour. The manner in which an organism responds to a STIMULUS. See also ETHOLOGY.

behaviourism. A school of experimental psychology which holds that the proper subject for psychological investigation is the prediction and control of BEHAVIOUR and no other, so dismissing such issues as consciousness, mind and free will. Basic drives are defined operationally (e.g., hunger or thirst in terms of the length of time without food or water).

beheaded river. See RIVER CAPTURE.

bel. A unit of sound volume equal to 10 DECIBELS.

Belemnoidea. See DECAPODA.

bench-mark. See BIOLOGICAL BENCH-MARKING.

beneficial use. A use of the environment, or some part of it (e.g., recreation, agriculture, water storage, etc.), that benefits a human population and therefore should be protected so that it may continue.

beneficiation. The separation of a valuable mineral from GANGUE and COUNTRY ROCK in order to effect a low-cost, artificial concentration of the ore. Beneficiation involves such processes as crushing, magnetic separation and FROTH FLOTATION.

Benioff zone. An inclined plane of earthquake foci (*see* FOCUS) which DIPS beneath some continental margins and ISLAND ARCS.

Bennetitales. *See* GYMNOSPERMAE.

benthic. Applied to organisms living close to the bottom of a lake or sea. *See also* BENTHOS, DEMERSAL.

benthon. *See* BENTHOS.

benthos (benthon). The bottom of a sea or lake anywhere from the high-water mark down to the deepest level; the organisms living there. The abyssobenthos refers to an ocean floor at great depths; the phytobenthos is the part of a lake or sea bottom covered with vegetation; the potamobenthos is a river bottom; the geobenthos that part of the bottom of a freshwater lake that does not support rooted vegetation.

bentonite. A very fine CLAY formed by the alteration of volcanic ASH deposits. It is used in water softeners, to reduce seepage in mines and channels, and as a major constituent in drilling muds.

benzene (C_6H_6). The simplest AROMATIC hydrocarbon, found in COAL TAR. It is used extensively as an industrial solvent, in laboratories and in the manufacture of styrene, lacquers, varnishes and paints. It is a highly flammable narcotic liquid and is also a CARCINOGEN.

benzene hexachloride. *See* LINDANE.

benzene ring. The conventional representation of BENZENE as a hexagonal ring of carbon with alternate double and single bonds.

benzoapyrene (benzpyrene). A CARCINOGEN found in tars and produced in the course of the manufacture of ASPHALT, tar and other organic products.

Berberis vulgaris (barberry). A hedgerow shrub that occurs locally in small quantities in the UK. It harbours a stage in the life cycle of the fungus *Puccinia graminis*, which causes black rust on wheat and other

GRAMINEAE. *See* RUST FUNGI.

Bergeron–Findeisen mechanism. An explanation for the initiation of some kinds of rain, where the upper parts of cloud are frozen. The vapour pressure over supercooled water (*see* SUPERCOOLING) exceeds that over ice at the same temperature, so that if ice and water particles are present in a cloud together the droplets evaporate and the ice crystals grow. This mechanism supposes that after the initial growth the ice crystals continue to grow by ACCRETION as they fall through the clouds of droplets, melting below the cloud.

Bergmann's rule. As the mean temperature of the environment decreases (e.g., with increasing latitude) the body size of warm-blooded animal species or subspecies tends to increase.

bergschrund. A fissure formed during warm weather between the steep head wall of a CIRQUE and the glacial ice filling the cirque.

berg winds. Warm winds blowing from mountains, especially well known in coastal Natal, South Africa. Sometimes their temperature indicates recent adiabatic descent of the air from above the height of the mountain (*see* ADIABATIC LAPSE RATE). Berg winds are a type of FÖHN WIND, not associated with cloud or rain.

beri-beri. A disease caused by a dietary deficiency of vitamin B_1 (*see* THIAMINE). It is common among people subsisting on a diet in which polished rice is the staple food.

berkelium. *See* ACTINIDES.

Bermuda high. A high-pressure region near Bermuda, evident in pressure charts averaged over long periods because of the frequent occurrence of ANTICYCLONES there. It is part of the subtropical high-pressure belt and the cause of the SARGASSO SEA.

Berriasian. *See* NEOCOMIAN.

berry. A fruit, usually containing many seeds with a PERICARP consisting of a skin (epicarp), a succulent mesocarp and a membranous endocarp surrounding the seeds. Examples include tomato, grape and cucumber. *Compare* DRUPE.

beryl. The mineral beryllium aluminosilicate $(Be_3Al_2(SiO_3)_6)$; the major ore mineral of BERYLLIUM, found in PEGMATITES. Gem-quality beryl is known as aquamarine if it is blue and emerald if it is green.

beryllium (Be). A hard, poisonous metallic element used in the production of very hard, corrosion-resistant, non-ferrous alloys, in X-ray tubes and in the nuclear industry as a MODERATOR. If inhaled, beryllium has been known to cause malignant growths (beryllosis) in the lungs of workers and some residents near factories, and its use is now strictly controlled. The element is found in the atmosphere in minute traces, and it is converted into a radioactive isotope in the STRATOSPHERE by COSMIC RAYS. $A_r = 9.0122$; $Z = 4$; SG 1.85; mp 128°C.

beryllosis. A disease of the lungs caused by inhaling particles of beryllium oxide.

Bessemer process. A steel-making process, named after its inventor, Sir Henry Bessemer (1813–98). Steel was made by this process from about 1860 to the 1960s. A large vessel (the converter) is charged with 25–50 tonnes of molten pig iron, and air is blown through holes in its base. The oxygen in the air oxidizes the iron, with the iron oxide formed dissolving and then oxidizing silicon, manganese and carbon impurities. The metallic oxides form a slag, and the carbon is removed as carbon monoxide, some of which burns to form carbon dioxide. The vessel is tilted, and while its contents are being poured into a large ladle a manganese alloy is added which combines with the remaining iron oxide, so removing it. Other compounds are added to assist the deoxidizing of the (by now) steel, then ANTHRACITE is added to bring the carbon content to the desired level.

best practicable means. The concept applied to the control of air pollution in the UK by the ALKALI INSPECTORATE, whereby the best possible level of control is attained within constraints imposed by the technological and economic capabilities of the industrial plant in question.

beta diversity (habitat diversification). Diversity that results from competition between species, leading to more precise ADAPTATION to the HABITAT as a whole and thus narrowing the range of tolerance to other environmental factors.

beta mesosaprobic. *See* SAPROBIC CLASSIFICATION.

beta-naphthylamine (ß-naphthylamine). An organic compound formerly used in the manufacture of dyestuffs, whose use is now banned in many countries, including the UK, because it is a powerful bladder CARCINOGEN.

beta-particles (ß-particles). Electrons emitted by radioactive decay. They have moderate penetrative power and can damage living tissue. *See also* IONIZING RADIATION

Beta vulgaris. A species of CHENOPODIACEAE of very varied habitat which includes the beetroot, sugar-beet, chard, mangold-wurzel and the wild sea-beet from which all the cultivated forms were derived. Most forms are biennial and store sugar in a swollen root.

Betula (birch). A genus of trees and shrubs (family: Betulaceae) of northern temperate regions and extending to the northern limit of trees. Birches bear catkins and have hard wood used for making shoes, charcoal, etc.

Bewsey's method. A weather-forecasting technique, named after its inventor, a British meteorologist, that is based on observation of the orientation of FALLSTREAKS from ice clouds. If these lie at right angles to the wind direction at the level at which they occur, with the tail pointing in a direction backed from the wind, development is indicated. Typically, the approach of a WARM FRONT is revealed, and rain is to be expected.

bezoar (*Capra aegragus*). An animal with large, scimitar-shaped horns, that inhabited Persia in prehistoric times and is believed to be the wild ancestor of domesticated goats.

BHC. Benzene hexachloride (*see* LINDANE).

Bhopal. A town in India where, on 3 December 1984, an accident at a pesticide factory owned by Union Carbide India released a cloud of METHYL ISOCYANATE, killing about 2500 people and injuring about 200000.

B horizon. *See* SOIL HORIZONS.

bias. In statistics, a systematic distortion, as distinct from RANDOM ERROR, which may distort on one occasion but balances out overall.

biennial. (1) Applied to any phenomenon that recurs at two-yearly intervals. (2) A plant (e.g., carrot) that completes its life cycle from seed to seed in two years. Food is stored in the first year and is used to produce seeds in the second. *Compare* ANNUAL, EPHEMERAL, PERENNIAL.

big bang. (1) The event that is believed to have initiated the formation of the universe. (2) An explosion large enough to produce a pressure oscillation of the order of 1 millibar or more at a distant point on the Earth (e.g., the eruption of KRAKATOA in 1883). Large NUCLEAR FUSION (thermonuclear) explosions are comparable.

big bud. *See* GALL.

bilateral symmetry. The condition in animals and flowers where there is only one plane of symmetry (i.e. one-half of the individual is the mirror image of the other). Most free-moving animals have similar right and left halves. Bilaterally symmetrical flowers (e.g., sweet pea, *Lathyrus odoratus*) are usually termed zygomorphic. *Compare* ACTINOMORPHIC.

bilharzia. *See* SCHISTOSOMA.

biliproteins. *See* PHYCOBILINS.

billow clouds. Clouds that are arranged in several parallel bars in close proximity in a layer. They are formed either by overturning due to CONVECTION, the cloud being cooled on top by radiation into space, the arrangement being due to WIND SHEAR, or by the overturning of unstable waves due to the generation of shear at a very stable layer when it is tilted, as in flow over a mountain.

binaural. The ability to hear with two ears for orientation or the electronic simulation of this.

binocular vision (stereoscopic vision). The arrangement in which both eyes face forward, so that each eye views the same scene from a slightly different angle. The pair of two-dimensional images is merged by the brain to form a single image with the impression of three dimensions. Binocular vision is found in primates and predators, to whom judging distance is particularly important.

binomial nomenclature (binomial system). The present scientific method of naming organisms, using pairs of Latin words, conventionally written in italics, the first (with a capital initial letter) denoting the genus and shared by the organism's closest relatives, the second (with a small initial letter) denoting the species. The name (or initials) of the person responsible for describing and naming the species should follow, in Roman characters. For example, the heath violet is *Viola canina* Linnaeus (or Linn. or L.). If varieties or subspecies have been described a third Latin name is added (e.g., *Viola canina* ssp. *canina* and *Viola canina* ssp. *montana*). *See also* CLASSIFICATION, LINNAEUS.

binomial system. *See* BINOMIAL NOMENCLATURE.

bioassay. The quantitative measurement, under standardized conditions, of the effects of a substance on an organism or part of an organism.

biochemical oxygen demand (BOD). The amount of oxygen used for biochemical oxidation by a unit volume of water at a given temperature and for a given time. It is used as a measure of the degree of organic pollution of water. The more organic matter the water contains, the more oxygen is used by microorganisms. *Compare* CHEMICAL OXYGEN DEMAND.

biochore. A subdivision of a BIOCYCLE, comprising a group of similar BIOTOPES (e.g., desert, including sandy and stony desert biotopes, or forest, including coniferous, deciduous and other forest biotopes).

biocides. Agents that kill living organisms. Sometimes the term is used as a synonym for PESTICIDES.

bioclastic. Composed of fragmental organic remains. *See also* CLASTIC.

bioclimatology. The scientific study of the relationship of living organisms to climate.

biocoen. All the living components of the environment. *Compare* ABIOCOEN.

biocoenology. *See* SYNECOLOGY.

biocoenosis. (1) A community of organisms occupying a BIOTOPE. (2) An assemblage of fossils consisting of the remains of organisms that once lived together. *Compare* THANATOCOENOSIS.

biocontrol. *See* BIOLOGICAL CONTROL.

biocycle. A subdivision of the BIOSPHERE. Those usually recognized are land, sea and fresh waters. *See also* BIOCHORE.

biodegradation. The breakdown of substances by microorganisms (mainly aerobic bacteria). Many manufactured substances are readily biodegradable, but others (e.g., ORGANOCHLORINES and 'hard' DETERGENTS) are much more resistant to bacterial action. *See also* ACTIVATED SLUDGE, ALKYL SULPHONATES.

biodynamic farming. Farming according to principles laid down by Rudolf Steiner (1861–1925). The principles are similar in many ways to those underlying ORGANIC FARMING, but in addition they relate farm operations to phases of the Moon and make use of very small quantities of various preparations.

bioengineering. *See* BIOTECHNOLOGY.

biogas. Any combustible gas, but usually one in which METHANE is the main constituent, produced by the fermentation of organic wastes.

biogenetic law. *See* PALINGENESIS.

biogeochemical anomaly. The concentration of a minor or trace element in the soil, determined by chemical analysis of plants. *Compare* GEOBOTANICAL ANOMALY.

biogeochemical cycles. The global circulation of elements between land, sea and the atmosphere, involving living organisms at one or more stage. *See also* CARBON CYCLE, HYDROLOGICAL CYCLE, NITROGEN CYCLE, PHOSPHORUS CYCLE.

biogeographical province. An area of the Earth's surface defined by the species of fauna and flora it contains. The concept is taxonomical (*see* TAXONOMY) rather than ecological (*see* ECOLOGY) and is usually based on the number of ENDEMIC species found in the area.

biogeosphere. The outer part of the Earth's crust (LITHOSPHERE) as far down as life exists.

biological amplification (biological magnification). The concentration of a persistent substance (e.g., ORGANOCHLORINE compound) by the organisms of a FOOD CHAIN, so that at each successive trophic level the amount of the substance relative to the BIOMASS is increased.

biological bench-marking. The use of plant or animal species to measure pollution (e.g., of LICHENS to measure SULPHUR DIOXIDE concentration) based on assessments (bench-marks) of population level and fitness, against which changes can be evaluated. *See also* BIOLOGICAL INDICATOR, BIOLOGICAL MONITORING.

biological complex. *See* BIOPLEX.

biological control (biocontrol). The control of pests by the use of other living organisms. The dramatic reduction in the rabbit population in Australia by the introduction of MYXOMATOSIS virus and the control of the citrus scale insect in California by the introduction of an Australian ladybird (ladybug) beetle (foiled later by the use of insecticides) were successful examples. An advantage of this type of pest control is that, once established, the control is self-perpetuating. It is also specific and does not pollute the environment. Another technique of biological control involves the introduction of large numbers of sterile (irradiated) males, resulting in the laying of infertile eggs. This has been successful in some parts of the USA in controlling the screw worm fly (*Callitroga macellaria*).

biological husbandry. *See* ORGANIC FARMING.

biological indicator. A species or organism that is used to grade environmental quality or change. *See also* BIOLOGICAL BENCH-MARKING, BIOLOGICAL MONITORING.

biological magnification. *See* BIOLOGICAL AMPLIFICATION.

biological monitoring. The direct measurement of changes in the biological status of a HABITAT, based on evaluations of the number and distribution of individuals or species before and after a change. *See also* BIOLOGICAL BENCH-MARKING, BIOLOGICAL INDICATOR.

biological shield. A thick (usually 3–4 metres) wall, usually made of concrete, surrounding the core of a NUCLEAR REACTOR, designed to absorb neutrons and gamma-radiation for the protection of personnel.

bioluminescence. The production of light of various colours by living organisms (e.g.,

some bacteria and fungi, glow-worms and many marine animals such as fishes, CEPHALOPODA and PROTOZOA). Luminescence is produced by a biochemical reaction, which is catalyzed by an enzyme; it produces no heat. In some animals (e.g., glow-worms) the light is used as a mating signal; in others it may be a protective device. In deep-sea forms luminous organs may serve as lanterns.

biomass. The total weight of the organisms constituting a given trophic level (*see* FOOD CHAIN) or inhabiting a defined area.

biomass fuel. A carbon-based fuel derived from plants that were living recently (e.g., wood) as opposed to a FOSSIL FUEL (e.g., coal). Crops may be grown for use as fuel. Alternatively, wastes from food or fibre crops may be used. The fuel may be burned directly or converted into a more convenient form (e.g., ETHANOL or METHANOL).

biome. A major ecological COMMUNITY of organisms, occupying a large area (e.g., tropical rain forest). *See also* FORMATION.

biometeorology. The scientific study of the relationship between living organisms and weather.

bioplex (biological complex). A system in which the waste products of each stage are used as raw materials for a succeeding stage, the whole system forming a cycle.

bioseston. All the living matter floating or swimming in water. *Compare* ABIOSESTON. *See also* SESTON.

biosphere. That part of the Earth and its atmosphere in which organisms live.

biosphere reserve. An area of land or coast that has been designated by IUCN and UNESCO (as part of the MAN AND THE BIOSPHERE PROGRAMME) as being of international importance for conservation, study and sustained development. In 1985, there were 243 such sites in the world (36 in Africa, 50 in

North America, 20 in South America, 30 in Asia, 78 in Europe, 17 in the USSR and 12 in Oceania) with a total area of almost 155 million hectares.

biostimulant. A substance that stimulates the growth of aquatic plants. *See also* EUTROPHICATION.

biostratigraphic zone. *See* INDEX FOSSIL.

biosystematics. The study of the biology of populations, especially in relation to their evolution, variation, reproductive behaviour and breeding systems.

biota. The flora and fauna of an area.

biotechnology (bioengineering).   The employment of biochemical processes on an industrial scale, most notably recombinant DNA techniques, to produce drugs or (by means of fermentation) bulk foodstuffs for humans or livestock, sometimes by the recycling of wastes. *See also* GENETIC ENGINEERING, NOVEL PROTEIN FOODS, MICROBIAL METALLURGY.

biotic. Living or biological in origin. The term is applied to components of an environment. *Compare* ABIOTIC.

biotic factors. Influences on the environment that result from the activities of living organisms.

biotic index. A rating used in assessing the quality of the environment in ecological terms. Rivers can be classified according to the type of invertebrate community present in the water using a biotic index which is largely an indication of the amount of dissolved oxygen present, this in turn being a measure of the level of organic pollution. Very clean water, holding a wide variety of species including pollution-sensitive animals (e.g., stonefly and mayfly nymphs), has a high biotic score. As pollution increases, oxygen levels decrease, and the more sensitive species disappear. Badly polluted water, in which only a few tolerant species (e.g., red midge larvae and annelid worms) can sur-

vive, together with a few animals that breathe air at the surface, has a very low biotic score.

biotic potential. An estimate of the maximum rate of increase of a species in the absence of competition from predators, parasites, etc. This is usually very large, but in fact it is never achieved in natural surroundings because natural selection operates.

biotic province. A major ecological region of a continent (e.g., the Hudsonian Province, which covers most of Alaska and Canada).

biotic pyramid. The graphic description of the trophic levels within a stable FOOD CHAIN, essentially a histogram laid on its side with each bar centred. Since a high proportion of the energy derived from food is used to sustain the metabolism of the organism that consumes it, the number of individuals at each level decreases (e.g., one lion kills about 50 zebra a year). Thus the histogram resembles a pyramid, with primary producers at the base, then primary consumers (herbivores), secondary consumers (omnivores and some carnivores) and tertiary consumers ('top' carnivores). Animals at higher levels are often larger.

biotic succession. That component of a SUCCESSION in which the composition of COMMUNITIES is controlled by interactions among different species rather than by physical characteristics.

biotin. A water-soluble, nitrogenous acid that is classified as one of the B group of VITAMINS, although at one time it was known as vitamin H. Biotin is involved in the formation of fats and the utilization of carbon dioxide in many animals. It may be essential for humans.

biotite. *See* MICA.

biotope. A HABITAT that is uniform in its main climatic, soil and BIOTIC conditions (e.g., a sandy desert).

biotype. A naturally occurring population that consists of individuals of the same genetic make-up.

birch. *See* BETULA.

bird ringing (banding). A method of marking birds with numbered bands fastened around one leg, enabling information to be obtained about their movements, life spans, etc. In the UK, the main scheme is run by the BRITISH TRUST FOR ORNITHOLOGY, which trains people to catch birds (by means of mist nets, etc.), to make observations on age, weight, etc. and to ring them before release. About 500 000 birds are ringed in the UK each year; the recovery rate is about 11 000 per year. *See also* DUCK DECOY POND.

birds. *See* AVES.

bird song. A form of communication between birds, developed strongly in PASSERIFORMES. The elaborate song of a male bird is used mainly to mark its TERRITORY and is often delivered from a vantage point (song post). Song is often the only certain means of separating some closely related species of warbler (e.g., willow warbler and chiffchaff) in the field. Sub-song is an extremely quiet form of song the significance of which is not clear.

birth rate. The number of births within a population in a given period of time, usually one year.

bisexual. *See* HERMAPHRODITE.

bit. A contraction of *bi*nary digi*t*; the unit of information, represented by a 1 or 0 in binary notation. It is used in information theory and in modelling. *See also* DIGITAL COMPUTER, MODEL.

biting lice. *See* MALLOPHAGA.

bittern. (1) The liquid remaining after the crystallization of sodium chloride (common salt) from sea water; a source of iodine, bromine and magnesium. (2) A marsh bird belonging to the order CICONIIFORMES.

bitumen. A general geological name for various solid and semi-solid HYDROCARBONS. The term was defined in 1912 by the American Society for Testing Materials (ASTM) to include all hydrocarbons that are soluble in carbon disulphide, but current usage is generally more restrictive.

bituminous coal. A COAL between lignite and ANTHRACITE in rank that burns with a smoky, luminous flame.

bivoltine. Applied to organisms that produce two generations a year.

bl. *See* BARREL.

black body. (1) An ideal body which would, if it existed, absorb all of the radiation falling upon it and reflect none, making it a perfect absorber of solar energy. (2) A body that radiates a spectrum appropriate to its temperature and by which its temperature may be measured. Thus daylight due to the Sun is black-body radiation at 3000K. Snow and cloud are black bodies radiating in the infrared.

black diamond (carbonado). A crystalline form of carbon, similar in some ways to DIAMOND. Because of its great hardness it is a valuable abrasive.

black earth. *See* CHERNOZEM.

black jack. Zinc blende. *Compare* BLUE JOHN.

black frost. Clear ice that is not white, like hoar frost, snow, etc. It sometimes forms on wet roads at night. It also forms in arctic seas when supercooled rain (*see* SUPERCOOLING) accumulates on the rigging of ships. The consequences of this can be disastrous.

black mustard. *See* BRASSICA.

black rust. *See* BERBERIS VULGARIS, RUST FUNGI.

black sand. A coarse, SEDIMENTARY concentrate of heavy minerals formed by water and/or wind action. The heavy minerals are commonly a mixture of MAGNETITE, ILMENITE and HAEMATITE, with lesser amounts of RUTILE, CASSITERITE, GOLD, MONAZITE, ZIRCON, CHROMITE, GARNET and the FERROMAGNESIAN MINERALS. Black sands are PLACER deposits.

black smoke. Smoke produced by carbon particles released by HYDROCARBON CRACKING followed by sudden cooling. *Compare* BROWN SMOKE.

bladderworm (cysticercus). A larval stage of some tapeworms (CESTODA) that inhabits the tissues of an intermediate host and consists of a bladder containing an inverted scolex ('head'). The bladderworm matures when it is eaten, along with the tissues of the intermediate host, by the definitive host. Bladderworms of the hydatid worm *Echinococcus granulosus*, whose adult stage infests carnivorous mammals (e.g., dogs), form large multiple cysts which can cause great damage to their hosts, including humans.

blanket bog (blanket mire). An area, often very extensive, of acid peatland, found in constantly wet climates, characteristic of broad, flat, upland WATERSHEDS. It develops where drainage is impeded and the soil is acid. Blanket bog is ombrogenous (i.e. maintained by water from precipitation, not by ground water). *See also* RAISED BOG.

blanket mire. *See* BLANKET BOG.

blastema. A mass of undifferentiated cells that will develop into an organ in an animal. The regeneration of a lost part often begins with the development of a blastema.

blast furnace. A furnace for the production of iron from iron ore, constructed from refractory bricks contained by steel plates. A charge of ore, coke and limestone ($CaCO_3$) is introduced from above, sometimes together with sinter pellets (*see* SINTER PLANT). Molten iron and SLAG are removed from the bottom of the furnace. The product is pig iron or cast iron.

blastula. An early stage in the development of an animal embryo, usually consisting of a hollow ball of cells.

bleaching powder. One of a number of chemical agents that can bleach or destroy natural colours, so rendering substances white. Formerly, chlorinated lime (i.e. calcium hydroxide more or less saturated with chlorine) was used. This has been largely displaced by elemental chlorine and by sodium hypochlorite (the most common domestic bleach).

bleicherde A grey, bleached layer of a PODZOL. *See also* SOIL HORIZONS.

blepharoplast. In certain flagellate PROTOZOA, a basal granule embedded in the cytoplasm of the FLAGELLUM. *See also* CENTRIOLES, CILIA.

blizzard. Strictly, a storm of snow blown up from the surface; more loosely, any snowstorm accompanied by strong wind.

block heating. *See* DISTRICT HEATING.

blocking. The mechanism whereby an ANTICYCLONE remains stationary and CYCLONES move around it. It appears to block storms in their tracks.

blood corpuscles. The cells that circulate in the blood of animals. Erythrocytes (red blood cells), present in vertebrate blood, contain HAEMOGLOBIN. In humans they are biconcave discs without nuclei, about five million in each cubic millimetre of blood being the normal count. Leucocytes (white blood cells) are present in vertebrates and invertebrates. The three types of leucocytes present in mammals are: lymphocytes, small amoeboid, but non-phagocytic (*see* PHAGOCYTE) cells, which make or carry antibodies; monocytes, large phagocytic cells which enter the tissues and ingest invading microorganisms; and polymorphs (polymorphonuclear leucocytes or granulocytes), phagocytic cells with lobed nuclei and granular cytoplasm, which also enter tissues to attack invaders.

blood group. A group of people or animals all of whom bear the same ANTIGENS on their red blood cells (*see* BLOOD CORPUSCLES). If blood from different groups is mixed the blood cells clump together because the antigens (agglutinogens) react with agglutinins in the blood plasma. In humans there are four main blood groups: A, B, AB and O (the last contains neither agglutinogen A or B). The proportions of the groups are not constant in populations throughout the world. If agglutinogens A or B are absent, the corresponding agglutinins are present. Other blood antigens including the Rh (rhesus) factor occur, but their related ANTIBODIES are not normally present in plasma.

blood plasma. The liquid part of blood. In vertebrates it contains proteins (e.g., fibrinogen, responsible for clotting), dissolved food substances (e.g., glucose), inorganic salts (e.g., sodium chloride, sodium bicarbonate) and excretory products.

blood platelets (thrombocytes). Minute bodies, possibly cell fragments, that are present in the blood of mammals. They produce an enzyme activator, thrombokinase, responsible for initiating the clotting of blood.

blood serum. The liquid that separates from clotted blood. It is similar to BLOOD PLASMA, but lacks the ability to clot.

blood sugar. *See* CARBOHYDRATES.

bloodworms. (1) The aquatic larvae of certain non-biting (chironomid) midges, which contain HAEMOGLOBIN and are tolerant of organic pollution. (2) Sludgeworms or river worms, mud-dwelling oligochaete worms (*see* OLIGOCHAETA) (e.g., *Tubifex*), which contain haemoglobin and are tolerant of organic pollution. (3) Red bristleworms (*see* POLYCHAETA), which occur on muddy shores.

bloom. A sudden discoloration of water caused by the explosion of a population of ALGAE (algal or phytoplankton bloom) producing a green colour or of DINOFLAGELLATES producing a red colour. A bloom may occur at any time when conditions are favourable, but is more common in spring and early summer when primary production (*see* PRODUCTION) increases ahead of the growth in consumer populations. Blooms may also be caused by the addition of plant nutrients to water (e.g., phosphates or nitrates in sewage or fertilizer run-off from farm land), and they are characteristic of EUTROPHICATION in fresh water.

blow-by. Gases that blow past the piston rings and into the crankcase of an internal combustion engine, from which they are removed through a tube and fresh air introduced into the crankcase through a filter. Crankcase emissions are now controlled in California, Australia and elsewhere, the gases being carried back to the intake manifold by the low (less than 1 atmosphere) pressure in the induction system. Uncontrolled, crankcase emissions, consisting of unburned fuel and exhaust gases, are a major source of hydrocarbon emissions.

blowdown. An occasion when a substantial, but limited area of trees is blown down in a short time. Such incidents are rare; most occur on the lee side of mountains and are due to the strong winds in the trough of MOUNTAIN WAVES.

blowing-down. The operations of discharging part of the water in a boiler while still under steam to remove concentrated solids.

blow-out. (1) The catastrophic failure of an oil well, allowing oil to gush uncontrolled from the well until the situation can be remedied by capping (i.e. sealing) it. (2) A deflation basin excavated in sand or other easily eroded REGOLITH. Blow-outs can be initiated in partly stabilized sand dunes by the killing of marram grass and other binding agents by people walking on them.

blue asbestos. *See* ASBESTOS.

blue–green algae. *See* CYANOPHYTA.

blue ground. Altered and brecciated (*see*

BRECCIA) IGNEOUS rock which occurs in pipes within CRATONS, composed mainly of the UL-TRABASIC rock KIMBERLITE, with a range of XENOLITHS including ECLOGITE. Some blue grounds contain diamonds.

Blue John. A decorative variety of FLUORITE consisting of alternating bands of blue and yellow crystals, mined principally in Derby-shire, England, where it is possible that the miners who first worked it named it light heartedly to contrast with 'black jack' (zinc blende), mined locally as a zinc ore.

Blueprint for Survival. A document written by Edward Goldsmith, Robert Allen, Michael Allaby, John Davoll and Sam Lawrence, which was first published in January 1972, as a special issue of *The Ecologist* and later as a book, which was translated into many languages. It outlined the reasons for the non-sustainability of eco-nomic growth based on a constant expansion of industrial activity due to increasing con-sumption of resources and energy, and sug-gested strategies whereby a 'no-growth' society might be created. It attracted much publicity.

blue schists. *See* SCHIST.

blue sky. RAYLEIGH SCATTERING of light by atmospheric particles that are small com-pared to the wavelength of light. The blue sky is strongly polarized in directions seen at right angles to the Sun, indicating that it is single scattering. Blue sky is seen in shad-ows when the air is viewed from above (e.g., from an aircraft or in dark-sided mountain valleys).

blue whale. *See* BALAENOIDEA.

BNFL. *See* BRITISH NUCLEAR FUELS PLC.

BOD. *See* BIOCHEMICAL OXYGEN DEMAND.

body louse. *See* ANOPLURA

boehmite (A1O(OH)). One of the major ore minerals of ALUMINIUM. It is one of the main constituents of BAUXITE and LATERITE.

bog. An area of wet acid PEAT in which grow characteristic plants (e.g., *Sphagnum* mosses; sundews, *Drosera* species) and bog myrtle (*Myrica gale*). Bogs form on badly drained ground, the lack of oxygen in the waterlogged soil preventing the decomposi-tion of dead plants, and they may cover large areas as BLANKET BOG. RAISED BOG may de-velop over valley bog or FEN. The term bog is sometimes used more widely to include alkaline valley bog as well as acid peatland.

boghead coal. A sapropelic (containing finely divided plant remains, in this case mostly algal and fungal matter) COAL rich in KEROGEN.

bog soil. Brown or black peaty material, sometimes metres thick, lying over buried PEAT (or in the case of half-bog soil over mottled mineral soil). Its accumulation is due to the slow decomposition of organic matter in waterlogged conditions. *See also* GLEY.

boiler. A pressure vessel designed to pro-duce vapour from liquid by the application of heat.

boiling water reactor (BWR). A NUCLEAR REACTOR that uses enriched URANIUM as fuel, boiling light water (i.e. ordinary water as opposed to HEAVY WATER) as a coolant and light water as a MODERATOR. Steam from the reactor is fed directly to turbines and then back into the reactor.

bolson. A low area or basin that is com-pletely surrounded by higher ground. A bolson drains centripetally into a PLAYA.

bolster eddy. An EDDY sometimes found at the foot of a cliff or steep slope, with air rotating in the corner, driven by an upslope wind.

bomb. A large piece of molten lava ejected into the air from a volcano. During flight and on impact, bombs develop characteristic forms giving rise to such descriptive names as cow-dung bombs, bread-crust bombs, ribbon bombs and spindle bombs.

bomb calorimeter. Apparatus in which the heat produced by a reaction is measured, the reaction being made to occur in a closed vessel. It is used in analyses to determine the CALORIFIC VALUE of a material.

Bombyx mori (silkworm moth). The commercial silk wormmoth, whose larvae are usually raised on a diet of mulberry leaves (although they can be induced to survive on other fare). Each larva spins a cocoon of about 300 metres of silk fibre from its salivary secretions. The pupa is usually killed before it can mature and break from the cocoon (breaking the fibre in the process) and the silk is unwound. About 50 000 cocoons are unwound to yield 1 kilogram of silk thread.

bone. A vertebrate skeletal tissue consisting of cells regularly arranged in a matrix of COLLAGEN fibres and bone salt (a complex compound containing calcium and phosphate). Channels permeating the matrix connect the cells and contain blood vessels and nerves. *See also* CARTILAGE.

bone conduction. The means by which sound can reach the inner ear and be heard without travelling through the air in the ear canal.

booklice. *See* PSOCOPTERA.

bora. A cold, KATABATIC WIND from the highlands, characteristic of Trieste and northern Yugoslavia.

book. *See* LUNG.

borax. *See* BORON.

Bordeaux mixture. A mixture of copper sulphate, quicklime (*see* CALCIUM OXIDE) and water long used as a FUNGICIDE in European vineyards. This or similar mixtures containing inorganic copper are used to control potato blight, fruit scab and other diseases of horticultural crops.

Boreal. A dry climatic period in northern Europe from about 7500 to 5500 BC, during which summers were warm and winters cold. The dominant vegetation was pine, birch and hazel woodland.

boreal. (1) Applied to a climate zone with short, warm summers and snowy winters. (2) Applied to northern coniferous forests growing in a boreal climate.

boric acid. *See* BORON.

bornite (Cu_5FeS_4). A copper iron sulphide mineral. It is a major ore of copper and is found in HYDROTHERMAL veins and in the zone of SECONDARY ENRICHMENT.

boron (B). An element; a brown powder or yellow crystals that occurs as borax and boric acid. It is used for hardening steel and in the production of enamels and glasses. In addition, it is employed in the control of rods of NUCLEAR REACTORS because of its property of absorbing slow neutrons. It is an essential nutrient for plants. A_r=10.811; Z=5; SG (brown powder) 2.37; (yellow crystals) 2.34; mp 2300°C.

Bos primigenius. *See* AUROCHS.

boss. (1) In North American usage, STOCK. (2) A steep-walled stock that is roughly circular in plan. Some small bosses may be volcanic PLUGS.

Botanical Society of the British Isles (BSBI). The senior botanical society of the UK, founded in 1836 and responsible for the production of such works as the *Atlas of the British Flora*.

botryoidal. Having a form resembling that of a bunch of grapes.

bottle gas. *See* LIQUEFIED PETROLEUM GAS.

bottom load. *See* BED LOAD.

bottom sets. The gently sloping layers of fine-grained sediment at the seaward end of a DELTA.

botulism. A disease caused by the

bacterium *Clostridium botulinum*, which can cause many wildfowl deaths during hot summers when water levels drop. The disease also affects other animals, including humans, and is often fatal.

boulder. A block of rock more than 25 centimetres in diameter.

boulder clay (till). An unsorted (*see* SORTING) and usually unstratified glacial deposit of ROCK FLOUR, often containing coarser, ice-transported material ranging in size from SAND to BOULDERS. It is a component of GLACIAL DRIFT.

boundary layer. The layer of fluid which moves more slowly than the main stream because of its proximity to a rigid boundary or, more generally, applied to properties of the fluid other than velocity (e.g., pollution, temperature, etc.). The layer partaking significantly of properties caused by the presence of the boundary.

Bovidae (cattle, sheep, goats). A large family of ARTIODACTYLA which have four-chambered stomachs and horns that are not shed (*compare* ANTLERS).

BP. Years before present; present taken conventionally to be 1950 AD.

Bq. *See* BECQUEREL.

Br. *See* BROMINE.

Brachiopoda (lamp shells). A small phylum of animals with a shell of two valves – dorsal and ventral – superficially resembling bivalve molluscs (*see* MOLLUSCA). They are marine, solitary and usually live attached to the sea bottom. They were much more plentiful in PALAEOZOIC and MESOZOIC times.

Brachycera (short-horned flies). A suborder of the DIPTERA that includes horseflies and gadflies (Tabadinae), the females of which are blood-suckers, robber flies, some of which catch insects in flight, and bee flies, the adults of which hover and feed in flowers.

bracken (*Pteridium aquilinum*). A widespread fern (*see* FILICALES) that is often a troublesome weed on acid grasslands and heaths because it is not eaten by most sheep or rabbits. It recovers rapidly after burning. In the UK it is a common dominant in the field layer of woods on acid soil.

brackish. Applied to water that contains some salt in solution, but less than is contained in seawater. Brackish waters are usually regarded as those containing 0.5–30 parts per thousand sodium chloride (the average salinity of seawater is 35 parts per thousand).

bract. A modified leaf beneath a flower or below a branch of an inflorescence.

bracteole. A small or secondary BRACT.

braided stream. A wide, shallow stream that divides repeatedly between islets of ALLUVIUM and then rejoins, giving it a braided appearance.

brain. An anterior enlargement of the central nervous system in an animal. The development of a brain is correlated with the aggregation of sense organs at the anterior end. *See also* CEREBELLUM, CEREBRAL HEMISPHERES, MEDULLA.

Branchiopoda. A subclass of the CRUSTACEA that includes water fleas (Cladocera, e.g., *Daphnia*), brine shrimps and fairy shrimps, all of which have leaf-like thoracic appendages. Most live in fresh water and commonly reproduce by PARTHENOGENESIS.

Branchiura (carp lice). A small group of CRUSTACEA that are temporary ectoparasites (*see* PARASITISM) of freshwater and marine fishes. They have a large, flattened head with sucking mouthparts and swim well by means of four pairs of thoracic limbs.

brass. Any of a series of alloys consisting basically of copper and zinc.

Brassica. A genus of plants (family: CRUCIFERAE), many of which are cultivated for their stems, roots, leaves or seeds. *B. nigra* is black mustard, *B. oleracea* cabbage, *B. campestris* turnip and *B. napus* rape. Rape is grown for its seeds from which oil is extracted.

Brassicaceae. *See* CRUCIFERAE.

braunerde. *See* BROWN FOREST SOIL.

bread-crust bomb. *See* BOMB.

breakwater. A wall built offshore to protect a beach or harbour from wave action.

bream zone. *See* RIVER ZONES.

breccia. A rock composed of angular fragments of pre-existing rock, predominantly more than 2 millimetres in diameter. Breccias can be of sedimentary origin (implying little transport), TECTONIC (fault breccia) or volcanic (vent breccia).

breeder reactor (fast breeder reactor). A NUCLEAR REACTOR that produces more fuel than it consumes (if, in operation, it is set to consume more fuel than it produces it is called an incinerator). The fuel is plutonium-239 (^{239}Pu), which releases fast (i.e. high-energy) neutrons. The reactor uses no moderator to slow neutrons, but the core is surrounded by a blanket of non-fissile uranium-238 (^{238}U). By bombardment with fast neutrons ^{239}Pu is produced in the blanket from the ^{238}U. The ^{238}U is obtained in the first place from the spent fuel from conventional thermal reactors; the ^{239}Pu can be used to fuel thermal reactors or, if the nuclear industry is expanding, other breeder reactors. The word fast, often used in the name of such reactors, refers to their use of fast neutrons, not to the rate at which they produce plutonium, which is slower than that of most thermal reactor designs (where the plutonium is produced in the fuel itself and almost all of it decays as it contributes to the overall reaction). If the plutonium is extracted from spent thermal reactor fuel, it is possible to obtain enough from one reactor to charge a breeder reactor with fuel after 17 years of operation, and a breeder reactor can breed a second breeder reactor after a further 31 years. The breeder reactor thus extends the energy value of the initial thermal reactor fuel by about 100 times and reduces the overall amount of plutonium for disposal or storage (by burning it), but the development of a breeder programme depends upon the pre-existence of thermal reactors. Most existing breeder reactors use molten sodium as a heat transfer liquid by which energy is carried to water for the production of steam to power turbine generators, liquid sodium being more stable than steam under the conditions within the reactor. This may make breeder reactors inherently safer than thermal reactors. There are prototype breeder reactors in the UK, France, Germany and the USSR.

breeze. A light to moderate wind, force 2–6 on the BEAUFORT SCALE (6–50 km/h, 4–31 mph). Land and sea breezes are caused by temperature differences between land, which warms rapidly by day and cools rapidly at night, and sea, which warms and cools more slowly. Similar breezes occur inland and are not part of the general motion of air over a large area.

brigalow forest. Forest in which the dominant species is the brigalow (*Acacia harpophylla*), covering large areas in Australia.

brimstone. *See* SULPHUR.

brine shrimps. *See* BRANCHIOPODA.

bristle-cone pine (*Pinus aristata*). A species of trees that live to great ages and are used in the investigation of past climates. One living specimen has been dated by DENDROCHRONOLOGY as being 4600 years old, making it the oldest living organism so far known. The oldest redwood (*Sequoia sempervirens*) so far dated is about 3230 years old. *See also* TAXODIACEAE.

bristletails. *See* APTERYGOTA.

bristleworms. *See* POLYCHAETA.

British Nuclear Fuels plc (BNFL). The publicly owned UK company responsible for the reprocessing of spent fuel from nuclear reactors and the disposal of the resultant HIGH-LEVEL WASTES. It operates and is based at the Sellafield complex in Cumbria. (The L in the initials BNFL stands for Limited; under modern company law BNFL is a public limited company, i.e. plc, and the Limited is no longer used.)

British thermal unit (Btu). The energy required to raise the temperature of 1 pound of water through 1 degree Fahrenheit, usually specified as from 39 to 40°F, and in SI units equal to 1059.52 joules, or the mean, being the energy needed to raise the temperature from 32 to 212°F divided by 180 and equal to 1055.179 joules. The Btu is used widely in the USA; in the UK the THERM is used, mainly in relation to gas supply. Both units are tending to be replaced by SI units.

British Trust for Ornithology (BTO). A society founded in 1932 to further the scientific study of birds in the field. Its work includes running the BIRD RINGING scheme and organizing counts of particular species.

brittle stars. *See* OPHIUROIDEA.

brl. BARREL; a volumetric unit used for crude oil.

broadleaved trees. Trees belonging to the DICOTYLEDONEAE, some of which are evergreen (e.g., holly), especially in low latitudes; in higher latitudes (e.g., the UK) most are DECIDUOUS. *See also* CONIFERALES, HARDWOODS.

Brocken spectre. A GLORY observed on the Brocken, a mountain in East Germany.

broiler. A system for the intensive, semi-industrial production of poultry meat, in which birds are housed indoors, but not caged, and are fed and watered by semi-automated systems. Stocking densities are usually very high. Effluent disposal often presents problems. *Compare* BATTERY.

bromine (Br). An element that is a

brownish–red, corrosive liquid at room temperature, used as a raw material in the manufacture of pharmaceuticals and dyestuffs, in the production of ANTIKNOCK ADDITIVES, as a CATALYST, as a component of photographic materials and as an ingredient of chlorofluorobromine (CFB) compounds used in fire extinguishers. In this case, it is suspected of being implicated in the chemistry of the OZONE LAYER. A_r=79.909; Z=35; bp 58.8°C.

bronchial diseases. Diseases of the upper respiratory (bronchial) system, involving the excessive production of mucus, and stimulated by smoking or air pollution. The diseases are chronic because the system becomes adapted to and dependent on the stimulant for mucous production and on cleansing the system by coughing because of the increased viscosity. Infections are less easily prevented in a bronchitic respiratory system, and coughing can cause structural damage to the alveoli (air sacs) in the lungs, leading to emphysema in which the damaged part is filled with mucus and rendered permanently useless.

bronze. Any of a number of alloys of copper and tin.

brood parasite. *See* PARASITISM.

Brower, David. *See* FRIENDS OF THE EARTH.

brown algae. *See* PHAEOPHYTA.

brown coal. *See* COAL.

brown earth. *See* BROWN FOREST SOIL.

brown forest soil (braunerde, brown earth). A dark brown, friable soil with no visible layering, although there is a lighter soil below the surface. The soil is well aerated, and the amount of organic matter gradually decreases with depth, down to a CALCAREOUS parent material whose high calcium carbonate content helps to retard LEACHING. The soil has high agricultural potential. *See also* SOIL CLASSIFICATION, SOIL HORIZONS.

brownian motion. The random movement

of molecules or small, suspended particles in a fluid due to impacts with other molecules or particles.

brown ores. Soils leached (*see* LEACHING) of the more soluble minerals, but retaining rich amounts of iron oxides and hydroxides, particularly GOETHITE, which can be mined. Such ores have declined in importance with the increased mining of BANDED IRONSTONES, but they may constitute a major resource for the distant future.

brown podzolic soil. An acid forest soil comprising a surface layer of litter over a dark, greyish-brown, organic and mineral soil, with a pale leached (*see* LEACHING) layer beneath. *See also* GREY–BROWN PODZOLIC SOIL, SOIL CLASSIFICATION, SOIL HORIZONS.

brown smoke. Smoke produced by volatile, tarry substances emitted when COAL is burned at a low temperature. *Compare* BLACK SMOKE.

brown soils. Soils similar to CHERNOZEMS, but occurring in warmer areas where rainfall is lower. The grasses, which form the natural vegetation, are shorter than those on a chernozem. *See also* SOIL CLASSIFICATION.

brunizem. *See* PRAIRIE SOIL.

brush discharge. An almost continuous release of electric charge to the air from prominent objects when a strong electric field, (e.g., due to a thunder-cloud) is present.

bryocole. An animal that lives among moss (e.g., some TARDIGRADA).

Bryonia dioica. See CUCURBITACEAE.

Bryophyta. The liverworts (Hepaticae), hornworts (Anthocerotae) and mosses (Musci), all of which are small plants, mostly terrestrial (although many prefer damp surroundings), attached to the SUBSTRATE by RHIZOIDS. Mosses and some liverworts have stems and leaves, the rest are thalloid (*see* THALLUS). The plant is a gametophyte and liberates motile male GAMETES which fertilize solitary egg cells housed in flask-shaped organs (archegonia). A capsule (sporophyte), dependent on the gametophyte, grows out and produces SPORES which give rise to new plants.

Bryoza. *See* POLYZOA.

BSBI. *See* BOTANICAL SOCIETY OF THE BRITISH ISLES.

Btu. *See* BRITISH THERMAL UNIT.

bubble bursting. The chief mechanism for the projection of tiny water droplets from the sea surface, which are quickly dried and remain in the air as sea salt AEROSOL.

bubble policy. Where limits are placed on the amount of a particular pollutant a factory may discharge, an accommodation with the regulatory agency whereby stricter standards are imposed on a particular part of the factory in return for a relaxation of standards in another part. The bubble policy was introduced in 1980 as a means of increasing the efficiency of pollution control while reducing its cost.

budding. (1) *See* GEMMATION. (2) The grafting of a bud on to a plant.

buffer. A solution (e.g., in seawater and many fluids found in animal and plant bodies) that resists changes in pH if an acid or alkali is added. The solution contains a weak acid or BASE together with a salt of the acid or base. When more acid or base is added, it reacts with the solution and is removed, with the result that there is little change in the pH. Lakes fed with water that crosses fertile land and sedimentary rock (e.g., many UK lakes) are often well buffered because of the carbonic acid and carbonates they contain; lakes in more sparsely vegetated regions developed on hard IGNEOUS rocks (e.g., many Scandinavian lakes) are more poorly buffered and more susceptible to chemical pollution (e.g., from ACID RAIN).

bugs. *See* HEMIPTERA.

bulb. A modified underground shoot with a much shortened stem, bearing fleshy, food-storing leaf bases or scale leaves (e.g., onion, daffodil). A bulb is an organ of perennation (*see* PERENNIAL) and vegetative propagation, enclosing the next year's bud.

bumpiness. The rough ride experienced by aircraft travelling through the up- and downcurrents of atmospheric convection. A glider or balloon may avoid it by remaining in an upcurrent; a fast aircraft may experience only rapid vibrations.

bunodont. Applied to teeth that have separate conical elevations (cusps) on the crown. Such teeth are typical of mammals that subsist on a mixed diet (e.g., pigs, humans). *Compare* LOPHODONT, SELENODONT.

bunt. *See* SMUT FUNGI.

Bunter. A stage in the Lower TRIASSIC System; most of the RESERVOIR ROCKS in the southern North Sea gas fields are Bunter sandstones.

buoyancy. The upward force exerted on a body (including a body of fluid) that is immersed in a fluid. *See also* ARCHIMEDES' PRINCIPLE.

Bureau of Land Management. The branch of the US Department of the Interior that is responsible for the administration of land owned by the Federal Government. It also issues permits for the grazing of livestock on public lands.

burette. A graduated glass column with a stopcock at the lower end, used in chemical analysis by titration.

burn-up. (1) The quantity of fissile material destroyed in a NUCLEAR REACTOR by NUCLEAR FISSION or by neutron capture, as a percentage of the original quantity present. (2) The heat obtained from a unit mass of any fuel.

bush layer. *See* LAYER.

butadiene $(CH_2CHCHCH_2)$. A major HYDROCARBON raw material mainly for the production of butadiene–styrene (synthetic) rubbers, but also for water-based latex paints, various plastics, etc. At room temperature it is a heavy, colourless gas, but liquefies at -4.4°C (and at 25°C under 2.8 atmospheres pressure). It is extremely reactive, making it difficult and dangerous to handle.

butte. An isolated, flat-topped hill formed when horizontal bedding is worn away at the sides. *See also* KOPJE, MESA.

Buys Ballot's law. The law which states that the wind circulates an area of low pressure in an anticlockwise direction and high pressure in a clockwise direction in the northern hemisphere (and in the opposite directions in the southern hemisphere), so that for a person whose back is to the wind, low pressure is to the left, and high pressure to the right. The law does not apply close to the equator or in violent squalls, or where buildings or other obstructions deflect the wind, altering its direction unpredictably. The law is named after the Dutch meteorologist Buys Ballot (1817–90).

BWR. *See* BOILING WATER REACTOR.

by. Billion (i.e. 10^9) years.

by-pass valve. A valve arranged so as to cause the fluid it controls to flow past some part of its normal path (e.g., to allow a liquid to avoid a filter through which it usually passes).

byssinosis. A disabling lung disease caused by the inhalation of cotton dust during cotton manufacture.

byte. In information processing, an amount of information equal to eight BITS.

C

C. (1) *See* CARBON. (2) *See* COULOMB.

c. *See* CENTI-.

Ca. *See* CALCIUM

cabbage. *See* BRASSICA.

Cactaceae (cacti). A family of DICOTYLEDONEAE; XEROPHYTES that are found mainly in tropical America. The leaves are usually much reduced. The fleshy, water-storing stems are covered in a thick cuticle and bear spines, which retard TRANSPIRATION, promote the formation of dew at their tips and protect the plant from the heat of the Sun and from grazing. *OPUNTIA* bears edible fruit and is grown as the food plant of the cochineal beetle. A few cacti provide timber, and some are used as hedge plants.

cacti. *See* CACTACEAE.

Cactoblastis cactorum. *See* LEPIDOPTERA.

caddis flies. *See* TRICHOPTERA.

cadmium (Cd). An element; a soft, silvery–white metal found associated with zinc ores and in the rare mineral greenockite. It is used in the manufacture of fusible alloys and pigments, in electroplating, and for making control rods in NUCLEAR REACTORS. Cadmium pollution was blamed for the Japanese itai-itai disease. $A_r = 112.40$; $Z = 48$; SG 8.642; mp 320.9°C.

caecilians. *See* APODA

caenogenetic. Applied to a special feature present in the EMBRYO of an animal which adapts that animal to the particular stage it has reached in its development, but which will not be present in the adult (e.g., the embryonic membranes of a vertebrate).

caesium (Cs). An element; a highly reactive, silvery–white metal, used in photoelectric cells and as a CATALYST. Compounds are rare. The radioactive isotopes (^{134}Cs and ^{137}Cs) are likely components of accidental radioactive discharges (e.g., from CHERNOBYL). $A_r = 132.905$; $Z = 55$; SG 1.87; mp 28.5°C.

caffeine. An alkaloid derived from tea and coffee, in which drinks it is the principal stimulant. It is also produced synthetically.

Cainozoic. *See* CENOZOIC.

cairngorm. QUARTZ that contains small amounts of impurities, staining it brown.

cake urchins. *See* ECHINOIDEA.

Calamites. *See* EQUISETALES.

calc-alkaline. Applied to IGNEOUS rocks in which the dominant FELDSPAR is calcium-rich. Such rocks tend to contain calcium-rich FERROMAGNESIAN MINERALS (e.g., HORNBLENDE and AUGITE). *Compare* ALKALINE.

calcareous. Containing calcium carbonate ($CaCO_3$).

calcareous grassland. *See* GRASSLAND.

calcic. Applied to SOIL HORIZONS of secondary carbonate accumulation, which are not less than 15 centimetres thick and whose calcium carbonate content exceeds 15 percent.

calcicole (calciphile). A plant that grows best on CALCAREOUS soils (e.g., old man's beard, *Clematis vitalba*; rockrose, *Helianthemum chamaecistus*; wayfaring tree *Viburnum lantana*). Compare CALCIFUGE.

calciferol. *See* VITAMIN D.

calcifuge (acidophile). A plant that grows best on acid soils (e.g., ling, *Calluna vulgaris*; bracken, *Pteridium aquilinum*). Compare CALCICOLE.

calcimorphic soil. A soil containing an excess of lime (*see* CALCIUM HYDROXIDE).

calcination. Strong heating in air (e.g., to convert metals to their oxides).

calciphile. *See* CALCICOLE.

calcite ($CaCO_3$). A very abundant CARBONATE MINERAL; the main constituent of LIMESTONES, MARBLES and CARBONATITES, and also a GANGUE mineral in some HYDROTHERMAL deposits. Calcite is an important CEMENT in many SEDIMENTARY ROCKS.

calcium (Ca). An element; a soft, white metal that tarnishes quickly in air. It occurs very widely as calcium carbonate ($CaCO_3$) and has many industrial uses. It is an essential MACRONUTRIENT for plants and animals. $A_r = 40.08$; $Z = 20$; SG 1.55; mp 845°C.

calcium carbonate ($CaCO_3$). A common mineral, widely distributed as the main constituent of chalk and LIMESTONE, and used as a filler in paints, rubbers and plastics, as a pigment in whiting, in the manufacture of putty, as an ingredient in polishes and in medicine.

calcium chloride ($CaCl_2$). A drying agent, often used in laboratories in U-shaped tubes to remove water from a gas stream.

calcium cycle. The global circulation of CALCIUM. Plants take up calcium from the soil, it passes through the FOOD CHAIN within the ECOSYSTEM and returns to the soil. Losses

by LEACHING enter aquatic ecosystems and are made good in terrestrial ecosystems by the capture of calcium from surface water, by organisms (whose bodies contain calcium) migrating from neighbouring ecosystems, or from underlying rock. Chalk and LIMESTONE rocks are predominantly of biological origin, being the accumulated, insoluble residues of former marine organisms. *See also* OOLITE.

calcium hydroxide (slaked lime, $Ca(OH)_2$). Agricultural lime used to raise the soil pH, also used in the purification of sugar, glass production and in the manufacture of plasters and mortars, which is produced by the action of water on CALCIUM OXIDE.

calcium oxide (quicklime, CaO). A chemical of very wide industrial use, having a high affinity for water to give CALCIUM HYDROXIDE, and used in metallurgy, paper manufacture, petroleum processing and the food industries.

caldera. A large (several kilometres in diameter), pit-like depression within a surrounding wall of volcanic material, formed by the collapse of the roof of a MAGMA chamber into the chamber itself.

caliche. *See* HARDPAN.

californium. *See* ACTINIDES

Callovian. A stage of the JURASSIC System.

Calluna. *See* ERICEAE.

callunetum. A COMMUNITY of plants dominated (see DOMINANT) by ling or heather (*Calluna vulgaris*).

calomel (mercurous chloride). *See* LAWN SAND.

caloric requirement. The amount of energy an animal needs to maintain its normal functions. The energy is provided by the oxidation of food. When fully oxidized, 1 gram of CARBOHYDRATE (e.g., glucose) liberates 3.74 Calories (i.e. 3.74 kilocalories =

3740 calories = 15.6 kilojoules); 1 gram of PROTEIN liberates 4.1 Calories (17.1 kilojoules), and 1 gram of FAT liberates 9.3 Calories (38.9 kilojoules). The minimum daily caloric intake for a human varies according to age, sex and occupation, but the average is approximately 2400 Calories (10 megajoules).

calorie (gram calorie). The amount of heat required to raise the temperature of 1 gram of water from 15 to 16°C. A Calorie, or kilocalorie, (i.e. 10^3 gram calories) is the unit by which the energy value of food is commonly measured, although in all uses these units are being replaced by the joule (1 joule = 4.182 calories).

calorific value. The number of units of heat obtained by the complete combustion of a unit mass of a substance. This is the most important characteristic of a municipal or industrial waste that is to be disposed of by burning and also provides a measure by which fuels may be compared. The gross calorific value includes the LATENT HEAT captured when water vapour released during combustion is condensed; the net calorific value omits latent heat.

calyx. *See* FLOWER.

cambic. Applied to a SOIL HORIZON that is changed either in structure or in mineral content.

cambium. *See* MERISTEM.

Cambrian. The oldest period of the PALAEOZOIC Era, usually taken as beginning some time between 570 and 610 million years ago. The name also refers to the rocks formed during the Cambrian Period; these are called the Cambrian System, and in the UK they are divided into four series, probably laid down over 70–80 million years. Starting with the oldest, these are the Comley, St David's, Merioneth and Tremadoc. Geologists in most other countries put the Tremadoc into the ORDOVICIAN, so making the Cambrian shorter. The series are identified by the different species of fossil trilobites they contain.

Camellia (Thea). A genus of shrubs and small trees (family: Theaceae), many of which are cultivated. *C. sinensis* is the tea plant, whose leaves contain CAFFEINE, and which is an economically important crop in India, China, Japan, etc., where the young shoots are picked, fermented (except for green tea) and dried.

Campaign for Lead-Free Air (CLEAR). A UK voluntary organization that campaigns to reduce human exposure to lead and that urged strongly the introduction of lead-free petrol (gasoline) in Britain and the rapid phasing out of leaded petrol.

Campaign for Pesticide Reform. A US voluntary organization, representing 41 consumer and environmental organizations and the National Agricultural Chemicals Association, that seeks to achieve the safer manufacture and use of pesticides.

Campanian. *See* SENONIAN.

Campbell–Stokes sunshine recorder. An instrument that records sunshine by focusing it through a glass sphere onto a sensitive chart.

Canadian deuterium–uranium reactor (Candu). A NUCLEAR REACTOR that uses natural URANIUM as a fuel, HEAVY WATER as a MODERATOR and as a coolant heavy water under a pressure of 90 atmospheres inside tubes that also contain the fuel and run through a tank containing the moderator. A Candu reactor is in operation at Pickering, Ontario.

Canadian pondweed. *See* HYDROCHARITACEAE.

candela (cd). The SI unit of luminous intensity, being the intensity perpendicular to a surface of 1/600000 square metres of a BLACK BODY at the temperature of freezing platinum (2042K) at a pressure of 101325 newtons per square metre.

Candu. *See* CANADIAN DEUTERIUM–URANIUM REACTOR.

cane sugar. *See* CARBOHYDRATES.

Cannabiaceae (Cannabidaceae). A family of DICOTYLEDONEAE that contains only two genera — *Humulus* and *Cannabis*. *Humulus lupulus* (hop) is a perennial climbing herb widely cultivated for its cone-like INFLORES-CENCES, used to flavour beer. *Cannabis sativa* (hemp) is cultivated in temperate and tropical regions for its fibre, and for the drug (known variously as ganja, marijuana, charas, bhang, pot, etc.) contained in its resin.

Cannabis. *See* CANNABIACEAE.

Canna edulis. *See* ACHIRA.

cannel coal. A sapropelic coal (*see* BOGHEAD COAL) rich in KEROGEN. The organic material appears to be finely divided vegetable matter, spores, algae and fungi.

canopy cover. The percentage of the ground that is covered when a polygon drawn about the extremities of the undisturbed canopy of each plant is projected upon the ground and the areas of all such projections within a given area are added.

cantharophily. Pollination by beetles.

CAP. *See* COMMON AGRICULTURAL POLICY.

cap cloud. A cloud that remains stationary in a strong wind on or over a mountain peak. *See also* WAVE CLOUD.

capillary flow. (1) The ascent or descent of a liquid within a tube of very small diameter due to the relative attraction between molecules of the liquid and molecules of the material composing the tube. (2) The movement of water upwards through soil spaces above the WATER TABLE as a result of pore surface attraction. The smaller the pore spaces the greater the height to which the water will rise.

capillary moisture. Water held in pores around soil particles by surface tension forces. *See also* FIELD CAPACITY, HYGROSCOPIC MOISTURE.

Capra aegragus. *See* BEZOAR.

Caprimulgidae (nightjars). A family of nocturnal birds (order: Caprimulgiformes) that feed on insects caught in flight. Eggs are laid on bare ground. One American species hibernates.

cap rock. The relatively impermeable upper seal for an underground RESERVOIR containing CRUDE OIL or NATURAL GAS.

Capsicum. *See* SOLANACEAE.

Capsidae (capsid bugs). A large family of plant bugs (suborder: Heteroptera), some of which are serious pests in orchards (e.g., the apple capsid, *Plesiocoris rugicollis*). A few are predatory and play an important role in the control of pests (e.g., *Blepharidopterus angulatus*, which feeds on the red spider mite).

capsule. (1) A dry fruit formed from more than one carpel (*see* FLOWER), which opens to liberate its seeds (e.g., poppy). (2) The swollen terminal portion of the SPORANGIUM of a moss or liverwort, inside which the SPORES are formed. (3) An envelope of CONNECTIVE TISSUE, gelatinous material, etc., surrounding an animal organ, egg, bacterial cell, etc.

capture of droplets. A phenomenon in which a large hailstone or ice crystal falls faster than smaller cloud droplets and grows by ACCRETION. It is also significant on some trees (e.g., pine) whose small leaves, when enveloped in cloud, accrete particles from the cloud and add materially to the rainfall in the area. Sometimes a rain gauge placed beneath a tree collects more than one on open ground. *See also* RIME.

Caradocian. The fourth oldest series of the ORDOVICIAN System in the UK.

carat. (1) The unit of weight for jewels, equal to 200 milligrams. (2) The standard of purity for gold alloys; 24 carat corresponds to pure gold, 22 carats to 22 parts of gold and 2 parts of alloying metal, etc.

carbamates. A group of organic compounds with a wide range of biological activity, used as herbicides, fungicides or insecticides. They have a low toxicity to mammals and persist in the soil for only a short time. They include carbamates (e.g., CIPC, CARBARYL, BARBAN, ASULAM), also thiocarbamates and dithiocarbamates (e.g., DIALLATE, TRIALLATE, ZINEB, MANEB, METHAM-SODIUM).

carbaryl (sevin). A CONTACT INSECTICIDE, earthworm killer and growth regulator of the CARBAMATE group, used to kill earthworms in turf and to control such insects as winter moth caterpillars and earwigs, and for fruit thinning in apples. It is harmful to bees and fish, but is not very toxic to mammals.

carbohydrates. Organic compounds with the general formula $C_x(H_2O)y$. Glucose (dextrose) is a six-carbon atom (hexose) monosaccharide (a simple sugar). It is made during PHOTOSYNTHESIS and is the principal energy source for metabolic processes in plants and animals; it is the blood sugar of vertebrates. Galactose is also a hexose sugar, differing only slightly in configuration from glucose. Ribose is a five-carbon atom (pentose) monosaccharide and is an important constituent of nucleic acids (*see* RNA). Sucrose (cane sugar) is a 12-carbon atom disaccharide formed by the combination of the (six-carbon atom) monosaccharides glucose and fructose. It is found in plants, but not usually in animal tissues. Lactose (milk sugar) is a 12-carbon atom disaccharide formed by combination of the monosaccharides glucose and galactose and is a constituent of mammalian milk. Maltose (malt sugar) is a 12-carbon atom disaccharide, each molecule a compound of two glucose molecules, formed as a breakdown product during the digestion of starch. It occurs in germinating seeds. Starch is a polysaccharide, a combination of many monosaccharide molecules, made during photosynthesis and stored as starch grains in many plants. Glycogen (animal starch) is a polysaccharide made up of glucose units, stored by fungi and animals, notably in the liver and muscles of vertebrates. Cellulose is a long-chain polysaccharide made up of glucose units and a fundamental constituent of plant cell walls. Pectin is an acid polysaccharide, also a constituent of plant cell walls.

carbon (C). An element that occurs in several allotropic forms (*see* ALLOTROPY) including DIAMOND and GRAPHITE, and as amorphous carbon (lamp-black, etc.). Owing to its valency of four, carbon forms compounds with many other elements and is able to form very large molecules that are the chemical basis for life. $A_r = 12.011$; $Z = 6$; SG 3.51 (diamond), 2.25 (graphite), 1.8–2.1 (amorphous carbon); mp 3550°C.

carbon, activated. *See* ACTIVATED CARBON.

carbon-14 dating. *See* RADIOMETRIC AGE.

carbonaceous. (1) Coal-like. (2) Rich in organic compounds and therefore a potential source rock for HYDROCARBONS.

carbonado. *See* BLACK DIAMOND.

carbonate minerals. Minerals that contain the carbonate group (CO_3^{2+}). Since a limited substitution can occur between Ca, Mg, Fe^{2+}, Mn and Zn, many minerals exist with compositions intermediate between calcite ($CaCO_3$), dolomite ($CaMg(CO_3)_2$), siderite ($FeCO_3$), ankerite ($FeMg(CO_3)_2$), magnesite ($MgCO_3$), rhodocrosite ($MnCO_3$) and smithsonite ($ZnCO_3$). Other carbonate minerals include aragonite ($CaCO_3$), which is a polymorph (*see* POLYMORPHISM) of calcite and malachite ($CuCO_3.Cu(OH)_2$). Many living organisms produce hard parts of aragonite or calcite with differing amounts of magnesium substituted for calcium; after death many varieties change to a more stable form or suffer preferential LEACHING.

carbonatite. An intrusive or EXTRUSIVE rock composed mainly of one or more of the carbonate minerals CALCITE, DOLOMITE or ANKERITE, and typically a very rare associate of alkaline igneous INTRUSIONS. Carbonatites have suites of minor elements dissimilar to those of LIMESTONES, and some are sources of niobium and tantalum ores. Niobium (called columbium in the USA) finds increasing use in NUCLEAR REACTORS and is produced almost wholly from carbonatites.

carbon black. Finely divided forms of carbon made by the incomplete combustion or thermal decomposition of various HYDRO-CARBONS. The fine particles are a severe pollution nuisance in areas where they are produced industrially. The form of carbon used in printing ink is usually carbon black. It has been proposed to be of use in controlling weather and climate, by covering snow surfaces with carbon black to absorb sunshine and so cause the snow to melt, but this is unlikely to be practicable because the carbon black itself would be covered and obscured by fresh falls of snow.

carbon cycle. The global circulation of carbon through living organisms, water and the atmosphere. Carbon atoms from carbon dioxide are incorporated into organic compounds formed by green plants during PHOTOSYNTHESIS. These compounds are eventually oxidized during RESPIRATION by the plants that made them, or by consumers of those plants or carnivores feeding on the consumers, and on the death of organisms the carbon in their tissues is oxidized during decomposition, thus returning the carbon to the atmosphere as carbon dioxide. Carbon dioxide held between soil particles also dissolves in soil water, reacts with other substances, eventually drains into rivers and thus carbon is carried to the sea where it becomes incorporated in the shells of marine animals, which in turn may eventually form sedimentary rocks (e.g., LIMESTONES), a long-term store of carbon most of which was once in the atmosphere. Carbon dioxide is introduced to the carbon cycle from volcanoes.

carbon dioxide (CO_2). A minor constituent of the atmosphere, comprising about 0.4 percent; a compound essential to living organisms (*see* CARBON CYCLE). Since the burning of coal, oil and gas began on a large scale the atmospheric content has increased, lead-

ing to fears of widespread climatic changes (*see* GREENHOUSE EFFECT). The atmospheric content of carbon dioxide varies somewhat with the seasons, and the gas is slightly soluble in seawater, the oceans containing a much larger total amount than the atmosphere.

Carboniferous. The fifth oldest period of the PALAEOZOIC Era, usually taken as beginning some time between 345 and 350 million years ago and lasting about 90 million years. The name also refers to rocks formed during the Carboniferous Period, which are called the Carboniferous System, in Europe divided into two series the Lower (Dinantian) and Upper (Silesian). These series are subdivided into stages: the Dinantian into the Tournaisian, Visean and Namurian; the Silesian into the Westphalian and Stephanian. In North America the Carboniferous is replaced by the Mississippian and Pennsylvanian, which are usually ranked as systems.

carbon monoxide (CO). A colourless gas found in trace quantities in the natural atmosphere and produced by incomplete combustion, notably in motor vehicles and cigarettes. Carbon monoxide can form a stable compound with blood haemoglobin (*see* CARBOXYHAEMOGLOBIN), but is harmless in small doses, although lethal in large doses. When domestic (town) gas contained carbon monoxide it was the most common means of suicide (natural gas contains no carbon monoxide). The natural level in the blood is about 0.5 percent saturation, due to the decomposition of old red blood corpuscles. In dense traffic the blood concentration rarely rises above 2–3 percent. Smokers frequently achieve levels of 4–8 percent blood saturation, which produces a slight intoxicating sensation similar to alcohol; chain smokers often achieve 9–10 percent. Such elevation of blood carbon monoxide causes no known permanent harm, except in patients with overloaded or diseased pulmonary or cardiac function. (J.B.S. Haldane once raised his blood saturation by carbon monoxide to 40 percent with acute symptoms of ANOXIA, but no permanent effects.)

carbon tetrachloride (CCl_4). A widely-used industrial solvent, best known as a dry-cleaning agent, but now largely replaced by other compounds because of its toxicity. It is an irritant whose vapour may cause dizziness and headache. Prolonged exposure may lead to liver and kidney damage. It is also a CARCINOGEN.

carboxyhaemoglobin (COHb). The stable compound formed between CARBON MONOXIDE and blood HAEMOGLOBIN. Haemoglobin has a greater affinity for carbon monoxide than for oxygen, so combines with it preferentially, and since the resulting compound is stable the ability of the blood to transport oxygen is reduced by the presence in it of carbon monoxide.

carboxyl. The group –COOH, which forms part of many organic molecules.

carboxylic acid. One of the organic acids in which a carbon fragment (radical) is attached to a carboxyl (–COOH) group. The carboxylic acids include formic acid (HCOOH), acetic acid (CH_3COOH), proprionic acid (CH_3CH_2COOH), butyric acid (($CH_3)_2CH CH_2COOH$), valeric acid ($CH_3(CH_2)_3COOH$), lauric acid ($CH_3(CH_2)_{10}COOH$), myristic acid ($CH_3(CH_2)_{12}COOH$), palmitic acid ($CH_3(CH_2)_{14}COOH$), stearic acid ($CH_3(CH_2)_{16}COOH$) and benzoic, phthalic and terphthalic acids formed from BENZENE RINGS joined to carboxyl groups.

carboy. A large glass bottle, often enclosed in a wicker or steel basket, traditionally used for storing chemicals, although its use in industry is decreasing.

carcinogen. A cancer-producing substance (e.g., some hydrocarbons such as benzo[*a*]-pyrene, vinyl chloride monomer, cutting oils, as well as asbestos and radioactive substances). In many cases the carcinogen does not cause cancer directly, but initiates a series of reactions whose end result, or one of whose products, triggers the aberrant cell behaviour which develops into cancer.

carcinoma. A synonym for cancer.

Caribbean Action Plan. *See* CARTAGENA CONVENTION.

Caribbean Floral Region. *See* CENTRAL AMERICAN FLORAL REGION.

Carica papaya (pawpaw). A small tree (family: Caricaceae) cultivated in warm countries for its edible fruit. The milky juice is used in digestive salts, and meat may be tenderized by being wrapped in the leaves, which brings about partial digestion.

carnallite ($KCl.MgCl_2.6H_2O$). An EVAPORITE mineral; one of the major sources of potassium, an essential plant nutrient. *See also* SYLVITE.

carnassial. *See* CARNIVORE.

carnelian. *See* CHALCEDONY.

Carnivora. *See* CARNIVORE.

carnivore. (1) Any flesh-eating animal or plant (e.g., sundew); a secondary consumer in a FOOD CHAIN. (2) A placental mammal belonging to the order Carnivora (e.g., dog, cat, bear, hyaena, otter, stoat, badger, mongoose, raccoon, etc.), comprising the suborder Fissipedia of land-dwelling carnivores, and walruses, seals and sea-lions, comprising the suborder Pinnipedia of aquatic carnivores. Carnivora have large canine teeth and, typically, carnassials, teeth specially adapted for cutting and formed from the fusion of premolars or molars.

carnotite ($K_2(UO_2)_2(VO_4)_2.3H_2O$). A mineral that is a hydrated oxide of uranium and vanadium found in SEDIMENTARY ROCKS. It is a major ore mineral of uranium and vanadium.

Carnot's principle. In a perfectly reversible engine powered by heat (i.e. an ideal engine in which there is no friction, no waste of heat and which consequently can be run backwards to covert motion into exactly the same amount of heat as would generate that amount of motion) the output of the engine is a function solely of the difference in temperatures of the bodies between which heat is transferred. This principle makes it possible to calculate the theoretical maximum efficiency of any fuel-burning engine. The principle was devised by the French physicist Nicolas Léonard Sadi Carnot (1796–1832).

carotene. *See* CAROTENOIDS.

carotenoids. A group of orange, yellow and red pigments, including carotene and xanthophylls, found in plants (e.g., in leaves, carrot roots, some fruits and flowers) and in CYANOPHYTA, XANTHOPHYTA and some bacteria and fungi. They absorb light and can therefore assist in PHOTOSYNTHESIS, although they are not essential for this process. Carotene is made into VITAMIN A by vertebrates.

carpals. *See* PENTADACTYL LIMB.

carpel. *See* FLOWER.

Carpinus betulus (hornbeam). A tree (family: Carpinaceae) of northern temperate climates and native to southern England. It bears bunches of nutlets, each with a leaf-like, three-lobed wing. Hornbeam is often dominant as a COPPICE shrub in oakwoods in south-east England. The timber is not much used.

carp lice. *See* BRANCHIURA.

carr. FEN woodland dominated by plants such as alder (*Alnus*) and willows (*Salix*).

carrying capacity. (1) The maximum number of species an area can support during the harshest part of the year, or the maximum BIOMASS it can support indefinitely. (2) The maximum number of grazing animals an area can support without deterioration. (3) The level of use, at a given level of management, a natural or manmade resource can sustain without an unacceptable degree of deterioration of the character and quality of the resource (e.g., the maximum level of recreational use, in terms of numbers of people and types of activity, that can be accommodated before the ecological value of the area declines).

Cartagena Convention. An international agreement signed by representatives of 27 countries at Cartagenas de Indias, Colombia, in March 1983, to protect the marine environment of the Caribbean. The Convention resulted from the UNEP REGIONAL SEAS PROGRAMME and called for the establishment of a $1.5 million Caribbean Action Plan.

cartel. An association of independent undertakings in the same or related branches of industry that aims to improve or stabilize conditions of production or sale for the mutual benefit of its members. To be strong, a cartel must control a large proportion (usually 75 percent or more) of the production of a commodity and also its distribution.

cartilage. Flexible vertebrate skeletal tissue consisting of groups of cells lying in a matrix containing COLLAGEN. It occurs on the ends of bones, in the PINNA of the ear, in discs between vertebrae and forms most of the skeleton in embryos. *See also* CHONDRICHTHYES.

Caryophyllaceae. A widespread family of DICOTYLEDONEAE, most of which are herbs, including the campions, pinks, chickweeds, stitchworts, pearlworts and sand spurreys.

caryopsis. The fruit of grasses; an ACHENE in which the ovary wall is fused to the seed coat.

cascade. A repetitive system for separation and purification, in which components of a mixture gradually become more clearly separated as the process is repeated. The principle is used in many industrial processes, including the separation of radioactive isotopes.

case hardening. A method for producing a hard surface layer on a non-hardening type of steel by heating the steel in contact with certain carbon compounds. Carbon is absorbed to produce a thin layer of high-carbon steel which can then be heat-treated.

casein. The principal PROTEIN in milk and cheese, but with many industrial uses (e.g., in the production of medicines, foodstuffs, paints, glues, plastics, synthetic fibres, etc.).

casparian strip. *See* ENDODERMIS.

Caspian Sea. *See* KARA-BOGAZ-GOL.

cassava. *See* MANIHOT.

cassiterite (SnO_2). A mineral that is the major source of tin, found in acid IGNEOUS rocks (particularly PEGMATITES), contact metamorphic zones (*see* CONTACT METAMORPHISM) and in HYDROTHERMAL deposits, but much of the world's production comes from PLACER deposits. Tin is used mainly in tin plating, where a thin layer of tin acts as an anticorrosion coating for mild steel, but it is also an ingredient of many alloys (e.g., solder, bearing metal, gunmetal, bronze, bell metal and pewter) and industrial chemicals.

caste. An individual anatomically specialized for a particular function within an insect colony. For example, in a honeybee colony there are three castes: fertile females (queens); sterile females (workers); males (drones). *See also* POLYMORPHISM.

castellatus. Turret-shaped CUMULUS clouds, often present in lines; a very unstable form of ALTOCUMULUS, individual turrets having lifetimes of 5–10 minutes.

Castillo de Bellver. A Spanish oil tanker that caught fire and eventually broke in two as it was rounding the Cape of Good Hope on 6 August 1983, with a cargo of 260 000 tons of Gulf crude bound for Spain. Soot from the fire covered farm crops and sheep on shore, but an oil slick 43 kilometres (27 miles) long and 11 kilometres (7 miles) wide was carried northwards by the Benguela Current and did not contaminate beaches.

casting. The making of a pottery or metal object by pouring material into a mould. The mould is usually of plaster for clay casting and of sand for iron casting.

castles in the air. *See* FATA MORGANA.

castor oil plant. *See* EUPHORBACEAE.

casual plant. *See* ADVENTIVE PLANT.

catabolism (katabolism). The breaking down by organisms of complex molecules into simpler ones with the liberation of energy. *Compare* ANABOLISM.

catadromy (katadromy). The migration of some fish (e.g., freshwater eel) from rivers to the sea for spawning. *Compare* ANADROMY.

catalase. An iron-containing enzyme that breaks down hydrogen peroxide (H_2O_2) into oxygen and water.

catalysis. A process in which the rate of reaction is altered by the presence of an added substance (a CATALYST) which remains unchanged at the end of the reaction.

catalyst. (1) A substance whose presence alters the rate at which a chemical reaction proceeds, but whose own composition remains unchanged by the reaction. Catalysts are usually employed to accelerate reactions (positive catalysts), but retarding (negative) catalysts are also used. *See also* CATALYSIS. (2) *See* CATALYTIC REACTION.

catalytic converter. A device fitted to the exhaust system of a petrol-driven motor vehicle to reduce emissions of pollutants, especially of unburnt hydrocarbons, carbon monoxide and nitrogen oxides, by catalyzing chemical reactions to trap the pollutants. The device is very efficient, but must be replaced about every four years, and can be fitted only to engines burning lead-free petrol (gasoline).

catalytic cracking. The breaking of carbon–carbon bonds with the aid of a CATALYST. This is an essential process in the refining of petroleum. *See also* CRACKING.

catalytic reaction. A chemical reaction in which the amount of one of the substances involved (the catalyst) is not decreased, although its presence is necessary for the reaction to occur at all.

catalytic reforming. The use of heat, pressure and a CATALYST to isomerize (*see* ISOMER) HYDROCARBON molecules, thus varying their properties for particular uses.

catarobic. Applied to a body of water in which organic matter is decomposing fairly slowly and oxygen is not in short supply.

catastrophism. The theory that the history of the Earth and of life upon it is punctuated by a series of discontinuous events of great magnitude that effect profound changes. Most modern geologists adhere to UNIFORMITARIANISM, and biologists to the concept of evolution by natural selection (*see* DARWIN, CHARLES ROBERT), although mass extinctions of organisms have occurred in the past, and modern evidence suggests that at least some of them may have been the consequence of major impacts by large meteorites or comets, so a much-modified idea of catastrophism now finds some favour.

catchment (in US usage, watershed). The area from which a major river system or lake derives its water (i.e. the area drained by a particular river system or lake). The UK water authorities, responsible for the public water supply, are allotted areas based on natural catchments. One catchment is separated from a neighbouring catchment by a watershed (in US usage a divide).

catena. (1) A group of soils showing variations in type due to differences in topography or drainage, although they are derived from uniform or similar parent material. (2) A diagram illustrating such a distribution of soils.

Cathartidae. A family of New World VULTURES.

cathode. *See* ANION.

cation. A positively charged ION. *Compare* ANION.

catkin. A long, usually tassel-like INFLORESCENCE that contains either male or female flowers (e.g., as borne by hazel, birch, poplar, willow, oak).

cattle. *See* BOVIDAE.

Caudata. *See* URODELA.

cauliflory. In plants, the condition in which flowers (and therefore fruit) are borne on the stem (e.g., in cocoa, *Theobroma*, and coffee, *Coffea*).

caustic scrubbing. A process, used in the USA, for removing SULPHUR DIOXIDE from FLUE GASES by passing them through a solution of caustic soda (sodium hydroxide). The sulphur dioxide and sodium hydroxide react to produce sodium sulphite and sodium bisulphite, and the addition of lime (calcium carbonate) causes the precipitation of calcium sulphate (GYPSUM), leaving the water enriched in sodium carbonate, a harmless substance, present naturally in most mineral water, which can be diluted and then discharged into surface waters.

caustic soda (sodium hydroxide, NaOH). A strong alkali that causes severe burns if it comes into contact with flesh (caustic is from the Greek *kaustikos* meaning burning). It is manufactured in large quantities for use in an extremely wide range of industrial chemical processes.

Cavtat. A Yugoslav ship which sank in the Adriatic in 1974 carrying a cargo of drums containing TETRAETHYLLEAD. Recovery of the cargo was delayed by legal arguments, and the last of the drums was not recovered until April 1978, by which time some of them had begun to leak, although not sufficiently to cause serious pollution.

cayman. *See* CROCODILIA.

CCAMLR (Convention on the Conservation of Antarctic Marine Living Resources). *See* ANTARCTIC TREATY.

CCC. *See* CHLORAMQUAT.

Cd. *See* CADMIUM.

cd. *See* CANDELA.

celestine. *See* CELESTITE.

celestite (celestine SrSO$_4$). The mineral strontium sulphate, found chiefly in SEDIMENTARY ROCKS, but also in HYDROTHERMAL deposits. It is the major source of strontium, which is used to impart a crimson colour to fireworks and also finds uses in sugar refining.

cell. A very small unit of living matter (PROTOPLASM), bounded by a thin membrane (i.e. PLASMA MEMBRANE) and, in plants, surrounded by a CELL WALL usually made of cellulose. The protoplasm of most cells is divided into a nucleus and the cytoplasm, which contains various inclusions (PLASTIDS, VACUOLES, MITOCHONDRIA, RIBOSOMES, CENTRIOLES, LYSOSOMES, ENDOPLASMIC RETICULUM, GOLGI APPARATUS). Many microorganisms consist of a single cell. The cells of multicellular organisms (which in sexually reproducing forms are all derived from a single cell, the zygote) are specialized for particular functions and vary greatly in structure. *See also* COENOCYTE, PLASMODIUM, PROTOPLAST, SYNCYTIUM.

cell division. *See* MEIOSIS, MITOSIS.

cell membrane. *See* PLASMA MEMBRANE.

cellophane. A strong, flexible, transparent film made from cellulose nitrate or cellulose acetate and used as a wrapping material.

cell respiration. *See* RESPIRATION.

cell theory. The theory, advanced in 1838–9 by Matthias Jakob Schleiden and Theodor Schwann, that all organisms are composed of cells and their products, and that reproduction and growth are the results of cell division.

cellular convection. CONVECTION in a layer of uniform thickness, best demonstrated in very viscous fluids.

cellular respiration. *See* RESPIRATION.

cellulose. *See* CARBOHYDRATES.

cell vessels. *See* TRACHEA.

cell wall. The outer, supporting layer of a plant cell, made by the PROTOPLAST, and in higher plants, many ALGAE and a few FUNGI consisting largely of cellulose. The long-chain cellulose molecules are arranged in bundles, forming a network of strands separated by other polysaccharides including pectin. A middle lamella of pectic material forms across the centre of a dividing cell, and the wall is built on this from each side. It is subsequently thickened in most cells, but minute pits are left through which pass fine threads of CYTOPLASM (plasmodemata) connecting adjacent cells. Further thickening of the wall may occur, incorporating strengthening deposits of LIGNIN on SCLERENCHYMA, vessels and TRACHEIDS, or waterproofing deposits of cutin on epidermal cells and of suberin on CORK cells.

Celsius, Anders (1701–44). A Swedish astronomer who devised the temperature scale named after him (and also called, incorrectly, centigrade).

cement. (1) The mineral matter binding together the fragments in SEDIMENTARY ROCK. Common constitutents are SILICA (as QUARTZ or CHALCEDONY), CALCITE, DOLOMITE, HAEMATITE, LIMONITE, SIDERITE and GYPSUM. (2) A powder that sets hard after having been mixed with water. PORTLAND CEMENT is the material most commonly used. Cements contain burnt lime (quicklime, CaO) with silica (SiO$_2$), alumina (Al$_2$O$_3$), with small amounts of iron (as Fe$_2$O$_3$) and often magnesium oxide (MgO). Gypsum is usually added to slow the hardening process. Where these minerals occur in a natural rock in the required proportions, the rock is called cement rock. If the limestone clinker is ground more finely than for Portland the result is a rapid-hardening cement. High-alumina cement (aluminous cement) is made by melting bauxite ($Al_2O_3.2H_2O$) and limestone and grinding the clinker without gypsum to provide a cement that reaches its maximum hardness very quickly. Sulphate-resistant cement has a higher (not less than 3.5 percent) content of tricalcium aluminate ($3CaO.Al_2O_3$), hydrated aluminium compounds being very susceptible to sulphate attack. Glass fibre-reinforced cement is a mixture of glass fibres and cement which can be cast, pressed or injection moulded for tunnel linings, window frames, etc. *See also* CONCRETE, MORTAR.

cementation. Increasing the carbon content of steel by heating it in contact with carbon compounds, as in CASE HARDENING.

cement kiln. A long, slightly inclined rotary chamber in which the slurry is calcined (*see* CALCINATION) in the manufacture of PORTLAND CEMENT.

cement rock. *See* CEMENT.

Cenomanian. A stage of the CRETACEOUS System.

Cenozoic. (1) The era that followed the MESOZOIC; in some usages it was followed by the QUATERNARY Era (i.e. Cenozoic is synonymous with TERTIARY) and in others (e.g., that of the US Geological Survey) the Cenozoic continues to the present day (i.e., is composed of the Tertiary and Quaternary, both of which rank as PERIODS). (2) Refers to rocks formed during the Cenozoic Era.

census. A complete enumeration of a whole population with respect to specific variables, as distinct from sampling (taking a statistically representative sample).

centi-. A prefix used in conjunction with SI units to denote the unit x 10^{-2} (i.e. one-hundredth).

centipedes. *See* MYRIAPODA.

Central American Floral Region (Carib-

bean Floral Region). The region of the NEOTROPICAL REALM that comprises Central America, the southern tips of California and Florida, the islands of the Caribbean and the northern part of South America, including Guyana, Surinam, French Guiana and most of Colombia and Venezuela.

central eruption. A volcanic eruption from a central vent, as distinct from an eruption from a fissure. Central eruptions form cones of two major varieties—SHIELD VOLCANOES and STRATOVOLCANOES.

central nervous system (CNS). Nervous tissue that coordinates the activities of an animal. In most invertebrates the CNS consists of solid ventral nerve cords bearing swellings (ganglia). In vertebrates it is hollow and situated dorsally, inside the skull and vertebral column, and consists of the brain and spinal cord. Nerve cell bodies are aggregated in a CNS, and many junctions (synapses) are present between nerve cells, so that impulses coming in from sense organs and going out to effector organs can be coordinated finely. In higher animals, the brain is able to store information and is the 'seat of learning'. In vertebrates, the outer layer of the CNS consists of WHITE MATTER, composed largely of nerve fibres, and the inner layer (and the CEREBELLUM) is GREY MATTER, made up mainly of nerve cell bodies. *Compare* AUTONOMIC NERVOUS SYSTEM, PERIPHERAL NERVOUS SYSTEM. *See also* NERVE NET, NEURON.

Centre for Economic and Environmental Development. A UK environmentalist organization backed by conservation and professional bodies and industry.

centres of diversity. *See* VAVILOV, NIKOLAI IVANOVICH.

centres of origin. *See* VAVILOV, NIKOLAI IVANOVICH.

centrifugal force. The inertial force directed away from the centre of curvature that is evident in a body made to move in a curved path.

centrioles. Bodies situated in the CENTRO-SOME and revealed by electron microscopy to be very similar in structure to the kine-tosomes found at the bases of CILIA and fla-gella (*see* FLAGELLUM). Each centriole is a cy-linder whose walls are formed by a ring of nine groups of fibres. Centrioles are self-duplicating and form the poles of the spindle during cell division. *See also* MEIOSIS, MITO-SIS.

centrosome. (1) A small region of cyto-plasm that is present in animal cells situated near the nucleus, and contains a pair (usu-ally) of CENTRIOLES. (2) A synonym for centriole.

cephalic index. The breadth of the head expressed as a percentage of the length from front to back.

Cephalochordata. *See* ACRANIA.

Cephalopoda (Siphonopoda, octopuses, squids, cuttlefish, nautilus, (extinct) am-monites, etc.). A class of MOLLUSCA charac-terized by the possession of a well-devel-oped head bearing tentacles, complex eyes and a shell that is often reduced and internal. *See also* DECAPODA, OCTOPODA.

CEQ. *See* COUNCIL ON ENVIRONMENTAL QUALITY.

Ceratostomella ulmi. *See* DUTCH ELM DIS-EASE.

cercaria. *See* TREMATODA.

cerebellum. A part of the hindbrain of vertebrates, particularly well-developed in birds and mammals, that coordinates com-plex muscular movements.

cerebral cortex. *See* CEREBRAL HEMI-SPHERES.

cerebral hemispheres (cerebrum). A pair of swellings at the anterior end of the verte-brate brain, which are concerned largely with the sense of smell in primitive forms, but are very well developed in birds and mammals, and which are respon-sible for the general coordination of the ac-tivities of the animal. In higher verte-brates the outer layer (cerebral cortex) con-sists of GREY MATTER and is extensive and much folded. *See also* CENTRAL NERVOUS SYS-TEM.

cerebrum. *See* CEREBRAL HEMISPHERES.

cesspool. A storage tank for sewage that must be emptied at intervals. *Compare* SEP-TIC TANK.

Cestoda (tapeworms). A class of flatworms (phylum: PLATYHELMINTHES) the adults of which parasitize (*see* PARASITISM) verte-brates. The adult tapeworm has no gut or complex sense organs, but possesses a thick CUTICLE, with hooks and/or suckers at one end to attach it to the host's gut. Most pro-duce a chain of proglottides (segments), each containing reproductive organs, and it is this which gives the animal its character-istic ribbon-like appearance. The larval stage is passed inside one or more intermedi-ate hosts (e.g., rabbit, sheep, human or flea in the case of various dog tapeworms). *See also* BLADDERWORM.

Cetacea (whales, dolphins, porpoises). An order of aquatic, placental mammals (*see* EUTHERIA) that lack hair or (other than inter-nal and rudimentary) hindlimbs, but have blubber, front limbs modified as paddles and a transverse (i.e. horizontal to the body, rather than vertical as in fishes) tail fin. The hunting of whales for their flesh, oil, sperma-ceti and baleen ('whalebone') has threat-ened some species with extinction. The IN-TERNATIONAL WHALING COMMISSION has im-posed a moratorium on all hunting of whales except for 'scientific research'.

cf. Cubic feet, (ft^3). A unit used, for ex-ample, in measurements of HYDROCARBON RESERVES.

CFBR (Commercial Fast Breeder Reactor No. 1). A British BREEDER REACTOR.

CFC. *See* CHLOROFLUOROCARBONS.

CGIAR. *See* CONSULTATIVE GROUP FOR INTERNATIONAL AGRICULTURAL RESEARCH.

chaetae. *See* ANNELIDA, CHAETOPODA.

Chaetognatha (arrow-worms). A small phylum of PELAGIC animals with transparent, elongated bodies. They are often found in swarms among marine PLANKTON.

Chaetopoda. A class of ANNELIDA whose members possess bristles (chaetae). *See* OLIGOCHAETA, POLYCHAETA.

chain reaction. A situation in which one event causes a second, similar event. For example, when a raindrop, having grown by ACCRETION of cloud droplets to about 5 millimetres diameter is broken into several smaller droplets by aerodynamic forces, each of the small droplets grows by accretion and thus rain is produced by a chain reaction. In a NUCLEAR REACTOR, the release of neutrons from the fuel causes neutron bombardment of other fuel, leading to a further release of neutrons and further bombardment.

chalcedony. A CRYPTOCRYSTALLINE variety of SILICA, some varieties of which are semiprecious (e.g., agate, onyx, carnelian, jasper). Chalcedony occurs in sediments, as GANGUE in HYDROTHERMAL deposits, and as filling in amygdales (*see* VESICLES).

chalcocite. A copper sulphide mineral (Cu_2S), that is a major ore of copper and is found mainly in the zone of SECONDARY ENRICHMENT.

chalcopyrite $(CuFeS_2)$. A copper iron sulphide mineral that occurs widely, but is mined mainly from hydrothermal deposits (e.g., PORPHYRITIC COPPERS) and from stratiform deposits confined to particular horizons in organic-rich SEDIMENTARY ROCK and metamorphosed sediments (*see* METAMORPHISM). The latter deposits show evidence both of deposition of copper minerals from seawater during sedimentation and METASOMATISM during early DIAGENESIS.

chalk. (1) *See* CALCIUM CARBONATE. (2) A very fine-grained, friable pure LIMESTONE that forms extensive deposits of CRETACEOUS and Lower TERTIARY age. Most chalks are white.

chamaeophyte. *See* RAUNKIAER'S LIFE FORMS.

change of state. The transformation of a substance from one to another of the three states of matter: gas, liquid, solid.

channelized flow. *See* EROSION.

chaparral. A type of vegetation dominated by shrubs with small, broad, hard evergreen leaves, found in areas with a mediterranean climate.

char. The solid product of the DESTRUCTIVE DISTILLATION or carbonization of an organic material. The char from the destructive distillation of urban waste is sometimes used as a fuel.

Chara. *See* CHAROPHYTA.

characteristic impedence. A measure of the efficiency with which sound travels through a particular substance, expressed as the ratio of the SOUND PRESSURE LEVEL at a given point to the effective particle velocity. It is equal to the product of the density of the substance and the speed of sound within it, the result being a value in rayls. Air at 20°C and a pressure of 1000 millibars has a characteristic impedence of 408 rayls.

characteristic species. Plant species that, almost without exception, are localized within a given ASSOCIATION. They provide the most reliable floristic expression of the ECOLOGY of the group.

Charadriidae (waders). A family (order: CHARADRIIFORMES) of small or medium-sized birds with long legs, wings and bills. Most are shore birds that nest on the ground and are highly migratory, often occurring in flocks on passage (e.g., curlews, snipe, sandpipers, plovers, avocet, woodcock).

Charadriiformes. A large order of swimming or wading birds which includes the waders (family: CHARADRIIDAE), auks (family: ALCIDAE), gulls and terns (family: LARIDAE), skuas, oystercatchers and the stone curlew.

charcoal. Impure carbon obtained by heating CARBONACEOUS substances (e.g., wood) with little or no oxygen. It is used as a fuel and also as a deodorant, decolourant, absorbant and catalyst.

charcoal, activated. *See* ACTIVATED CARBON.

chard. *See* BETA VULGARIS.

Charophyta (stoneworts). A group of what are usually recognized as ALGAE, found in still or slow-moving fresh or BRACKISH water. The filamentous THALLUS bears whorls of branches, and the plant often becomes heavily encrusted with lime (e.g., *Chara*).

chelate. A chemical compound whose molecule has a ring structure, with a metallic ion attached by coordinate bonds to two or more atoms in the ring.

Chelonia (Testudines; turtles, terrapins, tortoises). An order of toothless reptiles in which the body is encased in a 'shell' consisting of two plates, in most groups with the thoracic vertebrae and ribs fused to the dorsal plate.

chemical engineering. Originally the branch of engineering concerned with the industrial manufacture of chemical products, but today many other products and industries fall within the compass of chemical engineering practice. In general, the term is applied to the process industries, covering a wide range of manufacture, including the conversion of industrial raw materials and production of foodstuffs.

chemical oxygen demand (COD). The weight of oxygen taken up by the organic matter in a sample of water, expressed as parts per million of oxygen taken up from a solution of boiling potassium dichromate in two hours. The test is used to assess the strength of sewage and trade wastes. *Compare* BIOCHEMICAL OXYGEN DEMAND.

chemoautotrophic (chemosynthetic). Applied to organisms that produce organic material from inorganic compounds using simple inorganic reactions as a source of energy. For example, *Thiobacillus* obtains energy by oxidizing hydrogen sulphide to sulphur; other bacteria utilize energy from the oxidation of ferrous salts to their ferric form. *See also* AUTOTROPHIC.

chemoreceptor. A sense organ (e.g., insect antenna, vertebrate taste bud, etc.) that is sensitive to the presence of chemical substances.

chemosynthetic. *See* CHEMOAUTOTROPHIC.

chemotaxis. *See* TAXIS.

chemotrophic. Applied to organisms that obtain energy from any source other than light. The energy may be obtained from simple inorganic reactions (*see* CHEMOAUTOTROPHIC) or (by HETEROTROPHIC ORGANISMS) from the oxidation of organic material. *Compare* PHOTOTROPHIC.

chemotropism. (1) An outdated synonym for chemotaxis (*see* TAXIS). (2) A growth response in plants in which the stimulus is a chemical concentration gradient. An example is the growth of a pollen tube through the STIGMA and style in response to substances contained in the OVULE or ovary wall.

Chenopodiaceae. A family of DICOTYLEDONEAE, mostly comprising herbaceous plants, mainly of arid regions and frequently adapted to tolerate salty soil (halophytic). The leaves are often reduced, fleshy or hairy. UK species include fat hen (*Chenopodium album*), oraches (*Atriplex* species), sea purslane (*Halimione portulacoides*) and samphires (*Salicornia* species). Economically useful members of the family include *BETA VULGARIS*, spinach (*Spinacia oleracea*) and some species of *Chenopodium* grown for leaves and seeds.

Chernobyl. The site in the Ukraine of a large complex of nuclear power plants where, on 26 April 1986, one of four operational RMBK REACTORS failed catastrophically as a result of experiments conducted on it during a routine maintenance shut-down. The reactor building was partly destroyed by explosions and fire, which released large clouds of radioactively contaminated material, amounting to an estimated total of 10^{16} becquerels. The fallout affected principally the western USSR, Poland and, to a lesser extent, parts of north-western Europe. One worker was killed by falling debris, a second by steam burns and a further 29 from injuries or radiation sickness. Estimates of long-term cancer deaths have varied from less than 1000 to about 75 000 over 50 years in the whole of Europe, but mainly in the USSR.

chernozem (black earth). A grassland soil found in subhumid to temperate areas, where a HUMUS remains near the surface and a black or brown surface layer grades downward through a lighter coloured layer to a layer where lime has accumulated. Tall grasses are the natural vegetation of chernozem. It is the soil of the Russian steppes (chernozem is the Russian for black soil), North American Great Plains and prairies, and the pampas of Argentina. *See also* SOIL CLASSIFICATION.

chert. A CRYPTOCRYSTALLINE form of SILICA, usually occurring as nodules or as a replacement material in SEDIMENTARY ROCKS. *See also* FLINT.

chestnut soils. Soils similar to CHERNOZEMS, but found in warmer, drier areas, where they support grasses shorter than those that grow on chernozems.

Chilopoda (centipedes). *See* MYRIAPODA.

Chimaera. A genus of fishes (e.g., king of the herrings, rabbitfish) belonging to the largely fossil group Holocephali, whose members have large, flat crushing teeth and the upper jaw fused to the skull.

chimaera (chimera). An organism composed of a mixture of cells with different genetic compositions. This condition results from chromosome changes in part of the organism during its development (*see* MOSAIC) or from grafting material from one plant on to another (the resulting plant being a graft hybrid).

chimney. A vertical passage through which a stream of gas is dispersed into the atmosphere. It is most often used to cool combustion products from domestic fires and from a wide range of industrial processes (e.g., power stations, cement works). Chimneys are typically made of brick or steel, and movement in them may be actively induced by fans, although being warmer and therefore less dense than its surroundings the gas stream tends to rise and so produce a natural draught.

china clay. *See* KAOLIN.

China man. *See* HOMO.

chinampas. A Mexican style of agriculture based on irrigation canals constructed to a rectangular grid pattern enclosing 'islands' to which silt dredged periodically from the canals, aquatic weeds and organic wastes of all kinds are added to supply nutrients, and which form cultivated beds. The water is flushed out and replenished from freshwater springs connected by aquaducts to the chinampas site, thus preventing SALINATION. The system was developed in Aztec times and was highly productive.

China Syndrome. The title of a novel and a film released at about the time of the THREE MILE ISLAND incident and fuelling the fears of antinuclear groups, referring to the worst accident that can be imagined concerning a nuclear reactor. It supposes that a total failure of all cooling and safety systems, and exposure of fuel elements leads to a rapid overheating and the consequent melting of the core. The intensely hot material then melts the floor of the building and descends (in the general direction of the Earth's core and thence through the Earth and by a further stretch of the imagination to China) until it encounters ground water, when its heat is dissipated, and there is a large explosion and major release of radioactive material to the atmosphere. Experience in nuclear accidents (e.g., CHERNOBYL, Three Mile Island) and extensive theoretical modelling of serious accidents indicate that when a core begins to melt down its material disperses, and the nuclear reaction is slowed. Thus the MELT-DOWN is self-limiting, and heat is dispersed in the surrounding core and the floor of the building, making the China syndrome appear impossible.

chinook. A warm wind from the west in Canada, similar to a FÖHN WIND, that descends the eastern side of the Rocky Mountains in winter and displaces cold air masses.

Chinook Arch. See ARCH CLOUDS.

Chiroptera (bats). A very large order of placental mammals (subclass: EUTHERIA) whose forelimbs are modified to form wings, covered with a membrane of skin supported by the elongated bones of the forelimbs and the second to fifth digits. There are two groups: the Microchiroptera comprising the small, mainly insectivorous species, and the Megachiroptera comprising the fruit-eating bats (flying foxes), although the two groups may not be closely related. Insectivorous bats are extremely numerous, and their predations exert a major control on the size of populations of flying insects. In the UK, bats are the most rigorously protected of all animals.

chi-squared test (χ^2 test). A statistical procedure that is widely used to test the agreement between a set of data derived from observation and a set of hypothetical figures.

chitin. A tough, flexible substance that forms much of the EXOSKELETON of ARTHROPODA and the bristles of ANNELIDA. Its long-chain molecules are partly polysaccharide (see CARBOHYDRATES), but they also contain nitrogen. The cell walls of many FUNGI contain a similar substance.

chloracne. A disfiguring skin rash caused by excessive exposure to DIOXIN. It is difficult to treat and has been known to persist for 15 years, although workers exposed to TCDD (a dioxin) at the Bolsover, Derbyshire, UK, factory of Coalite and Chemical Products Ltd in 1968 were free from chloracne after four years. Many cases of chloracne occurred as a result of the accident at SEVESO, Italy.

chloramquat (CCC). A growth regulator used to shorten and strengthen straw in oats and wheat.

chlorates. Salts containing the chlorate ion (usually ClO_3^-), but the term also includes hypochlorite (ClO^-), chlorite (ClO_2^-) and perchlorate (ClO_4^-). They are very strong oxidizing agents (see OXIDATION) and are used widely as disinfectants and bleaching agents, but they can form explosive mixtures with organic materials.

chlordane. A persistent ORGANOCHLORINE used as an earthworm killer on turf. Because of its side effects, its use in the UK is now limited. See also CYCLODIENE INSECTICIDES.

chlorinated hydrocarbons. See ORGANOCHLORINES.

chlorination. (1) The process of introducing one or more chlorine atoms into molecules of a compound, often using gaseous chlorine. (2) The application of chlorine to water, sewage or industrial wastes for disinfection or other biological or chemical purposes. Chlorine is extremely toxic to bacteria, but is rapidly deactivated in the process, especially on contact with protein.

chlorine (Cl). An element that is an essential precursor of many industrial chemicals. An extremely toxic gas, heavier than air, it combines with water (e.g., in the lungs if inhaled) to produce hydrochloric acid (HCl) and hypochlorite (see CHLORATES). It was used as a weapon in World War I. It is also used as a bleach and in water purification (see CHLORINATION). It occurs naturally as halite (common salt, NaCl), as chlorides of other metals and is manufactured almost entirely by the electrolysis of brine. $A_r=35.453$; $Z=17$.

chlorine demand. The amount of chlorine needed to kill all the pathogens in a sample of water.

chlorite. A group of SILICATE MINERALS closely related to MICAS that are found in low-grade metamorphic rocks (*see* METAMORPHISM) as an alteration product of IGNEOUS rocks and in sediments.

chloroethene. *See* VINYL CHLORIDE MONOMER.

chlorofluorocarbons (CFCs, freons). A range of compounds of carbon and halogens (principally chlorine and fluorine, but sometimes bromine), named freons by Du Pont Nemours and Co., which developed them, but also known as CFC followed by a number. CFC11 (CFCl$_3$) and CFC12 (CF$_2$Cl$_2$) are used mainly as propellants in AEROSOL SPRAYS; CFC21 (CFHCl$_2$) and CFC22 (CF$_2$HCl) are used as refrigerants in refrigerators, deep freezers and air conditioners. CFCs are also used in plastic foams, and those containing bromine are used in fire extinguishers. They are chemically very stable, their non-flammability and complete non-toxicity making them attractive in use. Released into the TROPOSPHERE some of them have a long life (11 years or more) and molecules may migrate across the TROPOPAUSE and eventually to the OZONE LAYER. There they are dissociated by ultraviolet radiation, releasing free atoms of chlorine which engage in a series of chemical reactions whose overall effect is to deplete the concentration of ozone. For this reason many countries have banned or restricted their manufacture and use. They are also very effective greenhouse gases (*see* GREENHOUSE EFFECT).

chloroform (trichloromethane, CHCl$_3$). A simple organochlorine compound used as a solvent in the plastics, rubber and resin industries, as well as an industrial solvent and raw material. It was formerly used as an anaesthetic, but was abandoned because it is dangerous (the difference between an anaesthetizing dose and a lethal one being small), can cause liver damage and may be a CARCINOGEN.

chloromethane (methyl chloride, CH$_3$Cl). A colourless gas manufactured industrially for use as a refrigerant similar to a CHLOROFLUOROCARBON (CFC) and as a local anaesthetic, but also produced naturally by certain fungi, most notably *Phellinus pomaceus*, which grows on rotting wood. Natural releases amount to about 5 million tonnes a year (compared with an estimated 26000 tonnes of CFCs). Chloromethane is believed to be the major contributor of stratospheric free chlorine involved in the chemistry of the OZONE LAYER.

chlorophenotone. *See* DDT.

Chlorophyceae. *See* CHLOROPHYTA.

chlorophyll. A green pigment, present in algae and higher plants, that absorbs light energy and thus plays a vital role in PHOTOSYNTHESIS. Except in Cyanophyta (blue–green algae), chlorophyll is confined to CHLOROPLASTS. There are several types of chlorophyll, but all contain magnesium and iron. Some plants (e.g., brown algae, red algae, copper beech trees) contain additional pigments that mask the green of their chlorophyll. *See also* BACTERIOCHLOROPHYLL.

Chlorophyta (Chlorophyceae). The green algae, the largest and most diverse division of ALGAE, occurring in fresh and saltwater and in damp places on land. Some are microscopically small, often able to move by means of flagella (*see* FLAGELLUM) and occur as single cells or as colonies. Others are filamentous (e.g., *Spirogyra*) or have a flattened THALLUS (e.g. *Ulva*, the sea-lettuce).

chloroplast. A cytoplasmic body (a PLASTID) of plant cells that contains CHLOROPHYLL and is the site of PHOTOSYNTHESIS. Each chloroplast contains numerous grana, assemblages of chlorophyll molecules which lie stacked like piles of coins. It is here that the light reactions of photosynthesis take place. The subsequent, dark reactions occur outside the grana, in the surrounding stroma, which is bounded by the membrane of the chloroplast.

chlorosis. A yellowing of the normally green parts of plants, due to the prevention of CHLOROPHYLL formation (e.g., by lack of light, iron or magnesium).

chlorpropham. *See* CIPC.

chlorthiamid. A soil-acting herbicide used for total weed control and also for selective weed control in orchards and forests. It is used to kill floating and submerged aquatic plants in still or slow-moving water.

Choanichthyes. A group of mostly extinct bony fishes, comprising the lungfishes (DIPNOI) and CROSSOPTERYGII, characterized by having nostrils that open into the mouth, and fleshy, lobed fins with a central bony axis. AMPHIBIA are believed to have evolved from the DEVONIAN Choanichthyes.

choline. A nitrogenous alcohol that was once believed to be necessary in the diet of many animals and was therefore classed as a vitamin. Provided the diet contains adequate amounts of METHIONINE and COBALAMINE (vitamin B_{12}), choline is probably not necessary, so placing in doubt its status as a vitamin, although it may be valuable in marginal diets. Where it is necessary, its absence leads to haemorrhage of the kidneys and the excessive deposition of fat in the liver. Choline is an important constituent of LECITHIN.

cholinesterase. *See* ACETYLCHOLINE.

Chondrichthyes (Elasmobranchii). The cartilaginous fishes, almost all of them marine, including sharks, dogfish, skates, rays, *CHIMAERA*, and many extinct forms. They have no bone, but their cartilage may be calcified. *Compare* OSTEICHTHYES.

chondriosomes. *See* MITOCHONDRIA.

chondrite. A type of stony METEORITE, usu-

ally containing small, spherical, mineral bodies (chondrules). *Compare* ACHONDRITE.

chondrule. *See* CHONDRITE.

Chordata. A phylum of animals characterized by the possession, at least in the early stages of development, of a notochord (a longitudinal supporting rod of vacuolated cells, *see* VACUOLE, enclosed in a firm sheath, lying just below the central nervous system), pharyngeal ('gill') pouches and a hollow, dorsal, nerve cord. The phylum includes the vertebrates (VERTEBRATA) and the more primitive Protochordata (extinct forms lacking skulls and vertebrae).

chorion. (1) One of the embryonic membranes of AMNIOTA; in MAMMALIA it forms much of the PLACENTA. (2) A non-cellular envelope surrounding certain ova (e.g., insect eggs).

C horizon. *See* SOIL HORIZONS.

chorology. The study of areas and their floral and faunal development.

CHP. *See* COMBINED HEAT AND POWER.

chromatid. One of a pair of strands formed by the duplication of a CHROMOSOME.

chromatin. The NUCLEOPROTEIN that makes up the CHROMOSOMES.

chromatography. A technique of chemical analysis that identifies substances by comparing the rate at which they move across a base material with the rates of movement of known substances. The technique has been widely used to analyze small quantities of ORGANOCHLORINES. There are two main types of chromatography. Partition chromatography involves the selective solution of the material between two solvents. In paper chromatography the rate is measured at which materials move along a piece of paper one end of which is immersed in a solvent. *See also* GAS CHROMATOGRAPHY.

chromatophore. (1) In plants, a chromoplast (a pigmented PLASTID). (2) In animals, a pigment cell. Some animals (e.g., prawns, octopuses, flounder, minnow, frogs, chameleons) are able to change colour by the concentration or spreading of the pigment granules within the chromatophores. This ability is under the control of HORMONES, and in fishes and chameleons it is also controlled by the nervous system. Cuttlefish are able to change colour rapidly because the chromatophores can be dilated by the contraction of radiating muscle fibres attached to the surface of the cells. *See also* MELANINS.

chromic acid (CrO_3). A bright orange, caustic material that reacts with organic matter; the name given is incorrect as it is not an acid. It is widely used in laboratories as the strongest cleaning agent for glassware, and industrially it is used in ceramic glazes, rubber and textile manufacture, chromium plating, as an oxidizing agent and generally for degreasing.

chromite ($FeCr_2O_4$). A mineral of the SPINEL group: the major ore mineral of CHROMIUM. Chromite ore occurs as layers in ULTRABASIC IGNEOUS bodies and in PLACER deposits, and more rarely in HYDROTHERMAL veins.

chromium (Cr). A hard, white metallic element that occurs as chrome–iron ore (*see* CHROMITE) and is used in the manufacture of stainless steel and, to a lesser extent, in chromium plating. $A_r = 51.996$; $Z = 24$; SG 6.92; mp 1890°C.

chromophores. Functional groups attached to HYDROCARBON RADICALS that produce the colours in dyestuffs. More generally, any process or thing that produces colour in a substance.

chromoplast. *See* CHROMATOPHORES.

chromosomes. Strands of genetic material. In animals and most plants they are situated in the cell nucleus and consist of DNA and PROTEIN, but in BACTERIA, CYANOPHYTA (blue–green algae), and VIRUSES they are made up of nucleic acid only (RNA in some viruses). Each species has a characteristic number of chromosomes (46 in humans), which are present in pairs. The members of a pair look alike (except for the SEX CHROMOSOMES) and associate during MEIOSIS. These pairs are homologous chromosomes. GAMETES and the cells of gametophytes contain the haploid number of chromosomes (only one from each pair); most other cells have the full (diploid) number. Each chromosome consists of a linear series of genes, the constant positions (loci) of which have been mapped in some species. All the body cells of an organism have an identical set of genes because the DNA in each chromosome is exactly replicated and during MITOSIS the identical halves of the chromosomes separate and form the daughter nuclei.

chrysalis. *See* PUPA.

Chrysophyta (Chrysophyceae). The golden–brown and orange–yellow algae; a diverse group of microscopically small algae which inhabit fresh and saltwater, many being planktonic (*see* PLANKTON). They contain CAROTENOID pigments and may be unicellular, colonial, filamentous or amoeboid.

chrysotile. A fibrous SERPENTINE worked (where the industry has not been banned) for the manufacture of ASBESTOS.

chylocaulus. Applied to plants whose stems are succulent (e.g., cactus). *See also* XEROPHYTE.

chylophyllous. Applied to plants whose leaves are succulent (e.g., stonecrops, *Sedum*). *See also* XEROPHYTE.

Ciconiiformes (herons, bitterns, storks, spoonbills, flamingos). An order of birds that are large, with long necks, legs and bills. Most of them feed in shallow water on fish and other aquatic organisms, making them highly sensitive to anything which contaminates water and may poison them directly or reduce the populations on which they depend for food.

cilia. Short threads of cytoplasm that project

from the surface of cells and beat in a regular sequence in a constant direction. They occur in many PROTOZOA and on ciliated EPITHELIUM in many multicellular animals (e.g., in the respiratory passages of many land animals, including humans). They have the effect of either moving fluid past the cell or of moving (i.e. rowing) the cell through a liquid medium. Many animals (e.g., bivalve molluscs, AMPHIOXUS) feed by filtering organic particles from a stream of water passed through the animal by means of cilia. True cilia occur in very few plants (e.g., the male gametes of CYCADALES). A cilium contains two central filaments surrounded by a ring of nine double ones, and it is anchored to a body (kinetosome) at its base. *Compare* FLAGELLUM.

Ciliophora. A class of PROTOZOA whose members move by means of CILIA and who possess a double nucleus (e.g., *Paramecium*, of the subclass Ciliata). *See also* INFUSORIA.

cilium. *See* CILIA.

cinders. (1) Solid particles produced by the burning of substances containing carbon that have ceased to burn, but still contain combustible material. (2) SCORIA 4–32 millimetres in diameter.

cinnabar (HgS). The mineral mercuric sulphide; the principal ore of MERCURY, found in near-surface HYDROTHERMAL deposits. Native mercury is often associated with cinnabar.

CIPC (chlorpropham). A herbicide of the CARBAMATE group, used to control germinating weeds and to prevent sprouting in stored potatoes.

circadian rhythm (diurnal rhythm). Rhythmic changes, with a periodicity of approximately 24 hours, that occur in plants and animals even when they are isolated from daily changes in the environment. Cycles of sleep and waking and the leaf movements of plants are examples of circadian rhythms.

circle of inertia. The circular track along which a theoretical particle would travel on a frictionless rotating Earth were it subject to no horizontal forces.

circumferential velocity. *See* TANGENTIAL VELOCITY.

circumhorizontal arc. A brightly coloured horizontal spectrum of colours, with red at the top, occurring in the sky at an altitude below 32° and below the Sun. It is seen when the Sun is rather more than 58° above the horizon and is caused by the Sun's rays entering vertical faces and leaving horizontal faces of hexagonal ice crystals with vertical axes.

circumnutation. *See* NUTATION.

circumpolar vortex. The system of westerly winds in high latitudes, particularly at altitudes between 1000 and 12000 metres, that circulate air around the pole.

circumscribed halo. A halo seen around the Sun that is tangent to the 22° halo at the top and bottom, but outside it at the sides. It is roughly elliptical when the Sun is high in the sky, but has a sagging shape when the Sun is low. It is caused by the passage of the Sun's rays through hexagonal ice crystals with horizontal axes.

circumzenithed arc. A brightly coloured circular arc, with red at the bottom, seen more than 58° above the Sun when the Sun is below 32°, centred on the zenith. It is caused by the Sun's rays entering the horizontal top and emerging from a vertical side of hexagonal ice crystals with vertical axes.

cirque (corrie, cwm). A steep, bowl-shaped hollow in a mountainous region made by glacial ice and often forming the head of a valley.

Cirripedia (barnacles). Sedentary, aquatic CRUSTACEA, usually with a body enclosed in calcareous plates and with feathery appendages used for gathering food particles from water. Barnacles have free-swimming larvae. Some species are parasitic.

cirrocumulus. Literally, a fibrous heap cloud, which is a contradiction in terms, but the name given to ALTOCUMULUS when it occurs at the same levels as fibrous-looking ice cloud and, perhaps because of its height, appears to be composed of small puffs.

cirrostratus. A layer of fibrous-looking cloud which covers a large part of the sky.

cirrus. Cloud with a fibrous appearance, always composed of ice crystals, compact masses of cloud being drawn out into 'fibres' as the crystals fall through the air without evaporating.

cistron. A length of DNA upon which a molecule of messenger RNA is built. The messenger RNA determines the sequence of AMINO ACIDS that go to form a particular PROTEIN. A cistron is usually regarded as being equivalent to a gene.

CITES (Washington Convention on International Trade in Endangered Species of Wild Flora and Fauna). A convention that seeks to provide protection for certain species (e.g., peregrine falcon, white-tailed eagle) against over-exploitation through international trade.

citric acid cycle (Krebs' cycle, tricarboxylic acid cycle). Part of the process of aerobic respiration, comprising a series of enzyme-controlled reactions, occurring in the MITOCHONDRIA, during which pyruvic acid (formed in GLYCOLYSIS) is broken down to carbon dioxide and water; ATP is built up.

citrine. QUARTZ made yellow by slight impurities.

Citrus. A genus of trees and shrubs (family: Rutaceae), including the lemon, lime, orange and grapefruit, that are widely cultivated in warm countries. The fruit is a hesperidium (a BERRY in which the fleshy part is divided into segments and the whole fruit is surrounded by a tough skin).

citrus scale insect. *See* BIOLOGICAL CONTROL.

city climate. The climate produced within a city which often differs from the climate in the surrounding countryside. Some of the changes resulting from the influences of the city are beneficial (e.g., the reduction in HUMIDITY due to rain running off roofs and paved surfaces, which dry quickly, the increase in night temperature due to the use of fuel and the trapping of air in streets, and the reduction in wind strength due to shading by small buildings). Others are adverse (e.g., the production of pollution, consequent reduction in sunshine and lowering of temperature and generation of FOGS and photochemical SMOG, and the increase of wind caused by tall buildings).

city farm. (1) An establishment within a city where some plant crops are grown on a small scale and where farm livestock is reared, mainly for educational purposes. (2) A theoretical concept in which all the wastes from an urban community are processed into forms that can be fed to fish or other livestock. *See also* BIOPLEX.

Civic Trust. A UK voluntary organization, formed originally to protect the nation's urban architectural heritage, but by extension involved in many urban environmental issues. It is a member of the COMMITTEE FOR ENVIRONMENTAL CONSERVATION, and was one of the founders of the EUROPEAN ENVIRONMENTAL BUREAU.

Cl. *See* CHLORINE.

cladistics. A method of classifying plants or animals on the basis of their similarities to one another. It is often used to indicate recent descent of contemporaneous forms from a common ancestor. In the modern sense the term is applied especially to the numerical calculation of degrees of similarity.

Cladocera. *See* BRANCHIOPODA.

cladode (phylloclade). A stem that is modified to resemble a leaf, in appearance and function (e.g., asparagus, butcher's broom).

cladogram. A tree-like diagram illustrating similarities among species and therefore degrees of relationship and evolutionary descent.

clarification. The removal of TURBIDITY and suspended solids by settling. Certain chemicals can accelerate the process through COAGULATION.

classification. Biological classification is based mainly on structural criteria and arranges organisms in a hierarchy of groups that reflect evolutionary relationships. Usually the smallest group is the species, although SUBSPECIES and VARIETIES may be recognized. A species is generally regarded as a group of organisms that resemble one another more closely than they resemble members of other groups and that form a reproductively isolated group whose members will not normally breed with members of another group. The common names of plants and animals often denote species. Similar species are grouped into genera (sing. genus), genera into families, families into orders, orders into classes, classes into phyla (sing. phylum) for animals and divisions for plants, and these into kingdoms. The systematic position of the European crested newt can be given thus:

kingdom	Metazoa
phylum	Chordata
subphylum	Vertebrata (Craniata)
class	Amphibia
order	Urodela (Caudata)
family	Salamandridae
genus	*Triturus*
species	*cristatus*

All names other than those for the species, subspecies and variety are given an initial capital; the names for genera, species and subspecies are written in italics. The system is flexible; as knowledge increases organisms are continually being reclassified. *See also* BINOMIAL NOMENCLATURE.

clast. *See* CLASTIC.

clastic. Applied to sediments made up from fragments (clasts) of parent rocks or minerals that have been deposited by mechanical transport. Sediments formed from organic remains are called bioclastic.

clathrate. A compound in which one kind of molecule is enclosed within the structure of another without there being any direct bonding between the two. At Michigan Technological University clathrates have been developed whose enclosed molecule is water, giving the lattice a melting point of 10–30°C according to the design used. Such clathrates, sealed in the exterior walls of buildings or similar locations, might be used for thermal insulation.

Claus kiln. An oil refinery unit for recovering sulphur from gases rich in hydrogen sulphide. The hydrogen sulphide is burnt in an insufficient supply of air, giving the reaction $2H_2S + O_2 \rightarrow 2H_2O + 2S$

Claviceps. *See* ERGOT.

clay. Sedimentary particles, mostly CLAY minerals, each of which is less than 2 micrometres (μm) in diameter.

clay minerals. Stable SECONDARY MINERALS formed by the weathering of some primary minerals. They form the bulk of the silicate material in soil and many SEDIMENTARY rocks. Characteristically, they have a layered crystal structure with an average grain size of less than 2 micrometres in diameter. Examples include kaolinite (*see* KAOLIN), illite, MONTMORILLONITE and CHLORITE.

claypan. *See* HARDPAN.

Clean Air Act, 1970. A US federal law passed in order to strengthen the provisions of the AIR QUALITY ACT, 1967 by giving more specific powers to the ENVIRONMENTAL PROTECTION AGENCY (e.g., to specify the actual discharges to be permitted from a particular plant). This proved impracticable since it required states where existing discharges exceeded those permitted to ban all further industrial development, including modification of the offending plants. Amendments to the Act allowed the introduction of BUBBLE POLICIES.

CLEAR. *See* CAMPAIGN FOR LEAD-FREE AIR.

clear air turbulence. A small-scale motion in clear (i.e. cloudless) air, causing BUMPINESS to aircraft, and which may occur above cloud tops where convection from the ground surface is inoperative. It can be detected by radar and the forward scattering of radio waves, mainly because it is usually produced by unstable waves on stably stratified layers (*see* BILLOW CLOUDS). In its more intense forms it is notably associated with JET STREAMS, particularly their exit regions, and MOUNTAIN WAVES.

cleavage. (1) The tendency for a crystal to split along planes determined by the crystal structure. (2) The tendency for a metamorphic rock (*see* METAMORPHISM) to split along closely spaced parallel planes which are not usually parallel to BEDDING PLANES. Rock cleavage is developed by pressure and is usually accompanied by at least a partial recrystallization of the rock. (3) *See* SEGMENTATION.

cleidoic egg. The egg of a land animal (e.g., bird, reptile, insect) whose shell isolates it from the environment and allows only for gaseous exchange and a limited loss of water.

cleistogamy. Fertilization that occurs inside an unopened flower. For example, wood sorrel (*Oxalis*) and violets produce, after the showy flowers, small, unattractive flowers that never open and so are capable only of SELF-POLLINATION.

Clematis vitalba (traveller's joy, old man's beard). A perennial woody climber (family: Ranunculaceae), bearing large heads of plumed ACHENES, that grows in hedgerows and woodland edges. Its presence indicates that the soil is calcareous. *See also* CALCICOLE.

clepto-parasite. *See* PARASITISM.

climate. The long-term or integrated manifestation of weather.

climatic climax. A CLIMAX whose composition is controlled by the climate rather than any other factor (e.g., soil) and is in equilibrium with the climate of the area.

climatological station. A meteorological recording station where records are collected over many years for the purpose of studying climate.

climatological statistics. Statistics that describe the weather averaged over many years.

climatology. The scientific study of climates over long periods of time.

climax. A community of organisms that is in equilibrium with existing environmental conditions and forms the final stage in the natural SUCCESSION (e.g., oak woodland in much of lowland Britain, ash woodland on upland limestone, etc). *See also* FIRE CLIMAX, PLAGIOCLIMAX, POSTCLIMAX, PRECLIMAX, PROCLIMAX, SERCLIMAX, SUBCLIMAX.

Clinch River. A tributary of the Tennessee River, in Tennessee, USA, beside which was the site proposed for the first US BREEDER REACTOR. The building of the reactor was opposed by antinuclear groups and led to much political controversy. Work on the site was halted by President Carter under the terms of the NUCLEAR NON-PROLIFERATION ACT, and in the autumn of 1983 the Senate voted to end funding for the project, which was abandoned.

cline. A gradation of structural differences among the members of a species that is closely correlated with ecological or geographical distribution. Populations at each end of a cline may differ substantially from one another. An example is the coal tit (*Parus ater*) which has 21 recognized subspecies that tend to blend into one another and are distributed over Europe and parts of North Africa and Asia.

clinometer. An instrument for measuring angles in the vertical plane (i.e. inclinations from the horizontal) by means of a pendulum and calibrated scale, or a spirit level. Clinometers are used mainly for measuring the DIP of such geological features as BEDDING PLANES, FAULT PLANES, JOINTS and CLEAVAGE.

clint. A flat surface in a LIMESTONE PAVEMENT. Clints are separated by GRIKES.

clisere. A sequence of CLIMAXES set in motion by a major change in climate (e.g., glaciation).

clone. A group of organisms of identical genetic constitution (unless mutation occurs) produced from a single individual by ASEXUAL REPRODUCTION, PARTHENOGENESIS or APOMIXIS. *See also* ORTET, RAMET.

closed community. A COMMUNITY in which colonization is precluded because all the NICHES are occupied.

Clostridium botulinum. See BOTULISM.

Clostridium pasteurianum. A species of anaerobic nitrogen-fixing BACTERIA responsible, along with other nitrogen-fixing bacteria such as *Azotobacter*, for the gradual increase in the nitrogen content of unmanured grassland. *See also* NITROGEN FIXATION.

cloud. Water present in the atmosphere in the form of suspended droplets that have condensed on condensation nuclei (*see* CONDENSATION NUCLEUS). The formation of cloud requires that moist air be cooled. The main cooling mechanism is decompression caused by the upward movement of air, the base of cloud (*see* CLOUD BASE) being determined by the limit of vertical convection. This process forms CUMULUS cloud in unstable air. In stable air, heat may be lost by radiation until the water vapour condenses into stratus cloud. Continuous layers of cumulus-type clouds may result from TURBULENCE.

cloud amount. The proportion of the sky covered by cloud. It is usually measured in oktas or in eighths of the sky occupied by cloud, and reported as a single number. The number 9 indicates that the sky is obscured by FOG.

cloud base. The height of the lowest part of a cloud, or layer of cloud, usually with special reference to low cloud (e.g., scud or lifting FOG), a layer of STRATOCUMULUS or CUMULUS for which the base represents the CONDENSATION LEVEL).

cloud bow. An arc in the sky caused, like a RAINBOW, by the refraction of light by spherical cloud droplets. *See also* FOG BOW.

cloud chamber. A vessel in which air may be expanded adiabatically (*see* ADIABATIC) to produce cloud. In physics laboratories the AITKEN NUCLEI are removed by repeated condensation and sedimentation. The tracks of ALPHA-PARTICLES can then be made visible because the ionization they produce forms CONDENSATION NUCLEI. The cloud chamber was the first instrument to be used to observe tracks of particles caused by radioactivity.

cloud generation. The production of clouds, which is almost wholly caused in natural clouds by the ADIABATIC cooling due to the ascent of air, the exceptions being STEAM FOG, radiation FOG and aircraft CONDENSATION TRAILS.

cloud height. *See* CLOUD SEARCHLIGHT.

cloud searchlight. A light used to point a beam vertically and measure cloud height by observing the elevation of the illuminated spot. It can be used by day, employing audiofrequency modulation or a monochromatic source.

cloud seeding. The introduction of dry ice (solid carbon dioxide), sea salt or silver iodide crystals into cloud to promote rainfall. *See also* ARTIFICIAL RAIN.

cloud shadows. Shadows cast on the ground by clouds. They are the major cause of any reduction in sunshine and because most computer models of large-scale climatological phenomena are unable to predict accurately the formation of cloud in moist air, cloud shadows are the principal unknown quantity in all physical theories of climate change.

cloud streets. Clouds arranged in long, parallel lines, usually along the direction of the WIND SHEAR, which frequently differs little from that of the wind. They are caused when CONVECTION is limited to a uniformly deep layer over a wide area. *Compare* CONVECTION STREETS.

clubmosses. *See* LYCOPODIALES.

Club of Rome. An 'invisible college', founded in April 1968, to foster understanding of the varied, but interdependent components making up the global system in which we all live, to bring that new understanding to the attention of policy makers and the public, and so to promote new policy initiatives and action. The Club consisted originally of 30 scientists, educators, economists, humanists and civil servants from 10 countries, led by Dr Aurelio Peccei. Membership was restricted and could not exceed 100, but it grew quickly to a membership of 70, from 25 countries. The Club initiated a project on The Predicament of Mankind, whose best-known result was the computer modelling of the world conducted at the Massachusetts Institute of Technology, funded by the Volkswagen Foundation and published in 1972 as *The Limits to Growth*. The Club's second report, *Mankind at the Turning Point*, was published in 1975.

Clupeidae. A family of bony fishes, including the herring (*Clupea*), sprat, pilchard, anchovy and shad, all of which are important commercially.

Cnidaria (sea-anemones, corals, jellyfish, comb jellies, etc.). A phylum of aquatic, mostly marine animals, that are usually radially symmetrical. They have simple, two-layered bodies with only one opening to the gut (the coelenteron) and characteristically they bear sting cells (NEMATO-BLASTS) on tentacles. There are two basic types of individual — the POLYP and the MEDUSA. Colonial forms are common. *See also* ACTINOZOA, COELENTERATA, HYDROZOA, SCYPHOZOA.

cnidoblast. *See* NEMATOBLAST.

CNS. *See* CENTRAL NERVOUS SYSTEM.

Co. *See* COBALT.

coagulation. The process of converting a finely divided or colloidally dispersed (*see* COLLOID) suspension of a solid into particles of such a size that a reasonably rapid settling occurs. *See also* FLOCCULATION.

coal. Carbonaceous fossil fuel of worldwide distribution that occurs in stratified accumulations. Coals form two series: the humic coals, which, with an increase in carbon content (i.e. RANK) pass from PEAT to lignite to BITUMINOUS COAL to ANTHRACITE; the sapropelic coals (e.g., CANNEL COAL, BOGHEAD COAL, OIL SHALE). These consist of finely divided vegetable material (e.g., spore cases, algae, fungal matter). Coal is the most abundant of fossil fuels. *See also* COAL GAS.

coalescence. The formation of large drops out of smaller ones. Cloud droplets do not coalesce unless there is a fair spread of sizes, causing collisions due to different fall speeds. It is thought not to be a significant factor in rain generation among droplets, all of which are less than 20 micrometres in diameter.

coal gas. A mixture of combustible gases obtained by the DESTRUCTIVE DISTILLATION of coal, a process that produces a heavily contaminated waste gas and waste water with a high content of organic substances. By-products of gas manufacture are COKE, AMMONIA and many organic compounds (*see* COAL TAR). The process has declined in importance during the 20th century.

coal tar. A black, viscous liquid obtained by the DESTRUCTIVE DISTILLATION of COAL. It is a major raw material for pharmaceuticals, dyes, solvents and other organic compounds. It contains several organic CARCINO-GENS, as do many of the by-products from the manufacture of heating coals or oils (e.g., CUTTING OILS). *See also* COAL GAS.

coastal effects. Differences in the weather experienced near a coast compared with the weather inland, which may be caused by temperature differences, altitude differences or differences in AERODYNAMIC ROUGHNESS. The presence of cliffs may increase wind speeds. Vegetation may be affected by sea spray, and snow may lie for a shorter time due to the presence of salt or to warmer sea air. FOGS may be more prevalent near coasts than over open sea because of the upwelling of cold weather.

cobalamine (vitamin B_{12}). A cobalt-containing VITAMIN required by many animals, including humans, for normal cell division (*see* MEIOSIS, MITOSIS). It is synthesized by the bacterium *Bacillus subtilis*, an inhabitant of the gut in many animals. Cobalamine deficiency in humans causes pernicious anaemia.

cobalt (Co). A hard, silvery–white, magnetic metallic element that occurs in combination with sulphur and arsenic. It is used in alloys and also to produce a blue colour in glass and ceramics. It is an essential MICRONUTRIENT. A_r=58.9332; Z=27; SG 8.9; mp 1480°C.

cobalt-60. A radioactive isotope of COBALT used as a TRACER, an X-ray source and in radiation therapy. It is a fallout product of nuclear explosions and the medium of the 'doomsday machine' an atomic weapon encased in cobalt-59, the stable isotope, which is converted to cobalt-60 by the explosion to become a toxic product of a 'dirty' bomb and spread widely by fallout.

cobaltite (CoAsS). One of the major ore minerals of cobalt; cobalt arsenic sulphide found in HYDROTHERMAL veins. The other major sources of cobalt are LINNAEITE, PYRITE and cobaltiferous LATERITE.

cob nut. *See* CORYLUS AVELLANA.

coccidiosis. A disease of mammals and birds, often endemic in wild populations, caused by parasitic SPOROZOA which invade internal organs (e.g., the liver).

Coccinellidae. *See* COLEOPTERA.

coccus. Any spherical bacterium (*see* BACTERIA).

cochlea. The small, snail-shaped structure that forms part of the inner ear in mammals and contains the sound receptors.

cockles. *See* LAMELLIBRANCHIATA.

cockroaches. *See* DICTYOPTERA.

cocksfoot. *See* GRAMINEAE.

cocktail party effect. A degree of hearing loss in which the victim finds it difficult to concentrate on and so to hear one sound amid a background of generally similar sounds (i.e. to separate signal from noise).

coconut palm. *See* COCOS NUCIFERA.

Cocus nucifera (coconut palm). A characteristic plant of tropical marine islands, with a large, single-seeded fruit (DRUPE) capable of floating in seawater for long periods while remaining viable. The plant provides many of the basic necessities of life for inhabitants of tropical areas, and its products are widely exported, making it one of the world's most economically important species. The fruit provides food for humans and cattle, and the oil is used in the manufacture of soap and margarine. Its outer layer yields a valuable fibre (coir). The wood from the stem provides useful timber. The leaves are woven for thatching, basket making, etc., and the leaf stalks are used as stakes. The apical bud is edible, and the flower stalk is tapped for its juice, which yields sugar or is fermented to provide an alcoholic drink (toddy) or vinegar.

COD. *See* CHEMICAL OXYGEN DEMAND.

code, genetic. *See* DNA, RNA.

codon. A sequence of three adjacent nucleotides in DNA or the corresponding RNA that specifies a particular AMINO ACID.

coelacanth. *See* CROSSOPTERYGII.

Coelenterata. Animals that have a single body cavity (the COELENTERON). The name was formerly given to a phylum comprising the CNIDARIA and CTENOPHORA, but these are now regarded as phyla in their own right, and the name Coelenterata has fallen from use, although it is sometimes used as a synonym for Cnidaria.

coelenteron. The single body cavity in animals formerly classed as COELENTERATA. It serves as a gut, and its walls form a skeleton. It has only one opening, which can be closed by a sphincter.

coelom. The main body cavity in animals. It is large in VERTEBRATA, ECHINODERMATA and many ANNELIDA, and in these it forms the cavity around the gut. In ARTHROPODA and MOLLUSCA, it is small, the main body cavity being a haemocoel, which is an expanded part of the blood system.

coelomic. Applied to structures, fluid, etc. contained in the COELOM of animals.

coen. (1) All the components of an environment. (2) An abbreviation for COENOSIS.

CoEnCo. *See* COMMITTEE FOR ENVIRONMENTAL CONSERVATION.

coenocline. A sequence of natural COMMUNITIES associated with an environmental gradient.

coenocyte. A mass of protoplasm containing many nuclei, found in fungi and some green algae. *See also* PLASMODIUM, SYNCYTIUM.

coenosis (coen). A random assemblage of organisms that have common ecological requirements, as distinct from a COMMUNITY.

coenospecies. A group of species among which fertile HYBRIDS may occur.

coenzyme. An organic substance (e.g., ATP, CYTOCHROME) that plays a part in an enzyme-controlled reaction or group of reactions without being used up. Many coenzymes are continually changed chemically and are then reconstituted by enzymes (e.g., ATP is converted to ADP, which is then rebuilt to ATP; cytochromes are oxidized then reduced again).

coevolution. The evolution of two or more closely associated species in which changes in one species are complemented by changes in others. The close relationship between certain flowering plants and the insects, birds or bats that pollinate them results from coevolution and some prey–predator relationships may also have coevolved.

COHb. *See* CARBOXYHAEMOGLOBIN.

coincidence. In acoustics, an agreement in wavelengths between sound waves and oscillations caused in structures by them, which causes sound to be transmitted through the structure with great efficiency. Conversely an avoidance of coincidence increases the effectiveness of sound insulation. When a body is subjected to air movement (e.g., sound waves) it will tend to be displaced, but if it is held under tension (e.g., part of a building) it will oscillate as it is displaced and then returns to its former position. This oscillation has a wavelength characteristic of the structure, and bending waves move within it. If the length of these waves coincides with that of the sound waves striking the structure they will be amplified and transmitted through the material very efficiently. There is a critical frequency below which coincidence cannot occur, but sound insulation is reduced at all higher frequencies. This critical frequency is determined by the flexibility of the structure and its thickness, so that the effectiveness of sound insulation increases as it is made thicker and more rigid.

coir. *See* COCOS NUCIFERA.

coke. A porous, brittle solid fuel that contains about 80 percent carbon. It is made in coke ovens from COAL and is also obtained as a residue from the manufacture of COAL GAS.

col. (1) A pass in the shape of a saddleback between adjacent mountain peaks. (2) The region between two high or two low areas of atmospheric pressure, where the pressure gradient is small and therefore winds are light. Birds ready to migrate often mistake a col for the centre of an ANTICYCLONE, whose fair weather extends over a large area.

cold-bloodedness. *See* POIKILOTHERMY.

cold front. The FRONT (boundary) of an advancing cold AIR MASS, at which the cold (denser) air usually undercuts the adjacent warm (less dense) air, so lifting it and generating rain. *See also* ANA-FRONT, KATA-FRONT.

cold low. A centre of low atmospheric pressure caused by a strong vertical uplifting of air, in mid or low latitudes accompanied by strong CONVECTION, but often with little precipitation.

cold pool. A cold mass of air whose uplifting causes a COLD LOW.

cold trough. A projection of air at low pressure with cold air above it. It causes vigorous CONVECTION and the formation of CUMULUS-type clouds.

Coleoptera (beetles). One of the largest orders of insects, with a very wide distribution. The forewings form horny coverings (elytra) over the hind part of the body, and the hind wings are membranous, reduced or absent. Some beetle larvae are active and predacious, some are caterpillar-like, whereas others are maggot-like. Some beetles are serious pests (e.g., the cotton boll weevil, *Anthonomus grandis*; death watch beetle, *Xestobium rufovillosu*; wireworms the larvae of certain click beetles). Others (e.g., the ladybirds, Coccinellidae) are mostly carnivorous and of great importance in the natural control of APHIDIDAE and SCALE INSECTS. *Novius cardinalis* is a coccinellid beetle used in the BIOLOGICAL CONTROL of the citrus scale insect.

coliform bacteria. A group of BACTERIA that are normally abundant in the intestinal tracts of humans and other warm-blooded animals and are used as indicators (being measured as the number of individuals found per millilitre of water) when testing the sanitary quality of water.

collagen. A fibrous PROTEIN that forms one of the main skeletal substances in many animals. In vertebrates it is a component of BONE, CARTILAGE and CONNECTIVE TISSUE. Tendon is almost pure collagen. When boiled, collagen yields gelatin. Ascorbic acid (*see* VITAMIN C) is essential for its synthesis.

Collembola. *See* APTERYGOTA.

collenchyma. A type of supporting tissue found in growing parts of plants (e.g., leaves and young stems). It consists of living cells in which parts of the walls, usually the corners, are thickened with cellulose. *See also* SCLERENCHYMA.

colloid. A substance (e.g., a gum or protein) which, when dispersed in a liquid, forms a mixture that behaves in many ways like a solution, but whose particles are much larger than those in a true solution, and too big to diffuse through a semipermeable membrane.

colluvium. A deposit that moves down a slope by MASS WASTING or CREEP.

colostrum. The milk, rich in proteins and antibodies, produced during the first few days after a mammal has given birth.

colour phase. An unusual, but regularly occurring colour variety of a plant or animal (e.g., the white variety of sweet violet).

Columbiformes (pigeons, doves, sandgrouse, etc.). An order of rather heavy grain- or fruit-eating birds, most of which are good fliers. The domestic pigeon is descended from the rock dove, found wild on mountains and sea cliffs. The dodo was a large, flightless member of the order which lived in Mauritius and was hunted to extinction by humans in the 17th century.

columnar joints. JOINTS that split up an IGNEOUS rock body, usually a SILL or LAVA flow, into columns which often have a hexagonal cross-section.

Colymbiformes. *See* PODICIPEDIDAE.

combe rock (coombe rock, head).   An unsorted (*see* SORTING), earthy deposit containing angular fragments that is formed as a result of SOLIFLUCTION in periglacial conditions. The name combe rock is used mainly in south-east England, the synonym head in south-west England and the USA.

combined heat and power (CHP). The use of cooling water from steam-generating power plants to provide water and space heating to nearby buildings. In theory, a CHP scheme can double the efficiency with which fuel is converted into useful energy. *See also* DISTRICT HEATING.

combined sewer. A sewer pipe that is designed to carry all household liquid wastes (sullage water, i.e. water used in washing, cooking, etc., plus sewage), as well as rain and storm water, as opposed to a sewerage system in which rain and storm water are carried separately from household wastes.

comb jellies. *See* CTENOPHORA.

combustion. The reaction of a substance with oxygen at high temperature with the production of heat.

combustion air. In incinerators, the air introduced by natural, induced or forced draught through the fuel bed and into the primary chamber.

Comley Series. The oldest series of the CAMBRIAN System.

command economy. A centrally planned national economy in which decisions regarding the production, distribution and marketing of goods is made at the centre of a command structure. Many, but not all, socialist economies are command economies.

commensalism. A close association between organisms of different species that benefits at least one partner and harms none. The association may be one in which the partners share a home (e.g., humans and house-martins) or it may be much closer (e.g., between a shark and a remora (sucking fish), which is transported, protected and gathers superfluous food from the shark). *Compare* AMENSALISM. *See also* PARASITISM, SYMBIOSIS.

commercial fast breeder reactor (CFBR 1).   A fast BREEDER REACTOR planned for commercial use in the UK.

comminutor. A machine that crushes coarse material to a finer particle size.

commiscum. A species (i.e. the set of all individuals that can exchange genes successfully with one another (thus can breed to produce fertile offspring).

Committee for Environmental Conservation (CoEnCo). A body, formed in 1969, made up of representatives of all the major UK national, non-governmental conservation bodies, covering interests in wildlife, archaeology, architecture, outdoor recreation, amenities, etc., with the purpose of facilitating the exchange of views and information among them and of coordinating their approach to national issues concerning more than one of them.

common agricultural policy (CAP). The framework on which is built the agricultural policy of the EUROPEAN ECONOMIC COMMUNITY. Its aim is to facilitate trade in agricultural products by establishing common levels of market support throughout the Community, measured in a notional currency unit whose value against the national currencies of member states is fixed, rather than float-ing, to prevent distortions in trade caused by the revaluation of currencies. The CAP guarantees prices paid to producers for many commodities, provides grants for many capital schemes and defines trading relation-ships in agricultural commodities with non-member states (i.e. third countries).

common land. Under English law, formalized in Norman times, but based on custom established in Saxon times, land that is owned privately, but on which certain persons have 'rights of common' (e.g., the right to graze livestock, cut peat for fuel, etc.), whether these rights may be exercised at all times or only during limited periods. Waste lands of a manor (i.e. uncultivated land) which are not subject to rights of common may also be regarded as common land provided such land is registered as such under the Commons Registration Act, 1965. The public has no automatic right of access to common land under common law; where such a right exists it is by virtue of a specific grant, convenant or statute. The Law of Property Act, 1925, S. 193, gave rights of public access, but with restrictions on certain activities (e.g., parking vehicles, lighting fires, etc.), to all commons that lie wholly or partly within urban areas. In rural areas the owner of a common may, if he wishes, apply the benefits of this section to the land by deed.

Commons, Tragedy of the. *See* TRAGEDY OF THE COMMONS.

Commons Registration Act, 1965. *See* COMMON LAND.

community. Any naturally occurring group of organisms that occupy a common environment. The term is a general one, covering groups of various sizes (e.g., ASSEMBLIES, CONSOCIATIONS and SOCIETIES). *Compare* COENOSIS. *See also* ASSOCIATION.

comparative advantage. The economic theory according to which individuals, societies or nations concentrate their economic activities in those fields for which

they are best fitted by reason of geography, resources, skills or level of development, since it is these activities that will prove most profitable to them.

compensation point. (1) The light intensity, or concentration of carbon dioxide in the air surrounding a plant, at which the rates of RESPIRATION and PHOTOSYNTHESIS in a plant are equal, so there is no net absorption or evolution of carbon dioxide or oxygen. (2) The depth in a sea or lake below which, because of low light intensity, plants use up more organic matter in respiration than they make during photosynthesis.

competent bed. A layer of rock that is relatively strong and flexes rather than flows during folding. Such beds, in a varied sequence, determine to a large extent the style of deformation. INCOMPETENT BEDS passively accommodate themselves.

competition. The struggle for existence that results when two or more species have requirements (for food, shelter, etc.) which exceed the available supply.

competitive exclusion principle. *See* GAUSE'S PRINCIPLE.

complemental males. Males, often reduced in size except for their reproductive organs, that live attached to female or HERMAPHRODITE animals of the same species (e.g., angler fish, some stalked barnacles).

complex. The meeting of several COMMUNITIES, in which each community is characterized by its own DOMINANTS, SUBDOMINANTS, etc., and these may or may not differ from one community to another (e.g., several different stages in a SERE may be found). The communities are related to one another by certain species that are shared.

Compositae. A very large and widespread family of mainly herbaceous DICOTYLE-DONEAE that includes dandelions, daisies, thistles, etc. The flowers are arranged in heads, often with a disc of tubular florets surrounded by strap-like ray florets. The massing of the flowers makes them very attractive to insects, which pollinate many at a single visit. The fruit is a CYPSELA and often bears a parachute of hairs which aids dispersal. The family contains few commercially useful plants, apart from decorative garden plants, but many troublesome weeds. However, Compositae are vitally important ecologically because of their relationships with insects.

composite volcano. *See* STRATOVOLCANO.

compost. A soil conditioner made by mixing organic wastes in a container in order to accelerate their decomposition.

compound interest. *See* EXPONENTIAL GROWTH.

conchoidal. Applied to a rock fracture that produces a curved surface, often with concentric waves.

concrete. A structural material made by mixing PORTLAND CEMENT, sand, crushed stone or gravel and water in predetermined proportions. In large structures it is often reinforced with steel rods to improve its tensile properties.

concrete cancer. A popular, but misleading name given to the ALKALI–AGGREGATE REACTION, which can weaken CONCRETE structures.

concretion. (1) A solid mass of matter formed by the cohesion or coalescence of its constituent particles. (2) A localized rock body in a SEDIMENTARY ROCK or PYROCLASTIC rock. The body is harder than the surrounding rock and is AUTHIGENIC.

condensate. HYDROCARBONS that are present in a RESERVOIR as a gas, but separate as a liquid when they are brought to the surface. Condensates resemble petrol (gasoline).

condensation. (1) The change of a vapour into a liquid, which occurs when the pressure of the vapour becomes equal to the maximum vapour pressure of the liquid at that temperature (i.e. the medium containing the vapour becomes saturated). (2) A chemical change in which two or more molecules react with the elimination of water or some other simple substance.

condensation level. The level to which a parcel of unsaturated air must ascend before it becomes saturated. In the case of convection clouds the CLOUD BASE coincides with the condensation level.

condensation nucleus. A solid particle in the atmosphere on to which water vapour condenses when the air is saturated, so forming a liquid droplet. Droplets will not form if air is cooled beyond saturation (i.e. below the DEW POINT) unless condensation nuclei are present. There are usually at least 10–100, and often 1000, nuclei per cubic centimetre of air, the higher number being more common close to cities, where the air cannot become supersaturated (*see* SUPERSATUR-ATION). *See also* CLOUD CHAMBER, FREEZING NUCLEI.

condensation sampling. A gas-sampling technique in which gas is passed through tubes immersed in refrigerants, thus condensing and therefore trapping various fractions of the gas mixture. The gases trapped depend on the temperature, which in turn depends on the refrigerant used.

condensation trail (contrail). A trail of CLOUD formed by the mixing of hot aircraft exhaust with the surrounding cold air. Trails persist only if they become frozen and the air is saturated for ice. *Compare* DISTRAIL. *See also* ICE EVAPORATION LEVEL.

conditional instability. The instability of cold air that comes into contact with a warmer land or sea surface, so that air warmed by the surface rises convectively. (*see* CONVECTION). This leads to the formation of CUMULUS and CUMULONIMBUS clouds and TURBULENCE. The situation is common in low latitudes; in temperate regions it occurs over land in summer and over sea in winter. The LAPSE RATE is intermediate between the dry ADIABATIC LAPSE RATE and the WET ADIABATIC

LAPSE RATE. The instability is conditional because the air will be stable if it is unsaturated, but unstable if it is saturated. If unsaturated air receives water evaporating from the surface, as it approaches saturation its stability will decrease, so it will be potentially unstable.

conditional reflex. A REFLEX action that has been modified by experience. The classic example is that studied by I.P. Pavlov (1849–1936). When shown food, a dog will salivate (an unconditioned reflex). If a bell is rung each time the food is shown, eventually the dog will salivate when the bell rings, even though it sees no food (a conditioned reflex).

conduction. *See* HEAT TRANSFER.

cone. (1) (strobilus) A compact group of reproductive organs found in conifers, cycads, horsetails and clubmosses. (2) A type of nerve cell present in the retina of most vertebrates and concerned with discrimination of fine detail, but not sensitive to very dim light. *Compare* ROD. (3) In a BLAST FURNACE, the steel cone suspended at the top and counterweighted so it sinks when the charge is tipped in and rises again afterwards, preventing the escape of gas. (4) The form taken by a STRATOVOLCANO.

cone of depression. A conical depression in the WATER TABLE immediately surrounding a well and caused by the abstraction of water from the well.

cone-sheet. A funnel-shaped DYKE, usually surrounding an IGNEOUS INTRUSION.

confluence. The flowing together of STREAMLINES so as to increase the speed of flow rather than causing an increase in the vertical depth of the stream. *Compare* CONVERGENCE.

conformable. Applied to rock strata that lie one above another in a parallel order.

congelifluction. The MASS WASTING of permanently frozen ground. *See also* CONGELIFRACTION, CRYOTURBATION, GELIFLUCTION.

congelifraction. The splitting of rocks by frost.

congeliturbation. *See* CRYOTURBATION.

congeneric. Belonging to the same genus.

conglomerate. A SEDIMENTARY ROCK composed of rounded grains, most of which are more than 2 millimetres in diameter.

Coniacian. *See* SENONIAN.

Coniferales (conifers). An order of CONE-bearing plants which includes nearly all the present day GYMNOSPERMAE. Most are tall evergreen trees with needle-like (e.g., pines), linear (e.g., firs) or scale-like (e.g., cedars) leaves. They are characteristic of temperate zones and are the main forest trees of colder regions. They provide timber, resins, tars, turpentine and pulp for paper.

conifers. *See* CONIFERALES.

coning. The behaviour of a PLUME when it expands to form a roughly conical shape. The phenomenon occurs when the path of the plume is determined mainly by the momentum of the efflux, by buoyancy or by some combination of both.

connate water. Water that is trapped with a sediment as it is deposited and that remains there when the sediment becomes rock. Since most waters in deep subsurface reservoirs are chemically quite different from seawater, connate water often refers to the water existing in a RESERVOIR ROCK immediately prior to drilling.

connective tissue. A supporting or packing tissue of vertebrates, containing cells, white fibres of COLLAGEN and sometimes more elastic yellow fibres, lying in a jelly-like, polysaccharide-containing matrix (*see* CARBOHYDRATES). It is widely distributed, found beneath the skin and around muscles and many organs. *See also* ADIPOSE TISSUE.

consequent. Applied to rivers whose course is determined by the existing slope and pattern of the landscape. *Compare* SUBSEQUENT.

conservation. (1) The preservation or protection from decay or destruction of anything whose loss it is desirable to prevent. (2) The planning and management of resources so as to secure their wide use and continuity of supply while maintaining and possibly enhancing their quality, value and diversity. Resources may be natural or manmade (e.g., historic buildings, landscapes produced by farming). Nature conservation is the application of this concept to fauna, flora and physiographical features. (3) *See* SOIL CONSERVATION.

Conservation Area. Under planning law in England and Wales (Civic Amenities Act, 1967), an area of special architectural or historic interest, the character of which it is desirable to preserve or enhance. The designation of a Conservation Area is a public statement of intent to conserve by the planning authority; it is not in itself a proposal for specific action or a planning technique, but defines an objective and a set of problems for which supporting policies will be required.

conservation of energy. *See* THERMODYNAMICS, LAWS OF.

Conservation Society. A UK voluntary organization, founded in 1966, but disbanded in 1987, that studied all aspects of resource use and produced reports on them. It was instrumental in founding TRANSPORT 2000.

consociation. A climax community of plants dominated by one particular species (e.g., oak woodland dominated by *Quercus robur*).

consocion. A layer in which patches of different DOMINANT species alternate with one another.

conspecific. Belonging to the same species.

constancy. In ECOLOGY, the degree of frequency with which a particular species occurs in different STANDS or samples of the same ASSOCIATION.

constant. In ECOLOGY, a species that occurs in 95 percent or more of samples taken at random within a COMMUNITY.

constant-level balloon. A balloon that is designed to rise to a predetermined height and then remain at that height. Such balloons are often used to study the worldwide circulation of air. The altitude of the balloon is regulated by fixing its volume (e.g., by covering its envelope with an inelastic net so it can inflate as it rises only to a certain size) or by a servo device, but occasional large vertical movements of the air and changes in temperature in the balloon as it moves into and out of sunshine mean control of its height is only approximate.

constant-pressure map. A map that shows, as contours, the heights above sea level of a surface of constant atmospheric pressure in order to study the movement of air at high altitude.

Consultative Committee on the Ozone Layer. *See* CONVENTION TO PROTECT THE OZONE LAYER.

Consultative Group for International Agricultural Research (CGIAR). The agency, sponsored by the World Bank, FAO and United Nations Development Programme, that aims to coordinate international agricultural research.

consumer. *See* FOOD CHAIN.

consummatory act. An action (e.g., eating) that ends APPETITIVE BEHAVIOUR.

consumption residues. Wastes that result from the final consumption of goods or services, rather than from their production or distribution (which are termed production residues).

contact. The boundary surface between one body of rock and another or between different fluids in a subterranean RESERVOIR. Contacts between different rocks may be the result of sedimentation under varied conditions or of faulting (*see* FAULT), but the term is more specially used for the boundary between an IGNEOUS INTRUSION and the COUNTRY ROCK.

contact insecticide. An INSECTICIDE that kills without being eaten by the insect (e.g., by penetrating the CUTICLE or blocking the insect's spiracles, *see* STIGMA).

contact metamorphism. METAMORPHISM in the vicinity of an IGNEOUS body, forming a METAMORPHIC AUREOLE. The width of the aureole depends on the size of the INTRUSION and on the temperature of the intrusion and of the COUNTRY ROCK. Without METASOMATISM the chief effects are the baking of CLAY, the hardening of SHALE, the conversion of SLATE to SPOTTED SLATE or to HORNFELS, and the conversion of LIMESTONE to MARBLE. Very high pressures along the base of THRUSTS can also produce a metamorphosed ROCK-FLOUR called MYLONITE.

contact process. A method for producing SULPHURIC ACID, using a platinum CATALYST for the oxidation of SULPHUR DIOXIDE to sulphur trioxide.

containerization. The loading of goods into containers of standard sizes and shapes, which can be loaded and unloaded easily by standard equipment, thus making it cheaper to transport the goods by road, rail or water. An increasing proportion of goods in international trade are handled in this way.

contamination. The introduction to water or foodstuffs of substances containing toxins or live pathogens which constitute a hazard to human health.

contemporary city. An ideal city conceived by Le Corbusier (the Swiss architect Charles Edouard Jeanneret, 1887–1965) in which the railway station stood at the city centre, linked to subways and other transport facilities, with a helicopter link to the airport. The population was to be housed in sky-scraper blocks with hanging gardens, surrounded by parks in which were located restaurants, shops, theatres, etc. *See also* GARDEN CITY.

contest competition. A form of COMPETITION in which each successful competitor obtains all that it needs for survival or reproduction, but its rivals obtain insufficient or nothing.

continental climate. A climate characterized by one or more persistent features of the weather associated with the interior of large continents (e.g., hot summers, cold winters, short autumns and springs, rainy and dry seasons) and a general lack of moderation due to excess of features such as long frosts, tornadoes and deserts.

continental drift. The theory that the present continents result from the break-up of a larger continent and have moved independently to their present positions (and continue to move). The evidence for the existence of such a supercontinent comes from PALAEO-MAGNETISM, the distribution of OROGENIC BELTS, faunas, sedimentary FACIES and climatic belts, and the morphological fit of the continents. The theory has been superceded by those of SEA-FLOOR SPREADING and PLATE TECTONICS, which provide mechanisms for the movement of continents.

continental rise. The gently sloping region connecting the bottom of the CONTINENTAL SLOPE with the ABYSSAL PLAINS. The continental rise is well developed only in some areas and has a thick blanket of TURBIDITE fans and reworked sediments swept up by sea-bottom currents.

continental shelf. The comparatively shallow sea area surrounding the continents, extending seawards to the CONTINENTAL SLOPE, where there is a sharp increase in gradient from an average of 0.1 percent to about 4 percent. Shelf widths vary considerably, as does water depth at the shelf break (average 130 metres). For the purposes of establishing EXCLUSIVE ECONOMIC ZONES, an arbitrary distance of 200 nautical miles (370 kilometres) from the coast defines the legal continental shelf.

continental slope. The portion of the continental margin that begins at the outer edge of the CONTINENTAL SHELF and descends into the ocean deeps.

contingency. In statistics, the degree of dependence or independence between VARIABLES arranged in a complex classification.

continuous spectrum. A FREQUENCY analysis having components ranged continuously over the spectrum of wave frequencies.

continuous variations. *See* VARIATION.

contour farming. The performing of cultivations along lines parallel to contours, rather than across them, so reducing the loss of topsoil by EROSION, increasing the capacity of the soil to retain water (since each furrow is a small reservoir) and reducing the pollution of water by soil. *See also* CONTOUR STRIP CROPPING.

contour strip cropping. The growing of crops on strips that run parallel to contours in order to reduce EROSION. *See also* CONTOUR FARMING.

contractile vacuole. *See* VACUOLE.

contrail. *See* CONDENSATION TRAIL.

control. That part of an experimental procedure which is like the treated part in every respect except that it is not subjected to the test conditions. The control is used as a standard for comparison, to check that the outcome of the experiment is a reflection of the test conditions and not of some unknown factor.

controlled tipping. *See* LAND-FILL.

Control of Pollution Act, 1974. A UK Act of Parliament dealing with the disposal of wastes on land, water pollution, air pollution and noise, and in particular to the functions of water authorities.

control rods. In a NUCLEAR REACTOR, rods that can be raised or lowered into receiving holes in the core, among the canisters containing the fuel elements. The control rods contain substances (e.g., BORON) that absorb neutrons, and their presence or absence thus accelerates or decelerates the reaction. When all the control rods are fully inserted the reaction is reduced to a minimum (but does not cease altogether). Rods are capable of being lowered very rapidly (and automatically) in the event of overheating or other emergencies.

conurbation. A large area occupied by urban development, which may contain isolated rural areas, and formed by the merging together of expanding towns that formerly were separate.

convection. (1) The transfer of heat (*see* HEAT TRANSFER) through a fluid by the movement of the fluid. (2) The vertical movement of air, leading to cooling of the air. This is the most efficient atmospheric cooling process. Free convection occurs when air heated by conduction from a warm land surface rises, or when a mass of warm air rises above an adjacent mass of cold air due to the lateral movement of either air mass. Forced convection occurs when a moving air mass encounters mountains and is lifted regardless of its initial temperature; this can produce high rainfall. Orographic lifting occurs under very stable conditions when a slow-moving warm air mass moves gently upward over wedges of cooler air, producing typical winter rains which are light, steady and extensive. Flash floods in the southern USA are produced by the combined effects of horizontal convergence of air masses, upglide over cooler air and orographic lifting of unstable, moist maritime air. Penetrative convection occurs when the ground or sea surface is warmer than air immediately above it, as when cold air moves across a warm surface (a common situation in low latitudes). This produces CONDITIONAL INSTABILITY, leading to the formation of CUMULUS and CUMULONIMBUS clouds.

convection streets. The same phenomenon

as CLOUD STREETS, but without visible cloud. Sometimes it is visible in haze or haze tops, and it can be discovered by glider pilots who find upcurrents in parallel lines.

convective equilibrium. A state in which a vertically displaced parcel of air experiences no BUOYANCY force.

convective instability. A state in which a vertically displaced parcel of air experiences a BUOYANCY force in the direction of displacement, usually produced by heating of the ground by sunshine. *See also* CONDITIONAL INSTABILITY, WET ADIABATIC LAPSE RATE.

Convention for the Conservation of Antarctic Seals. *See* ANTARCTIC TREATY.

Convention for the Prevention of Marine Pollution from Land-Based Sources (Paris Convention). An international agreement signed in March 1974 by Denmark, France, Iceland, Luxembourg, The Netherlands, Norway, Spain, Sweden, the UK and West Germany to augment earlier agreements controlling the dumping of wastes at sea.

Convention on Long-Range Transboundary Air Pollution. *See* GENEVA CONVENTION ON LONG-RANGE TRANSBOUNDARY AIR POLLUTION.

Convention on the Conservation of Antarctic Living Marine Resources. *See* ANTARCTIC TREATY.

Convention on the Conservation of European Wildlife and Natural Habitats. An agreement among members of the EEC that came into force in June 1982, requiring member states to protect indigenous endangered species.

Convention on the Prevention of Marine Pollution by Dumping of Wastes and Other Matter (London Dumping Convention). The international agreement among 25 nations that is concerned with the effects of disposing of wastes at sea.

Convention on Wetlands of International Importance. *See* RAMSAR CONVENTION ON WETLANDS OF INTERNATIONAL IMPORTANCE.

Convention to Protect the Ozone Layer. An international convention, drawn up under the auspices of UNITED NATIONS ENVIRONMENT PROGRAMME and signed by 49 countries in March 1985, under which nations cooperate to monitor the OZONE LAYER, to note changes in it and to agree recommendations for action to mitigate any depletion of it. Its members meet regularly in the UN Consultative Committee on the Ozone Layer.

convergence. (1) In three dimensions, the compression of a fluid into a smaller volume (the opposite of DIVERGENCE), but in meteorology, more specifically, the horizontal convergence of an AIR PARCEL (e.g., at the ground), which necessitates its upward extension; this is not to be confused with CONFLUENCE. Convergence intensifies the vertical component of vorticity, and therefore is usually accompanied by falling pressure. (2) *See* CONVERGENT EVOLUTION.

convergent evolution (convergence). The evolution of two different groups of organisms so that they come to resemble one another closely (e.g., the marsupial and placental moles, which are both adapted for burrowing).

convivium. A population that is differentiated within a species and is isolated geographically. This is usually a SUBSPECIES or ECOTYPE.

cooling pond. (1) A large water tank in which irradiated fuel elements from NUCLEAR REACTORS are stored while their short-lived fission products decay. (2) An artificial lake or pond in which industrial cooling water is allowed to cool naturally.

cooling tower. A device for cooling water by the use of a flow of air drawn from outside and at the ordinary outside temperature.

cool-season plant. A plant that grows mainly during the cooler part of the year, usually the spring, but sometimes the winter.

coombe rock. *See* COMBE ROCK.

Copepoda. A group of small, freshwater and marine CRUSTACEA, some of which form a major part of the PLANKTON (e.g., *Calanus* on which herrings feed). Some are parasitic. Many have enlarged first antennae, used for swimming. *Cyclops* is common in ponds.

copper (Cu). A red, malleable and ductile metallic element, unaffected by water or steam, and the best conductor of electricity after silver (apart from superconductors). It occurs as the element and as chalcopyrite ($CuFeS_2$), cuprite (Cu_2O) and copper glance (Cu_2S). It is used in boilers, plumbing (where it is being replaced by plastics), electrical wire (where it is being replaced by satellites, in the case of submarine cables, and by optical fibres) and in many alloys (e.g., BRONZE, BRASS, GUNMETAL, etc.). A_r=63.54; Z=29; SG 8.95; mp 1084°C.

copper glance. *See* COPPER.

coppice. (1) To fell a tree or shrub close to the ground, causing several shoots to arise from ADVENTITIOUS buds on the stump. Coppicing was formerly a common practice in woodland management and is being revived in some places, mainly for conservation reasons (the clearance of trees encourages the growth of a wide diversity of herbs). *See also* COPPICE WITH STANDARDS. (2) A crop of coppice shoots obtained by coppicing.

coppice forest. *See* FOREST.

coppice with standards. A method of woodland management in which the woodland is severely thinned, leaving large, timber-producing trees (e.g., oak) at regular intervals (the standards), the spaces being occupied by coppiced (*see* COPPICE) underwood (e.g., hazel, *CORYLLUS AVELLANA*), which is cut near to ground level every 10–15 years. The standards draw up the newly coppiced underwood resulting in the production of numerous long, straight growths for stakes, hurdles, etc.

coprolite. A fossilized faecal pellet; coprolites are usually phosphatic.

coquina. A soft, porous LIMESTONE composed of shells and other BIOCLASTIC remains.

Coraciiformes. *See* ALCEDINIDAE.

coral. A colonial member of the CNIDARIA that secretes a calcareous skeleton. The larger, reef-building corals (Madreporaria) and the alcynonarian corals (e.g., the organ pipe coral, *Tubiflora*; precious coral, *Corallium*) belong to the ACTINOZOA. Hydrocorallinae (e.g., the stinging coral, *Millepora*) belong to the HYDROZOA, and unlike the actinozoan corals they produce a free-swimming MEDUSA stage.

corallines. *See* POLYZOA.

coral reef. A structure on the sea bed composed mainly of calcium carbonate and constructed by colonial CORALS. Coral reefs can form only in clear water, because light penetration is necessary to the photosynthesizing ZOOXANTHELLAE and algae that live symbiotically (*see* SYMBIOSIS) with the coral polyps, and the water must be warm. Reefs occur only where the sea temperature is between 20 and 28°C and sea depth is 10–60 metres. Because of their need for clear water, they are extremely vulnerable to pollution. Living corals are often brightly coloured, the colour being supplied by the microorganisms living on them. Reefs are classified according to their shape as: barrier reefs, which lie offshore and do not make contact with the shore; fringing reefs, which extend from the shore; atolls, which form around the edges of submerged volcanic craters and so are commonly circular, containing a lagoon; patch reefs, which are small, irregularly shaped reefs found between a barrier reef and the shore or inside an atoll lagoon. *See also* GREAT BARRIER REEF, BARRIER REEF MARINE PARK, REEF ECOSYSTEM.

Cordaitales. *See* GYMNOSPERMAE.

core. (1) That part of the Earth which lies

below the MANTLE. The radius of the core is approximately 3500 kilometres; the inner core (radius: about 1400 kilometres) is believed to be solid and composed predominantly of nickel and iron, surrounded by an outer core which behaves as a liquid. Movements within the core produce the Earth's magnetic field. (2) *See* NUCLEAR REACTOR.

core area. *See* HOME RANGE.

core city. A planning concept, departing radically from conventional planning, in which buildings are packed very densely. In some plans the total floor area provided is equal to the total land area of the city, in a continuous body of intense density and activity. Theoretically, a population of 20 million could be accommodated in a 16-kilometre (10-mile) radius. In most plans the floor space amounts to about 10 percent of the total area.

Corf. *See* CORIOLIS FORCE.

Coriolis force (CorF). The force apparently acting on a body that is moving with respect to fixed coordinates on the surface of the Earth. It acts at right angles to the direction of motion, is proportional to the speed and latitude of the moving body and is caused by the rotation of the Earth. It is important only in horizontal motion and deflects bodies to the right in the northern hemisphere and to the left in the southern hemisphere. It is named after the French engineer Gaspard Gustave de Coriolis (1792–1843).

cork (phellem). A protective, impermeable layer of dead cells produced by cell division in the phellogen (cork cambium), a secondary MERISTEM that arises in the CORTEX of woody plants. Cork replaces the EPIDERMIS of the young stem or root. In the cork oak (*Quercus suber*) the cork layer is exceptionally thick.

cork cambium. *See* CORK, MERISTEM.

corm. The swollen, food-storing base of a plant stem (e.g., crocus) that lies under-ground and bears buds and shrivelled leaf bases. A corm is an organ of perennation and VEGETATIVE PROPAGATION that lasts only one year, next year's corm arising on the top of the old one. *Compare* RHIZOME, TUBER.

cormorants. *See* PELECANIFORMES.

corn. (1) In British usage, any cereal crop. (2) In US usage, maize (*Zea mays*).

cornucopian premises. A term coined by Preston Cloud to describe five tenets which may underlie certain attitudes to problems of resource depletion, etc. The tenets are: (a) nuclear fuel will provide an inexhaustible supply of abundant energy; (b) new technology, combined with nuclear power, will enable much lower grades of resources to be exploited, so making reserves virtually inexhaustible; (c) the development of synthetics and substitutes will permit the replacement of resources that become scarce; (d) technological choices and development and the use of resources are governed only by economic considerations; (e) populations will stabilize of their own accord before crises occur. *See also* TECHNOLOGICAL FIX.

corolla. *See* FLOWER.

corona. The coloured rings seen around the Sun or Moon in tenuous or thin cloud in which the droplet size is fairly uniform.

corpus luteum. In a mammalian ovary, a temporary HORMONE-producing area that forms after the ripe egg has been shed. It produces PROGESTERONE and persists only if the egg is fertilized. Its formation and activity are promoted by hormones secreted in the PITUITARY GLAND.

corrasion. The abrasive mechanical EROSION of rocks by the action of particles carried by water, ice or wind.

correlation. The relationship between two VARIABLES, according to a specific range of values. *See also* ASSOCIATION.

corrie. *See* CIRQUE.

corrosion. The conversion of iron and other metals to oxides and carbonates by the action of air and water. The process commonly takes place at ambient temperatures and is fastest in the presence of seawater or a humid atmosphere. There are several ways to prevent corrosion, and some metals (e.g., aluminium) are naturally corrosion-resistant.

cortex. An outer layer, especially of a gland, the kidney, the brain or a plant stem.

cortisone. A HORMONE secreted by the adrenal cortex. One of its effects is to reduce local inflammation.

corundum (Al_2O_3). A mineral noted for its extreme hardness (9 on MOHS'S HARDNESS SCALE). Corundum occurs in some alkaline PEGMATITES, in some regionally metamorphosed rocks (*see* METAMORPHISM) and in PLACER deposits. Gem corundum is known as sapphire or ruby, but most corundum is used as an abrasive. EMERY is an impure form of corundum.

Corvidae. A family (order: PASSERIFORMES) of large omnivorous birds, some of which are pests of cereal crops, including the crow, magpie, jay, jackdaw, rook, raven and chough.

Coryllus avellana (hazel, cob nut). A deciduous shrub or small tree (family: Corylaceae); the common shrub layer (*see* LAYERS) DOMINANT of lowland oakwoods and sometimes of ashwoods in Britain. The plant, which is wind-pollinated, has pendulous male catkins and bud-like female INFLORESCENCES with projecting red STIGMAS. Its nuts have been eaten by humans throughout history, but the unreliability of its fruiting make it unlikely that the plant was ever cultivated for food on a large scale. It has been coppiced in woodlands to yield small timber. *See also* COPPICE, COPPICE WITH STANDARDS.

corymb. A broad, flattish INFLORESCENCE in which the outer flowers open first.

cosere. A series of SUCCESSIONS in the same place.

cosmic rays. A complex system of radiation that reaches Earth from outer space.

cost–benefit analysis. An economic technique for estimating the desirability of a proposed course of action, in which the advantages and disadvantages are listed, expressed in monetary terms and the totals compared. *See also* COST–BENEFIT RATIO.

cost–benefit ratio. The ratio of the gross costs of a proposed activity to the gross benefits, both costs and benefits being discounted over the life of the project at an annual rate of interest. The difference between them is the present value of the net benefit or net cost, and the ratio of one to the other is the gross cost–benefit ratio.

cotyledon. A seed leaf: part of a plant EMBRYO. MONOCOTYLEDONEAE have one, DICOTYLEDONEAE two. In some plants (e.g., beans) the cotyledons are large, fleshy and store food; in other (e.g., grasses) they absorb food from the ENDOSPERM and pass it to the growing embryo. Some cotyledons (e.g., sycamore) grow out of the ground, develop CHLOROPHYLL and carry out PHOTOSYNTHESIS.

cotype. *See* SYNTYPE.

coulomb (C). The derived SI unit of electric charge, being the quantity of electricity transferred by 1 ampere in 1 second. It is named after Charles Augustin Coulomb (1736–1806).

coumarin ($C_9H_6O_2$). (1) A white, crystalline aromatic compound found in sweet vernal grass and other plants, and the substance which gives hay its characteristic smell. It is used as a vanilla flavouring and in perfume. (2) *See* WARFARIN.

Council for Nature. A body founded in 1958 to represent all the UK local natural history societies. It was instrumental in founding the COMMITTEE FOR ENVIRONMENTAL CONSERVATION, into which it was absorbed.

Council on Environmental Quality (CEQ). The US federal agency, formed under the terms of the National Environmental Policy Act, 1969 (NEPA) for the enforcement of measures for environmental protection. The aim of the Act is to 'encourage productive and enjoyable harmony between Man and his environment; to promote efforts which will prevent or eliminate damage to the environment and biosphere and stimulate the health and welfare of Man; to enrich the understanding of the ecological systems and natural resources important to the Nation; and to establish a Council on Environmental Quality'. The Act also calls for ENVIRONMENTAL IMPACT STATEMENTS. *See also GLOBAL 2000 REPORT TO THE PRESIDENT*.

country park. In the UK, an area of land, or water and land, usually not less than 10 hectares (25 acres) in extent, that is designed to offer the public, with or without charge, opportunity for recreation pursuits in the countryside. Country parks are designated officially by the COUNTRYSIDE COMMISSION.

country rock. The body of rock that encloses an IGNEOUS INTRUSION or MINERAL vein.

Countryside Act, 1968. *See* COUNTRYSIDE COMMISSION.

Countryside Commission. The British official body, formed under the terms of the Countryside Act, 1968, to assume the functions of the earlier National Parks Commission and charged with keeping under review all matters relating to the provision and improvement of facilities for the enjoyment of the countryside in England and Wales (there is a separate Countryside Commission for Scotland, based in Perth). It oversees matters relating to NATIONAL PARKS, COUNTRY PARKS, long-distance routes, etc. Its members are appointed by the Secretary of State for the Environment and the Secretaries of State for Wales and Scotland. The Commission's headquarters are in Cheltenham, Gloucestershire.

county trusts for nature conservation. *See* NATURALISTS' TRUSTS.

coupled modes. In acoustics, modes of vibration that influence one another.

COV. Cross-over value (*see* CROSSING-OVER).

covar. In statistics, an abbreviation for covariance (i.e. varying together).

covariance. *See* COVAR.

covellite (CuS). A copper sulphide mineral: a major ore of copper, which is usually found in the zone of SECONDARY ENRICHMENT.

cover value. The amount of cover accounted for by a particular plant species within a QUADRAT according to various arbitrary scales. *See also* DOMIN SCALE.

cow-dung bomb. *See* BOMB.

cow-month. The amount of feed or grazing required to maintain a mature cow in good condition for 30 days.

coypu (*Myocastor coypus*). A large, plump herbivorous aquatic rodent introduced into England from South America for fur farming. It escaped in the 1930s and has established itself in the wild in eastern parts of England, where it causes damage to crops and river banks.

Cr. *See* CHROMIUM

cracking. A process that uses heat to decompose complex substances. Examples include the breaking down of methane (CH_4) to carbon (C) and hydrogen (H_2) or of ammonia (NH_3) to nitrogen (N_2) and hydrogen (H_2). *See also* HYDROCARBON CRACKING

crag-and-tail. A hill with a steep mass of highly resistant rock (crag) at one end and a streamlined tail of a more gently sloping body of softer rock or MORAINE. Crag-and-tails are the product of glacial erosion.

Craniata. *See* VERTEBRATA.

Crataegus (hawthorn). A genus of deciduous trees or shrubs (family: ROSACEAE), usually thorny, that are numerous in northern temperate regions. The two species native to Britain are *C. monogyna*, which is the commonest scrub DOMINANT on most soils other than wet peat and poor acid sand and is commonly planted to make field hedges in Britain, and *C. laevigata* (woodland hawthorn or Midland hawthorn) which is more tolerant of shade, but much less common. *C. laevigata* has broader leaf lobes than *C. monogyna* and usually two styles rather than one.

craton. A continental block that has been stable over a relatively long period of Earth history and has undergone only faulting or gentle warping.

creep. The plastic flow of solids under constant stress. *See also* SOIL CREEP.

creode. The pattern of development of a single organism from fertilized egg to its death and dissolution.

crepuscular rays. Literally evening rays. They are rays seen on account of shadows cast by cloud on haze, usually diverging from the Sun, which is obscured by cloud, but sometimes converging to the ANTISOLAR POINT at about sunset.

Cretaceous. The youngest period of the MESOZOIC Era, usually dated as beginning 135–136 Ma and lasting about 70 million years. The name also refers to the rocks deposited during the Cretaceous Period. These are called the Cretaceous System, which is divided into Lower and Upper Cretaceous. The Lower Cretaceous is divided into the Neocomian, Aptian and Albian Stages. The Upper Cretaceous is divided into the Cenomanian, Turonian, Senonian, Maastrichtian and, usually, Danian, although Danian is considered by some to be PALAEOCENE. The name Cretaceous is derived from the Latin *creta* (chalk), which characterizes the rocks of the period.

cretonism. *See* THYROXIN.

crickets. *See* ORTHOPTERA.

Crinoidea (sea lilies). A class of ECHINODERMATA that are abundant as fossils and have been present since the CAMBRIAN. Most adults are sedentary and stalked, with long, branched feathery arms bearing CILIA which set up feeding currents. The feather star (*Antedon*), the only surviving British genus, breaks from its stalk and swims by waving its arms.

critical frequency. *See* COINCIDENCE.

critical links. Organisms in a FOOD CHAIN that are responsible for energy capture and which play a critical role in the assimilation of nutrients and their subsequent release in forms available at higher trophic levels.

critical mass. The mass of fissile material that will sustain a CHAIN REACTION. At this mass, escaping neutrons have a sufficiently good chance of colliding with more atoms, rather than escaping, for the process to be self-sustaining. The critical mass for uranium-235 is a sphere about 80 millimetres in diameter, weighing about 10 kilograms.

critical organ. An organ within the human body that has a special capacity for concentrating within itself a specific radioisotope (e.g., the skeleton, which can concentrate strontium-90), or that is especially sensitive to radiation and that is studied with particular care when assessments are made of the effects of radiation on the body.

critical reaction. A very violent form of fighting behaviour that occurs when an animal is motivated strongly by fear, but cannot flee from the animal threatening it because of lack of space or overriding ties (e.g., those that forbid a mother from deserting her young). In this 'cornered' situation an animal will fight desperately.

crocidolite (blue asbestos). When asbestos was in use, a commercially important form of it. *See also* CHRYSOTILE.

Crocodilia. An order of large carnivorous

reptiles with powerful jaws, an elongated hard palate, numerous conical teeth, webbed feet, nostrils that can be closed and are located at the end of the snout, and heavy bony plates as well as scales covering the body. Crocodilians are archosaurs (ruling reptiles), the group that also includes the dinosaurs. The present-day crocodilians are the alligator, cayman, crocodile and ghavial.

crocodiles. *See* CROCODILIA.

Cro-Magnon man. *See Homo.*

crop milk (pigeon's milk). A secretion of the crop lining in male and female pigeons, on which the young are fed. Its production is stimulated by the LACTOGENIC HORMONE, secreted by the PITUITARY GLAND.

crop rotation. A farming method in which, season after season, each field is sown with a series of crop plants in a regular rotation, so that each crop is repeated at intervals of several years. Since plant species differ in their nutrient requirements and demand for water, a crop rotation minimizes the risk of depleting the soil of particular nutrients selectively or of GROUNDWATER at a particular level, so helping to maintain the fertility of the soil. LEGUMES are usually included in rotations because of the nitrogen-fixing bacteria in their root nodules.

cross-bedding (cross-lamination, false bedding). The arrangement of bedding or lamination within a BED, where it is at an angle to the main planes of stratification of the strata concerned. The laminae are the non-eroded parts of the layers of dunes and ripple marks formed in arenaceous (*see* ARENITE) sediments (e.g., sands, silts and ooliths, *see* OOLITE). The nature and orientation of cross-bedding are used in determining the way up of the strata and the conditions (e.g., sediment supply, current speed and direction, and, if an aqueous deposit, water depth) at the time of deposition.

crosscut. In mining, a horizontal tunnel that runs through the COUNTRY ROCK or ore at a large angle to the STRIKE of the ore body.

cross-fertilization. The fusion of gametes from two different individuals (not RAMETS from a CLONE) of the same species. *Compare* SELF-FERTILIZATION.

crossing-over. An equal exchange of material between the CHROMATIDS of HOMOLOGOUS CHROMOSOMES, which occurs during the pairing of the chromosomes at MEIOSIS. This causes a reassortment of ALLELOMORPHS (alleles) and increases the diversity of GAMETES. The frequency of crossing over between two genes at different loci (*see* LOCUS) on the same chromosome is the cross-over value (COV), which is proportional to the distance apart of the genes and is used to map their relative positions.

cross-lamination. *See* CROSS-BEDDING.

Crossopterygii. A group of primitive bony fishes (CHOANICHTHYES) that are now usually considered to be distinct from the lungfishes (DIPNOI) from which they differ (e.g. in having conical teeth). The Crossopterygii were believed to have been long extinct until the coelacanth (*Latimeria*) was discovered in 1939 off the coast of East Africa.

cross-over value. *See* CROSSING-OVER.

cross-pollination. The transference of pollen from the stamen of one plant to the stigma of another plant of the same species. *Compare* SELF-POLLINATION

crown of thorns starfish (*Acanthaster planci*). A starfish (*see* ASTEROIDEA) that reaches a diameter of 60 centimetres and that eats hard coral. It is associated with coral reefs in tropical waters, and an increase in its numbers since the early 1960s has led to serious damage to some reefs. Concern has been expressed about the effect these animals might have on the GREAT BARRIER REEF.

Cruciferae (Brassicaceae). A family of DICOTYLEDONEAE, most of which are herbaceous plants, chiefly of northern temperate regions. The flowers have four sepals alternating with four petals, which are usually white or yellow. Many are weeds of cultivation (e.g., charlock, *Sinapis arvensis*; shepherd's purse, *Capsella bursa-pastoris*). Others are food plants (e.g., cabbage, *Brassica oleracea*; turnip, *Brassica campestris*; white mustard, *Sinapis alba*; garden cress, *Lepidium sativum*; watercress, *Nasturtium officinale*).

crude oil. Liquid PETROLEUM in the form in which it is extracted and prior to refining. Crude oils range from colourless liquids as thin as petrol (gasoline) to viscous black ASPHALTS. Very dense crudes (i.e. those with low values on the API SCALE) may, exceptionally, underlie the water in an oil pool. Crude oil containing less than 0.5 percent sulphur is termed low-sulphur crude and is more valuable than the corresponding high-sulphur crude.

crust. The outermost solid shell of the Earth, defined by its composition and the properties of some seismic waves. The crust rests on the MANTLE, with the intervening boundary being called the MOHOROVICIC DISCONTINUITY (named after Andrija Mohorovicic, 1857–1636). The crust varies in thickness from an average of about 6 km in ocean areas to 35–70 kilometres in continental regions. In oceanic areas the crust is apparently young (JURASSIC or younger) and of basaltic composition (*see* BASALT), the top layer (layer 1) being sediment (e.g., TURBIDITES, OOZES, RED CLAY), the next (layer 2) being basaltic PILLOW LAVAS, and the lowest (layer 3) being an intrusion complex of parallel DOLERITE DYKES and GABBRO. In continental regions, sedimentary and metasedimentary rocks overlie a layer broadly composed of GRANODIORITE and its metamorphic equivalents, which in turn overlie layer 3. Continental crust underlies the CONTINENTAL SHELVES (e.g., Barents Sea) and also forms some submerged microcontinents (e.g., Rockall) which appear to be fragments of continental crust torn off during CONTINENTAL DRIFT.

Crustacea. A class of mainly aquatic

ARTHROPODA that includes crabs, shrimps, lobsters, crayfish, woodlice, water fleas, barnacles, etc. *See also* BRANCHIOPODA, BRANCHIURA, CIRRIPEDIA, MALACOSTRACA, OSTRACODA.

cryogenic system. A system in which a local temperature is produced that is lower than the surrounding temperature (e.g., refrigerated transport).

cryopedology. The scientific study of frozen ground.

cryophyte. A plant that grows on snow and ice. Cryophytes include ALGAE, FUNGI and MOSSES, and they may be sufficiently abundant to colour the snow.

cryoturbation. The disturbance of material by frost action, frost heaving and differential mass movements.

cryptic coloration. A colour pattern that protects an animal by making it resemble its surroundings closely. Many moths (e.g., the peppered moth, *Biston betularia*) and ground-nesting birds are camouflaged in this way.

cryptocrystalline. Composed of crystals too small to be resolved by the naked eye.

Cryptogamia. A large group of plants, comprising the THALLOPHYTA, BRYOPHYTA and PTERIDOPHYTA, the last of which are VASCULAR cryptogams.

Cryptophyta. A small division of variously pigmented unicellular ALGAE that live in marine and fresh waters. Most move by means of two unequal flagella (*see* FLAGELLUM). A few species are SAPROPHYTES or ingest food particles, but most are holophytic (*see* HOLOPHYTIC NUTRITION).

cryptophyte. *See* RAUNKIAER'S LIFE FORMS.

Cryptozoic. *See* PRECAMBRIAN.

cryptozoic. Applied to animals that live in crevices.

crystal. An homogeneous solid that possesses long-range, three-dimensional internal order.

crystalline. Composed of, or resembling, crystals; especially IGNEOUS and metamorphic rock (*see* METAMORPHISM).

Cs. *See* CAESIUM.

CS gas (*o*-chlorobenzylidene malonitrile). A fine powder used in riot control because of its property of inducing acute irritation to the eyes, to any skin abrasion and to the tissues of the respiratory passages, causing coughing and nausea. The powder is absorbed readily by soils and may find its way into surface waters, but the toxicity to fish of it and its two main hydrolysis products (*o*-chlorobenzaldehyde and malonitrile) is low.

CSM (corn, soya, milk). A mixture of 70 percent processed maize (corn), 25 percent soya protein concentrate and 5 percent milk solids that is used as a protein-enriched food in dealing with acute food shortages. *See also* NOVEL PROTEIN FOODS.

Ctenophora (comb jellies, sea-gooseberries, etc.). A small phylum of PELAGIC marine animals, formerly included with the CNIDARIA in the COELENTERATA, but that differ from cnidarians in having no NEMATOBLASTS and in showing a measure of bilateral symmetry (*see* BILATERALLY SYMMETRICAL). Their bodies are jelly-like and rounded, and they swim by means of comb-like rows of fused CILIA.

Cu. *See* COPPER.

cuckoos. *See* CUCULIDAE.

Cuculidae (cuckoos). A family (order: Cuculiformes) of long-tailed, slender-winged birds that have two toes pointing forwards and two backwards. Some are brood parasites (*see* PARASITISM) (e.g., the European cuckoo, *Cuculus canorus*) which lay eggs in the nests of small passerines (e.g., warblers, meadow pipit, pied wagtail, robin, linnet). Individual birds usually parasitize only one species, and their eggs resemble those of the host species.

Cucurbitaceae. A family composed mainly of tropical and subtropical, tendril-climbing plants, including the melons, cucumbers, marrows and squashes. Some species yield gourds and calabashes, and the fibrous fruit wall of *Luffa aegyptiaca* forms the loofah. *Bryonia dioica* (white bryony), a common hedgerow plant, is the only species native to Britain.

cuesta. An asymmetric ridge, one slope of which is steep and the other long, gentle and roughly parallel to the underlying BEDS.

Culicidae (mosquitoes and certain gnats). A large family of delicate flies (suborder: NEMATOCERA), most of which have scaly wings and females of which have a piercing blood-sucking proboscis. The larvae and pupae are aquatic. Species of *Anopheles* transmit malaria to humans, *Aedes (Stegmyia) aegyptii* transmits yellow fever, and *Culex fatigans* transmits the worm (*FILARIA*) that causes elephantiasis.

cullet. Broken scrap glass that can be collected and used again.

culm. The flowering stem of a grass (GRAMINEAE).

culm and gob banks. Hills, or banks, made from inferior fuels and wastes discharged from plants that process coal. Banks of culm (ANTHRACITE) and gob (bituminous waste) are often unsightly and present a fire hazard.

cultch. Materials, often shells and pebbles, pumped into the sea to provide sites for the growth of the larval stages of shellfish intended for eventual commercial exploitation.

cultivar. A plant variety that is found only under cultivation. It may be maintained by asexual propagation or controlled breeding.

cumulonimbus. A CUMULUS cloud that pro-

duces rain; a shower cloud. If it is sufficiently large its upper limits may be composed of ice crystals swept by high-altitude
winds into the shape of an anvil.

cumulus. A heap cloud, produced by buoyant upward CONVECTION. It has growing pro-
tuberances on top giving it a cauliflower-like
appearance.

cupola. (1) A vertical-shaft furnace used for
melting metals, as distinct from a BLAST
FURNACE in which the ore is melted. It is a
source of grit, dust and metallurgical fumes,
especially from the hot-blast type of cupola.
(2) A small, dome-like upward projection
from a BATHOLITH. Some STOCKS and BOSSES
may well be cupolas.

cuprite. *See* copper.

currare. *See* STRYCHNOS.

curie (Ci). The former unit of measurement
for radioactivity, equal to 3.7×10^{10} disinte-
grations per second, which is the intensity of
the radioactivity of radium, so that the curie
effectively compares the radioactivity of a
substance with that of radium. It is named
after Marie Curie (1867–1934), the Polish-
born, but French by marriage, physicist. The
curie has been replaced by the BECQUEREL
(Bq): $1 \text{ Ci} = 3.7 \times 10^{10} \text{ Bq}$.

curium. *See* ACTINIDES.

curlews. *See* CHARADRIIDAE.

Cuscuta (dodder). A genus of leafless,
rootless totally parasitic plants, lacking
CHLOROPHYLL, but having thread-like, twining stems bearing suckers which penetrate
the host plant and absorb food. *C. epithy-
mum*, the common British species, usually
parasitizes gorse and ling; the more local *C.
europea* is found mainly on stinging nettle.

cuticle. (1) A non-cellular covering to a
plant or animal, made by the EPIDERMIS, and
often in terrestrial forms serving to prevent
excessive loss of water as well as to protect
the organism. In higher plants, the cuticle is
made from waterproof cutin, a mixture of
compounds derived from fatty acids. In
invertebrates, it is made of CHITIN or a substance similar to COLLAGEN. It forms the
exoskeleton in ARTHROPODA, and in insects it
is covered with a waterproof layer of wax.
(2) More generally the outer protective layer
of dead cells in the skin of a terrestrial vertebrate.

cutin. *See* CUTICLE.

cut-off low. A CYCLONE (i.e. low) that is cut
off from the main sequence of cyclones
associated with a FRONT, the front having
long since been OCCLUDED so that no warm
AIR MASS is identifiable at ground level.

cutting oils. Petroleum products that contain additives to enable them to resist very
high pressures. They are used as lubricants
and to lubricate and cool machine tool cut-
ters.

cuttlefish. *See* DECAPODA.

cwm. *See* cirque.

cyanides. A class of very toxic compounds,
ingestion or inhalation of which may cause
death to mammals (although recovery from
sublethal doses is usually complete with no
after-effects). The cyanide ion (CN^-) is derived from hydrogen cyanide (HCN), of
which all other cyanides are salts. Its toxicity
derives from the ability of CN^- to form a
stable compound with the iron in HAEMOGLO-
BIN, so inhibiting the transport of oxygen by
the blood, and leading to asphyxiation.
Cyanides are used in the metal-plating industry, for the heat treatment of metals, in
some pesticides and in some disinfectants,
and they may survive in industrial wastes to
cause disposal problems. Once released into
the environment, however, they do not persist.

cyanogenic glycosides. *See* GLYCOSIDES.

Cyanophyceae. *See* CYANOPHYTA.

Cyanophyta (blue–green algae, Cyanophyceae). A group (division) of primitive, unicellular, colonial or filamentous PROKARYOTES, known formerly as the Myxophyta or Myxophyceae. Their cells have no nuclear membrane, MITOCHONDRIA or CHLOROPLASTS, but do contain CHLOROPHYLL and various other pigments, so they exhibit a variety of colours. They are widespread in marine and freshwaters and in soils, and occur even in hot springs. Some are symbiotic (e.g., in some LICHENS) and others fix atmospheric nitrogen. The name blue–green algae is no longer favoured since they are not necessarily blue or green, and they resemble BACTERIA more closely than algae. They are known from fossils to have been present on Earth 3000 Ma and are believed to have been the first organisms to produce oxygen on a large scale as a by-product of PHOTOSYNTHESIS.

cybernetics. The study of communication and control in systems.

cyborg. A fusion of a human and a machine within a single physical entity.

cycads. *See* CYCADALES.

Cycadales (cycads). An order of GYMNOSPERMAE, most of which are extinct, but whose living representatives are native to tropical and subtropical regions. The tall, unbranched stem bears a crown of fern-like leaves. Cycads produce cones and are the most primitive present-day seed-producing plants.

cycles per second. *See* FREQUENCY.

cyclodiene insecticides. A subgroup of the ORGANOCHLORINE insecticides which includes ALDRIN, CHLORDANE, DIELDRIN and ENDRIN.

cyclogenesis. The intensification of the rotation of air as a result of horizontal CONVERGENCE accompanied by upward motion and the formation of cloud. It is common at FRONTS where the warm air mass is lifted over the cold air.

cyclone. (1) A centre of atmospheric low pressure around which the air circulates cyclonically (i.e. in the same direction as the Earth). (2) *See* TROPICAL CYCLONE.

cyclone collector. A structure without moving parts in which the velocity of an inlet gas stream is transformed into a confined vortex from which suspended particles are driven to the wall of the cyclone body and collected. It is used widely as a pollution control for dusts.

cyclonic curvature. Curvature in the direction of flow around a CYCLONE (i.e. to the left in the northern hemisphere and to the right in the southern hemisphere).

cyclonic vorticity. Circulation in a cyclonic direction (i.e. in the same sense as the circulation of the rotating Earth). A parallel straight airstream possesses cyclonic vorticity if the speed decreases to the left across the stream. A stream with anticyclonic curvature may possess cyclonic vorticity.

Cyclops. *See* COPEPODA.

Cyclorrapha. A large suborder of DIPTERA that includes bluebottles, houseflies, hoverflies and fruit flies, and also frit flies whose larvae bore into crops and are serious pests, warble flies and bot flies, whose larvae are endoparasites (*see* PARASITISM) of livestock, and the tsetse fly which transmits the TRYPANOSOMIDAE responsible for sleeping sickness in humans and cattle in Africa. The Tachinidae are important parasites of insects, especially the larvae of LEPIDOPTERA. The larvae of Cyclorrapha have much-reduced heads, and the pupae are enclosed in barrel-shaped cases.

Cyclostomata (lampreys and hagfish). Present-day members of the AGNATHA, which are eel-like, scaleless, jawless and have no paired fins. They attach themselves to fish and feed by sucking their blood.

cyclostrophic force. The centrifugal force evident in the wind and caused by the curvature of the path of particles. It needs to be taken into account in calculating the GRADI-

ENT WIND only when the curvature is large.

cyclothem. A vertical sequence of SEDIMEN-TARY ROCK that occurs many times in a succession and denotes a repetition of physiographic and sedimentary events in a more or less constant order.

cyme. An INFLORESCENCE that branches repeatedly, with the oldest flower at the end of each branch.

Cyperaceae (sedges). A family of MONO-COTYLEDONEAE that are grass-like herbs of wet places and include the true sedges (*Carex*), bulrushes, clubrushes, cotton grasses, etc.

Cyprinidae (bream, carp, chub, dace, roach, rudd, tench, barbel, gudgeon, minnow). A large family of freshwater fishes with scales on their bodies but not on their heads.

cypsela. A single-seeded fruit, resembling an ACHENE, but formed from an inferior ovary and typical of members of the COMPOSITAE. The fruit of a dandelion is a cypsela.

cysteine. An AMINO ACID ($SHCH_2CH(NH_2)COOH$) with a molecular weight of 121.1.

cysticercus. *See* BLADDERWORM.

cystine. An AMINO ACID with the formula ($HOOCCH(NH_2)CH_2)_2S_2$) and with a molecular weight of 240.3.

cytochromes. A group of iron-containing proteins that play an important part in aerobic RESPIRATION, acting mainly as COENZYMES.

cytogenetics. The scientific study of heredity in relation to the physical appearance of the CHROMOSOMES.

cytokinins (phytokinins). Plant HORMONES that stimulate cell division (*see* MEIOSIS, MITOSIS). Their effects include the promotion of bud formation and germination, and they are thought to be associated with increased nucleic acid and protein metabolism. *See also* AUXINS, GIBBERELLINS, ZEATIN.

cytology. The scientific study of the structure and function of living cells.

cytolysis. Cell destruction, especially when brought about by the destruction of the PLASMA MEMBRANE.

cytoplasm. All the PROTOPLASM of a cell, except for the NUCLEUS.

cytoplasmic inheritance. The inheritance of characters through PLASMAGENES.

cytosine. *See* DNA, RNA.

cytotaxonomy. A method of classifying plants that is based on observation of the number, shape and size of CHROMOSOMES.

cytotype. A sample member of a population composed of individuals whose KARYOTYPE is identical, and different from that of other populations.

D

2,4-D (2,4-dichlorophenoxyacetic acid). A hormonal TRANSLOCATED HERBICIDE used to control many broadleaved weeds in cereals and grass.

D. *See* DEUTERIUM.

d. *See* DECI-.

da. *See* DECA-.

DAC. *See* DEVELOPMENT ASSISTANCE COMMITTEE.

dacite. A fine-grained CALC-ALKALINE IGNEOUS rock with a SILICA content between those of RHYOLITE and ANDESITE. Dacites contain QUARTZ PHENOCRYSTS and occur, with andesites, in large volumes in OROGENIC BELTS.

dalapon. A herbicide used for the control of grasses and to prevent water courses from becoming clogged by the growth of reeds and other monocotyledonous plants. *See also* MONOCOTYLEDONEAE, TRANSLOCATED HERBICIDES, TRANSLOCATION.

DALR. *See* DRY ADIABATIC LAPSE RATE.

dam. A structure built in order to restrict the flow of tidal or river water, or of semi-liquid mining or other industrial waste. Dams are commonly used to form artificial reservoirs or to provide hydroelectric power. *See also* LARGE DAM.

damage risk criterion. That noise level, as a function of FREQUENCY, waveform (i.e. pure tone or random noise), intermittency, etc., above which more than a specified degree of permanent hearing loss is likely to be sustained by a person exposed to it.

damaging UV. *See* ULTRAVIOLET RADIATION.

damping. The removal of energy from an oscillating system or particle by means of friction or viscous forces, the energy removed being converted to heat.

damselflies. *See* ODONATA.

dangerous waste. *See* SPECIAL WASTE.

Danian. A subdivision (usually ranking as a STAGE) of geological time, which is considered either uppermost CRETACEOUS or lowermost PALAEOCENE.

Danube Circle. An unofficial Hungarian environmental organization that in 1984 sought Austrian support in its efforts to prevent a Hungarian—Czechoslovakian hydroelectric scheme that would divert water from the Danube.

Daphnia. See BRANCHIOPODA.

dark minerals. In petrology, usually the FERROMAGNESIAN MINERALS which are present in IGNEOUS rocks.

dark reactions. *See* PHOTOSYNTHESIS.

Darrieus generator (Darrieus rotor). A vertical-axis AEROGENERATOR whose rotor blades consist of long, narrow aerodynamically efficient strips, designed to do mechanical work or generate electricity.

Darwin, Charles Robert (1809–82). The author of *The Origin of Species by Means of Natural Selection*, which revolutionized concepts of evolution by proposing a mechanism for it, and *The Descent of Man*, which advanced evidence for the evolution of humans from subhuman forms. Darwin collected evidence for his theories over many years, notably during his voyage as naturalist on *HMS Beagle*, when he studied the unique fauna of the Galápagos Islands. He and WALLACE, ALFRED RUSSELL, jointly published the first work proposing the theory of evolution by natural selection in 1858, and Darwin elaborated this in 1859 in his *Origin of Species*. The theory may be summarized as follows: organisms produce large numbers of offspring, but the overall numbers of a particular species remain relatively constant. Therefore a struggle for existence occurs among the offspring. Individuals of a species exhibit variation. These differences may confer advantages on certain individuals, increasing their chances of survival and reproduction. This results in the 'survival of the fittest' and implies the adaptation of the organism to its environment. The possession of advantageous variation is handed down to the offspring. (Darwin did not know how random variations were inherited, as the findings of MENDEL, GREGOR, had not come to his notice.) Thus, when conditions change or organisms spread to new areas, new forms will arise, each adapted to its own environment.

Darwin's finches. *See* ADAPTIVE RADIATION.

dawn. *See* SUNRISE.

day-degrees. The sum of the degrees of temperature above a threshold (e.g., a daily mean of 4°C) over a certain period (e.g., the growing season of a particular crop), usually in order to determine in advance whether temperatures in a particular place are suitable for the growing of a certain crop that has not been grown there previously.

dB. *See* DECIBEL.

dBA. *See* DECIBEL.

DBCP. *See* DIBROMOCHLOROPROPANE.

DCMU. *See* DIURON.

DDE. *See* DDT.

DDT (dichlorodiphenyltrichloroethane, chlorophenotone, dicophane).   A persistent ORGANOCHLORINE insecticide that was introduced in the 1940s and used widely because of its persistence (meaning repeated applications were unnecessary), its low toxicity to mammals and its simplicity and cheapness of manufacture. It became dispersed all over the world and, with other organochlorines, had a disruptive effect on species high in FOOD CHAINS, especially on the breeding success of certain predatory birds. DDT is very stable, relatively insoluble in water, but highly soluble in fats. Health effects on humans are not clear, but it is less toxic than related compounds (e.g., ALDRIN, DIELDRIN, LINDANE). It is poisonous to other vertebrates, especially fish, and is stored in the fatty tissue of animals as sublethal amounts of the less toxic DDE. Because of its effects on wildlife its use in most countries is now forbidden or strictly limited.

deadly nightshade. *See* SOLANACEAE.

deamination. The removal of an amino (–NH_2) group from a molecule. In mammals, AMINO ACIDS are deaminated by ENZYME action in the liver and kidneys, leaving carbon compounds that may be used in RESPIRATION. The waste product of deamination – ammonia – is converted to the less harmful UREA in the liver.

deca- (da). The prefix used in conjunction with SI units to denote a quantity equal to the unit x 10.

Decapoda. (1) A suborder of CEPHALOPODA, whose members possess 10 tentacles, and including cuttlefish, squids and ancestral MESOZOIC forms (Belemnoidea). *Compare* OCTOPODA. (2) *See* MALACOSTRACA.

deci- (d). The prefix used in conjunction with SI units to denote the unit x 10^{-1}.

decibel (dB). A unit used to measure the intensity of sound, on a logarithmic scale based on measurements of sound intensity in watts per square metre and related to a reference 10^{-12} W/m², which is the intensity of the quietest sound perceptible to the human ear. Because the scale is logarithmic (i.e. $\log_{10}$) each doubling of intensity increases the decibel value by three. The scale is then weighted according to three further systems, designated A, B and C (of which A is the most commonly used, to give the dBA unit) to reduce the response of measuring instruments to very high and very low sound frequencies and to emphasize those within the range that is audible to humans. Some sound meters have a further (D) weighting, to measure perceived noise (PNdB), often used in assessing aircraft noise. On the dBA scale, the rustle of leaves is 10 dBA, a quiet office 40, an alarm clock at 1 metre distance 80, a Saturn rocket lifting off at 300 metres distance 200. One decibel is equal to one-tenth of a bel (although the bel is rarely used). The unit is named after Alexander Graham Bell (1847–1922).

deciduous. Applied to an organism that sheds certain parts readily or regularly, or to those parts themselves. Deciduous trees shed all their leaves at a particular season (usually autumn) each year (*compare* EVERGREEN, TROPOPHYTE). Deciduous teeth are the first (milk teeth) of the two sets possessed by most mammals and are shed. There are fewer grinding teeth in the first set, but otherwise the two sets are similar (*see* DENTAL FORMULA). Certain fish (e.g., mackerel) have deciduous scales, which are shed readily, especially if the fish rubs against a solid object.

Declaration on the Human Environment. The declaration, agreed at the 1972 United Nations Conference on the Human Environment, which set out the common attitude of all signatory nations to environmental issues. *See also* UNITED NATIONS ENVIRONMENT PROGRAMME.

decomposer. *See* REDUCER.

decomposition. The separation of complex organic substances into simpler compounds.

decurrent. Projecting downwards, below

the point of attachment, as in a leaf blade continued as a wing running down the stem.

deep sea. An imprecise term usually restricted to that part of the ocean beyond the CONTINENTAL SHELF. *See also* ABYSSAL, BATHYAL.

deficiency disease. An illness caused by the lack of an essential food substance (e.g., scurvy caused by lack of VITAMIN C).

definitive host. *See* HOST.

deflation. The picking up and removal of loose material by the wind.

defoliant. A herbicide designed to remove leaves from trees and shrubs or to kill plants. Examples include 2,4-D and 2,4,5-T, which disturb hormonal balances in plants and so induce metabolic disorders.

deforestation. The permanent removal of forest and undergrowth.

deformation. Any change in the original form or volume of a rock body produced by TECTONIC forces. Deformation can be a contraction or extension and can be produced by folding, faulting or solid flow.

degenerative disease. An illness caused by the deterioration of organs or tissues, rather than by infection. *See also* DISEASES OF CIVILIZATION.

degreaser's flush. *See* DEGREASING.

degreasing. The process of removing grease, oils and dirt from machine parts by dipping them into a tank containing a degreasing agent, commonly an organic solvent (e.g., trichloroethylene (*see* TRICHLOROETHANE) known as 'tri' or 'trike'). Exposure to trike followed by the consumption of alcohol can cause a skin inflammation called degreaser's flush. Trike is also a suspected CARCINOGEN.

dehiscent. Applied to fruits that open to release seeds (e.g., gorse, poppy). *See also*

CAPSULE, FOLLICLE, LEGUME.

dehumidifier. A device incorporated in many air-conditioning systems to dry incoming air by passing it across a bed of a HYGROSCOPIC substance or through a spray of very cold water. *Compare* HUMIDIFIER.

dehydrogenases. ENZYMES that catalyze the removal of hydrogen from a substance. *See also* OXIDASES, RESPIRATION.

Delaney Clause. A clause added to the Food Additives Amendment to the US federal Food, Drug and Cosmetic Act, 1938, stating that 'no additive shall be deemed safe if it is found to induce cancer when ingested by man or animal'. It is interpreted to mean that no substance may be added to food if its administration to any experimental animal, in any quantity or over any period of time whatever, produces cancer.

delayed density-dependent. Applied to a situation in which the mortality among a HOST population depends on the population density of the host in successive generations, thus affecting the size of the parasite population.

delta. An accumulation of sediment at the mouth of a river. Conditions for delta building occur when the rate of deposition of sediment into the sea or lake exceeds the rate at which it can be removed. As sediment blocks river channels, new channels must be found for the water and its load. Three sets of bedding are usually observable in the structure of a delta. Bottomset beds of fine sediment are deposited farthest into the sea or lake. Foreset beds of coarser sediment are deposited at the mouth of the river and advance the delta into the sea or lake as they accumulate. Topset beds overlie the foreset beds and build up to sea or lake level. They are composed largely of alluvial sediment as in the FLOOD PLAIN. Deltas are typically triangular, with the apex upstream (the Nile Delta being the prototype), but other forms occur, notably the Mississippi Delta, which has a digital or bird's-foot form projecting into the sea.

-deme. A suffix used in experimental taxonomy to indicate a group of plants that exhibits clearly definable characteristics (e.g., a gamodeme is a group of individuals that are capable of interbreeding).

demersal. Applied to organisms living in the lowest part of a sea or lake, and sometimes used as a synonym for BENTHIC.

demetron-S-methyl. An organophosphorus SYSTEMIC INSECTICIDE and ACARICIDE used to control aphids and red spider mites on agricultural and horticultural crops. It is poisonous to vertebrates. *See also* ORGANOPHOSPHORUS PESTICIDES.

demographic transition. A transition in the pattern of increase in a human population from one characterized by high birth and high death rates to one with low birth and low death rates. It proceeds in stages. From the initial condition, in which high mortality means population size can be maintained only if birth rates are high, improved health care leads to a reduction in mortality, especially infant mortality. The high birth rate and relatively low death rate allows population size to grow, sometimes rapidly. Further improvements in health care, combined with greater educational and economic opportunities for women and generally greater prosperity, produce a situation in which most babies may be expected to survive to adulthood. Very large families become an economic burden to the parents, so birth rates begin to fall until the final condition is reached. The size of the population then stabilizes at its new level.

demography. The study of the age and sex structure, geographical distribution, rate of change of size, etc., of human populations.

denaturing. The addition of a noxious substance to render a product unfit for human consumption. Denaturing has been used to prevent wheat from being sold for human consumption when economic policies required it to be used for feeding livestock. Fish caught surplus to market requirements are often denatured and dumped at sea. Denaturing substances are often dyes. Methyl alcohol pyridine is added to industrial ethanol to denature it, producing methylated spirits, the drinking of which leads to extreme intoxication due to the impurity and the high concentration of ethanol.

dendritic. Many-branched, like a tree. The word is from the Greek *dendros* (tree).

dendritic crystals. Ice CRYSTALS commonly found in snow and characteristically formed when ice particles fall through supersaturated air (*see* SUPERSATURATION) and grow by SUBLIMATION. They are hexagonal and have many complex, but symmetrical branches, producing patterns that inspire much Christmas card art.

dendritic drainage. A drainage pattern in which the channels branch many times.

dendrochronology (tree ring dating). The scientific dating and investigation of historical climates through the study of differences between the successive annual growth rings of trees. Such differences result from the correlation between ring growth and climate, especially. in certain species. Using bristle-cone pines (*Pinus aristata*), the oldest living example of which is 4600 years old, and by correlating rings in living and dead wood in the same area, dendrochronologically dated wood up to 8200 years old can then be dated by carbon-14 dating (*see* RADIOMETRIC AGE) to check the carbon-14 date. This has shown that early radiocarbon dates were much too young; a calibration curve published in 1970 showed that 6000 BP determined by radiocarbon dating should be nearly 7000 BP. Consequently radiocarbon dates, and inferences from them, published before 1970 must be regarded with suspicion.

denitrification. The breakdown of NITRATES by soil bacteria (e.g., *Bacterium denitrificans*), resulting in the release of free nitrogen. This process takes place under ANAEROBIC conditions, such as are found in waterlogged soil, and it reduces soil fertility. *See also* NITROGEN CYCLE.

density. The mass of a unit volume of a substance.

density current. *See* TURBIDITE.

density-dependent. Applied to a limiting factor in the growth of a population that is dependent upon the existing population density (e.g., disease, reproductive rate, access to food).

density-independent. Applied to a situation in which the percentage mortality or survival of a species varies independently of population density.

dental formula. A conventional way of indicating the number and type of teeth present in any mammal, by listing the teeth on one side only of the upper and lower jaws, with those of the upper jaw displayed above those of the lower. The human formula is:

$$i\frac{2}{2}.c\frac{1}{1}.p\frac{2}{2}.m\frac{3}{3}$$

where i is incisors, c canines, p premolars and m molars. (Molars are not represented in the DECIDUOUS set.)

denticles. *See* DENTINE.

dentine. The bone-like substance that makes up the bulk of a tooth and lies inside the enamel. Ivory is composed of dentine. It also occurs in the tooth-like scales (denticles) of present-day cartilaginous and some fossil fishes.

denudation. The wearing away of the surface of the land by the combined effects of WEATHERING and EROSION.

deodorizer. In US usage, equipment for the removal of noxious gases and odours by combustion, ABSORPTION or ADSORPTION.

deoxyribonucleic acid. *See* DNA.

Department of the Environment (DoE). The UK government department that is responsible for a wide range of matters relating to land-use planning and the environment. It is headed by a Secretary of State and comprises the Ministries of Housing and Construction, Local Government and Development, and Transport Industries.

depression. In meteorology, a region of low pressure or CYCLONE.

derelict land. Land that has been damaged by extractive or other industrial processes and/or by serious neglect, that in its existing state is unsightly and that is incapable of reasonably beneficial use unless treated.

Dermaptera (earwigs). A small order of CRYPTOZOIC insects (division: EXOP-TERYGOTA) that bear a pair of forceps at the posterior end of the abdomen (used in a few species for seizing prey) and usually have well-developed hind wings folded under the short, leathery forewings. Earwigs eat plants and other insects, and are sometimes a nuisance in gardens.

dermis. The thick, inner layer of the skin in vertebrates. In mammals it is composed of CONNECTIVE TISSUE in which lie blood and lymph vessels, sense organs, nerves, fat cells, sweat glands and hair follicles (which are invaginations of the EPIDERMIS) with their erector muscles.

Dermoptera (flying lemurs). A small order of insectivorous PLACENTAL MAMMALS that glide by means of flaps of skin stretched between the limbs and tail.

derris (rotenone). An insecticide and acaricide used for the control of aphids, thrips, red spider mites, etc. It is harmful to fish, but not to mammals or birds, and it breaks down very rapidly after application. It is extracted from the root of *Derris elliptica* and other LEGUMINOSAE, and was the active ingredient of AL63, an insecticide used against lice in the 1939–45 war, before the introduction of DDT.

DES. *See* DIETHYLSTILBESTROL.

desalination. The extraction of fresh water from saltwater by the removal of salts, usually by DISTILLING.

Descent of Man, The. *See* DARWIN, CHARLES ROBERT.

desert. Any area in which one or more of the factors necessary to living organisms is in critically short supply so that the area is devoid of life or is very sparsely populated. Possible limiting factors include the number of days with temperatures high enough for plant growth, light, nutrient or water. On land, a desert will develop if evaporation exceeds precipitation, for whatever cause, leading to aridity and a consequent lack of vegetation. Evaporation rates vary according to temperature, but less than 25 centimetres of rain annually produces a desert in almost any temperature range. A semiarid area has a ratio of precipitation to evaporation that is less than one (i.e a deficiency of rainfall for the year as a whole). A true desert has one-half the precipitation that would separate semiarid climates from humid climates at that temperature range. *See also* ARID ZONE, DESERTIFICATION, DROUGHT.

desert crust. *See* DESERT VARNISH.

desertification. The process of desert expansion, observed in most of the major deserts, but most marked in the SAHEL. Deserts may expand as a result of natural climate change, but the process may be exacerbated (e.g., by overgrazing with livestock at the desert edge or by the clearance of vegetation adjacent to the desert).

desert pavement. A single layer of closely spaced stones, collected on the surface of silt and sand, and grading down to gravel, that is found in arid and semiarid areas.

desert rose. A coarsely crystalline mass of tabular GYPSUM or BARYTE crystals, found buried in deserts, and bearing a vague morphological resemblance to a rose. The crystals usually contain sand grains.

desert soils. Soils of arid regions where the

annual rainfall is generally less than 255 millimetres, although temperatures may vary from cool to hot. The vegetation is sparse and/or sporadic, due to the net deficiency of rainfall rather than an inherent lack of nutrients; consequently the organic layer is thin or even discontinuous. A pebble layer (*see* DESERT PAVEMENT) may accumulate at the surface. The leached layer (*see* LEACHING) is usually less than 150 mm thick, and characteristically there is a carbonate layer within 300 mm of the surface. *See also* SOIL CLASSIFICATION.

desert varnish (desert crust, patina). A hard, usually black coating of iron and manganese oxides, on the surface of rocks in deserts, probably formed by deposition from evaporating mineral-charged water drawn to the surface by CAPILLARY FLOW.

desiccator. A laboratory glass vessel, with a close-fitting lid and a chamber in the base for moisture-absorbant material, in which substances can be placed for the gradual extraction of water at ambient temperatures.

design rule. A rule that requires manufacturers to design products so they will conform to an environmental or other standard.

Desmidiaceae. A large group of unicellular, freshwater green algae (division: CHLOROPHYTA), which often form films on mud and aquatic plants. A desmid cell is usually composed of two symmetrical halves with sculptured or spinous walls.

desorption. The reverse of ABSORPTION or ADSORPTION.

desoxyribonucleic acid. *See* DNA.

destructive distallation. The DISTILLING of solids accompanied by their decomposition (e.g., of coal to produce coke and coal tar). Solid domestic refuse can be processed in this way by heating it in a retort without air at 500–1000°C, which reduces its weight by about 90 percent. Combustible gases, VOLATILE fluids, tar and CHARCOAL are produced, and the useless residue can be disposed of at LAND-FILL sites.

desulphurization. The removal from a substance (e.g., crude oil, coal, iron, a nonferrous metal or an ore) of sulphur or sulphur compounds by such processes as ELUTRIATION, FROTH FLOTATION and magnetic separation.

detention period. The average length of time for which a unit volume of fluid is retained in a tank during a flow process.

detergent. A SURFACE-ACTIVE AGENT used to remove dirt and grease from a surface. Soap is a detergent. Early synthetic detergents, containing alkyl benzene sulphonate (*see* ALKYL SULPHONATES) proved resistant to bacterial decomposition, causing foaming in rivers and difficulties at sewage-treatment plants. These 'hard' detergents were replaced in domestic use in the mid-1960s in Europe, North America and Australia by 'soft' biodegradable (*see* BIODEGRADATION) detergents containing straight alkyl chains. Problems remain, arising from the use of phosphate compounds (mainly sodium tripolyphosphate), which can cause EUTROPHICATION and for which no satisfactory substitute has been found.

determinant. (1) In genetics, a hereditary factor (*see* GENES, PLASMAGENES). (2) The part of an ANTIGEN molecule that combines with the corresponding ANTIBODY molecule.

detrital sediments. Sediments formed from fragments of pre-existing minerals and rocks and from the alteration products of rocks (e.g., CLAY MINERALS) that have been transported to the site of deposition and then compacted into SEDIMENTARY ROCKS. *Compare* CLASTIC.

deuteric. Applied to the alteration of IGNEOUS rocks by the action of VOLATILES derived from the MAGMA during the later stages of consolidation. Kaolinization, tourmalinization, greisening and serpentinization are examples of deuteric processes.

deuterium (D). An element; the ISOTOPE of

hydrogen with mass number 2 and atomic mass (Z) of 2.0147. Natural hydrogen contains 0.0156 percent deuterium and in water about one part in 5000 has hydrogen displaced by deuterium, giving deuterium oxide (D_2O, heavy water), which is used as a MODERATOR or coolant in some NUCLEAR REACTORS. Heavy water has a specific gravity of 1.1, freezes at 3.82°C and boils at 101.42°C.

Deuteromycotina. *See* FUNGI IMPERFECTI.

developed countries. *See* ECONOMIC DEVELOPMENT.

developing countries. *See* ECONOMIC DEVELOPMENT.

development. In meterorology, the generation of motion by BUOYANCY forces in the atmosphere, involving the ascent of warm air and a DIRECT CIRCULATION. It is an agent of CYCLOGENESIS.

development, economic. *See* ECONOMIC DEVELOPMENT.

Development Assistance Committee (DAC). The committee of the Organization for Economic Cooperation and Development (OECD) which is concerned with economic development and aid to developing countries.

Devensian. *See* ICE AGE.

Devonian. The fourth oldest period of the PALAEOZOIC Era, usually taken as beginning 400–425 Ma, and the rocks formed during the Devonian Period. These are called the Devonian System, which in Europe is divided into three series: Lower, Middle, Upper. The series are subdivided into stages: the Lower into the Gedinnian, Siegennian and Emsian; the Middle into the Eifelian and Givetian; the Upper into the Frasnian and Fammenian. The Devonian Period lasted 45–50 million years. Some geologists place the Downtonian Series into the Devonian rather than the Silurian. The non-marine FACIES of the Devonian in northern Europe is commonly called the Old Red Sandstone.

dew. Water vapour that condenses on to solid objects when the DEW POINT is reached, usually in the evening when surfaces cool more rapidly than the surrounding air. If surfaces cool to below freezing point HOAR FROST will form by SUBLIMATION.

dew bow. A RAINBOW phenomenon seen in dew drops on the ground, but rarely noticeable because of its weak intensity.

dew point. The temperature at which air becomes saturated with water vapour on being cooled. If the cooling is due to ADIABATIC expansion, cloud droplets are formed by further cooling; if the air is cooled by contact with a cold surface, condensation occurs on the surface as DEW. If cooling is due to mixing with colder, but unsaturated air, the mixture may be colder than its dew point, and then cloud is formed; this happens in FOGS and CONDENSATION TRAILS.

dew pond. A shallow pond, often with a puddled (*see* PUDDLING) clay bed, made in high pasture (e.g., chalk downland), in which water collects from precipitation and by condensation at night.

dew retting. *See* RETTING.

dextral fault. A strike–slip FAULT or a fault with a considerable component of strike–slip motion, in which the distant block shows relative displacement to the right when viewed from across the fault plane.

dextrose. *See* CARBOHYDRATES.

dia-. A prefix meaning through.

diabase. In North American usage, DOLERITE. In British usage, dolerite, older than the TERTIARY, that is so altered that few, if any, of the original minerals survive. The term is derived from continental European usage, but is considered obsolete in Britain. *See also* MICROGABBRO.

diacetylmorphine. *See* HEROIN.

diachronous. Applied to a rock unit that is

apparently continuous, but represents the development of the same FACIES at different places and at different times. The BED immediately above an UNCONFORMITY is usually diachronous.

diagenesis. The changes undergone by a sediment after deposition. These include changes caused by organisms, compaction and the resulting decrease in porosity, and changes related to the solution and deposition of minerals by CONNATE WATER and circulating water. Diagenesis grades into META-MORPHISM.

diageotropism. *See* GEOTROPISM.

diallate. A soil-acting herbicide of the thio-carbamate group (*see* CARBAMATES) used to control wild oats and blackgrass in brassica and beet crops. It can be irritating to the skin and is harmful to fish.

dialysis. The separation of smaller molecules from larger ones in a solution by means of a semipermeable membrane which allows the passage of the smaller molecules, but not the larger ones.

diamond. A mineral composed of the high-pressure form of carbon. Diamonds are found in ULTRABASIC pipes of KIMBERLITE and in PLACER deposits. Diamond is the hardest mineral known (10 on MOHS'S HARDNESS SCALE) and is chiefly used in abrasives. Gem-quality stones are cut for jewellry.

diapause. A dormant stage in the life cycle of some invertebrates, during which the metabolic rate is much decreased (e.g., HIBERNATION in insects). *See also* DORMANCY.

diapir. An INTRUSION that domes up the overlying layers, having cut through lower layers. Salt domes are examples of diapirs. Diapirs form where relatively dense material overlies less dense material, and the system is disturbed (e.g., by an earthquake); the denser material begins to sink, fracturing as it does so, and mushroom-shaped intrusions of less dense material rise through it.

diaspore. (1) One of the major ore minerals of aluminium, Diaspore (A1O(OH)) is one of the main constituents of BAUXITE and LATERITE. (2) (disseminule). Any part of a plant (e.g., SPORE, seed, TURION) that is dispersed and can give rise to a new individual.

diastases. *See* AMYLASES.

diastem. A minor break in a sedimentary sequence of rocks.

diastrophism. The process of large-scale DEFORMATION of the Earth's crust, producing continents, land masses, seas and ocean basins, and mountain ranges. Diastrophism is usually divided into orogenesis (mountain-building) and epeirogenesis (vertical movements without major crustal shortening).

diatomaceous earth (kieselguhr). A friable, siliceous deposit composed of the skeletal remains of diatoms (*see* BACILLARIOPHYTA) and used as an abrasive, a filtering medium, a filler, a physical insecticide and as a thermal and acoustic insulator.

diatoms. *See* BACILLARIOPHYTA.

dibromochloropropane (DBCP). A US pesticide, banned in 1977 because it was suspected of causing sterility and possibly cancer (based on research undertaken in the 1950s). It leaked and contaminated many private wells in California in 1982.

dicaryon. *See* DIKARYON.

dichlobenil. A soil-acting herbicide used for total weed control in land that is not intended for cropping, and for selective weed controls in orchards and forests. It is also used as an aquatic herbicide to kill floating and submerged plants in still or slow-moving water.

dichlorodiphenyltrichloroethane. *See* DDT.

dichloroethylene. A solvent, used in the

electronics industry and as a dry-cleaning fluid, suspected of causing cancer in humans.

2,4-dichlorophenoxyacetic acid. *See* 2,4-D.

dichlorprop. A hormonal TRANSLOCATED HERBICIDE used to control many broadleaved weeds in cereals.

dichlorvos. An organophosphorus insecticide and acaricide (*see* ORGANOPHOSPHORUS PESTICIDES) of short persistence, used domestically, in glasshouses and outdoors on fruit and vegetables for a rapid kill close to harvest. Resistant strains of aphids and red spider mites have appeared in some areas. It is possibly harmful to human health.

dichogamy. The maturation at different times of the male and female parts of a flower. This prevents SELF-POLLINATION. *See also* PROTANDROUS, PROTOGYNOUS.

dichotomous. Equally-forked.

dicophane. *See* DDT.

dicotyledon. *See* COTYLEDON, DICOTYLEDONEAE.

Dicotyledoneae. The larger of the two classes of flowering plants (ANGIOSPERMAE), in which the embryo has two (rarely more) COTYLEDONS. The leaves are mostly net-veined, and the VASCULAR BUNDLES in the stem usually contain CAMBIUM and are arranged in a ring. The flower parts are commonly in fours or fives, or multiples of these numbers. The class includes many trees as well as herbaceous plants. *Compare* MONOCOTYLEDONEAE. *See also* HORMONE WEED KILLERS.

Dictyoptera. An order of insects (division: EXOPTERYGOTA), formerly included in the ORTHOPTERA, that includes the swift-running, omnivorous cockroaches and the slow-moving mantids, whose forelegs are enlarged to grasp insect prey. Both groups have somewhat flattened bodies, long legs and usually leathery forewings covering membranous hind wings.

dieldrin. A CYCLODIENE INSECTICIDE that was used widely in the 1950s and 1960s. It is highly persistent, fat-soluble and thus becomes concentrated along FOOD CHAINS to produce adverse effects on species high on food chains (e.g., birds of prey). Its use is now forbidden or severely restricted in most countries.

diethylstilbestrol (DES). A synthetic oestrogenic hormone (*see* OESTROGENS) that was formerly used in livestock husbandry to accelerate growth and increase the proportion of lean meat to fat in beef cattle, but was implicated in cancer of the vagina in humans.

differential cooling. *See* DIFFERENTIAL HEATING.

differential heating. Heating by sunshine, when surfaces warm at different rates because of variations in the thermal capacities and conductivities of different materials (e.g., sand becomes hotter than solid rock), absorption over different depths (e.g., land becomes warmer than water), different ALBEDO (e.g., bare earth or dark surfaces become warmer than snow-covered surfaces) or cooling by evaporation (e.g., a wet surface or one covered with vegetation warms more slowly than a dry or bare surface). BREEZES, especially sea breezes and ANABATIC WINDS, are generated by differential heating. The complementary phenomenon of differential cooling is less spectacular because albedo and condensation effects are small, most bodies being equivalent to BLACK BODIES when radiating. Furthermore, heating at the bottom of the air produces motion that is more intense and over a greater depth than does cooling at the bottom.

differentiation. The development, during the growth of an organism or the regeneration of one of its parts, of different cells and organs from unspecialized cells. Differences in cell structure enable the cells or organs to perform different functions (e.g., the development of XYLEM and PHLOEM elements from cambium (*see* MERISTEM) in higher plants).

diffluence. The flowing apart of air particles with the consequent separation of STREAMLINES accompanied by deceleration, the motion being horizontal.

diffraction. The passage of waves around sharp edges that are not large compared with the WAVELENGTH. Diffraction of light around particles in the air produces the separation of wavelengths (e.g., in the CORONA and GLORY). The diffraction of sound waves can cause a SOUND SHADOW behind an acoustic screen.

diffraction analysis. A technique involving the DIFFRACTION of electromagnetic radiation or particle beams (e.g., X-rays, electrons or neutrons) to study the structure of matter, especially solids.

diffuse field. A sound field in which sound pressure is equal at every point, and sound waves are likely to be travelling in all directions.

diffusion. The spreading or scattering of a fluid (i.e. the process by which molecules intermingle as a result of their random thermal motion). *See also* EDDY DIFFUSION, MOLECULAR DIFFUSION.

digestion. The breaking down of complex food substances into simpler compounds, which can then be used in METABOLISM. Digestion is brought about by enzymes (e.g., lactase which is secreted by the small intestine of mammals and splits lactose into glucose and galactose).

digital computer. A device, originally mechanical, but now usually electronic, that is used in the solving of problems and in the construction of models. It accepts and processes information in the form of digits (i.e. encoded into a binary notation, as sets of zeros and ones), rather than with physical quantities directly (*compare* ANALOGUE COMPUTER) and can handle large amounts of information at high speed.

Digitalis. *See* SCROPHULARIACEAE.

digitigrade. Applied to animals that walk on the ventral surfaces of the digits only and not on the whole foot (e.g., cat, dog). *Compare* PLANTIGRADE, UNGULIGRADE.

dikaryon (dicaryon). A fungal HYPHA or MYCELIUM made up of cells, each of which contains two haploid nuclei (*see* CHROMOSOMES). *Compare* MONOKARYON.

dike. *See* DYKE.

dilution. The dispersal of a fluid (e.g., an effluent) within a much large receiving volume of another fluid.

diluvium. A general name for glacial deposits, which at one time were attributed to the Great Flood of Noah. The word persisted in literature as a synonym for glacial drift, but has fallen increasingly into disuse, and its literal meaning of flood deposit has been little used in its true sense. *Compare* ALLUVIUM.

dimethoate. A systemic organophosphorus insecticide and acaricide (*see* ORGANOPHOSPHORUS PESTICIDES, SYSTEMIC INSECTICIDES), poisonous to vertebrates, that is used to control red spider mites and insects (e.g., aphids) on agricultural land and horticultural crops. Strains of mites and aphids resistant to this chemical have appeared in some areas.

dimethyl sulphide. *See* ACID RAIN.

dimorphism. (1) Possessing two forms, as when a substance crystallizes in two forms. (2) The existence within a single species of two different forms (e.g., male and female individuals; POLYP and MEDUSA in some Cnidaria).

Dinantian. The lower CARBONIFEROUS, usually ranked as a series and comprising the Tournaisian, Visean and Namurian Stages.

dinitro-*o*-cresol. *See* DNOC.

dinitro pesticides. Compounds whose

molecules contain a dinitro group (i.e. two nitrogen atoms) and that are used as contact herbicides, fungicides and insecticides (e.g., DINOSEB, DINOCAP, DNOC). They are very poisonous to plants and animals, and may be harmful to humans in very small doses so their use is under review. They break down rapidly after application and so cause no delayed environmental contamination.

dinocap (DNOPC). A fungicide and acaricide of the dinitro group (*see* DINITRO PESTICIDES) used to control powdery mildew on horticultural crops and to suppress red spider mites. It can be irritating to the eyes and skin, and is harmful to fish.

dinoflagellates. A group of single-celled aquatic, mainly marine organisms with the characteristics of both plants (e.g., PHOTOSYNTHESIS) and animals (e.g., motility) and classed by some authorities as ALGAE and by others as PROTOZOA. The group includes the ZOOXANTHELLAE. When conditions are favourable their numbers may increase rapidly to produce 'red tides' (*see* BLOOMS) which can kill large numbers of fish and other animals and render molluscs extremely toxic to humans because of the water-soluble poison dinoflagellates produce.

dinoseb (DNBP, DNSBP, DNOSPB). A contact DINITRO PESTICIDE used to control many broadleaved weeds in leguminous crops, cereals, etc. It is very poisonous.

dioecious. Applied to organisms in which the male and female reproductive organs are borne on different individuals. *Compare* HERMAPHRODITE, MONOECIOUS, UNISEXUAL.

Diomedea exulans. See DIOMEDEIDAE.

Diomedeidae (albatrosses). A family of oceanic birds that engage in DYNAMIC SOARING. Since this generally leads to downwind drift, the largest albatrosses live in the southern hemisphere, where the pole may be circumscribed repeatedly without encountering land obstacles. Albatrosses live by scavenging surface material and breed on steep oceanic islands. The wandering albatross

(*Diomedea exulans*), the largest of all ocean birds with a mature wingspan of more than 3 metres, has difficulty in getting airborne in the absence of wind.

diorite. A course-grained, INTERMEDIATE, IGNEOUS rock that consists essentially of a plagioclase FELDSPAR which is more calcium-rich than the feldspar in GRANODIORITE, together with one or more FERROMAGNESIAN MINERALS (e.g., BIOTITE, HORNBLENDE, AUGITE). Diorites are the PLUTONIC equivalent of ANDESITE.

dioxin (2,3,7,8-tetrachlorodibenzo-*p*-dioxin, TCDD; $C_{12}H_4Cl_4O_2$). A by-product formed during the preparation of the herbicide 2,4,5-T, and sometimes produced by the incineration of chlorinated organic compounds. It may also occur naturally and is distributed widely in the environment, except locally in extremely low concentrations. Substantial amounts were released by the industrial accident at SEVESO in 1976. Exposure to high concentrations causes CHLORACNE. Although dioxin is suspected of causing more lasting damage, including CHROMOSOME malformation, there is no conclusive evidence for this.

dip. The angle between the greatest slope in a rock surface and the horizontal. The direction of dip is at right angles to the STRIKE. Dip is the complement of HADE.

diphyodont. Applied to an animal that has two sets of teeth: DECIDUOUS and permanent. It is a characteristic feature of mammals. *Compare* MONOPHYODONT, POLYPHYODONT.

diploblastic. Applied to animals in which the body wall consists of only two layers of cells – the ectoderm and endoderm (*see* GERM LAYERS) – separated by a layer of jelly (mesogloea). In many species there is a considerable invasion of the mesogloea by cells of the ectoderm and endoderm. *Compare* TRIPLOBLASTIC.

diploid. *See* CHROMOSOMES.

diplont. The diploid (*see* CHROMOSOMES)

stage in the life history of an organism; in almost all animals this is the whole life cycle apart from the GAMETES. The SPOROPHYTES of BRYOPHYTA, ferns, seed-bearing plants and some ALGAE are diplonts. In other algae and many FUNGI only the ZYGOTE is diploid. *Compare* HAPLONT.

Diplopoda. *See* MYRIAPODA.

Dipneusti. *See* DIPNOI.

Dipnoi. (Dipneusti, lungfish. A subclass of the bony fishes (class: OSTEICHTHYES) which first appeared in the DEVONIAN. The three living genera are air-breathing and inhabit tropical rivers, which dry up or become very stagnant. *Neoceratodus* is found in two Queensland rivers; the South American genus *Lepidosiren* and the African *Protopterus* can lie dormant in mud for at least six months. Dipnoi have characteristic broad toothplates for crunching food, which in modern forms consists of decaying vegetable matter and small invertebrates.

dip plating. A method for producing a thin coating of one metal on the surface of an object made from another metal by immersing the object in a solution of a salt or salts of the metal to be deposited. The process may pollute wastewater streams with metals.

Dipsacaceae (Dipsaceae). A family of dicotyledonous (*see* DICOTYLEDONEAE), mainly herbaceous plants that bear dense heads of flowers. Fuller's teasel (*Dipsacus fullonem*) has hooked bracts on the fruit heads, which were once used for raising the nap on cloth.

dip–slip fault. *See* FAULT.

dip slope. An inclined land surface that DIPS in the same direction as the underlying rocks and at approximately the same angle. The term is often applied to the back slope of a CUESTA.

Diptera (two-winged (true) flies). A large order of insects (division: ENDOPTERYGOTA) in which the hind wings are reduced to stumps and the larvae usually lack legs.

There are three suborders: BRACHYCERA, CYCLORRHAPHA, NEMATOCERA.

diquat. A contact herbicide used to control broadleaved weeds in many situations, including still and slow-flowing water, and to dry up foliage in order to facilitate harvesting of potatoes, clover seed, etc. It is harmful to mammals.

direct circulation. The sinking of cold air and ascent of warm air with consequent development (CYCLOGENESIS). The direct circulation at the entrance to a JETSTREAM provides the energy to accelerate the air. It is through the agency of the CORIOLIS FORCE that the circulation is converted into wind energy at right angles to the plane of the circulation.

directive. In the European Economic Community, an instruction based on agreements among ministers. It is issued by the Commission of the European Communities to member governments, and each of them should enforce it, if necessary by enacting appropriate national legislation.

directive evolution. *See* ORTHOGENESIS.

dirty. Applied to an arenaceous (*see* ARENITE) or rudaceous (*see* RUDITE) rock with a matrix of CLAY MINERALS.

disaccharides. *See* CARBOHYDRATES.

discharge. The volume of water flowing past a given point in a stream channel in a given period of time.

disclimax. (1) A SUBCLIMAX that endures for a long time and is prevented from reaching a full CLIMAX by human or other animal interference. (2) A modification or replacement of a true climax due to disturbance by humans or domestic livestock. *See also* PROCLIMAX.

disconformity. An UNCONFORMITY in which there is no angular divergence between the old and younger strata.

discontinuous distribution. A pattern of

distribution in which similar species are found in widely separated parts of the world. This is usually taken to indicate that the group is ancient and was once more generally distributed, but has become extinct over much of the original range.

discontinuous variation. *See* VARIATION.

diseases of civilization. A group of illnesses that occur more commonly among members of industrialized societies than among agrarian peoples and therefore may be associated with urban or industrial life. They include various cancers, diseases of the heart and circulation, certain digestive disorders, etc., many of them being DEGENERATIVE DISEASES.

dishpan experiments. Experiments in which the motion of a fluid is modelled in a rotating vessel that is heated from beneath or at an outside or inside vertical wall. The experiments have demonstrated the formation of a meandering JETSTREAM with a small number of waves and a CIRCUMPOLAR VORTEX relative to the vessel.

disinfectant. *See* DISINFECTION.

disinfection. The destruction of PATHOGENS by applying agents (disinfectants) such as CHLORINE.

disintegration. In nuclear physics, the disruption of the nucleus of an atom with the release of an ALPHA-PARTICLE or BETA-PARTICLE.

displacement activity. An apparently irrelevant action that is performed by an animal when it is stimulated to carry out two incompatible behaviour patterns (e.g., grooming activity performed by a rat when it is presented simultaneously with stimuli that normally elicit approach and flight).

display. A method of communication between animals during courtship, mating, defence of territory, etc. It involves showing off conspicuous features (e.g., peacock's tail, newt's crest, the red breast of a stickle-back), ritual performance of actions (e.g., the elaborate courtship dance of the great crested grebe) or producing sounds (e.g., birdsong, the 'drumming' of snipe).

disruptive coloration. *See* APATETIC COLORATION.

disseminule. *See* DIASPORE.

dissipation trail. *See* DISTRAIL.

dissolved load. The weathered (*see* WEATHERING) rock constituents carried in chemical solution by moving water.

dissolved oxygen. Oxygen molecules that are dissolved in water, usually expressed in parts per million (ppm). The presence of dissolved oxygen is vital to AEROBIC aquatic organisms because it is loosely held by the water and available to them for RESPIRATION. Normal saturation at 0°C is about 10 ppm, but it falls as temperature rises to about 6.5 ppm at 20°C and 5.5 ppm at 30°C, and the saturation point also depends on atmospheric pressure and on the chemical content of the water. In still water, oxygen dissolved from the atmosphere diffuses slowly through the lower levels of water, but in moving water the constant exposure to the air of unsaturated water usually leads to a higher dissolved oxygen content.

distilled water. Water of great purity, prepared by repeated DISTILLING and used in electrical conductivity measurements, etc.

distilling. The process of heating a mixture in order to separate its components by condensation of the volatile elements driven off by the heating.

distrail (dissipation trail). A clear lane in a thin layer of cloud caused by the downwash behind an aircraft, vortices following the aircraft forcing a line of clear air into the cloud. It often takes the form of a series of holes, which correspond to the blobs of a CONDENSATION TRAIL, and sometimes the shadow of a condensation trail on a cloud is imagined by observers to be a distrail.

distribution. Arrangement or pattern; statistically, the way in which variate values are apportioned.

district heating. A system that uses hot water from a single source (e.g., cooling water from a power station) to heat buildings nearby. Block heating is the heating of one or more apartment blocks or a shopping precinct from a central source. Group heating is the heating of a small group of houses from a central source. *See also* COMBINED HEAT AND POWER.

disulfoton. A systemic ORGANOPHOSPHORUS PESTICIDE used to control carrot fly and aphids in many crops. It is poisonous to vertebrates.

dithiocarbamate. *See* CARBAMATES.

ditocous. Producing two young at a time. *Compare* MONOTOCOUS, POLYTOCOUS.

diuresis. An increase in the volume of urine produced by the kidneys, usually because of an increase in the amount of liquid drunk.

diurnal. Daily: usually applied to events or cycles that repeat at daily intervals (*see* CIRCADIAN RHYTHM). The diurnal cycle of air pollution is of interest to those concerned with pollution control.

diuron (DCMU, DMU). A soil-acting herbicide of the UREA group, used for total weed control on land not intended for cropping and for the selective control of annual weeds in fruit crops and tree nurseries. Its effects can last for 12 months after application. It can be irritating to the eyes and skin and is harmful to fish.

divergence. In three dimensions, the rate of increase of unit volume of a fluid which, in the atmosphere, is approximately proportional to the vertical velocity because of the ADIABATIC expansion in the HYDROSTATIC PRESSURE field. In two dimensions, related to

horizontal motion as portrayed in meteorological charts, divergence at the ground is accompanied by descending motion above and by the dissipation of clouds. Horizontal divergence decreases vertical velocity and cyclonic rotation and is therefore accompanied by rising pressure.

divers. *See* GAVIIDAE.

divide. *See* CATCHMENT.

division. *See* CLASSIFICATION.

dizygotic twins (fraternal twins, non-identical twins). Twins produced as a result of the simultaneous fertilization of two ova. The twins are not genetically identical and may be of different sexes. *Compare* MONOZYGOTIC TWINS.

DMU. *See* DIURON.

DNA (deoxyribonucleic acid, desoxyribonucleic acid, thymonucleic acid). The principal material of inheritance. It is found in CHROMOSOMES and consists of molecules that are long unbranched chains made up of many nucleotides. Each nucleotide is a combination of phosphoric acid, the monosaccharide deoxyribose and one of four nitrogenous bases: thymine, cytosine, adenine or guanine. The number of possible arrangements of nucleotides along the DNA chain is immense. Usually two DNA strands are linked together in parallel by specific base-pairing and are helically coiled. Adenine links (by hydrogen bonding) only to thymine and guanine only to cytosine. Replication of DNA molecules is accomplished by separation of the two strands, followed by the building up of matching strands by means of base-pairing, using the two halves as templates. By a mechanism involving RNA, the structure of DNA is first transcribed and then translated into the structure of PROTEINS during their synthesis from AMINO ACIDS. *See also* GENES.

DNBP. *See* DINOSEB.

DNC. *See* DNOC.

DNOC (DNC, dinitro-*o*-cresol). A contact DINITRO PESTICIDE used to control broadleaved weeds and the overwintering stages of many insect and mite pests. It is extremely poisonous.

DNOPC. *See* DINOCAP.

DNOSBP. *See* DINOSEB.

DNSBP. *See* DINOSEB.

dodder. *See Cuscuta.*

dodo. *See* COLUMBIFORMES

DoE. *See* DEPARTMENT OF THE ENVIRONMENT.

dogger. (1) Rock strata deposited during Middle JURASSIC TIMES. (2) A large (BOULDER size) calcareous CONCRETION.

dog's mercury. *See* EUPHORBIACEAE.

doldrums. The region of small pressure gradients between latitudes 5°N and 5°S, where winds are light because the CORIOLIS FORCE is negligible. It is a region of widespread showers over the ocean.

dolerite. A medium-grained, basic HYPABYSSAL rock, equivalent to GABBRO and BASALT. Dolerites are called diabases in North America.

doline. A feature of KARST landscape consisting of funnel- or dish-shaped hollows from 2 to 300 metres deep, with varied outlines and floors filled with fallen rocks.

dolomite ($CaMg(CO_3)_2$). (1) A CARBONATE MINERAL that occurs in EVAPORITE deposits, as a replacement in LIMESTONES, as a CEMENT, as a GANGUE mineral in HYDROTHERMAL deposits and in CARBONATITES. (2) A rock consisting of a high percentage (usually 50 percent) of the mineral dolomite. To avoid confusion such rocks are sometimes called dolomite rock or dolostone.

dolomitization. The alteration of original CALCITE LIMESTONES by percolating magnesium carbonate solutions.

dolostone. *See* DOLOMITE.

dolphins. *See* CETACEA.

domestic wastes. Water-borne and other wastes from households, including sewage and sullage (i.e. water used for cooking, washing, etc.) water.

dominance frequency. The proportion of sampling units in which a particular species is most numerous.

dominant. (1) The characteristic, and often the tallest, species in a particular plant COMMUNITY. The dominant species is the one that exerts the greatest influence on the character of the community and may give it its name (e.g., oak in oakwood, reed in reed swamp. (2) A character which is the one of a pair of contrasted characters that is fully developed, whether the individual be a HETEROZYGOTE or HOMOZYGOTE. *Compare* RECESSIVE. (3) The leader in a group of animals (e.g., the most aggressive male in a herd of red deer during the rutting season).

Domin scale. A scale used to indicate the approximate percentage cover of individual plant species in a circumscribed area (e.g., a metre-square QUADRAT). The scale ranges from 10 (100 percent cover) to 1 (insignificant cover), with 8 representing 50–75 percent cover, 5 indicating 10–5 percent cover and 3 representing 1–5 percent cover.

Donora smog incident. An air pollution incident that occurred in Donora, Pennsylvania in October 1948, when FOG accumulated in very stable atmospheric conditions over a total of seven days before being washed down by rain. Of the total population (14 000), 42 percent suffered illness, 10 percent were seriously ill and 18 people died. The principal pollutants were believed to be SULPHUR DIOXIDE and particulate matter.

doomsday machine. *See* COBALT-60.

Doppler effect (Doppler shift). The apparent change in FREQUENCY of sound or electro-magnetic waves caused by the relative motion of the source and the observer. An approaching source emits waves, each of which begins from a point closer to the observer than the previous one, producing an apparent increase in frequency and consequent rise in the pitch of a sound or increase in blue light. A departing source emits waves whose frequency appears to decrease, lowering the pitch of a sound or increasing the amount of red light.

dormancy. A resting condition in which the growth of an organism is halted and its metabolic rate slowed. Dormancy may involve the whole organism, or only its reproductive bodies, and may be caused by unfavourable conditions or be part of a rhythmic cycle (e.g., winter dormancy in deciduous trees, regulated in some by PHOTO-PERIODISM; summer dormancy in daffodil bulbs). *Compare* AESTIVATION, DIAPAUSE, HIBERNATION

dormin. *See* ABSCISIN.

dorsal. (1) Applied to the part of an animal or organ that is at, or nearest to, the back; in most species this is directed upwards. In bipedal vertebrates the dorsal side is directed backwards, and in bony flatfishes the upper side is anatomically lateral, not dorsal (i.e. they swim on their sides). A few invertebrates move about with the dorsal surface downwards (e.g., back-swimming water boatmen, *Notonecta*) or lie on their sides (e.g., freshwater shrimps, *Gammarus*). (2) *See* ABAXIAL.

dorsiventral. Applied to leaves of DICOTY-LEDONEAE that lie more or less horizontally and whose upper and lower sides show differences in structure. *Compare* ISOBILATERAL. *See also* MESOPHYLL.

dose equivalent. *See* RADIATION DOSE EQUIVALENT.

double recessive. An individual that is homozygous (*see* HOMOZYGOTE) in respect of a particular RECESSIVE gene.

doubling time. *See* EXPONENTIAL GROWTH.

doves. *See* COLUMBIFORMES.

downdraught. A descending current of air, which may occur beside buildings where pollution from a chimney may be brought to ground level, and beneath rainstorms in which the rain cools the air by evaporation into it, so causing the air to descend and spread out at ground level.

downland vegetation. The vegetation characteristic of the chalk downs of southern England, comprising short turf rich in flowering herbs, formed originally by intensive and exploitive grazing by livestock which depleted the fertility of the land and so denied any advantage to aggressive plant species that might otherwise have become dominant. Until recently the SUCCESSION was arrested because of grazing by rabbits and sheep. Since MYXOMATOSIS drastically reduced the rabbit population and sheep farming declined, scrub has spread to the detriment of many attractive herbs. *See also* GRASSLAND.

Downtonian. The youngest series of the SILURIAN System in the UK; equivalent to the Pridolian elsewhere in Europe. The Downtonian is placed in the DEVONIAN System by some geologists.

downwash. *See* WING-TIP VORTICES.

dowsing. The detection of underground water or other substances or objects by feeling the motion of a split stick (often of hazel) or wires held in the hands. Used by experienced persons the technique has a long history of successful application, but attempts to investigate its principles have invariably failed and these are not understood.

Dracunculus medinensis. *See Filaria.*

drag. *See* AERODYNAMIC DRAG.

dragonflies. *See* ODONATA.

Dragon reactor. An experimental NUCLEAR

REACTOR, funded by 10 nations and built at Winfrith, Dorest, UK to investigate the principles of HIGH-TEMPERATURE GAS-COOLED REACTORS.

drainage basin. The land area from which water drains to a river or lake. *See also* CATCHMENT.

drainage morphometry. The study of drainage patterns.

dreikanter. A type of VENTIFACT in which sand blown by winds from three directions has produced three facets, each at right angles to the wind direction. *Compare* EINKANTER.

drift. (1) Superficial deposits caused by wind, ice or water, especially a deposit of wind-blown sand in the lee of a gap between two obstacles. Drift editions of geological maps show such deposits in colour, in contrast to solid editions, which ignore drift and show underlying rocks. (2) *See* GLACIAL DRIFT. (3) *See* DRIVE. (4) *See* CONTINENTAL DRIFT. (5) An ocean current.

drive. (1) (drift) In mining, a horizontal tunnel or opening, lying in or close to the ORE BODY, and parallel to the STRIKE of the ore body. (2) In animal behaviour, a state of activity and responsiveness to stimuli that normally leads to the satisfaction of a need (e.g., sex drive, hunger drive).

drizzle. Falling drops of water that are carried significantly by air motion, having diameters of less than about 0.5 millimetres. It is a form of soft rain, usually produced in clouds less deep than those producing larger raindrops, and sometimes produced close to the ground in dense fogs.

drop sonde. A radiosonde device (*see* BALLOON) that is released at high altitude from a balloon or aircraft and descends on a parachute to obtain soundings of the air at the levels through which it passes.

Drosera. See DROSERACEAE.

Droseraceae. A family of DICOTYLEDONEAE, all of which are insectivorous herbs, mostly found in acid bogs. Sundew (*Drosera*) catches insects by means of sticky tentacles on the rosettes of its leaves. Once trapped, the insect is pressed down on to the leaf blade by the bending of the tentacles, which secrete a protein-digesting enzyme. Because of its insectivorous mode of nutrition, sundew is able to live in very poor soil.

Drosophila. A genus of small, yellowish or brownish fruit flies (suborder: CYCLORRAPHA) that are much used in genetic research because of their short life cycle and the large chromosomes (*see* MEGACHROMOSOMES) in the salivary glands of their larvae.

drought. A long period of unusually low rainfall that leads to the parching of ground and the withering of vegetation. The term is not precise, being defined to suit the region in which it occurs, and used only if the condition is regarded as abnormal and people are unprepared for it. Thus a drought can develop in a few weeks in areas where normally rainfall is distributed evenly throughout the year and in months where the onset of a rainy season is delayed; in desert regions it may be regarded as a succession of unusually dry years.

drumlin. A smooth, oval hill of GLACIAL DRIFT (usually BOULDER CLAY), characteristically with one end that is blunter than the other in plan and has a steeper slope. A drumlin field, with many drumlins, is sometimes called basket-of-eggs topography.

drumlin field. *See* DRUMLIN.

drupe. A single-seeded fruit in which the PERICARP consists of a skin (epicarp), a thick, usually fleshy, middle layer (mesocarp), and a stony endocarp enclosing the seed (e.g., plum, cherry). A coconut is a drupe with a fibrous mesocarp; a blackberry is a collection of small drupes (drupelets). *Compare* BERRY.

drupelet. *See* DRUPE.

dry adiabatic lapse rate (DALR). The ordinary ADIABATIC LAPSE RATE, but called dry to distinguish it from the WET ADIABATIC LAPSE RATE.

dry ice. Solid CARBON DIOXIDE, which sublimes (*see* SUBLIMATION) at temperatures above -72°C. It is therefore a conveniently portable refrigerant, used by itinerant ice cream vendors, etc. As it sublimes, the condensation of water vapour in the air chilled by it forms cloud, a phenomenon much used for theatrical effects. *See also* ARTIFICIAL RAIN, CLOUD SEEDING.

dry impingement. A process in which particulate matter carried by a gas stream is pushed against a retaining surface which may be coated with an adhesive.

drying agents. Substances that remove water (e.g., calcium oxide, silica gel).

dry rot. *See* BASIDIOMYCETES.

dry-weather flow. The rate of flow of liquid through a sewer, or of water through a river channel, in dry weather.

dry weight rank method (DWR). A technique for estimating the percentage contribution each plant species makes to the total yield of a pasture.

duck decoy pond. A star-shaped pond with curving arms (pipes) ending in traps, into which ducks are enticed, usually by the intermittent sight of a dog specially trained to move around reed screens placed at intervals along the pipes. Decoy ponds were formerly much used for the commercial trapping of ducks, but few are now in existence. One in the Cambridgeshire Fens, UK is used extensively to trap ducks for ringing (*see* BIRD RINGING).

ducks. *See* ANATIDAE.

ductility. The capacity of metals for cold flow, which is accompanied by progressively increasing resistance to such flow, called work hardening. The ductility of metals makes possible the drawing of wire, cold pressing and similar operations.

ductless gland. *See* ENDOCRINE ORGAN.

dugong. *See* SIRENIA.

dun. *See* EPHEMEROPTERA.

dune stabilization. The prevention of the migration of SAND DUNES by erecting fences to trap moving sand, planting grasses to bind sand among their roots and to supply humus, and finally to plant other crops, including trees, or to allow natural vegetation to develop. Such techniques were first practised in Japan in the 17th century, were developed in Europe in the 18th century and are now practised widely.

dunite. An ULTRABASIC rock consisting entirely or almost entirely of OLIVINE.

Duplicidentata. *See* LAGOMORPHA.

duramen. *See* HEARTWOOD.

durilignosa. A plant COMMUNITY that consists of broadleaved SCLEROPHYLL forest and bush.

duripan. *See* HARDPAN.

durum. A variety of winter- or spring-sown wheat (*Triticum durum*) whose flour contains more gluten than that from most wheats, making it the variety best suited for making pasta. It is grown in the mediterranean region, the USSR, Asia and North and South America, especially in relatively arid areas. *See also* GRAMINEAE.

dusk. *See* SUNSET.

dust. Solid particles (1–1100 micrometres in diameter) that are carried into the atmosphere, from which they settle by gravity. Most originate naturally, from wind-blown soil, fires, etc., but about one-third of the dust in the air over much of the inhabited world is the result of human activity. Excessive inhalation of dust can cause injury to the respiratory system.

dust bowl. A large agricultural region of the central USA that experienced prolonged low rainfall in the 1930s, when the soil was ploughed or bare for other reasons: the prairie grasses which had once consolidated and protected the surface had been cleared. Dry soil was blown away by the wind, some areas losing 60–90 centimetres of topsoil. More generally, the term is applied to an area prone to this type of damage anywhere in the world.

dust burden. The weight of DUST suspended in a unit volume of a medium (e.g., FLUE GAS), expressed in grams per cubic metre at standard temperature and pressure.

dust collector. A device for collecting and so removing DUST from the exhaust gases of an industrial process. It may work by sedimentation, inertial separation (e.g., by impaction or impingement), precipitation (*see* electrostatic precipitator) or filtration (*see* electrostatic filter).

dust devil. A rotating convection current .made visible because of the particles that it contains, which have been carried by the whirlwind off the ground and into the air, where they ascend in the vortex.

dustlice. *See* PSOCOPTERA.

dust storm. A storm of DUST blown up from the ground when wind speed exceeds a critical value (commonly 24–48 kilometres per hour) which depends on the specific gravity, size, shape and dampness of the surface particles and their availability. There are two main types of dust storms. A haboob is local and is associated with a thunderstorm or CUMULONIMBUS cloud from which rain has begun to fall. The rain evaporates before reaching the ground, and the dust is blown into the air with the appearance of smoke, having a bulge at the leading edge and a slope at the upper surface. Dust may be carried to 1500–1800 metres or even higher. In contrast, the storm, known in Egypt as a khamsin and in Libya as a gibleh, covers a wide area and is associated with an area of low atmospheric pressure. In air that is thermally unstable hot dust and sand particles rising rapidly may warm the air surrounding them and increase the instability.

Dutch elm disease. A disease of elms (*Ulmus*) caused by the fungus *Ceratostomella ulmi* and spread by the elm bark beetle (*Scolytus scolytus*). The disease has been endemic in Britain at least since 1927, when a major outbreak began, and may have occurred at various times in the remoter past. It is called Dutch because the disease was first identified in Holland (in 1918) by Dutch scientists.

DUV. Damaging ultraviolet (*see* ULTRAVIOLET RADIATION).

dwale. *See* SOLANACEAE.

DWR. *See* DRY WEIGHT RANK METHOD.

dy. A type of lake bottom sediment that is composed largely of peaty plant detritus mixed with a gelatinous precipitate of iron salts. It is found in OLIGOTROPHIC lakes.

dyke. (1) (dike) A tabular IGNEOUS intrusion that is discordant (i.e. it cuts across the BEDDING PLANES in SEDIMENTARY ROCKS or across the FOLIATION in metamorphic rocks (*see* METAMORPHISM). (2) A ditch used for drainage. (3) A low wall or bank used to prevent water from invading low-lying land.

dynamic soaring. A technique in which a bird maintains or increases its air speed by flying across wind gradients, either behind wave crests or in the general increase of wind in the friction layer. It is not relied upon significantly by birds flying over land, but is undoubtedly important to the larger sea birds, particularly to the wandering albatross (*Diomedea exulans*), which remains airborne for many hours and allegedly days.

dynamic stability. In the atmosphere, in the presence of factors that induce TURBULENCE (notably WIND SHEAR), a condition in which small perturbations of the flow do not tend to grow. *Compare* STATIC STABILITY.

dynamometer. An instrument for measuring power (e.g., of an engine) or to assess the rate of emission of motor vehicle exhausts under test conditions.

Dynophyceae. *See* PYRROPHYTA.

dysphotic zone. The zone of water in a sea or lake that lies between the EUPHOTIC ZONE and APHOTIC ZONE. It is subject to dim light and usually spans depths of approximately 100–600 metres.

dystrophic. Applied to freshwater bodies that are deficient in calcium, very poor in dissolved plant nutrients, especially nitrates, and that are therefore unproductive. Such waters are typical of acid peat areas and have bottoms covered with undecomposed plant remains harbouring a poor fauna. The water is usually stained brown with peat. *Compare* EUTROPHIC, MESOTROPHIC, OLIGOTROPHIC.

E

ear. A sense organ of vertebrates that is both a sound receptor and an organ by which the animal is made aware of its movements and its position in relation to gravity; the sense of hearing may be absent in some fishes and reptiles. Different kinds of receptor cells in the inner ear are stimulated by vibrations initiated by sound waves, by movements of liquid caused by angular acceleration or by movements of otoliths (granules of carbonate) in response to gravity. The auditory ossicles (malleus, incus and stapes in mammals) are small bones that transmit vibrations of the ear drum to the inner ear.

Early Stone Age. *See* PALAEOLITHIC.

earthquake. A series of shock waves that are generated by a transient disturbance within the Earth's crust or mantle. The point of origin of the earthquake is called the focus, and the point on the Earth's surface directly above the focus is called the epicentre. The shock waves are classified as body waves (P WAVES and S WAVES), which travel within the Earth, and surface waves (L waves and R waves). The damage, caused by the surface waves, is classified on the MODIFIED MERCALLI SCALE (using Roman numerals), based on local structural damage which depends on the nature of the underlying soil and bedrock. Earthquake magnitudes are given using the Richter scale (using Arabic numerals), which is based on the amplitude of the largest trace recorded by a standard SEISMOGRAPH 100 kilometres from the epicentre, with 7 being regarded as a major earthquake; an earthquake of magnitude 8 probably releases about 30 times more energy than one of magnitude 7. Many of the deaths associated with earthquakes occur after the shock and are due to TSUNAMIS, landslides, fires, epidemics because of polluted water, exposure, etc. Earthquake prediction is a growing science based on such phenomena as changes in magnetic fields, pressure in wells, underground electrical currents, seismic velocities, amounts of radon underground and the build-up of stress in the Earth. The observation of animal behaviour also contributes since many animals seem able to sense imminent earthquakes and may leave the area.

Earth Resources Technology Satellite (ERTS-1). An unmanned Earth-orbiting satellite, developed by NASA and launched from the USA in 1972. It was equipped to scan the surface of the Earth and to obtain information pertaining to natural resources and the environment. ERTS-1 made 14 orbits each day, scanning overlapping strips 160 kilometres wide and taking 18 days to cover the whole of the Earth's surface. It produced images for different wavebands from blue to infrared, which could be combined, and which were used mainly for the mapping of resources, inaccessible areas of the world and such ephemeral phenomena as crop diseases and movements of polluted water. The programme was highly successful, leading to demands for more satellites of similar type; in 1975 the ERTS programme was expanded and renamed LANDSAT.

Earthscan. A news and information agency, supported and partly funded by the United Nations Environment Programme, and with close links to the International Institute for Environment and Development. It commissions original articles on environmental matters, which it sells as features to newspapers and magazines in many countries and publishes the findings of studies of major environmental problems. Its head office is in London.

Earth's shadow. The darkness that can be seen rising up the eastern sky just after sunset in suitable conditions of haze.

Earthwatch. A worldwide programme to monitor trends in the environment, established under the terms of the DECLARATION ON THE HUMAN ENVIRONMENT. It obtains data from a series of monitoring stations, and its activities are coordinated by the UNITED NATIONS ENVIRONMENT PROGRAMME.

earthworms. *See* OLIGOCHAETA.

earwigs. *See* DERMAPTERA.

East African Floral Region. The part of the PALAEOTROPIC REALM that comprises Africa from north of Lake Victoria south to southern Mozambique and extending westward to include southern Angola.

easterly wave. A sinuosity in the tropical easterly trade-winds, associated with increased rainfall and thought to be connected with the birth of hurricanes.

ecad. A form of a plant that has been modified by its habitat by the development of non-heritable characteristics.

ecdysis. Moulting; the periodic shedding of the outer covering of the body, especially in ARTHROPODA, AMPHIBIA and REPTILIA. In vertebrates, the sloughing of the outer EPIDERMIS is under the control of pituitary and thyroid HORMONES. In insects, the shedding of the hard EXOSKELETON, which usually occurs only in immature stages, is under the control of the hormone ecdysone, produced by glands in the first thoracic segment. In Arthropoda, an increase in body size can occur only between the time the old exoskeleton is shed and the new one hardens.

ecdysone. *See* ECDYSIS.

ecesic. The establishment of a colonizing plant species.

echidna. *See* MONOTREMATA.

Echinodermata. A phylum of marine invertebrate animals, most of which exhibit five-rayed symmetry as adults. The skin bears CALCAREOUS plates. The COELOM is intricate and large, with extensions into the many tube feet which protrude from the body surface. The PELAGIC larvae have affinities with those of the HEMICHORDATA. *See also* ASTEROIDEA, CRINOIDEA, ECHINOIDEA, HOLOTHUROIDEA, OPHIUROIDEA.

Echinoidea (sea-urchins, heart urchins, cake urchins, sand dollars). A class of spiny, armless, cushion-shaped or discoidal ECHINODERMATA whose skeletal plates are joined to form a rigid EXOSKELETON. Sea-urchins travel by means of their moveable spines and ten meridional rows of tube feet. They feed largely on seaweed, but also ingest mud and detritus. They are either ciliary feeders (*see* CILIA) or obtain food by means of their tube feet or, most commonly, use a complicated jaw apparatus (Aristotle's lantern).

Echiuroidea. *See* ANNELIDA.

echo. Reflected sound, which reaches the observer after a time interval long enough for it to be perceived as a separate sensation. The principle is used in ECHO LOCATION to determine the direction and distance of an object and in echo sounding (*see* SONAR) used to measure the distance between an instrument on board a ship and the sea bottom or to identify underwater objects. Radar works on a similar principle, but uses electromagnetic radiation transmitted as a beam. Radar is an acronym for *r*adio *d*etection *a*nd ranging.

echo location. A method used instrumentally (*see* SONAR) and by some animals (e.g., bats, porpoises) to locate and identify objects by emitting sounds, usually very high-pitched, and perceiving their ECHOS.

echo sounding. *See* ECHO, ECHO LOCATION, SONAR.

Echynorhynchus proteus. *See* ACANTHOCEPHALA.

eclipse plumage. *See* MOULT.

ecliptic. The plane of the Earth's orbit about the Sun.

eclogite. A fairly coarse-grained metamorphic rock (*see* METAMORPHISM) with the chemical composition of basic IGNEOUS rock, but with essential magnesium-rich GARNET and sodium-bearing PYROXENE, which are indicative of crystallization or recrystallization at high pressure and temperature. Eclogites are found as XENOLITHS in BLUE GROUND and in some metamorphic belts.

ecocatastrophe. A disaster threatening the quality of life of a community or population, or even leading to many human deaths, caused by excessive environmental damage.

ecocline. A CLINE, or gradient of ECOSYSTEMS, associated with an environmental gradient. *Compare* GEOCLINE.

ecodeme. *See* -DEME, ECOTYPE.

ecological balance (balance of nature). The condition of equilibrium among the components of a natural COMMUNITY such that their relative numbers remain fairly constant and their ECOSYSTEM is stable. Gradual readjustments to the composition of a balanced community take place continually in response to natural ecological SUCCESSION and to alterations in climatic and other influences. By removing or introducing plants or animals, by polluting the environment, by destroying habitats and by rapidly increasing their own numbers, humans can cause major changes, some of which may be irreversible.

ecological capacity. *See* CARRYING CAPACITY.

ecological factor. Any environmental factor that influences living organisms.

ecology. The study of the relationships among living organisms and between those organisms and their non-living

environment. *See also* AUTECOLOGY, SYNECOLOGY.

economic conservation. The management of natural resources, or the environment, so as to sustain a regular yield of a commodity at the highest level feasible.

economic development. The historical process whereby a country changes its economic base from one relying mainly on agriculture and the provision of raw materials, to the industrial processing of materials, to the provision of services and, finally, to a reliance on high-technology industries and the obtaining and disseminating of information. The distinction between developing and developed economies is arbitrary, but in a developing economy primary industries (agriculture, forestry, mining, etc.) are likely to provide 50 percent or more of all employment and up to 70 percent of export earnings. In a developed economy primary industries are likely to provide 20 percent of all employment or less, and a relatively small proportion of export earnings. *See also* DEMOGRAPHIC TRANSITION, PRIMARY ECONOMY, QUATERNARY ECONOMY, SECONDARY ECONOMY, TERTIARY ECONOMY, THIRD WORLD.

economic efficiency. The relationship between the monetary cost of attaining stipulated ECONOMIC ENDS and the monetary value of those ends, often measured as the cost per unit output. Provided product quality is not sacrificed, the lower the unit cost the greater the efficiency. The concept is useful, but has been much criticized for undervaluing social and environmental costs to which it may be difficult to attach a monetary value (e.g., increasing unemployment as unit labour costs are reduced, environmental pollution, etc.).

economic ends. The objectives of economic activity, both quantitative and qualitative. Economics is not concerned with the nature of the ends, but only with their number and relative importance.

economic entomology. The study of insects with particular reference to pests of agricultural crops and the control of their populations.

economic growth. The annual rate of change in the GROSS NATIONAL PRODUCT, usually expressed as the percentage change from the previous year, or as an index, with the value 100 being allotted to a particular past year.

economizer. A device for transferring heat from FLUE GASES to boiler feed water, thus increasing the efficiency of the heating system.

ecoparasite. *See* PARASITISM.

ecospecies. One or more ECOTYPES in a single coenospecies.

ecosphere. The biosphere, together with all the ECOLOGICAL FACTORS that act upon organisms.

ecosystem. A community of interdependent organisms together with the environment they inhabit and with which they interact, and which is distinct from adjacent communities and environments. For example, a pond with the species inhabiting it is distinct from the surrounding land, and an oakwood from, say, surrounding farmed land (the farmed land forming another ecosystem). An ecosystem may include humans, but if the living community and its environment are very large (e.g., the savannah grassland of Africa with the plants and animals, including humans, living on it) it is more usefully regarded as a BIOME.

ecotone. A transitional zone between two ECOSYSTEMS (e.g., where woodland borders grassland the edge of the wood and the edge of the grassland). Ecotones typically support species derived from the ecosystems bordering them as well as species found only in the ecotone, so they tend to be richer in species than the adjacent ecosystems.

ecotype (ecodeme, ecospecies). A subspecific group (regarded by some as a distinct species) that is adapted genetically to a particular habitat, but which can interbreed with other ecotypes of the same species or COENOSPECIES without loss of fertility.

ectoblast. *see* GERM LAYERS.

ectoderm. *See* GERM LAYERS.

ectoparasite. *See* PARASITISM.

ectoplasm. (1) (ectoplast, plasmalemma) In a plant, the external PLASMA MEMBRANE lying just outside the cell wall. (2) In an animal, the cell CORTEX; the outer layer of CYTOPLASM which in many cells (e.g. ova, PROTOZOA) is semi-solid (a GEL) and relatively free of granules and ORGANELLES. *Compare* ENDOPLASM.

ectoplast. *See* ECTOPLASM.

Ectoprocta. *See* POLYZOA.

ectotrophic. *See* MYCORRHIZA.

ecumenopolis. The ultimate city, occupying most of the Earth's land surface and accommodating all of its population, that would result from the continued growth of the human population. The concept was enunciated by C.A. Doxiadis, who defined 15 spatial units, starting with the human individual, proceeding to the large city and thence to the metropolis, conurbation, megalopolis and finally ecumenopolis.

edaphic factors. Those chemical, physical and biological characteristics of the soil which affect an ECOSYSTEM.

eddy. A current in a fluid that moves in a direction contrary to that of the main stream, often with a rotary motion.

eddy diffusion. The movement of a bulk quantity of one substance through another, to give mixing with local variations of concentration. It is the most important mixing process in the atmosphere. *Compare* MOLECULAR DIFFUSION. *See also* DIFFUSION.

Edentata (edentates; South American anteaters, sloths, armadillos). An order of PLA-

CENTAL MAMMALS whose teeth are much reduced or absent. The name was used formerly to include the African aardvark and the pangolins (scaly anteaters) of Africa and Asia, but similarities between these and the South American animals are only superficial, and they are not closely related, being now placed in the orders PHOLIDOTA and TUBULIDENTATA.

EEB. *See* EUROPEAN ENVIRONMENTAL BUREAU.

EEC. *See* EUROPEAN ECONOMIC COMMUNITY.

eel-grass. *See* ZOSTERACEAE.

eelworms. Free-living and plant parasitic NEMATODA (e.g., potato root eelworm, sugar-beet eelworm).

EEZ. *See* EXCLUSIVE ECONOMIC ZONE.

effective. Applied to quantities (e.g., effective sound pressure) to mean ROOT-MEAN-SQUARE VALUE.

effective height of emission. The height above ground at which rising waste gases are estimated to spread horizontally. This is higher than the top of the chimney, owing to the upward momentum and buoyancy of the gases.

effector. An animal organ or cell ORGANELLE that carries out movement (e.g., muscles, CILIA), secretion (e.g., glands) or other actions (e.g., CHROMATOPHORES, NEMATOBLASTS) in response to stimuli.

effluent. Generally, any fluid emitted by a source. More specifically, a waste fluid (usually liquid) produced by an agricultural or industrial process.

effluent charge. A charge levied against a pollutor for each unit of EFFLUENT discharged into public water. The charge may be general or variable, according to the nature of the waste being discharged and the absorptive capacity of the receiving water, and may be levied at all times or only when deteriorating

conditions are deemed to warrant it.

effluent standard. The maximum amount of a specified pollutant an EFFLUENT is permitted to contain. In the case of gaseous discharges, effluent standards are usually known as emission standards. *See also* ENVIRONMENTAL QUALITY STANDARDS.

egg membrane. The membrane that surrounds the ovum of an animal. For example, in a bird's egg, the film membrane surrounding the yolk (ovum) is the vitelline membrane, secreted by the ovum itself. The shell and the membrane immediately inside it are also egg membranes, secreted by the oviduct. In insects, the CHORION surrounding the eggs is a membrane secreted by the ovary.

egocentric. Applied to behaviour that favours the survival of the individual exhibiting it. *Compare* ALTRUISTIC.

EGR. *See* EXHAUST GAS RECIRCULATION.

EIA. *See* ENVIRONMENTAL IMPACT ASSESSMENT.

Eichhornia. *See* WATER HYACINTH.

Eifelian. The fourth oldest stage of the DEVONIAN System in Europe.

einkanter. A pebble that has been abraded by wind-blown sand (i.e. VENTIFACT) to produce one facet, at right angles to the prevailing wind. *Compare* DREIKANTER.

einkorn. A primitive wheat, first domesticated in the Near East and south-western Asia, probably about 11 000 years ago. Domesticated diploid (*see* CHROMOSOMES) einkorn (*Triticum monococcum*) was derived from the wild *T. boeoticum*, which occurs in several forms. *See also* GRAMINEAE.

einsteinium. *See* ACTINIDES.

EIS. *See* ENVIRONMENTAL IMPACT STATEMENT.

ekistics. The study of human settlements. It involves research into and knowledge of architecture, engineering, town planning, sociology, etc.

Ekman spiral. The spiral traced out by the velocity vector in the ocean as depth increases or in the atmosphere as height increases. It is caused by the interaction of the drag force between the air and ocean or ground, the CORIOLIS FORCE, and the shear stress (*see* WIND SHEAR) between layers of air or water.

Ekofisk Bravo 14. An oil platform in the Norwegian sector of the North Sea that blew out on 22 April 1977, releasing 20 000 tonnes of oil before it was sealed on 30 April. The oil dispersed before reaching any coastline, and there was little pollution.

elaioplast. A PLASTID in which fat is stored.

Elasmobranchii. *See* CHONDRICHTHYES.

elastic limit. The maximum stress that can be obtained in a structural material without causing permanent deformation.

elastic pavement. *See* PAVEMENT.

elbow of capture. *See* RIVER CAPTURE.

ELC. *See* ENVIRONMENT LIAISON CENTER.

El Chichón. A volcano in Mexico that erupted in March 1982 and continued to emit ash until May. The cloud penetrated the STRATOSPHERE and spread around the tropics, reducing insolation by 5–10 percent, then spread north and south into temperate latitudes. It produced a small worldwide climatic cooling.

electric field. In the atmosphere, a field that in fine weather has a typical value of about 200 volts per metre. There are great variations in the neighbourhood of thunderstorms, and the field is reduced by CONVECTION and rain.

electrodeposition bath. *See* ELECTROREFINING.

electrometer. An instrument used to measure the atmospheric electrical field. The first electrometer was probably the one used in 1766 by Horace Benedict de Saussure, who measured the change in potential with time between a conductor and the surface above which it was raised by a measured distance. Modern instruments also measure the potential in free air, between two balloons at different altitudes and may accelerate the measurement by using a small amount of radioactive material to ionize (*see* IONIZATION) the air immediately surrounding the instrument (then called a collector).

electron capture detector. A scientific instrument, invented by James E. Lovelock, that detects the presence of chemical substances in the atmosphere with great sensitivity. It detected the presence of ORGANOCHLORINE insecticides, providing Rachel Carson with the information she needed to write *Silent Spring*, as well as CHLOROFLUORCARBONS and other pollutants.

electrorefining. A process for removing impurities from a metal by making the crude metal the anode in an electrodeposition bath, where the required pure metal is deposited on the cathode.

electrostatic field. A region in which a stationary, electrically charged particle would be subjected to a force of attraction or repulsion as a result of the presence of another stationary electric charge.

electrostatic filter. A device in which the application of a static electric charge to a filter improves the efficiency with which it collects small particles.

electrostatic precipitator. A device that separates particles from a gas stream by passing the carrier gas between two electrodes across which a high voltage is applied. The particles pass through the electric field, become charged and migrate to the oppositely charged electrode. Electrostatic precipitators are very efficient collectors of extremely small particles and are used widely in the cement industry, etc.

elements. *See* PERIODIC TABLE OF ELEMENTS.

elephantiasis. *See* FILARIA.

elm bark beetle. *See* DUTCH ELM DISEASE.

Elmo Bumpy Torus. An experimental device for research into NUCLEAR FUSION that uses a magnetic container of a different shape from that in the TOKAMAK or conventional torus (doughnut-shaped) design.

El Niño Southern Oscillation (ENSO). An occasional climatic phenomenon that often begins in December (*El Niño* means the Christ child) and affects the west coast of tropical South America. Reduction in the south-east trade-winds allows warm surface water to flow eastward across the Pacific, then southward along the South American coast, inshore of the cold, northward-flowing Peru Current. It prevents the upwelling of nutrient-rich cold bottom water and so reduces biological productivity. Similar phenomena may also affect the South Atlantic and Indian Oceans.

elutriation. The separation of lighter from heavier particles by means of a stream of fluid. In separating powders, a stream of air may be used, directed upwards. The grain sizes of sediments can be separated and analyzed mechanically by passing them through currents of water of different velocities. The principle is also applied to the separation of lighter from heavier material in domestic refuse.

eluviation. The physical transport of insoluble soil particles (usually CLAY MINERALS) in water from upper to lower SOIL HORIZONS. The similar movement of soluble salts and minerals is called leaching.

elvan. Among miners and quarrymen in Cornwall, UK, a MICROGRANITE. Elvans are often used as aggregate and building stone.

elytra. *See* COLEOPTERA.

emagram. A thermodynamic diagram, having temperature and the logarithm of pressure as its coordinates, which is used for plotting and analyzing atmospheric soundings of temperature and humidity.

Embioptera (embiids). A small order of insects (division: EXOPTERYGOTA) that have soft, flattened bodies. All females and some males are wingless. They produce silk from glands in the front legs and make webs and tunnels in crevices in the soil, beneath bark, etc., where they live, often in small groups.

embryo. An organism in the process of developing from a fertilized or parthenogenetically activated (*see* PARTHENOGENESIS) ovum. In animals, the embryonic stage terminates with birth or hatching from the embryonic membranes. In seed plants, a well-developed embryo consists of a bud (PLUMULE), a root (RADICLE) and one or more COTYLEDONS. The embryonic stage terminates at the germination of the seed.

embryo sac. The megaspore (*see* SPORE) of a flowering plant, contained within the OVULE. The embryo sac is a large cell, containing several nuclei when it is mature. One of these is the egg nucleus, another the primary endosperm nucleus. At fertilization, one male nucleus from the pollen grain fuses with each of them producing a zygote (which develops into an EMBRYO) and an endosperm nucleus (which gives rise to the endosperm). The contents of the mature embryo sac represent the female gametophyte (*see* ALTERNATION OF GENERATIONS).

emerald. *See* BERYL.

emery. A naturally occurring abrasive composed of a mixture of the CORUNDUM, MAGNETITE and SPINEL. Emery occurs in some regionally metamorphosed rocks (*see* METAMORPHISM). It is used as an abrasive and in non-skid road surfaces and non-slip paints.

emission. *See* IMMISSION.

emission standards. *See* EFFLUENT STANDARD.

emmer. A primitive wheat, first domesticated in the Near East and south-west Asia about 11 000 years ago. Domesticated emmer (*Triticum dicoccum*) was probably derived from wild emmer (*T. dicoccoides*), which in turn was derived from the crossing of wild EINKORN (*T. monococcum*) and a species of goat-grass (*Aegilops*). *See also* GRAMINEAE.

emphysema. *See* BRONCHIAL DISEASES.

Emsian. The third oldest stage of the DEVONIAN System in Europe.

enamel. (1) The smooth, hard outer layer of the crown of the tooth of a vertebrate or of the exposed part of the scales (denticles) of present-day cartilagenous fishes and some fossil fishes. Enamel consists largely of crystals of a calcium phosphate–carbonate salt. (2) A glass, often coloured, applied to a metal surface by melting on to it, at high temperature, a powdered substance mixed with a FLUX, which makes it bond to the metal. The enamel protects the metal from CORROSION and adds decoration. (3) To apply an enamel finish to a metal surface.

enation. An outgrowth on a leaf caused by a local multiplication of cells in response to an infection or damage.

endangered. *See* RARITY.

Endangered Species Act, 1983. The US federal law that protects designated species and their habitats, requiring any development to be halted if it can be shown to endanger a species or subspecies throughout all or a significant part of its range.

endemic. (1) Confined to a given region and having originated there. (2) Of pests or organisms that cause disease, occurring continuously in a given area.

endergonic. Applied to a chemical reaction that requires an input of energy (e.g., PHOTOSYNTHESIS). *Compare* EXERGONIC .

end moraine. *See* MORAINE.

endoblast. *See* GERM LAYERS.

endocarp. *See* PERICARP.

endocrine organ (ductless gland). A gland (e.g., adrenal, pituitary) that produces a HORMONE.

endocytosis. The ingestion of particles or fluid by a cell. *See also* PHAGOCYTOSIS, PINOCYTOSIS

endoderm. *See* GERM LAYERS.

endodermis. The innermost layer of the CORTEX of roots and of the stems of some higher plants. Typically, the cells of the endodermis each have a complete band of waterproofing suberin (*see* SUBERIZATION) and LIGNIN (the casparian strip) laid down in radial and transverse walls.

endogamy (inbreeding). Sexual reproduction between closely related individuals (i.e. individuals that have at least some genes in common). If continued intensively this leads to pure lines. *Compare* EXOGAMY.

endogenous. Growing inside a plant or animal. *Compare* EXOGENOUS.

endometrium. The glandular lining of the uterus of a mammal. It undergoes cyclical thickening and regression during the OESTROUS CYCLE.

endomitosis. A doubling of the chromosome number of a cell without subsequent cell division. This may occur repeatedly, giving large multiples of the original chromosome number. Endomitosis occurs especially in flowering plants and insect tissues. *See also* ALLOTETRAPLOID, AUTOPOLYPLOID, POLYENERGID, POLYPLOID.

endoparasite. *See* PARASITISM.

endoplasm. (1) (endoplast) All the CYTO-PLASM except for the internal PLASMA MEMBRANE of a plant cell. (2) The inner part of the cytoplasm that, in many cells (e.g. PROTOZOA, ova), is more fluid than the outer layer and contains many granules and ORGANELLES. *Compare* ECTOPLASM.

endoplasmic reticulum (ergatoplasm). A complex network of membrane-bound channels that ramify through the CYTOPLASM and probably form a conducting system within the cell. By means of electron microscopy it has been shown that the endoplasmic reticulum may join up with the nuclear membrane and the GOLGI APPARATUS and that its membranes are often lined with RIBOSOMES.

endoplast. *See* ENDOPLASM.

Endoprocta. *See* POLYZOA.

Endopterygota (Holometabola). A large division of winged insects (subclass: PTERYGOTA) in which the immature stages (larvae) are very different from the adults (e.g., houseflies, butterflies, beetles). The transformation of larva into adult (i.e. META-MORPHOSIS) is a drastic remodelling which occurs at the pupal (*see* PUPA) stage. *Compare* EXOPTERYGOTA.

endoskeleton. A skeleton that lies inside the body. Vertebrates have a bony endo-skeleton. *Compare* EXOSKELETON.

endosperm. The food-storing tissue outside the EMBRYO in a seed. In endospermic seeds (e.g., wheat, pine, castor oil) some of the endosperm remains until germination. In non-endospermic (exalbuminous) seeds (e.g., pea, bean) the endosperm is absorbed by the embryo before the latter is fully developed.

endosulfan (thiodan). An ORGANOCHLORINE insecticide and acaricide used to control mites and insects pests such as aphids. Because of its side effects, use of this chemical is now restricted in many countries, including the UK.

endothelium. A single-layered sheet of flattened cells that lines the blood and lymph vessels and the hearts in vertebrates. Endothelium is derived from the mesoderm (*see* GERM LAYERS). *Compare* EPITHELIUM.

endothermic. Applied to a chemical reaction, process or exchange in which heat is absorbed. *Compare* EXOTHERMIC.

endotrophic. *See* MYCORRHIZA.

endozoochore. A seed, spore, etc. that is disseminated by being carried within the body of an animal.

endrin. A persistent CYCLODIENE INSECTICIDE that is much more toxic to vertebrates than other ORGANOCHLORINE compounds (e.g., DDT). Endrin is no longer approved for agricultural use in the UK.

energy budget. A record of the flow of energy through a system. Applied originally to ECOSYSTEMS, energy budgeting is now applied to industrial processes as a means of measuring the efficiency with which energy is used and as a complement to conventional financial budgeting.

energy flow. The passage of energy through the trophic levels of a FOOD CHAIN. Energy, almost all of it from sunlight, is trapped by the AUTOTROPHIC organisms of the first trophic level. Because much energy is dissipated during respiration, about 90 percent of the available chemical energy is lost each time energy is transferred from one trophic level to the next higher one.

Engel's law. *See* INCOME ELASTICITY OF DEMAND.

enology (oenology). The study of the production, history and use of wines and of the cultural associations with wine drinking.

ENSO. *See* EL NIÑO SOUTHERN OSCILLATION.

enteric bacteria. Bacteria that inhabit the human gut, including those that may cause disease. *See also* COLIFORM BACTERIA, PATHOGEN.

Enteropneusta. *See* HEMICHORDATA.

enthalpy. The heat content of a body or system, usually given by the formula

$$H = U + pV,$$

where H is the heat content, U the internal energy, p the pressure and V the volume.

Entisols. *See* SOIL CLASSIFICATION.

entoblast. *See* GERM LAYERS.

entoderm. *See* GERM LAYERS.

entomogenous. Growing parasitically on insects (e.g., fungi).

entomophily. Pollination by insects.

entropy. A measure of the degree of disorder within a SYSTEM. The term is derived from thermodynamics; heat passes from warmer to cooler bodies, thus becoming dispersed more generally with time, so that within a closed system eventually all heat will be distributed evenly. Since organization represents the local concentration of energy, its even distribution represents disorganization. As order decreases, entropy increases. This is summed up in the second law of thermodynamics (*see* THERMO-DYNAMICS, LAWS OF).

environment. (1) The physical, chemical and biotic conditions surrounding a living organism. (2) *See* INTERNAL ENVIRONMENT.

Environment, Department of. *See* DE-PARTMENT OF THE ENVIRONMENT.

Environmental Defense Fund. A US coalition of lawyers and environmentalists formed to monitor environmental issues and, when necessary, to take legal action in protection of the environment.

environmental disasters. The table (see page opposite) below lists the major pollu-tion and other environmental incidents that have occurred in recent years. They are arranged alphabetically with dates and brief details. Each incident is described more fully under its own entry.

environmental forecasting. The technique of predicting the environmental consequences of proposed developments, especially as these may impinge on public health or welfare.

environmental geology. The application of geological data and principles to the solution of problems created by human occupancy or other activity (e.g., the geological assessment of the effects of mineral extraction, the construction of SEPTIC TANKS, EROSION of land surfaces, etc.).

environmental impact assessment (EIA). The identification and evaluation of the environmental consequences of a proposed development and of the measures intended to minimize adverse effects. The EIA was introduced first in the USA, where it is a legal requirement of development permissions in most states, and versions of it, adapted to national legal conditions, have been adopted in many other countries. Impacts are identified and listed according to their significance. The technique is largely subjective, since not all impacts can be predicted, not all cause-and-effect relationships are understood, and decisions on the relative importance of each impact are a matter of opinion. Thus the system invites manipulation, and claims for its scientific objectivity are dubious. *See also* ENVIRONMENTAL IMPACT STATEMENT.

environmental impact statement (EIS). A written report, based on detailed studies carried out in the course of an ENVIRON-MENTAL IMPACT ASSESSMENT, that describes the environmental consequences of a course of action in order to assist in decision-making. In many countries organizations planning new projects are required by law to conduct such studies and produce an EIS, which can then be examined critically, sometimes in public.

Incident	Year	Place	Details
Amoco Cadiz	1978	French coast	Oil spill, severe pollution
Bhopal	1984	Bhopal, India	Industrial accident, 2500 people killed, 200 000 injured
Castillo de Bellver	1983	Cape of Good Hope, South Africa	Oil tanker fire, soot contamination on shore
Cavtat	1974–8	Adriatic	Ship sank with a cargo of tetraethyllead that later leaked
Chernobyl	1986	Ukraine	Explosion at a nuclear reactor, 31 people killed, widespread contamination
Donora	1948	Pennsylvania	Air pollution, 18 deaths, 5900 people ill
Ekofisk Bravo 14	1977	North Sea	Blow-out on an oil platform releasing oil, little coastal pollution
Esso Bernicia	1979	Shetland	Tanker collided with jetty, severe contamination, bird deaths and harm to sheep
Irene Serenade	1980	Greece	Oil tanker sank, severe coastal pollution
Ixtoc 1	1979	Mexico	Oil well blew out, released 3 million barrels of oil
London	1952, 1962	London	Smog lasting 4 days in 1952, 5 days in 1962, 4000 people died in 1952, 700 in 1962
Love Canal	1978	Niagara, New York	Leak from industrial waste dump, more than 200 families evacuated
Meuse Valley	1930	France	Air pollution, 60 people died, hundreds ill, many cattle slaughtered
Minamata	1953–60	Japan	Mercury poisoning due to industrial pollution, 43 deaths, many injuries, teratogenic consequences
Poza Rica	1950	Mexico	Air pollution by hydrogen sulphide, 22 deaths, 320 people made ill
Sangana	1980	Nigeria	Oil well blew out into Niger River, contaminating water supplies and fish
Seveso	1976	Italy	Factory explosion releasing dioxin, 700 people evacuated, livestock destroyed, crops burnt
Tanio	1980	English Channel	Oil tanker sank, broke in half and leaked a large amount of oil, causing severe pollution to beaches in Britanny
Three Mile Island	1979	Pennsylvania, USA	Nuclear reactor failed, small radioactive release, no injuries to the public
Times Beach	1982	Missouri, USA	Dioxin contamination from wastes, town evacuated and declared unfit for human habitation
Torrey Canyon	1967	Scillies, UK	Oil tanker sank, world's first major oil spill, contaminated beaches
Vaiont	1963	Italy	Rockslide filled a reservoir, 3000 people killed

environmental protection. That part of resource management which is concerned with the discharge into the environment of substances that might be harmful, or that might have harmful physical effects (e.g., noise, the release of radiation), and with safeguarding BENEFICIAL USES.

Environmental Protection Agency (EPA). The federal agency of the US government, established in 1970, that is responsible for dealing with the pollution of air and water by solid waste, pesticides and radiation, and with nuisances caused by noise.

environmental quality standards. The maximum limits or concentrations of pollutants that are permitted in specific media (e.g., air or water). In the USA, standards are based on estimates of those maxima which, with an allowance for safety, present no hazard to human health (primary standards) or to public welfare (secondary standards), and may take the form of EFFLUENT STANDARDS or relate to the content of products (e.g., additives or pesticide residues in food, phosphates in detergents).

environmental resistance. The restriction of population growth by the interaction of ECOLOGICAL FACTORS.

Environment Liaison Center (ELC). An organization formed under the auspices of the UNITED NATIONS ENVIRONMENT PROGRAMME and based in Nairobi. It acts as a clearing house for information among more than 3000 non-governmental environmental organizations.

enzyme (zymoprotein). An organic CATALYST. Enzymes are unstable PROTEINS or protein-containing compounds and are usually inactivated by heat. They control the many chemical reactions of metabolism, each enzyme being responsible for only one or a very limited range of reactions. An enzyme acts by combining with a specific substance (substrate) and activating it very rapidly, so that it undergoes a chemical change, simultaneously losing its combination with the enzyme. Enzymes each work best at a spe-

cific pH, and some are dependent for their action on the presence of COENZYMES. *See also* AMYLASES, ATPASE, DEHYDROGENASES, DIGESTION, OXIDASES, PROTEOLYTIC ENZYMES.

Eocene. A subdivision of the CENOZOIC Era, usually ranked as an EPOCH, following the PALAEOCENE and lasting from about 54 to 38 Ma. Some authors do not recognize the Palaeocene and consider the Eocene to follow the CRETACEOUS directly, and thus to last from about 65 to 38 Ma. Eocene also refers to rocks deposited during this time, which are called the Eocene Series.

eoclimax. (1). The peak during a period of dominance (*see* DOMINANT) of a fossil plant group. (2) The climax of the EOCENE Epoch.

eolian deposit. *See* AEOLIAN DEPOSIT.

eon (aeon). (1) All PHANEROZOIC time, or all Cryptozoic time (*see* PRECAMBRIAN). (2) In North American usage, one billion (i.e. 10^9) years.

eosere. The development of vegetation during an EON or ERA. A major development series within the CLIMATIC CLIMAX of a geological period.

EPA. *See* ENVIRONMENTAL PROTECTION AGENCY.

epeirogenesis. *See* DIASTROPHISM.

Ephedra. *See* GNETALES.

ephemeral. (1) Applied to any organism or phenomenon that is short-lived. (2) A plant that completes more than one life cycle, from seed to seed, in one year (e.g., groundsel).

Ephemeroptera (mayflies). An order of insects (division: EXOPTERYGOTA) in which the aquatic NYMPHS are long-lived (several years, their rate of development being governed by ECOLOGICAL FACTORS), whereas the non-feeding adults are short-lived. The adults have membranous wings and long tail filaments. A winged, subimago stage (the dun) occurs before the final moult yields the adult. All the stages are eaten in large numbers by freshwater fishes.

epiblast. *See* GERM LAYERS.

epicarp. *See* PERICARP

epicentre. *See* EARTHQUAKE.

epideictic display. A behaviour pattern used by an animal or group of animals to mark out a TERRITORY.

epidemiology. The study of epidemics and the patterns of the incidence of diseases.

epidermis. The outer layer of cells of an animal or plant. In many invertebrates and plants the epidermis is one cell thick and secretes a CUTICLE. In vertebrates, the epidermis is many cells thick, and in land-dwelling forms the outer layers are dead and KERATINIZED, protecting the body against excessive loss of water. *See also* EPITHELIUM.

epigamic character. An animal character, other than one associated with the reproductive organs, that is concerned with sexual reproduction (e.g., bird song, the distinctive coloration of many male birds and fish).

epigeal. Above ground. (1) Applied to the type of germination in which the COTYLEDONS grow out above ground (e.g., sycamore). *Compare* HYPOGEAL. (2) Applied to animals that live above ground.

epigene. Applied to processes that operate in and on the ground.

epilimnion. The warmer uppermost layer of water, which lies above the THERMOCLINE in a lake, and which is subject to disturbance by wind. *Compare* HYPOLIMNION. *See also* THERMAL STRATIFICATION.

epiorganism (superorganism, supraorganism). An entity (e.g., a bee colony or stand of vegetation) that is made up of a group of individual organisms.

Epipalaeolithic. *See* MESOLITHIC.

epipedon. The organic and leached (*see* LEACHING) upper layer of the soil. *See also* PEDON, SOIL HORIZONS.

epipelagic. Applied to organisms that live in the surface waters of the sea, down to about 100 metres depth (i.e. within the PHOTIC ZONE), this zone merging with the MESOPELAGIC zone below. *Compare* ABYSSOPELAGIC, BATHYPELAGIC.

epiphyte. A plant that grows on the outside of another plant, using it for support only and not as a source of nutrients. An example is lichen on trees.

epistatic gene. A GENE whose presence prevents the expression of another non-allelic gene. *Compare* HYPOSTATIC GENE. *See also* ALLELOMORPH.

epithelium. (1) In a plant, a layer of secretory cells that surrounds an intercellular cavity or canal (e.g., resin canal in coniferous plants). (2) In an animal, a sheet of cells that covers any free surface (e.g., the outer body surface, lining of the body cavity, gut lining, etc.). Epithelia may be one or many cells thick and may be ciliated (*see* CILIA) (e.g., lining of the respiratory passages in vertebrates). The EPIDERMIS of vertebrates is stratified epithelium, the lowest layer actively dividing and giving rise to more superficial layers, the outer layers consisting of dead flattened cells. *Compare* ENDOTHELIUM.

epithermal. Applied to ORE deposits that have formed from an ascending, essentially aqueous solution within about 1000 metres of the surface and in the approximate temperature range of 50–200°C. Most such ores occur as VEIN fillings, STOCKWORK, pipes and irregularly branching fissures. *Compare* HYPOTHERMAL, MESOTHERMAL.

epizoite. An animal that lives attached to another animal, but is not a parasite (*see* PARASITISM) (e.g., a limpet on a crab shell).

epoch. A unit of geological time, such that

two or more epochs constitute a period. Epochs are subdivided into ages.

equation of state. The equation ($p = R \rho T$) that relates the pressure (p), the gas constant (R), the density (ρ) and the absolute temperature (T) of a gas, and which embodies the gas laws of Charles and Boyle.

equilibrium level. The height at which a parcel (e.g., an AIR PARCEL) experiences no BUOYANCY force in a stably stratified fluid medium.

equilibrium population. *See* STABLE POPULATION.

Equisetales. An order of cone-bearing PTERIDOPHYTA that were numerous in the CARBONIFEROUS, when tree-like forms (e.g., *Calamites*) were present. The few living forms (horsetail, *Equisetum,* is the only genus) are herbaceous plants with furrowed photosynthesizing stems, bearing whorls of branches and scale-like leaves.

Equisetum. See EQUISETALES.

era. The second largest division (after the EON) of geological time. The PALAEOZOIC, MESOZOIC and CENOZOIC are eras and, on some schemes, so is the short QUATERNARY. Eras are divided into periods.

eradication. The complete extinction of a species throughout its RANGE.

erg. (1) A desert region of shifting sand. *Compare* HAMADA, REG. (2) A unit of work or energy, being the work done by a force of 1 dyne through 1 centimetre. In SI units 1 erg $= 10^{-7}$ joules.

ergatoplasm. *See* ENDOPLASMIC RETICULUM.

ergosomes. *See* RIBOSOMES.

ergosterol. A precursor of VITAMIN D, which is present in the skin of some animals, including humans and which is converted to vitamin D by ULTRAVIOLET RADIATION.

ergot (*Claviceps*). A genus of fungi (ASCOMYCETES) that infests the ovaries of cereals and grasses, gradually replacing the grains with black, dense, banana-shaped masses of interwoven hyphae (*see* HYPHA). These drop from the plant in the autumn and remain dormant in the soil throughout the winter. The fungus is very poisonous and was responsible for the holy fire or St Anthony's fire in medieval times. Ergotamine is extracted from it for use as a drug to constrict blood vessels.

ergotamine. *See* ERGOT.

Erica. See ERICACEAE.

Ericaceae. A cosmopolitan family of woody DICOTYLEDONEAE, mainly shrubs, that form ecologically important COMMUNITIES on peaty soil and in swamps. The family includes heaths (*Erica*), ling (*Calluna*), cowberry, bilberry, cranberry, etc. and the strawberry tree (*Arbutus*), the only British member of the family that is not CALCIFUGE. The roots of many species (e.g., heaths) have endotrophic MYCORRHIZAS.

ericfruticeta. *See* ERICILIGNOSA.

ericilignosa (ericifruticeta). Vegetation that is dominated by heaths (*Erica* species).

erosion. The breakdown of solid rock into smaller particles and its removal by wind, water or ice. As WEATHERING, erosion is a natural geological process, but more rapid soil erosion results from poor land-use practices, leading to the loss of fertile TOPSOIL and to the silting of DAMS, lakes, rivers and harbours. There are three classes of erosion by water. (a) Splash erosion occurs when raindrops (with an average speed of about 9.14 metres per second) strike bare soil, causing it to splash, as mud, to flow into spaces in the soil and to turn the upper layer of soil into a structureless, compacted mass that dries with a hard, largely impermeable crust. (b) Surface flow occurs when soil is removed with surface run-off during heavy rain. (c) Channelized flow occurs when a flowing mixture of water and soil cuts a channel, which is then deepened by further scouring. A minor erosion channel is called a rill, a larger channel a gully. Wind erosion may occur on any soil whose surface is dry, unprotected by vegetation (to bind it at root level and shelter the surface) and consists of light particles. The mechanisms include straightforward picking up of dust and soil particles by the airflow and the dislodging or abrasion of surface material by the impact of particles already airborne.

erosion of thermals. The mechanism whereby the diluted exterior of a thermal, rising in a stably stratified AIR MASS, is removed to find its own EQUILIBRIUM LEVEL, while the interior, warmer air continues to rise, itself to be eroded at a higher level.

erratic. A glacially deposited piece of rock that is different in composition to the BEDROCK beneath it. Erratics are used to reconstruct the movements of ice sheets.

error. In statistics, the difference between observed and expected values, usually caused by chance.

ERTS-1. *See* EARTH RESOURCES TECHNOLOGY SATELLITE.

erythrism. A colour variation in animals (especially birds) in which chestnut red replaces black or brown.

erythrocytes (red blood cells). *See* BLOOD CORPUSCLES.

escape. An organism that was formerly cultivated or in captivity and which has established itself in the wild (e.g., the COYPU in East Anglia, UK). *See also* FERAL.

esker. A long, narrow, usually sinuous ridge of sand or gravel that has been deposited by meltwater from a glacier or ice sheet.

essential amino acids. *See* AMINO ACIDS.

essential elements. *See* TRACE ELEMENTS.

essential minerals. The mineral constitu-

ents of a rock that are necessary to its nomenclature. An essential mineral need not be a major constituent. *Compare* ACCESSORY MINERAL.

Esso Bernicia. A British oil tanker which collided with a jetty at the Sullom Voe oil terminal, Shetland Islands, on 1 January 1979, releasing more than 1100 tonnes of heavy fuel oil and causing severe contamination. About 3700 seabirds and 50 sheep were killed, and a further 2000 sheep were badly oiled.

esters. Compounds derived by replacing the hydrogen in an organic acid by an organic RADICAL (e.g., acetic acid (CH_3COOH) becomes the ethyl ester ethyl acetate ($CH_3COOC_2H_5$)). Many esters are liquids with a pleasant smell and are used in flavourings. Many vegetable and animal fats and oils are esters.

estrogens. *See* OESTROGENS.

estrous cycle. *See* OESTROUS CYCLE.

etesian winds. *See* MISTRAL.

ethanal. *See* ACETALDEHYDE.

ethanoic acid. *See* ACETIC ACID.

ethanol (ethyl alcohol, C_2H_5OH). An alcohol traditionally produced by fermentation, but industrially more usually by the hydrolysis of ethene (ethylene, C_2H_4). It is the basis of alcoholic drinks and is used industrially as a solvent. In some countries (e.g., Brazil, USA) it is used as a fuel for vehicles. A mixture of pure ethanol and water is produced by distillation of the end product of yeast fermentation. This can be used as a fuel directly in engines modified to take it. The removal of the water produces a fuel that can be mixed with petrol (gasoline).

ethene. *See* ETHANOL.

Ethiopian Region. (1) The ZOOGEOGRAPHICAL REGION, forming part of ARCTOGEA, and consisting of Africa south of the Sahara and the south-western part of the Arabian peninsula. (2) (floral region) The part of the PALAEOTROPIC REALM that comprises Ethiopia and the south-western tip of the Arabian peninsula.

ethnobotany. The study of the uses to which plants and plant products are put by peoples of different cultures.

ethnocentrism. The evaluation of cultural traits in terms of a particular culture that is assumed, often unconsciously, to be superior (e.g., the view that a culture cannot be considered important unless it produces artifacts similar to those produced by one's own culture).

ethnozoology. The study of the uses to which animals and animal products are put by peoples of different cultures.

ethogram. A complete catalogue of the behaviour of an animal under conditions as natural as possible and of the contexts in which it occurs. *See also* MOTOR PATTERN.

ethology. The study of the behaviour of animals in their natural environment.

ethyl acrylate ($C_8H_8O_2$, $CH_2{:}CHCOOCH_2CH_3$). A foul-smelling, colourless liquid that boils at 101°C. It is sometimes employed in the manufacture of acrylates used in the plastics industry (e.g., in Perspex), when it can contaminate waste gases. It can be removed from gases by carbon filters or by passing the gases through a cooling pond and containing the condensed ethyl acrylate beneath a thick surface layer of foam, but should even a small amount escape it can be smelled for a considerable distance downwind.

ethyl alcohol. *See* ETHANOL.

ethylene. *See* ETHANOL.

ethyne. *See* ACETYLENE.

etiolation. A condition that is produced by growing a green plant in the dark. The plant fails to develop CHLOROPHYLL and so is yellow. It has weak, elongated stems, and its leaves are smaller than normal.

–etum. A suffix used in ecology to indicate a plant community dominated by a particular genus or species (e.g., fagetum, a woodland dominated by *Fagus sylvatica*, the beech).

Eucalyptus. The genus comprising the trees most characteristic of Australia (family: Myrtaceae). It includes gums, stringy bark and iron bark. Many species are now cultivated widely in warm climates. The leaves contain oil glands. Some species attain great size. They grow rapidly, and many species yield valuable timber, oils and gum. *See also* MALLEE SCRUB.

eucaryotic. *See* EUKARYOTIC.

eudominant. A DOMINANT that is more or less peculiar to a particular CLIMAX (e.g., beech or chestnut in their respective COMMUNITIES).

eugenics. The study of genetics based on the view that the evolutionary future of humans may be susceptible to guidance by the selective breeding of persons possessing characteristics that are considered desirable and by preventing the possessors of 'undesirable' characteristics from breeding. This view can be traced back to ancient times, and MALTHUS, THOMAS ROBERT, and DARWIN, CHARLES ROBERT, were influenced by it, but it was propounded more formally by Francis Galton, a cousin of Darwin, in 1869. The English Eugenics Society was founded by Galton in 1907, and the American Eugenics Society in 1905 by Madison Grant, Henry H. Laughlin, Irving Fisher, Fairfield Osborn and Henry Crampton. In the 1920s and 1930s eugenics laws were passed in several European countries and in 27 US states, permitting the sterilization of 'defective' persons. These laws were later repealed, partly over revulsion at their enthusiastic imitation in Nazi Germany, but also because of doubts about the relative contributions to characteristics, especially behavioural ones, of inheritance and environment (the NATURE–NATURE controversy) and regarding definitions of 'desirable' traits.

eugeosyncline. A GEOSYNCLINE with a thick sequence including volcanic rock, developed offshore of a CRATON and its MIOGEOSYNCLINE.

Euglenophyta. A division of unicellular, mostly freshwater algae whose members move by means of flagella (*see* FLAGELLUM) and lack a rigid cell wall. Some forms possess CHLOROPLASTS, others are colourless. Even the photosynthesizing forms require a supply of vitamin B$_{12}$ (*see* COBALAMINE). The group is also regarded as belonging to the Protozoa (Euglenoidina; class: Flagellata), thus they are considered to be animal rather than plant. *Euglena viridis* is often abundant in stagnant water that contains large amounts of nitrogenous organic matter. *See also* EUTROPHICATION.

euhaline. Applied to full-strength seawater or to water of equivalent salinity (i.e. to water containing about 35°/oo salt). Coastal waters have a salinity of about 30°/oo. *Compare* FRESH WATER, MESOHALINE, OLIGOHALINE, POLYHALINE.

eukaryotic (eucaryotic). Applied to cells in which there is a nuclear membrane separating the CYTOPLASM from the nuclear material, and to organisms composed of such cells. The cytoplasm contains the membrane-bounded ORGANELLES (e.g., MITOCHONDRIA, PLASTIDS) and the CHROMOSOMES consist of DNA and protein (NUCLEOPROTEIN). Apart from BACTERIA and CYANOPHYTA, the cells of all organisms are eukaryotic. *Compare* PROKARYOTIC.

Euphausiacea. *See* MALACOSTRACA.

Euphorbia. a genus of plants (family: EUPHORBIACEAE) that include the spurges in temperate regions, but in the tropics include species of trees (12 species have been identified in Brazil), related to the rubber tree (*HEVEA BRASILIENSIS*), whose sap is an emulsion of HYDROCARBONS resembling crude oil,

but lacking sulphur and other ingredients that are emitted as pollutants from the burning of petroleum.

euphotic zone. The upper zone of a sea or lake into which sufficient light can penetrate for active PHOTOSYNTHESIS to take place. This zone can be up to 100 metres deep and is the upper part of the PHOTIC ZONE.

Euphorbiaceae. A family of mainly tropical DICOTYLEDONEAE with unisexual flowers. Many of the plants are XEROPHYTES, often cactus-like and most are poisonous. They include economically important plants (e.g., cassava, *MANIHOT*; rubber trees, *Hevea* and *Sapium*; the castor oil plant, *Ricinus*; *EUPHORBIA*). Most Euphorbiaceae are trees and shrubs, but all the UK species are herbs. These include dog's mercury (*Mercurialis perennis*), frequently the DOMINANT plant of the field layer (*see* LAYERS) in old woodland, and the spurges.

euploid. Applied to an organism whose body cells contain an exact multiple of the haploid number of CHROMOSOMES. Each chromosome is represented the same number of times as the rest, so the organism is genetically balanced. *Compare* ANEUPLOID, NULLISOMIC, POLYPLOID.

Euratom (European Atomic Energy Community). One of the European Communities to which all EEC members belong, established by treaty, and charged with setting down basic standards for the protection of workers and the public from ionizing radiation.

European ash. *See FRAXINUS EXCELSIOR.*

European Atomic Energy Community. *See* EURATOM.

European Economic Community (EEC). A group of initially six, but now of 12 countries (Belgium, Denmark, France, Greece, Ireland, Italy, Luxembourg, Netherlands, Portugal, Spain, UK and West Germany) formed to promote free trade among members, but with wider social and political objectives, including the construction and implementation of coordinated policies for environmental improvement and the conservation of species, habitats and natural resources. *See also* DIRECTIVE.

European Environmental Bureau (EEB). An independent international organization, based in Brussels, consisting of large, non-governmental national conservation or environmentalist bodies from EEC member countries. The Bureau advises the Commission of the European Communities on environmental matters and acts as a clearing house for information between the Commission and its member bodies.

Euro-Siberian Floral Region. The region of the HOLARCTIC REALM that comprises the whole of Europe from southern Scandinavia south to northern Spain and Asia north of the Caspian Sea to northern Japan.

euryhaline. Applied to organisms that are able to tolerate a wide range of saline conditions (i.e. a wide variation of OSMOTIC PRESSURE) in the environment. *Compare* STENOHALINE.

eurythermous. Applied to organisms that are able to tolerate a wide range of temperatures in the environment. *Compare* STENOTHERMOUS.

eurytopic. Applied to organisms whose distribution is widespread. *Compare* STENOTOPIC.

eustatic. Applied to worldwide and simultaneous changes in sea level (e.g., from isostatic adjustments (*see* ISOSTASY) and the melting of ice sheets and glaciers).

Eutheria (placental mammals). A subclass that contains most of the present-day mammals apart from the MARSUPIALIA and MONOTREMATA. *See* ARTIODACTYLA, CARNIVORE, CETACEA, CHIROPTERA, DERMOPTERA, EDENTATA, HYRACOIDEA, INSECTIVORE, LAGOMORPHA, PERISSODACTYLA, PHOLIDOTA, PINNIPEDIA, PRIMATES, PROBOSCIDEA, RODENTIA, SIRENIA, TUBULIDENTATA.

eutrophic (polytrophic). (1) Applied to waters that are rich in plant nutrients and therefore highly productive, the large number of planktonic organisms (*see* PLANKTON) sometimes rendering them cloudy. (2) Applied to lakes in which the lower layer of cold water (i.e. HYPOLIMNION) becomes depleted of oxygen in summer through the decay of organic matter. *Compare* DYSTROPHIC, MESOTROPHIC, OLIGOTROPHIC. *See also* EUTROPHICATION.

eutrophication. The enrichment of a water body (usually of still or slow-flowing fresh water) with plant nutrients (e.g., by the input of organic material or by surface run-off containing NITRATES and PHOSPHATES). Eutrophication may happen naturally, but it is often a form of pollution. It leads to an increase in the growth of aquatic plants and often to algal BLOOMS, which may smother higher plants, reduce light intensity, produce toxins which kill fish and, through the aerobic DECOMPOSITION of organic matter, deoxygenate the water, causing the death of many aquatic animals and higher plants. Eventually the accumulation of organic matter may raise the bed of a lake until it becomes marsh, then dry land. *See also* EUTROPHIC, SAPROBIC CLASSIFICATION.

evaporation. The change of phase from liquid to gas, the energy being supplied as LATENT HEAT from the surrounding medium. The absorption of latent heat accounts for the chill associated with the evaporation of water from a surface and is exploited by animals, which sweat to cool the skin, and by plants, which secrete water at leaf surfaces (*see* EVAPOTRANSPIRATION).

evaporation pond. A body of seawater that is enclosed to allow the water to evaporate and so leave behind a deposit of salt.

evaporimeter. An instrument for measuring the rate of evaporation of water from a surface.

evaporite. A rock composed of minerals that have been precipitated from aqueous solution as a result of the evaporation of the water. Examples include oolitic limestones (*see* OOLITE), GYPSUM, ANHYDRITE, HALITE, SYLVITE and CARNALLITE.

evapotranspiration. The combined EVAPORATION of water from the soil surface and TRANSPIRATION from plants.

evergreen. Applied to a plant that bears leaves throughout the year (e.g., Scots pine, yew, holly, ivy), i.e. plants that shed their leaves, but not at any particular season of the year and not all at the same time. Many evergreens have leaves that remain on the plant for longer than a year, the old leaves persisting until new ones appear. *Compare* DECIDUOUS.

evolution. The process of cumulative change, usually gradual. In living organisms evolution occurs in successive generations and has led to the development of different species and subspecies from a common ancestor. *See also* DARWIN, CHARLES ROBERT.

evolution of the atmosphere. The change that is believed to have occurred and whose most recent outcome is the present atmosphere of the Earth. The present composition of the atmosphere is almost wholly of biological origin and has been fairly stable for 2–3 billion years, having been transformed from a reducing one, containing mainly compounds of carbon and hydrogen, into an oxidizing one, containing mainly nitrogen and oxygen, by the activities of living organisms (mainly microorganisms). The scavenging of carbon (e.g., to form LIMESTONES) has maintained a surface climate varying within only a small temperature range, and biological mechanisms have also controlled TURBIDITY (e.g., natural photochemical SMOG). It is probable that human activity is now perturbing this composition (*see* GREENHOUSE EFFECT). *See also* GAIA HYPOTHESIS.

exalbuminous. *See* ENDOSPERM.

excepted land. In UK planning, as defined by the COUNTRYSIDE COMMISSION, land that is included within an ACCESS AGREEMENT or ACCESS ORDER, but from which the public is excluded.

excitation. A forced variation in pressure, position or some similar quality, whereby energy is added. The excitation of a particle transfers it from its ground state to a higher energy level, the excitation energy being the difference in energy between the two states.

exclusive economic zone (EEZ). A concept proposed and then developed at the United Nations Conference on the Law of the Sea whereby coastal states assume jurisdiction over the exploration and exploitation of marine resources (e.g., fish, seabed minerals, etc.) in their adjacent sections of CONTINENTAL SHELF, defined arbitrarily to be a band extending 320 kilometres (200 miles) from the shore, with median lines agreed between states separated by seas less than this distance wide.

exclusive species. A species that is confined completely, or almost so, to one COMMUNITY.

excretion. The process by which an organism rids itself of the waste products of metabolism. Wastes may be stored (e.g., nitrogenous compounds in the fat body of insects and in the hollow wing scales of some butterflies). The elimination of the carbon dioxide and water which are by-products of respiration is also a form of excretion, but the term is particularly applied to the removal from the body of nitrogenous products (e.g., UREA, AMMONIA, etc.) by the kidneys in vertebrates and by nephridia (*see* NEPHRIDIUM), MALPIGHIAN TUBULES and other organs in invertebrates.

exergonic. Applied to a chemical reaction that gives out energy (e.g., respiration). *Compare* ENDERGONIC.

exfoliation. The separation of scale-like layers from a mineral or from an exposed or soil-covered rock. The process may be due to a variety of causes, including alternate heating and cooling of the rock, alteration of layers of minerals (especially by water) and pressure release as a result of erosion.

SPHAEROIDAL WEATHERING is one type of exfoliation.

exhaust gas recirculation (EGR). A technique to improve the combustion of pollutants normally emitted by an internal combustion engine and thereby reduce pollutant emissions.

Exocoetus. *See* FLYING FISHES.

exodermis. The outer layer of the CORTEX in a non-woody root, composed of cells impregnated with waterproofing suberin, which replaces the piliferous (hairy) layer, behind the root hair region, where suberin withers.

exogamy (outbreeding). Sexual reproduction between individuals that are not closely related, so retaining heterozygosity (*see* HETEROZYGOTE). Crossing between two genetically distinct lines may lead to HYBRID VIGOUR. *Compare* ENDOGAMY.

exogenous. Growing on the outside of a plant or animal. *Compare* ENDOGENOUS.

Exopterygota (Hemimetabola, Heterometabola). A large division of winged insects (subclass: PTERYGOTA) in which the young stages (NYMPHS) are similar to the adults, apart from being sexually immature and having wings that are incompletely developed or absent (e.g., dragonflies, mayflies, grasshoppers, earwigs, cockroaches, etc.). There is no pupal state (*see* PUPA), and the transformation of the nymph into an adult (*see* METAMORPHOSIS) is a gradual process, accomplished by a series of moults. *Compare* ENDOPTERYGOTA.

exoskeleton. A rigid support that covers the outside of an animal's body or lies in the skin. The exoskeleton of ARTHROPODA is a cuticle containing CHITIN, which overlies the EPIDERMIS. Some vertebrates (e.g., armadillo, tortoise) have exoskeletons of bony plates covered with KERATIN and lying on the skin. *Compare* ENDOSKELETON.

exosphere. The outermost layer of the Earth's atmosphere, lying beyond the IONOSPHERE, in which air density is such that a molecule moving directly outward has a 50 percent chance of escaping, rather than colliding with another molecule.

exothermic. Applied to a chemical reaction in which energy, as heat, is released. *Compare* ENDOTHERMIC. *See also* EXERGONIC.

exotic. (1) Applied to a species found in a region to which it is not native. (2) In geology, a synonym for AUTOCHTHONOUS.

explantation. *See* TISSUE CULTURE.

explosion wave. The wave that travels outward from an explosion to accommodate the extra volume produced by the explosion. Explosion waves travel as SHOCK WAVES near the source, then as spherical SOUND WAVES, but at distances beyond a few tens of kilometres they assume the form of external or sonic gravity waves in the atmosphere.

explosive. A substance that undergoes a rapid chemical change on heating or detonation, with the production of great heat and a large volume of gas.

exponential growth (compound interest, geometric growth). The increase in a value over a period by a fixed percentage of the original value, such that the increase in each period is equal to the original value plus the interest accumulated in the preceding period. Thus, although the rate of increase remains constant, the amount of increase is greater in each successive period. The time required to double the original value is called the doubling time and for exponential growth is approximately equal to 70 divided by the percentage rate of growth. For example, a growth (interest) rate of 10 percent will double the value in about seven periods. *Compare* LINEAR GROWTH.

exponential reserve index. *See* STATIC RESERVE INDEX.

exsiccata (exsic). Herbarium material that has been preserved by drying and dispersed to other centres for reference purposes.

externality. In economics, a cost or benefit attributable to an economic activity that is not reflected in the price of the goods or services being produced. Thus damage to the environment may not be counted as a cost (or environmental protection as a benefit) in production. It is an aim of the POLLUTER PAYS PRINCIPLE to require polluters to meet the cost of avoiding pollution or of remedying its effects, so internalizing the externalities.

external respiration. *See* RESPIRATION.

exteroceptor. A sense organ that detects stimuli originating from outside the organism that possesses it. *Compare* INTEROCEPTOR.

extinction. *See* RED DATA BOOK.

extracellular. Applied to processes that occur outside the cells of an organism (e.g., digestion in the rumen of a ruminant's gut, or around the hyphae of fungi).

extraclinal. *See* TOPOTYPE.

extraneous. Closer to the periphery than to the centre.

extrusive. Applied to IGNEOUS rocks formed from MAGMAS which flowed out at the surface of the Earth. *Compare* INTRUSIVE.

eye. A sense organ that detects light. An ocellus is a simple eye found in many invertebrates. A compound eye, found in CRUSTACEA and many insects, is a collection of units (ommatidia) each of which forms an image.

eye of the storm. The centre of a TROPICAL CYCLONE, where the pressure is lower than could be produced by the warming of air from below. Thus the centre is filled with air drawn down from above, and air from the stratosphere descends to sea level, with some mixing and moistening in the lowest few kilometres. This descending air is cloud-free, except sometimes for a layer of cloud at a height of 1 kilometre or less. It occupies the rain-free centre whose width is 10–50 kilometres. The wind at the centre is relatively calm, but surrounding it are the strongest winds and heaviest rain of the cyclone.

eye spot. *See* STIGMA.

F

F. (1) *See* FARAD. (2) *See* FLUORINE.

F₁. *See* FIRST FILIAL GENERATION.

f. *See* FEMTO-.

facies. The sum total of the characteristics of a rock (e.g., its texture, mineral or organic composition, form and structure) from which its environment of deposition and subsequent history can be deduced.

Factories Inspectorate. *See* HEALTH AND SAFETY EXECUTIVE.

facultative parasite. *See* PARASITISM.

faecal *Streptococcus*. A group of bacteria that are normally abundant in the intestinal tracts of warm-blooded animals other than humans and are indicators of the contamination of water by the FAECES of these animals.

faeces (feces). Material voided via the anus from the alimentary canal, consisting mainly of indigestible food residue, along with bacteria. In some animals excretory products are expelled in the faeces. The droppings of birds consist of faeces covered by whitish, semi-solid nitrogenous waste expelled from the urinary system.

fagetum. *See* -ETUM.

Fagus sylvatica (beech). A tall tree (family: Fagaceae) of northern temperate regions that often forms homogeneous forests. It is the characteristic DOMINANT of chalk and soft LIMESTONE in south-east England. Beech bears CATKINS and pairs of NUTS enclosed by prickly, woody, four-valved cupules. Beechwoods have a peculiar ground flora, which includes white helleborine (*Ceph-*

alanthera damasonium) and truffle (*Tuber aestivum*). Beech trees are useful for hedging and yield a hard timber.

fairy shrimps. *See* BRANCHIOPODA.

Falconidae. *See* FALCONIFORMES.

Falconiformes (Raptores). The birds of prey that hunt by day. They have sharp, hooked beaks, strong talons and powerful flight. Many soar on updraughts. The families represented in the British Isles are the Pandionidae (the fish-eating osprey), Falconidae (e.g., peregrine, hobby, merlin, kestrel), all of which have pointed wings and narrow tails, and Accipitridae, which includes eagles (large birds with broad wings and tails), vultures (eagle-like, but mostly with naked heads and necks), buzzards (medium-sized with broad wings and tails), harriers (long, broad wings and long tails), kites (long, angular wings and long, forked tails) and accipiters or birds hawks (e.g., goshawk and sparrowhawk), with short, rounded wings and long tails.

falling pressure. The most familiar prognostication of increasing cloud and the possibility of rain, produced when an upward movement of air, which may lead to cloud formation, produces a convergence of air at lower levels. The convergence increases the rotation of air near the ground with decreasing pressure at the centre.

fallout. The descent of airborne solid or liquid particles to the ground, which occurs when the speed at which they fall due to gravity exceeds that of any upward motion of the air surrounding them. Water falls as precipitation. The extent of particulate air pollution (e.g., by dust, soot, radioactive particles) can be estimated by measuring the mass and rate at which particles accumulate at receptors on the ground.

fallout front. The lower boundary of a region of FALLOUT. When the fallout front from a rain shower reaches the ground the rain begins.

fallstreak. A streak of falling cloud particles, almost always of ice particles which are not evaporating (*see* VIRGA).

fallstreak hole. A hole in a layer of cloud that is composed of supercooled water droplets, caused by the local freezing of some of the droplets and their conversion to FALLOUT, often in FALLSTREAK form, by the BERGERON–FINDEISEN MECHANISM. *See also* BEWSEY'S METHOD.

false bedding. *See* CROSS-BEDDING.

family. *See* CLASSIFICATION.

Fammenian. The youngest stage of the DEVONIAN system in Europe.

fanning. The behaviour of a PLUME when, in stable air, gases reach their EQUILIBRIUM LEVEL quickly and move horizontally, with some meandering, but very little vertical mixing, to produce a thin, but concentrated layer of pollutant that may come into direct contact with hillsides or tall buildings.

FAO. *See* FOOD AND AGRICULTURE ORGANIZATION OF THE UNITED NATIONS.

farad (F). The derived SI unit of capacitance (ability to store an electrical charge), being the capacitance of a capacitor between the plates of which there appears a potential difference of 1 volt when it is charged with 1 coulomb of electricity (i.e. farads are ampere-seconds per volt). The unit is named after Michael Faraday (1791–1867).

farmyard manure (FYM). A mixture of bedding straw and animal excrement that is used, either raw or after COMPOSTING, as a FERTILIZER and soil conditioner in agriculture

and horticulture. FYM is produced when livestock is housed in buildings or in yards, straw is provided as litter and fresh straw is used to cover old straw as this becomes soiled and trampled. FYM provides plant MACRONUTRIENTS and MICRONUTRIENTS and improves the structure of soils.

fast breeder reactor. *See* BREEDER REACTOR.

fat (lipid, lipoid). A compound that is formed from an alcohol with a FATTY ACID. A neutral (true) fat consists of glycerol combined with a fatty acid (e.g., oleic acid in the case of olive oil). The term lipid may be used more widely for any substance (e.g., sterols, steroids, carotene) that can be extracted from tissue by using a fat solvent such as ether, but is not necessarily a compound containing fatty acid(s). Fats are commonly stored as a source of chemical energy in the bodies of animals and in seeds and are also important as carriers of the fat-soluble VITAMINS. Fatty substances occur in suberin and cutin (*see* CUTICLE), and are important constituents of PROTOPLASM (e.g., LECITHIN).

Fata Morgana. A SUPERIOR IMAGE caused by a pressure anomaly, whereby the atmospheric pressure increases with height for a very short distance above the surface of cold water. Under perfect conditions the mirage can transform cliffs or distant houses into great castles, partly in the air and partly beneath the sea. The name refers to the fancied reflection of the submarine palace of Morgan le Fay (Fata Morgana), the fairy sister of King Arthur, in the Strait of Messina.

fat hen. *See* CHENOPODIACEAE.

fatty acid. A carboxylic acid with, attached to the carboxyl group, a sidechain of carbon atoms to which hydrogen atoms are attached: an essential ingredient of neutral FATS. There is usually an odd number of carbon atoms in the sidechain, commonly 15 or 17. If the carbon chain carries as many hydrogen atoms as it is capable of doing the fatty acid is said to be saturated. If there are fewer than the maximum number of hydrogen atoms, the fatty acid is unsaturated. If there are two or more unsaturated sites on the chain the fat is polyunsaturated.

fault. A fracture in the Earth along which there has been displacement parallel to the FAULT PLANE. Faults are classified by the relative motions of the faulted blocks as: (a) dipñslip faults (e.g., NORMAL FAULTS, REVERSE FAULTS and THRUSTS) in which the movement is parallel to the DIP of the fault plane; (b) strike–slip faults (e.g., wrench, tear and TRANSFORM FAULTS) in which movement is parallel to the STRIKE of the fault plane; (c) oblique slip faults in which movement is at an appreciable angle to both dip and strike. The upper surface above a fault is called the hanging wall, the lower surface the footwall. The vertical displacement of a point perpendicular to the fault plane is called the throw, the corresponding horizontal displacement the heave, and the horizontal displacement along the fault plane is called the strike slip.

fault breccia. A rock composed of broken fragments lying along a FAULT PLANE.

fault gouge *See* FAULT PLANE.

fault plane. The surface movement of a FAULT. Material along an active fault plane can become broken off (to form fault breccia) and ground up to a powder (a fault gouge), which can become fused to a MYLONITE. The surfaces can become polished and striated; these striations are called slickensides.

fauna. The animals of a particular region or period of time.

faunal regions. *See* ZOOGEOGRAPHICAL REGIONS.

Fe. *See* IRON.

feces. *See* FAECES.

feedback. In systems theory, the informational response to a cause which may inhibit further repetition of the cause (negative feedback), and therefore has a stabilizing effect on the system. An example is the feeling of satiation, which reduces an animal's motivation to continue eating. Positive feedback, which encourages the cause to continue, tends to produce vicious spirals (e.g., wage–price inflation, whereby higher wages cause prices to rise, so causing wages to rise again).

feeder reservoir. *See* RESERVOIR.

feedlot. An area of small pens in which beef cattle are fattened prior to slaughter. Food, mainly grains, is brought to the animals. The concentration of large quantities of sewage from feedlots presents serious disposal problems. Feedlots are common in the USA, but rare in Europe.

feldspar (felspar).   The most abundant group of rock-forming SILICATE minerals. Feldspars are grouped into plagioclase and alkali feldspars. Plagioclase feldspars range in composition from albite ($NaAlSi_3O_8$) to anorthite ($CaAl_2Si_2O_8$) to orthoclase ($KAlSi_3O_8$). Alkali feldspar is characteristic of the ALKALI IGNEOUS rocks, whereas plagioclase of varying composition is found in igneous rocks ranging from ACID to BASIC. Feldspars are abundant in metamorphic rocks (*see* METAMORPHISM) and in ARKOSES.

feldspathoids. A group of SILICATE minerals, all of which contain sodium and/or potassium and are related to FELDSPAR, but contain relatively less silica. Two of the simpler feldspathoids are nepheline ($NaAlSiO_4$) and leucite ($KAlSi_2O_6$). Feldspathoids occur in ALKALI IGNEOUS rocks, though never in rocks containing QUARTZ.

felsic.   Applied to the light-coloured minerals, FELDSPAR, FELDSPATHOIDS and QUARTZ (silica) in IGNEOUS rocks. *Compare* MAFIC.

felsite.   An IGNEOUS rock which is fine-grained to CRYPTOCRYSTALLINE, of ACID to INTERMEDIATE composition, and which may or may not be PORPHYRITIC. Because of this imprecision, felsite is best used as a convenient field term. Felsites are important sources of AGGREGATE and roadstone.

felspar. *See* FELDSPAR.

femto- (f). The prefix used in conjunction with SI units to denote the unit x 10^{-15}.

fen. An area of waterlogged PEAT which is alkaline or only slightly acid. Typical fen plants include reed (*Phragmites communis*), reed canary grass (*Pholaris arundinaceae*), ragged robin (*Lychnis flos-cuculi*), meadowsweet (*Filipendula ulmaria*), great spearwort (*Ranunculus lingua*) and fen sedge (*Cladium mariscus*), all of them tall herbs. Poor-fen vegetation develops over slightly acid peat and is intermediate between basic (alkaline) fen and bog. The richness of their flora and of the animals (especially insects) it supports means fens are of great scientific and educational interest and value, but they are under constant pressure from demands to drain them because, once drained, they provide highly productive agricultural land. *Compare* BOG.

fenitrothion. An ORGANOPHOSPHORUS PESTICIDE; an insecticide used to control aphids and caterpillars in fruit crops, moths and weevils in peas and leatherjackets in cereals. It is also used to control beetle pests in grain stores. It is poisonous to vertebrates.

fentins. A group of fungicides; organic compounds containing tin. They are used to control diseases such as potato blight, and are poisonous to vertebrates.

feral. Applied to animals that were once domesticated, but have established themselves in the wild (e.g., cat, goat and the rock dove which was formerly the domestic pigeon).

fermentation. The breakdown of organic substances by organisms with the release of energy; especially the ANAEROBIC breakdown of CARBOHYDRATES (e.g., glucose) by YEASTS and BACTERIA to form carbon dioxide and ETHANOL or other organic compounds.

fermium. *See* ACTINIDES.

Fernau glaciation (Little Ice Age). A climatic oscillation that led to a cooling of about 1°C over much of western Europe from about 1590 to 1850, causing the extension of glaciers. The change in climate amounted to much less than an ice age despite its popular name and is named more correctly after the Fernau Glacier, Austria. This glacier ends in the Bunte Moor peat bog, and its advances and retreats are recorded in layers in the bog by the formation of peat, which can be dated radiometrically (*see* RADIOMETRIC DATING) during ice-free periods, and by the deposition of morainic sand during periods of glaciation.

ferns. *See* FILICALES.

Ferrel's law. The law which states that in the northern hemisphere the wind is deflected (by the CORIOLIS FORCE) to the right. The American meteorologist W. Ferrel (d. 1891) deduced this on theoretical grounds some time earlier than Buys Ballot, who later acknowledged Ferrel's prior claim to the discovery.

ferromagnesian mineral. A rock-forming SILICATE mineral that contains essential iron and/or magnesium (e.g., OLIVINE, AUGITE, HORNBLENDE, BIOTITE).

Fertile Crescent. The approximately crescent-shaped area in the Near East, bounded by the rivers Tigris and Euphrates, that was the site of the earliest Near Eastern civilizations and one of the areas in which agriculture was first practised by members of what became HYDRAULIC CIVILIZATIONS. At its height, the Fertile Crescent supported a population density equal to that of much of modern Europe.

fertilization. The fusion of two GAMETES to form a ZYGOTE; the essential process of sexual reproduction, which results in the bringing together of an assortment of GENES from two haploid nuclei.

fertilizer. Any substance that is applied to land as a source of nutrients for plant growth. It may be a waste that is being recycled (e.g.,

FARMYARD MANURE, crop residues or compost) or produced industrially (e.g., compounds of ammonium or phosphorus). The main fertilizer constituents are nitrogen (N), phosphorus (P) and potassium (K) (giving the informal name NPK). Fertilizer values are commonly given as weights or percentages of the total content of N, P or K nutrients. Calcium (Ca), although required by plants and needed to reduce soil acidity, is not usually regarded as a fertilizer. Excessive use of salts that are highly soluble in water (e.g., NITRATES) can cause contamination of water, which may be serious in extreme cases (*see* METHAEMOGLOBINAEMIA), or its over-enrichment (*see* EUTROPHICATION). Excess nitrates may also lodge in the tissues of certain plants, and excesses of any one nutrient may lead to nutritional imbalances in crops. *See also* MACRONUTRIENTS, MICRONUTRIENTS.

fescues. *See* GRAMINEAE.

fetus. *See* FOETUS.

fibre. (1) In a plant, an elongated cell whose walls are thickened with cellulose (see CARBOHYDRATES) or LIGNIN, and which gives mechanical support to the plant. Fibres are extracted from many plants and used in the manufacture of such products as paper, hemp ropes (from *Cannabis sativa*) and linen (from *Linum usitatissimum*). *See also* COLLENCHYMA, SCLERENCHYMA. (2) In an animal, strands of material such as the COLLAGEN fibres found in CONNECTIVE TISSUE, TENDON and CARTILAGE of vertebrates. Fibres are made by irregularly shaped cells called fibroblasts or fibrocytes. *See also* MUSCLE, NEURON.

fibreglass. A manufactured, non-flammable fibre made from glass, which is used for insulation and, bonded with resin, as a strong, light construction material. Fibreglass is resistant to most chemicals and to solvents. Its fibres penetrate skin easily and can have adverse effects on the lungs if inhaled as a dust.

fibroblasts. *See* FIBRE.

fibrocytes. *See* FIBRE.

fibrous root system. A root system without a tap root, consisting of a tuft of roots of fairly equal diameter which bear many smaller, lateral roots (e.g., grasses, groundsel, strawberry). *Compare* TAP ROOT SYSTEM.

Ficus. A genus (family: MORACEAE) of trees, shrubs, climbers and EPIPHYTES of warm regions, whose flowers, lying inside a globular or pear-shaped receptacle, are pollinated by fig-wasps which lay eggs inside the ovaries. *F. carica* is the fig tree. *F. elastica*, the indiarubber tree, is stout, with large buttress roots and usually starts life as an epiphyte. It yields rubber when tapped. *F. benghalensis* is the banyan, which has large aerial roots supporting the branches.

fidelity. The degree to which a system reproduces accurately at its output the characteristics of the signal impressed on its input.

field. (1) The natural surroundings in which organisms or rocks are found and where they may be studied; hence field work. (2) An area of workable mineral riches (e.g., coal field, oil field).

field capacity. The greatest amount of water it is possible for a soil to hold in its pore spaces after excess water has drained away.

field layer. *See* LAYERS.

fig tree. *See* FICUS.

Fijian Floral Region. The part of the PALAEOTROPIC REALM that comprises the islands of Fiji.

filament. *See* FLOWER.

Filaria. A genus of NEMATODA, some of which cause serious diseases in humans. *F. (Wuchereria) bancrofti* lives in the blood and lymph vessels, causing elephantiasis. The worms are transmitted to new hosts by the mosquito *Culex fatigans*, inside which they pass through an essential stage of development. *F. (Dracunculus) medinensis*, the Guinea worm, grows to a length of 1 metre. It lives under the skin and discharges its eggs through ulcers into water. The intermediate host is a small aquatic crustacean, *Cyclops*, which may be swallowed in drinking water.

Filicales (ferns). A large group of PTERIDOPHYTA whose members produce SPORES on the underside of the leaves or on special leaf segments. The leaves are typically large and compound. Most ferns have RHIZOMES, but some (e.g., *Cyathea*, tropical and subtropical tree ferns) have aerial stems several metres high. Most species are terrestrial, but a few (e.g., *Azolla*) are aquatic. *See also* BRACKEN.

fiord. *See* FJORD.

fire algae. *See* PYRRHOPHYTA.

fire climax (pyroclimax). A CLIMAX community in whose development and maintenance fire is an important factor. In parts of the USA, for example, there are extensive stands of longleaf pine trees, whose growing tips are well insulated against heat and whose wood is not readily combustible. Forest fires destroy other species, thereby providing more space and supplying nutrients (as ash) for the pines, which flourish. Repeated at fairly regular intervals, extensive fires tend to suppress trees by destroying them before they become fully established, while encouraging fast-growing grasses and herbs. Thus fires are thought to have played an important part in the development of the major grassland areas of the world (e.g., the prairies of North America, pampas of South America, steppes of Asia, savannah of Africa).

fire storm. A wind storm caused by a large, intensely hot fire in a partly confined space (e.g., at ground level in a city). The heat causes expanding air to rise rapidly, creating an area of very low pressure into which air is drawn. If the air pressure is sufficiently low the centre of the fire will behave much like the centre of a TROPICAL CYCLONE. The inrushing air supplies oxygen to sustain combustion and carries combustible material that is exposed to radiated heat and ignites spontaneously.

firn. *See* NÉVÉ.

first filial generation (F_1). The first generation of offspring produced in a genetical experiment by crossing the parental generation (P_1). Crossing members of the F_1 generation gives the F_2 (second filial) generation.

first trophic level. *See* FOOD CHAIN.

First World. *See* THIRD WORLD.

Fish and Wildlife Service (FWS). The branch of the US Department of the Interior that is concerned with the protection of wildlife and its management as a resource. The FWS is responsible for administering much of the US legislation dealing with wildlife conservation.

fish lice. *See* MALACOSTRACA.

fissile. (1) Capable of being split along closely spaced, parallel planes. (2) Applied to ISOTOPES of elements that are capable of undergoing NUCLEAR FISSION upon impact with a slow neutron.

fission. (1) Asexual reproduction by splitting into two parts. Binary fission (splitting into two) occurs in many unicellular organisms (e.g., BACTERIA, PROTOZOA) and a few multicellular ones (e.g., corals). Multiple fission occurs in SPOROZOA. (2) *See* NUCLEAR FISSION.

fission energy per unit mass of uranium. The energy released by the fission of 1 gram atom (235 g) of uranium-235, which is 12×10^{12} J or 3.3×10^6 kWh. *See also* NUCLEAR FISSION, NUCLEAR REACTOR, URANIUM.

fissure. A fracture in a rock with displacement perpendicular to the break.

fissure eruption. A volcanic eruption in which LAVA or PYROCLASTIC material are extruded from a linear vent.

fitness. The response of a population of organisms to natural selection, measured in terms of the number of offspring produced in relation to the number needed to maintain a constant population size.

fixation. (1) In a soil, any process by which a nutrient essential for plant growth is changed from a soluble form that can be absorbed by plant roots to an unavailable and often insoluble form. (2) *See* NITROGEN FIXATION.

fixative. A chemical substance (e.g., FORMALDEHYDE or ETHANOL) that is used to preserve cells, ideally with as little distortion of them as possible.

fixed action patterns. MOTOR PATTERNS that are extremely stereotyped (e.g., the invariable use of the beak by the greylag goose to retrieve eggs that have rolled from the nest). *See also* ETHOGRAM.

fjord (fiord). A long, steep-sided inlet of the sea in a mountainous region, caused by the over-deepening of a valley by a glacier. Characteristically, fjords have a shallower BAR at their seaward end.

flag. *See* FLAGSTONE.

Flagellata (Mastigophora). A class of PROTOZOA whose adult members swim by means of flagella (*see* FLAGELLUM). They include plant-like forms (e.g., EUGLENOPHYTA), which usually contain CHLOROPLASTS and so are able to photosynthesize, and animal-like forms with HOLOZOIC NUTRITION. Examples include *Trypanosoma* (*see* TRYPANOSOMIDAE), a parasite that causes sleeping sickness in Africa, and *Trichonympha*, a wood-digesting SYMBIONT that lives in the guts of termites. *See also* PYRRHOPHYTA.

flagellum. A thread of CYTOPLASM, capable of lashing movements, that projects from a cell. Flagella are usually few in number and do not move in unison (*compare* CILIA). They are responsible for the motility of many microscopic organisms (e.g., GAMETES, SPORES and BACTERIA), and they also occur in some larger organisms, such as sponges (*see* PORIFERA) and CNIDARIA. Apart from the bacterial flagellum, which consists of a single filamént, the structure of a flagellum resembles that of a cilium. SPIROCHAETES resemble free-living flagella, and some biologists believe that flagella may have originated from a SYMBIOSIS between free-living forms and other cells.

flags of convenience. *See* PANHONLIB GROUP.

flagstone (flag). A SANDSTONE or sandy LIMESTONE that splits along BEDDING PLANES into slabs.

flame photometry. An analytical technique in which a substance is heated strongly in a flame, arc or high-voltage spark, thus exciting its atoms, which emit electromagnetic radiation whose spectrum is diagnostic for particular elements. Some 70 elements can be identified in this way.

flamingo. *See* CICONIIFORMES.

flash colours. Brightly coloured parts of an animal that are exposed suddenly to confuse a predator (e.g., certain patterns on the wings of butterflies and moths).

flat-plate solar collector. *See* SOLAR COLLECTOR.

flatworms. *See* PLATYHELMINTHES.

flax. *See* LINUM.

fleas. *See* APHANIPTERA.

flint. A CRYPTOCRYSTALLINE variety of SILICA with a CONCHOIDAL FRACTURE. Flint and CHERT are often used synonymously, but many geologists reserve flint for the irregular, siliceous nodules found in chalk, where the most abundant organic remains are sponge spicules.

flocculation. A process of contact and adhesion whereby the particles of a dispersed substance form large clusters or the aggregation of particles in a COLLOID to form small lumps, which then settle out. For example, when river water carrying electrically charged colloidal clays mixes with seawater, which contains electrically charged particles in solution, lumps are formed which settle as silt, often to form mudbanks. *Compare* AGGLOMERATION, COAGULATION.

floccus. Fleece-like clouds; evaporating CASTELLATUS.

flood. An unusual accumulation of water above the ground caused by high tide, heavy rain, melting snow or rapid run-off from paved areas. Many rivers have natural FLOOD PLAINS. Although floods may be caused by immediate run-off from urban areas, they also occur naturally. The most serious floods are experienced in coastal areas when low atmospheric pressure leads to a combination of heavy rain and consequent increased river flow, as well as onshore winds and waves, and these coincide with a high tide.

flood basalt. An accumulation of BASALTIC LAVA, many thousands of square kilometres in extent, with the individual flows themselves being very extensive, which implies a high rate of extrusion of very runny lava from FISSURES.

flood plain. A relatively level part of a river valley, adjacent to the river channel, formed from sediments (*see* ALLUVIUM) deposited by the river during periods of flooding or as river MEANDERS cut into the leading edge of the channel, deposit silt at the trailing edge, and so advance across level ground.

flora. (1) The plants of a particular region or period of time. (2) A descriptive list of plant species found in a particular place or period of time, often with a key to their identification.

floral formula. A conventional code to describe the number and arrangement of the parts of a FLOWER. The floral formula for a wallflower (*Cheiranthus cheiri*) is $K_{2+2}C_4A_{2+4}G_{(2)}$. This shows that the flower has a calyx (K) composed of four free sepals,

an inner and an outer pair, a corolla (C) of four free petals, an androecium (A) of six stamens, two outer and four inner ones, and a gynoecium (G) of two fused carpels, ( ) indicating that the carpels are united and the gynoecium is superior.

floral kingdom. *See* FLORAL REALM.

floral realm (floral kingdom). A large region of the world distinguished by the similarity of the plants growing within it; the highest hierarchical level recognized in the geographical grouping of plants. A realm is subdivided into regions, which are subdivided into provinces (domains), which are subdivided into districts. Four realms are generally recognized: the HOLARCTIC REALM, NEOTROPICAL REALM, PALAEOTROPICAL REALM, AUSTRAL REALM.

floret. *See* FLOWER.

florology. The study of the genesis, life and development of vegetative FORMATIONS.

florula. The plants of a small confined area (e.g., a pond).

flow banding. *See* FLOW STRUCTURE.

flower. The reproductive shoot of ANGIOSPERMAE. In a typical flower the axis (receptacle) bears: (a) the calyx, composed of sepals, the outermost organs, usually green and leaf-like, which protect the flower in the bud; (b) the corolla, composed of petals, often brightly coloured and attractive to insects; (c) the androecium, the male parts of the flower composed of stamens (microsporophylls) each of which consists of a stalk (the filament) bearing an anther which produces pollen; (d) the central gynoecium (pistil), the female parts of the flower composed of one or more carpels (megasporophylls), each of which consists of an ovary containing an OVULE or ovules and bearing a style ending in a stigma. The stigma is the surface on which the pollen grains are received. The carpels may be separate, as in the buttercup, or united, as in the wallflower (*see* FLORAL FORMULA). The calyx and corolla are accessory flower parts, together making up the perianth. A tepal is a unit of the perianth, the term being used especially when the perianth is not distinctly differentiated into sepals and petals (e.g., the tulip). In wind-pollinated flowers (e.g., grasses, poplars) and some plants relying on pollen or nectar to attract insects (e.g., willows) the perianth is reduced or absent. Some plants (e.g., hazel) have male and female organs in separate flowers. A floret is a small individual flower within a compound flower head (e.g., dandelion, *see* COMPOSITAE).

flowers of tan. *See* MYXOMYCOPHYTA.

flow structure. (1) A texture of IGNEOUS rocks, especially LAVAS in which the flow lines are shown by the alignment of prismatic crystals or elongated inclusions and by alternating bands of different minerals or crystal size (flow banding). (2) Sedimentary structures that result from subaqueous flow.

flue. A passage for conveying hot gases or for combustion. *See also* CHIMNEY, FLUE GAS, FLUE GAS DESULPHURIZATION, FLUE GAS SCRUBBER.

flue gas. Any hot waste gas, usually from a combustion process. *See also* FLUE, FLUE GAS DESULPHURIZATION, FLUE GAS SCRUBBER.

flue gas desulphurization. A technique for reducing emissions of SULPHUR DIOXIDE from industrial plants that burn coal and especially from coal-fired power stations. It involves passing the waste gases through lime. Sulphur dioxide reacts with lime to produce calcium sulphate (gypsum). The lime is obtained by kilning limestone, releasing carbon dioxide, the limestone being obtained by quarrying.

flue gas scrubber. Equipment for removing FLY ASH and other materials from the products of combustion by means of sprays or wet baffles, which also reduce the temperature of the effluent.

fluidized bed. A system for burning solid carbonaceous fuel (e.g., coal, organic ref-

use) efficiently and at a relatively low temperature, thus minimizing the emission of pollutants. The fuel is crushed to very small particles or a powder and mixed with particles of an inert material or of crushed LIMESTONE or DOLOMITE. The mixture is fed into a bed through which air is pumped vertically upwards, agitating the particles so they behave like a fluid. The forced circulation of air and the small size and separation of fuel particles ensures efficient burning. If the purpose of the fluidized bed is to raise steam, the water pipes are passed directly through it rather than being confined to the sides as in a conventional boiler. This allows the fuel to burn at a lower temperature because the water is heated more directly. The lower temperature reduces, possibly to zero, the production of NITROGEN OXIDES, and the more efficient combustion reduces emissions of HYDROCARBONS. If limestone or dolomite is mixed with the fuel it reacts with SULPHUR DIOXIDE and reduces its emission by about 90 percent. Full-size demonstration fluidized bed plants are in operation at several power stations in the USA and elsewhere.

flukes. *See* TREMATODA.

fluorescence. The emission of visible light by materials that are subjected to electromagnetic radiation of short wavelength (e.g., X-rays, ultraviolet or violet light). The name is derived from fluorspar (*see* FLUORITE), since some varieties of this mineral exhibit the phenomenon due to the ultraviolet light present in visible daylight. Geologically the property of fluorescence is used in prospecting for SCHEELITE, and in the rough identification of the API SCALE crude oil in drilled samples of rock.

fluoridation. The addition to public water supplies of FLUORIDES in order to prevent dental caries or to delay its onset. Many water supplies contain fluorides naturally, and a concentration of 1 ppm is considered optimal. The aim of fluoridation is to increase the fluoride content to the optimum level. The measure has aroused great controversy, although the consensus of medical opinion probably favours it.

fluorides. Compounds containing FLUORINE, the lightest and most reactive of the HALOGENS. Fluorides occur widely in nature, but are either dispersed at low concentrations or in stable compounds from which they are not released readily, but their ubiquity means they may be released into the atmosphere in higher concentrations than occur naturally in the course of a number of industrial processes. Such processes include the manufacture of cement and bricks, the refining of aluminium and the processing of slag from electric furnaces and from steel manufacture where a fluorite flux is used. As atmospheric pollutants fluorides are harmful to plants and mammals, and there have been many cases of serious economic damage to crops and livestock downwind of emission sources.

fluorine (F). An element; a pale, yellowish–green gas resembling CHLORINE, but more reactive. It occurs naturally as FLUORITE and cryolite. Organic fluorine compounds have a number of industrial uses (*see* CHLOROFLUOROCARBONS). A_r=18.9984; Z=9.

fluorite (fluorspar, CaF_2). A mineral found mainly in SEDIMENTARY ROCK, in ACID IGNEOUS rocks and as a GANGUE mineral. Its main uses are in the chemical industry, particularly in the manufacture of HYDROFLUORIC ACID, and in ceramics. A little is still used as a FLUX in steel making.

fluorocarbons. *See* CHLOROFLUOROCARBONS.

fluorosis. A disease in ruminants caused by the over-consumption of compounds containing fluorine, often as a result of air pollution which deposits FLUORIDES on vegetation, which is then grazed. Fluorosis produces mottling and weakening of the teeth, as well as excessive thickening of bones.

fluorspar. *See* FLUORITE.

flush. An area of boggy vegetation (*see* BOG) along a seepage line on sloping ground.

fluvial. Pertaining to the actions of rivers.

fluvioglacial deposits. *See* OUTWASH DEPOSITS.

fluviomarine. Pertaining to the action of both rivers and sea. It is used particularly of sediments deposited at the mouths of large rivers.

flux. (1) The rate of a flow of fluid or radiation across an area. In the case of a vector quantity (i.e. where direction and speed of flow are both specified) the flux is calculated as the rate of flow past a line crossing the area at right angles to the direction of flow multiplied by the area, and provides a value for electric intensity, magnetic intensity, etc. In nuclear physics, the flux is the number of particles per unit volume and their average velocity. (2) A substance added to assist the fusion of two substances (e.g., in metal working). (3) A morbid or excessive discharge of blood, excrement, etc; formerly the name for dysentery.

fly ash. Finely divided particles of ASH that are entrained in FLUE GASES resulting from the combustion of fuel or other material. The particles of ash may contain incompletely burned fuel and other pollutants.

flying fishes. Fishes belonging mainly to the family Exocoetidae, and a small number of unrelated fishes that are able to leave the water at speed and glide for some distance through the air, probably as a means of escaping from predators. The most common species, *Exocoetus volans*, which is about 25 centimetres long, is able to sustain a gliding flight by spreading its enlarged stiff pectoral fins and aiding its forward motion by beating rapidly with its tail fin, which is lifted when sufficient speed has been attained. The smaller pelvic fins are spread to make a second pair of 'wings'. Some fishes have been known to cover 400 metres (440 yards), taking about 43 seconds to do so.

flying foxes. *See* CHIROPTERA.

flying lemurs. *See* DERMOPTERA.

flysch. A rapidly alternating marine succession of SANDSTONES and SHALES, of late CRETACEOUS to early TERTIARY age, found on both flanks of the Alps. The sediments were derived from erosion of the rising mountain chain during DEFORMATION and were themselves later deformed. The term has also been used for many broadly similar sequences of GREYWACKE and shale. *Compare* MOLASSE. *See also* DIASTROPHISM.

foam. A gas in liquid dispersion, formed by vigorous mixing. As air bubbles burst above the sea the fragmentation of the thin films containing them is a major contributor of the small droplets that enter the air and evaporate, leaving airborne salt particles. *See also* SPRAY.

focus. *See* EARTHQUAKE.

FOE. *See* FRIENDS OF THE EARTH.

Foehn wind. *See* FÖHN WIND.

foetus (fetus). A mammalian EMBRYO during its later stages of development, after the main features have become recognizable.

fog. Visible moisture in the atmosphere that reduces horizontal visibility to below 1000 metres. Fog is caused by the cooling of relatively warm, moist air when it encounters a land or sea surface that is colder, so reducing the temperature of the air in immediate contact with the surface to below the DEW POINT. Advection fog is caused by the movement of moist, warm air across a colder surface as part of a general weather pattern. Radiation fog is caused by the cooling of the land surface at night by radiation, so that although the land and air temperature are equal during the day, at night the land cools more rapidly than the air. *Compare* MIST, SMOG. *See also* FOG SHOWERS, FROZEN FOG, PEA SOUP FOG, SEA FOG, STEAM FOG, VALLEY FOG.

fog bow. An arc, usually white because of the wide range of drop sizes, analogous to a RAINBOW, but seen in fog. Physically it is the same as a CLOUD BOW.

fog showers. A phenomenon characteristic of isolated high mountains that lie above the CLOUD BASE and are, from time to time, engulfed in passing CONVECTION CLOUDS. When supercooled (*see* SUPERCOOLING), the cloud droplets are captured as rime or glazed frost on wires, twigs and other small objects.

Föhn wind (Foehn wind). A warm wind descending from the mountains, originally in the European Alps, where physiological and psychological discomfort is produced by the big changes in temperature, humidity and wind strength that follow its onset. It is warm and dry because of its ADIABATIC descent. This is often attributed wrongly to the LATENT HEAT released when rain falls from the air following its ascent on the other side of the mountains. In fact the effect is due mainly to the blocking of a cold AIR MASS by the mountain range so that the air previously near the mountain top descends the lee slope. Very strong winds often occur in the troughs between MOUNTAIN WAVES, and the sky is characteristically filled with WAVE CLOUDS. *See also* CHINOOK.

folacin. *See* FOLIC ACID.

fold. A bend in strata or in any planar structure in rocks or minerals. Folds are classified geometrically by the DIPS of their LIMBS and attitude of their AXIAL PLANES. Folds are described as open if the interlimb angle is more than 70° and tight if it is less than 30°. Folds are said to plunge if their axes are not horizontal. Folds can also be classified by their mode of formation. *See also* ANTICLINE, ISOCLINE, MONOCLINE, RECUMBENT FOLD, SYNCLINE, TERRACE.

foliation. Layering in rocks that is caused by parallel orientation of minerals (*see* SCHISTOSITY) or bands of minerals. The texture is characteristic of such metamorphic rocks (*see* METAMORPHISM) as SLATE, PHYLLITE, SCHIST and GNEISS, and is also found in IGNEOUS rocks that flowed during cooling.

folic acid (folacin, pteroylglutamic acid, xanthopteryl-methyl-*p*-aminobenzoylglutamic acid). One of the B group of VITAMINS, belonging to a group of related substances that are essential in the diet of many animals. Folic acid deficiency in humans may lead to some types of anaemia, and because of a poorly understood relationship between folic acid and COBALAMINE, folic acid deficiency may also be implicated in pernicious anaemia. Folic acid is synthesized by intestinal bacteria, but in amounts insufficient for dietary needs.

follicle. (1) A dry fruit formed from a single carpel which splits down one side only to liberate its seeds (e.g., larkspur). (2) *See* HAIR FOLLICLE.

follicle-stimulating hormone (FSH). A HORMONE secreted by the pituitary gland of vertebrates that stimulates the growth of the OOCYTE and OVARIAN FOLLICLES in females and the development of spermatozoa (*see* SPERMATOZOON) in males. *See also* GONADOTROPIC HORMONES.

food additives. *See* ADDITIVES, DELANY CLAUSE.

Food and Agriculture Organization of the United Nations (FAO). One of the first of the UN specialist agencies to be formed and one of the largest, with headquarters in Rome. Its aim is to increase food production and availability among those sections of the world population where hunger is prevalent. It is closely associated with the GREEN REVOLUTION. It conducts many field projects, initiates and cooperates in research and seeks to reform world food-trading policies to the advantage of the poor. In addition to agriculture and trade, its operations extend into fisheries, forestry and nutrition. *See also* HIGH-YIELDING VARIETIES, INDICATIVE WORLD PLAN FOR AGRICULTURAL DEVELOPMENT.

food chain. A number of organisms that form a series through which energy is passed. At the base of the chain (the producer level or first trophic level, T_1) there is always a green plant or other AUTOTROPH which traps energy, almost always from light, and produces food substances, thereby making energy available for the other (consumer) levels. At the second trophic level (T_2) is a herbivore (primary consumer). At subsequent trophic levels there are small, then larger carnivores (secondary consumers). For example, a food chain might comprise unicellular algae → *Daphnia* → dragonfly nymph → smooth newt → grass snake. In a balanced community the BIOMASS of a lower trophic level is always higher than that of the succeeding level because at each stage a large amount of energy is dissipated during respiration. About 90 percent of the available chemical energy is lost each time energy is transferred from one trophic level to the next higher one. SAPROPHYTES are present at all consumer levels. Any natural community will have many interlinked food chains, making up a food web or a food cycle. *See also* ENERGY FLOW.

food cycle. *See* FOOD CHAIN.

food vacuole. *See* VACUOLE.

food web. *See* FOOD CHAIN.

fool's gold. *See* PYRITES.

footwall. The lower side of an inclined FAULT or VEIN, or the ore limit on the lower side of an inclined ORE BODY.

foraminifera. *See* RHIZOPODA.

forbs. All herbaceous plants, except for grasses and those resembling grasses (e.g., sedges).

forced convection (mechanical turbulence). The transport of a fluid (e.g., air) across the main stream, with consequent mixing induced by TURBULENCE, as the fluid moves over an uneven surface. *Compare* FREE CONVECTION.

forced oscillation (vibration). An oscillation that is maintained by the application of a fluctuating energy supply. *See also* NATURAL FREQUENCY.

foreset beds. Inclined layers of sediments

that are deposited on the advancing edge of a DELTA or on the LEE slope of an advancing SAND DUNE.

foreshock. A relatively small earthquake that precedes a larger EARTHQUAKE by a few days or weeks, and originates near or at the focus of the larger earthquake.

foreshore. The shore zone (*see* SHORE ZONATION) that is covered only by exceptionally high spring tides. Its vegetation is sparse and includes such specialized plants as sea rocket (*Cakile maritima*).

forest. (1) An extensive area of woodland that is either managed (*see* COPPICE) or maintained for the production of timber, etc. Coppice forest consists of trees derived from coppice shoots (produced from trees cut near the ground) and root suckers. High forest is mature woodland, usually composed of tall trees derived from seeds, their tops forming a closed canopy. The adjectives broadleaved, coniferous or the name of a particular tree (e.g., pine) are applied to woodlands in which at least 80 percent of the canopy consists of the trees the name describes; the canopy of mixed forest contains at least 20 percent of trees other than the DOMINANT species. Rain forest is evergreen forest growing in regions of high rainfall, where the dry season is short or absent. EPIPHYTES and climbers are abundant. The term is often used in the restricted sense of tropical rain forest, but rain forest can occur outside the tropics. Monsoon forest occurs in regions with a well-marked rainy season, and in some the trees are deciduous, losing their leaves for at least part of the dry season. (2) In Britain, an area that was originally unenclosed, not necessarily wooded, and was preserved for hunting (e.g., New Forest, Dartmoor Forest, deer forests of the Scottish Highlands).

form. (1) The smallest subgrouping of plants, based on trivial characteristics (e.g., colour of petals). (2) Loosely, various minor or informal groupings of animals. (3) The resting place of a hare, where it lies concealed by vegetation.

formaldehyde (methanal, HCHO). A colourless gas that is suspected of being carcinogenic (*see* CARCINOGEN) and that may be emitted by urea formaldehyde, which is widely used as foam insulation for cavity walls. Where walls are permeable (e.g., wood) formaldehyde may enter buildings.

formation. (1) The largest natural vegetation type (e.g., tropical rain FOREST) or the plants of a land BIOME. (2) A BED, or collection of beds, of a distinct rock type that can be traced over a considerable area of country. A formation is the basic mapping unit, subdivided into members or beds. Several formations may form a group.

formation type (formation class). A group of geographically widespread COMMUNITIES of similar physiognomy and life form, and related to major climatic and other environmental conditions.

fosse. (1) A depression separating two terminal MORAINES or an OUTWASH PLAIN and a terminal moraine. (2) In French usage, an ocean deep.

fossil. Remains or traces of an organism that have been preserved in the Earth's crust by natural processes. Fossils may be formed in a number of ways. The hard parts may be replaced by another mineral (e.g., PYRITES, SILICA). Animal shells and wood may be 'petrified' in this manner. Internal or external casts of the organism may form in the rock. Carbon residues may accumulate after organic decomposition. Impressions of soft parts may be left in fine-grained sediment. Traces of the activity of the organism may be preserved (e.g., tracks and burrows of animals, COPROLITES, root impressions from plants). Fossils are usually found in SEDIMENTARY ROCKS, but recognizable, albeit distorted fossils may be found in metamorphic rocks if the METAMORPHISM has not been too intense. Occasionally whole animals may be preserved if conditions are favourable (e.g., woolly mammoths in frozen ground in Siberia, insects preserved in amber). The term fossil is sometimes extended to include inorganic remains of geological age (e.g., fossil sand dune, fossil beach). It is also extended, still less formally, as living fossil, to living species that survive long after all their close relatives have become extinct (e.g., coelacanth is a living fossil fish).

fossil fuel. A fuel derived from ancient organic remains (i.e. PEAT, COAL, CRUDE OIL, NATURAL GAS, TAR SANDS, OIL SHALES). Fossil fuels result from the incomplete decomposition of organic material, Since the rate at which these fuels are consumed exceeds that of their formation, they are considered to be NON-RENEWABLE RESOURCES. They also represent a long-term storage of carbon removed from the atmosphere whose oxidation is completed by combustion, which returns it as carbon dioxide. *See also* CARBON CYCLE, GREENHOUSE EFFECT.

fossil turbulence. Inhomogeneities of temperature and humidity which remain in the air after the motion that produced them has subsided and the density, although not the temperature and humidity, has become uniform. Such inhomogeneities cause the scattering of radio waves and the formation of lumpy clouds when the air is made to ascend.

Foucault's pendulum. A pendulum hung from the ceiling which, swinging in a plane in space, appears to move in a rotating plane because the Earth is rotating. It is named after the French physicist J.B.L. Foucault (1819–68) and is often to be seen in museums of science.

Fourth World. *See* THIRD WORLD.

fracture. (1) A break in a mineral that is not along a CLEAVAGE plane. The broken surface may be described as even, uneven, HACKLY, splintery or CONCHOIDAL. (2) Breakage in a rock under stress, occurring at a point known as its ultimate strength.

fracture zone. The zone along which faulting (*see* FAULT) has taken place. The term is used more particularly for the linear zones of ridges and troughs approximately perpendicular to mid-oceanic RIDGES, which they

offset. Such zones are the topographic expression of TRANSFORM FAULTS.

fragipan. *See* HARDPAN.

Franklin. *See* GORDON AND FRANKLIN.

Franklin, Benjamin (1706–90). An American physicist who demonstrated the electrical nature of lightning by flying kites in thunderstorms. He also invented the forerunner of the wood-burning stove.

Frasnian. The sixth oldest stage of the DEVONIAN system in Europe.

fraternal twins. *See* DIZYGOTIC TWINS.

Fraxinus excelsior (European ash). A deciduous tree (family: Oleaceae) that has pinnate leaves, black buds and winged fruit (keys). It is especially common on calcareous soils, where it may form woods, and is frequently found in oakwoods. It yields valuable timber.

free acceleration test. A test for measuring exhaust emission from vehicles. The engine is accelerated rapidly, in neutral gear, and the exhaust gases are fed through a smoke meter, which analyzes them.

free convection. The motion and mixing of a fluid that is induced by BUOYANCY forces. *Compare* FORCED CONVECTION.

free field. In acoustics, a region in which no significant reflections of sound occur.

freemartin. A sterile, partly HERMAPHRODITE female hoofed animal, whose peculiarities result from the fusion of its placental circulation with that of its twin brother.

free progressive wave. A theoretical wave propagated in an infinite medium.

freestone. A SANDSTONE or LIMESTONE that does not split in one direction, but can be cut and dressed equally well in any direction.

freezing level. The height of the 0°C ISO-

THERM in the atmosphere. Although cloud droplets in rising air may remain super-cooled (*see* SUPERCOOLING) liquid above this height, descending ice particles begin to melt at it, although large hail may reach the ground unmelted.

freezing nuclei (ice nuclei). Atmospheric particles on to which water freezes. Normally water vapour condenses into liquid, even at temperatures well below freezing point, unless a freezing (or ice) nucleus is present. Ice nuclei are of mineral origin and possess a crystal structure similar to that of ice, so that within a supercooled (*see* SUPERCOOLING) droplet they initiate the formation of an ice crystal with an efficiency that increases the lower the temperature. Silver iodide and lead iodide crystals initiate freezing at about -5 and -7°C, respectively, but naturally occurring nuclei begin to function at below -10°C. Below -40°C spontaneous freezing occurs without the presence of freezing nuclei. The most efficient of all freezing nuclei are ice crystals themselves, which may grow by ACCRETION as they fall through air containing supercooled droplets.

freons. *See* CHLOROFLUOROCARBONS.

frequency. (1) The number of times a vibrating system or particle completes a repetitive cycle of movement within a given period of time. The derived SI unit of frequency is the hertz (Hz) (named after Heinrich Hertz, 1857–94), which is equal to 1 cycle per second. (2) The average number of statistical units found in an area. In practice this may be expressed as the percentage of total samples or QUADRATS in which a species occurs.

fresh water. Water whose SALINITY is less than 0.5°/oo.

friction layer. The BOUNDARY LAYER, within which friction occurs, between two layers of air with different wind speeds. At the ground surface this friction is manifested as SHEAR STRESS and at higher levels it causes TURBULENCE in bands a few hundred metres or more thick.

friction velocity (u_*). The relationship between the air density (ρ) and shear stress (T) such that

$$u_* = T/\rho$$

Friends of the Earth (FOE). An environmentalist organization formed in the USA in 1970 by David Brower, formerly executive director of the SIERRA CLUB. A branch opened in London in 1971, and the first UK director was Graham Searle. There are now branches or affiliated groups in 30 countries. Within countries it operates through independent, self-financing local groups, of which there are about 250 in Britain. Friends of the Earth mounts publicity campaigns to draw attention to particular issues, usually those concerning pollution, wildlife conservation and resource exploitation, and to exert political pressure for reform.

fringing reef. *See* CORAL REEF.

frogs. *See* ANURA.

front. A narrow transitional zone between AIR MASSES, named after the incoming air mass. *See also* COLD FRONT, OCCLUDED FRONT, WARM FRONT.

frontal analysis. The analysis of weather charts by marking the positions of the FRONTS between different AIR MASSES.

frontal slope. The inclination of the surface of a FRONT to the horizontal (α), which is usually given in terms of gravitational acceleration (g), the Coriolis parameter (f), the temperature (T), the discontinuity in temperature at the front (ΔT) and the discontinuity in GEOSTROPHIC WIND(ΔV_G) by

$$\Delta V_G = (\Delta T g)/(Tf)\sin \alpha$$
to a good degree of approximation.

frontal structure. The configuration of a FRONT which in nature is not always a sharp discontinuity, but more a zone of transition from one AIR mass to the other, sometimes with more than one fairly sharp demarcation. COLD FRONTS are more often sharply defined than WARM FRONTS, and on passage the transition may occur within a few minutes.

frontal wave. The wave-like perturbation at the frontal surface (usually the polar front in temperate latitudes) of a warm AIR MASS as pressure falls and cyclonic rotation of the air commences. As the pressure continues to fall, the wave-like deformation increases and near the centre the front becomes narrower because the cold air mass moves faster than the warm air. The wave then moves rapidly along the front, with the centre of low pressure at its crest, until the cold air overtakes the warm air to produce an OCCLUDED FRONT.

frontal zone. The broad band of weather associated with a FRONT, or the zone of transition from one AIR MASS to another when the front is not sharply defined.

frost. *See* DEW.

frost hollow. A relatively small, low-lying area that is subject to frequent and severe frosts because of the accumulation of cold air at night. Typically the frost is most severe where hills shade the ground from afternoon sunshine.

frost point. The temperature to which air must be cooled for frost to begin to form on solid surfaces. It is best measured by the temperature to which a solid surface must be raised at in order for the ice (frost) to evaporate. The frost point is higher than the DEW POINT by an amount which increases from zero at 0°C to about 3.5°C at -40°C.

frost wedging. The process by which the expansion of freezing water in pores or fissures in rocks shatters the rocks.

froth flotation. The separation of a mixture of finely divided minerals by mixing them in a froth of oil and water, so that some float and others sink.

frozen fog. A FOG of cloud composed of ice crystals. When supercooled (*see* SUPERCOOLING) fog begins to freeze, FALLOUT is generated by the BERGERON–FINDEISEN MECHANISM, so that frozen fogs usually clear quickly by fallout from the almost motionless air. Supercooled fogs have been dispersed by seeding to cause freezing.

fructose. *See* CARBOHYDRATES.

fruit. The ripened ovary of a flower. It contains the seed or seeds. *See* ACHENE, BERRY, CAPSULE, CARYOPSIS, CYPSELA, DRUPE, FOLLICLE, HESPERIDIUM, LEGUME, NUT, POME, SAMARA, SCHIZOCARPIC, SILIQUA.

fruit flies. *See* DROSOPHILA.

frutescent (fruticose). Shrubby.

fruticeta. FOREST composed mainly of shrubs.

fruticose. *See* FRUTESCENT.

FSH. *See* FOLLICLE-STIMULATING HORMONE.

fucoxanthin. A brown pigment found in certain primitive plants. It masks the green colour of CHLOROPHYLL, and any other pigments that may be present, giving the plant a brown or olive green colour. *See also* PHAEO-PHYTA.

fuel cell. A device for generating electrical power on a small scale in which a fuel (usually hydrogen) is positively ionized (*see* IONIZATION) and an oxidizing (*see* OXIDATION) substance negatively ionized to create a flow of electrons. The device is quiet, simple, does not produce significant amounts of waste heat and is fairly efficient.

fuel efficiency. The proportion of the potential heat of a fuel that is converted into useful energy.

fuel element. In a NUCLEAR REACTOR, one of the containers of fissile material ('fuel') that is inserted into the core.

fulgurite. A tube of glassy rock produced when lightning fuses together grains in loose sand or more compact rock.

fuller's earth. (1) An ARGILLACEOUS rock or CLAY with strong powers of absorption for water, colouring matter, grease and some oils. This property is due to the presence of MONTMORILLONITE clay. Fuller's earth was formerly used for degreasing (i.e. fulling) fleeces. (2) A stratigraphic name for a Middle JURASSIC rock unit in the UK that contains substantial beds of fuller's earth (see above).

fuller's teasel. *See* DIPSACACEAE.

fumarole. A volcanic vent that emits only gases, which are at a temperature higher than that of the atmosphere. The most abundant product is usually steam and from this minerals are deposited. Sulphur and chloride minerals, especially ammonium chloride, are the most common, but many metallic minerals have been found.

fume. Solids in the air that have been generated by the condensation of vapours, SUBLIMATION or chemical reactions. The particles are less than 1 micrometre in diameter and are often metals or metallic oxides, which may be toxic.

fumigation. (1) A rapid increase in air pollution close to ground level, which sometimes leads to very high concentrations of pollutants for an hour or more. The phenomenon occurs when a nocturnal temperature INVERSION has caused pollutants to accumulate aloft. In the morning, the warming of the ground initiates mixing upcurrents of air, and these bring down the pollutants held by the inversion, so preventing their escape upwards. The TURBULENCE gradually draws clean air from above the inversion into the lower layers, so diluting the pollutants. (2) A technique used to apply pesticides (especially fungicides) as a FUME in an enclosed space (e.g., a glasshouse). The area is then sealed for a time before workers are allowed to enter.

function. (1) The rate of biological energy flow through an ECOSYTEM (i.e. the rates of production and respiration of the populations in the COMMUNITY). (2) The rate at which materials or nutrients are cycled (i.e. the rate at which the BIOGEOCHEMICAL CYCLES proceed). (3) The biological or ecological regulation of species by the environment (e.g., PHOTOPERIODISM) and the regulation of the environment by organisms (e.g., NITROGEN FIXATION). (4) In statistics, a quantity that varies as a result of variations in another quantity.

fundamental frequency. The FREQUENCY with which a periodic function reproduces itself; the first harmonic.

fundamentalists. In the West German Green Party (*see* GREENS), the faction that opposes forming alliances with other parties because they are not prepared to compromise Green policies, especially those dealing with nuclear power, energy and defence. *Compare* REALISTS.

fundamental particles. Particles that cannot be demonstrated to contain simpler units.

Fungi. The kingdom that includes toadstools, mushrooms, MILDEWS, YEASTS, etc. They are either unicellular or composed of masses of fine filaments (hyphae) and reproduce by means of SPORES. None contains CHLOROPHYLL, and all are HETEROTROPHIC. They were formerly regarded as plants, which they resemble superficially, but now they are classified as distinct from, and not closely related to, plants or animals. Many (e.g., mildews, SMUT FUNGI, RUST FUNGI) cause plant diseases, and a few cause animal diseases (e.g., ringworm). Some can cause the decay of timber (e.g., dry rot) and of food, but many are beneficial because of the large part they play in the decomposition of organic matter in the soil. Certain antibiotics are produced by culturing fungi (e.g., penicillin is obtained from *Penicillium*), and others (e.g., yeasts) are used in the preparation or preservation of food and in brewing. *See also* AGARICS, ASCOMYCETES, BASIDIOMYCETES, FUNGI IMPERFECTI, PHYCOMYCETES.

fungicide. A chemical compound used to control fungal diseases. *See* BORDEAUX MIX-

TURE, CARBAMATES, DINITRO PESTICIDES, FENTINS, ORGANOMERCURY FUNGICIDES.

Fungi Imperfecti. FUNGI in which a sexually reproducing (perfect) stage is unknown. This makes them difficult to classify, and conventionally they are grouped as a subdivision (Deuteromycotina) of their own.

funnel cloud. A cloud that appears in the core of a tornado (*see* BATH PLUG VORTEX) or WATER SPOUT because of the low pressure. It resembles the air core of a bath plug vortex in shape, and its outline is approximately the ISOBAR of the CONDENSATION LEVEL.

furan (C_4H_4O). A compound fairly insoluble in water, but readily soluble in alcohols, ether and acetone, belonging to a group of organic compounds (furans) that occur naturally in wood oils, seed husks and other plant materials. Furans, especially furfural (C_4H_3O CHO), are used as bactericides and in the manufacture of perfumes and flavourings and in synthetic fibres. Furans are toxic to humans.

furfural. *See* FURAN.

furnace. Any container in which a material is heated to a very high temperature.

furze. *See* LEGUMINOSAE.

fusion reactor. A NUCLEAR REACTOR whose energy is derived from the fusion of two atoms (DEUTERIUM, TRITIUM, LITHIUM or some combination of these) to form one HELIUM atom, with the release of energy. The operating temperature is in the region of 100 x $10^{6}°C$, and the fuel is contained as a plasma in a magnetic field. Although deuterium is relatively easily available it requires much higher operating temperatures than the rarer lithium, and therefore the first commercial fusion reactors are likely to use lithium. Fusion reactors are still in the advanced experimental stage, but may provide significant amounts of electricity by the middle of the 21st century.

FWS. *See* FISH AND WILDLIFE SERVICE.

FYM. *See* FARMYARD MANURE.

G

G. *See* GIGA-.

gabbro. A coarsely crystalline BASIC IGNE-
OUS rock. It consists essentially of calcium-
rich plagioclase FELDSPAR and PYROXENE,
with or without OLIVINE. *Compare* DOLERITE,
BASALT.

gadflies. *See* BRACHYCERA.

Gadidae. A large and economically important family of marine fishes that includes the
cod, whiting and haddock.

gaging station. *See* GAUGING STATION.

Gaia hypothesis. The idea, proposed by
James E. Lovelock, Lynn Margulis and others, that on any planet supporting life the
living organisms respond to environmental
conditions and in doing so modify the environment, making it more hospitable to themselves. On Earth, the biota regulate the
global climate and drive the BIOGEOCHEMICAL
CYCLES.

galactose. *See* CARBOHYDRATES.

galena (PbS). The mineral lead sulphide,
which accounts for most of the world's production of LEAD. Galena usually occurs with
SPHALERITE in HYDROTHERMAL deposits, in
sedimentary stratiform deposits and in deposits formed by the replacement of LIME-
STONE by METASOMATISM. Many of the sedimentary ores are of PRECAMBRIAN age and
have undergone METAMORPHISM. Galena is
also an important source of silver, which it
contains as an impurity.

gall. An abnormal growth of plant tissue that
is produced by the plant as a response to

mechanical injury or to the invasion of
insects, mites, eelworms, fungi, bacteria or
viruses. Many are caused by gall-wasps
(ACULEATA) (e.g., robin's pincushion on wild
roses, and on oak oak-apples, spangle galls,
currant galls and marble galls). Big bud of
blackcurrant is caused by a mite; witches'
broom on birch by a fungus.

Galliformes (game birds). The order of
birds that includes grouse, ptarmigan, caper-
caillie, partridges, pheasants, quails, turkeys
and peacocks. These are mainly grain-eat-
ing, heavy-bodied, ground-nesting birds,
capable of only short, rapid flights. The
cocks are usually more colourful than the
hens.

gall-wasps. *See* ACULEATA.

galvanize. To coat iron or steel with zinc,
either by immersion in a bath of molten zinc
or by deposition from a solution of zinc
sulphate, to give protection against corrosion.

game birds. *See* GALLIFORMES.

gamete (germ cell). (1) A haploid (*see*
CHROMOSOMES) cell or nucleus that fuses with
another during FERTILIZATION to produce a
ZYGOTE, which develops into a new plant or
animal. Usually the two gametes are
different: the female one (ovum) is non-
motile with a large amount of cytoplasm; the
male one (spermatozoon in animals,
spermatozoid, or antherozoid in many
plants, male nucleus in seed plants) is motile,
usually by means of a FLAGELLUM, and small.
In some ALGAE, FUNGI and PROTOZOA, similar
gametes (isogametes) are produced. (2) A
cell similar to a female gamete, but usually

diploid, that develops by PARTHENOGENESIS into a new individual.

game theory. The theory that relationships within a community can be represented as a contest in which each participant seeks an advantage. The simulation of such relationships, often as a computer program, helps to determine winning strategies and this leads to a greater understanding of ecological and social relationships. *See also* ZERO SUM GAME.

gametocyte. A cell which gives rise to a GAMETE or gametes by MEIOSIS. *See also* OOCYTE, SPERMATOCYTE.

gametophyte. *See* ALTERNATION OF GENERATIONS, EMBRYO SAC.

gamma-BHC. *See* LINDANE.

gamma-rays. Electromagnetic radiation at the high-energy end of the spectrum, with wavelengths of 10 nanometres or less. Gamma rays are similar to X-rays, but are of shorter wavelength and less penetrative power. They cause IONIZATION along their track. Gamma-rays are received from cosmic sources, but are absorbed strongly in the atmosphere, radiation exposure approximately doubling for every 1500 metres increase in height. They are also emitted from some radioactive materials and from atomic transformations that occur in nuclear reactors and the detonation of nuclear weapons.

gamodeme. *See* -DEME.

ganglion. A discrete mass of tissue that contains NEURON (nerve cell) bodies.

gangue. The waste minerals in an ore. The term is essentially economic as material which is gangue in one mine may, in higher concentrations or under different economic conditions, be a valuable component in another. Common gangue minerals in HYDROTHERMAL veins are QUARTZ, tourmaline, CHLORITE, FLUORITE, HAEMATITE, PYRITES, CHALCEDONY, DOLOMITE and CALCITE.

gannets. *See* PELECANIFORMES.

gannister. A fine-grained, arenaceous (*see* ARENITE), siliceous rock that underlies some coal seams.

garden city. A town planning concept devised by Sir Ebenezer Howard (1850–1928) that would extend to all town dwellers the advantages of suburban conditions. The garden city was to be surrounded by countryside, and expansion was to be achieved by developing new garden cities on the other side of the GREEN BELT, leading to clusters of cities grouped around a central city. Several garden cities (e.g., Welwyn Garden City) were built in Britain. Howard was influenced by the city of Adelaide, Australia, which retained parklands on its northern side and expanded beyond them to form North Adelaide.

garnet. A group of minerals with the general formula $A_3B_2Si_3O_{12}$, where A can be iron, magnesium or calcium and B can be iron, aluminium or chromium. Garnets have no CLEAVAGE and are hard, and are therefore used as abrasives. Different garnets are found in a wide range of rocks.

garnierite. A nickeliferous SERPENTINE mineral $(H_4Ni_3Si_2O_9)$ that occurs in LATERITES developed from ULTRABASIC IGNEOUS rocks. Garnierite is one of the major ores of nickel, which is used in ALLOYS and stainless steel. Garnierite is called nouméite in New Caledonia. *See also* PENTLANDITE.

gas cap. An accumulation of NATURAL GAS above an oil POOL.

gas chromatography. An analytical technique for separating mixtures of volatile substances. The sample is placed on a separating column and is washed through with an inert gas. The column selectively retards, and thus separates, the substances. Packing with absorbent material coated with relatively non-volatile material gives gas–liquid chromatography; without liquid coating gives gas–solid chromatography. The technique provides quantitative results on small samples and is used widely (e.g., in PALYNOLOGY).

gas-cooled fast breeder reactor (GCFBR). A BREEDER REACTOR that uses gas as a coolant; it is less developed than the LIQUID METAL FAST BREEDER REACTOR, but is considered a possible alternative to it.

gas field. *See* OIL FIELD.

gas–liquid chromatography. *See* GAS CHROMATOGRAPHY

gas/oil ratio (GOR). The ratio of oil to gas in a produced CRUDE OIL, expressed as standard cubic feet of gas per BARREL of oil. Values may vary from less than 100 to several thousand.

gas pool. *See* POOL.

gas–solid chromatography. *See* GAS CHROMATOGRAPHY.

gas thermometer. A thermometer in which the expanding and contracting substance is gaseous.

Gastropoda. A large class of MOLLUSCA, most of which have a single shell, usually spirally coiled. The Prosobranchia (Streptoneura) are mainly marine (e.g., winkles, whelks, sea limpets); the Opisthobranchia (e.g., sea-slugs) are marine and many have the shell reduced or absent; the Pulmonata (e.g., land and pond snails, slugs) have lungs and live mainly on land or in fresh water.

gastrula. An animal EMBRYO at the stage of development that follows the BLASTULA. In the gastrula cells move about, forming the GERM LAYERS.

GATT. *See* GENERAL AGREEMENT ON TARIFFS AND TRADE.

gauging station (gaging station). A place at which the water flowing through a stream channel is measured. The surface level of the water, shape of the channel, stream velocity and amount of dissolved or suspended sedimentary matter are recorded. Data are provided for calculating the size and availability of the water resource, potential flood damage and stream pollution, and for projects connected with damming and irrigation schemes.

Gause's principle (competitive exclusion principle). The rule which states that two species whose ecological requirements are identical cannot exist together in the same habitat unless there is a superabundance of environmental resources (especially food).

gaussian distribution (normal distribution). A distribution that shows the maximum number of occurrences at or near to a centre or mean point, a progressive decrease in occurrences with increasing distance from the centre and symmetrical distribution of occurrences on both sides of the centre.

Gaviidae (divers). A family of diving birds (order: Gaviiformes) which includes the red-throated, black-throated and great northern divers. They inhabit open water, are clumsy on land and come ashore only to breed. They have webbed feet and feed mainly on fish.

GCFBR. *See* GAS-COOLED FAST BREEDER REACTOR.

GDP. *See* GROSS DOMESTIC PRODUCT.

Gedinnian. The oldest stage of the DEVONIAN System in Europe.

geese. *See* ANATIDAE.

Geiger counter (Geiger–Müller counter). An apparatus for counting charged particles by means of the IONIZATION they produce. It is hand-held and is thus a basic tool of the nuclear scientist and for those searching for radioactive substances.

Geiger threshold. The lowest voltage which, when applied to a GEIGER COUNTER, will produce pulses that in each case are of about the same size, irrespective of the number of primary ions produced.

gel. Material, often with a jelly-like appearance, that forms when a colloidal solution (*see* COLLOID) is allowed to stand. Gels may contain as little as 0.5 percent of solid matter, but their properties resemble those of solids more closely than those of liquids.

gelatin. A protein extracted from COLLAGEN that is used in medicine, biology, paper making, textile processing, food processing and the production of films and adhesives.

gelifluction. The MASS WASTING of thawed material over PERMAFROST. *Compare* CONGELIFLUCTION, SOLIFLUCTION.

gemmae. *See* GEMMATION.

gemmation. A form of asexual reproduction in plants and animals in which new individuals or members of a colony develop from groups of cells arising on the parent's body. In mosses and liverworts, small groups of cells (gemmae) become detached from the parent and then develop into new plants. In some CNIDARIA (e.g., *Hydra*) the new individuals develop while still attached to the parent. Gemmation in animals is usually called BUDDING.

GEMS. *See* GLOBAL ENVIRONMENTAL MONITORING SYSTEM.

genecology. The study of the genetics of plant and animal populations in relation to their environments.

gene exchange. Sexual reproduction within an ECOTYPE, species or COENSPECIES, which results in recombination of the parental genes.

gene flow. The movement of genes between populations as a result of sexual reproduction between members from each population.

gene frequency. The frequency with which a certain gene occurs in a population, compared with the frequency of all its ALLELOMORPHS.

General Agreement on Tariffs and Trade (GATT). An intergovernmental agreement, drawn up initially in 1947 and now involving about 80 nations, that aims to facilitate trade by removing restrictions, especially tariffs.

general circulation. The average, worldwide system of winds. Air movement is caused by differential heating of the Earth's surface and atmosphere and by the Earth's rotation, with topographic differences causing local variations. The distribution of alternating belts of high and low pressure between the equator and the poles causes a general flow of air from high- to low-pressure areas, which the Earth's rotation swings to the right in the northern hemisphere and to the left in the southern. BUYS BALLOT'S LAW states that winds flow around high- and low-pressure areas, and this produces the characteristic northern hemisphere north-east trade-winds, westerlies and north-easterlies, and the southern hemisphere south-east trade-winds, westerlies and south-easterlies moving from lower to higher latitudes in each case.

generation curve. The POPULATION DENSITY at a given stage of development (e.g., individuals of a particular age) plotted on a graph against the generation number over a sequence of generations.

genes. Physical units of inheritence, which are transmitted from one generation to the next and are responsible for controlling the development of characteristics in the individual receiving them. A gene is a short length of CHROMOSOME. Structural genes determine the sequence of AMINO ACIDS in the synthesis of PROTEINS (e.g., ENZYMES). Regulative genes control the activities of structural genes. *See also* ALLELOMORPHS, CISTRON, MUTATION, PLASMAGENES.

genetic drift. A change in the genetic composition of a population that occurs by chance and not as a result of natural selection.

genetic engineering. A popular name for techniques (e.g., using recombinant DNA) whereby the genetic structure of biological material is modified to produce a desired result. *See also* BIOTECHNOLOGY.

Geneva Convention on Long-Range Transboundary Air Pollution. An international convention, drawn up under the auspices of the United Nations, that came into force in March 1983. It calls for collaboration in research into air pollution and the exchange among signatories of information regarding pollutants, especially new ones.

genome. The genetic material characteristic of a particular species.

genotype. (1) The genetic constitution of an organism. *Compare* PHENOTYPE. (2) A group of organisms whose genetic constitution is identical. (3) The type species of a genus.

genus. *See* CLASSIFICATION.

geo-. A prefix meaning Earth, from the *Ge* or *Gaia* of Greek legend, who formed the Earth from chaos. The prefix is attached to such words as geology (study of the Earth), geography (depiction of the Earth), etc.

geobenthos. *See* BENTHOS.

geobotanical anomaly. The indication of enrichment or depletion of particular elements in the soil according to the presence or absence of certain plant species, or gross physical changes in plants (e.g., the 'copper flower', which grows in soils containing 100–5000 ppm of copper). *See also* BIOGEOCHEMICAL ANOMALY, INDICATOR SPECIES.

geobotany. The study of the global distribution of plants.

geochemical anomaly. The local enrichment or depletion of an element in soil or rock. *See also* GEOCHEMICAL DISPERSION, PRIMARY.

geochemical dispersion, primary. The dispersion of elements at depth within the Earth.

geochemical dispersion, secondary. The dispersion and redistribution of elements at or near the Earth's surface.

geocline. A CLINE associated with a geographic gradient. *Compare* ECOCLINE.

geode. A hollow, roughly globular body in a rock, lined with crystals projecting inwards.

geological time. The system whereby the history of the Earth from the time of its formation is divided into a chronology of definable episodes. The full history is divided into eras, subdivided into periods, and these are subdivided further into epochs. When formally naming a unit of geological time the initial letter of each word is capitalized (e.g., Devonian Period). The earliest eras belong to the PRECAMBRIAN, which is now usually divided into the Archaean and Proterozoic, together accounting for the first four billion years of Earth history, up to the start of the CAMBRIAN. The Precambrian has also been called the Cryptozoic (meaning hidden life), although evidence has now been discovered of Precambrian life forms more than three billion years old. PHANEROZOIC (meaning evident life) Eras account for time from the base of the Cambrian to the present day. The Precambrian Eras are followed by the PALAEOZOIC (meaning ancient life) Era, in which large numbers of life forms appeared, divided into the Cambrian (from about 570 Ma), ORDOVICIAN (from about 500 Ma), SILURIAN (from about 430 Ma), DEVONIAN (from about 395 Ma), CARBONIFEROUS (from about 345 Ma) and PERMIAN (from about 280 Ma) Periods. The MESOZOIC Era began about 225 Ma with the TRIASSIC Period, followed by the JURASSIC (from about 190–195 Ma) and CRETACEOUS (from about 136 Ma) Periods. The CENOZOIC Era, which includes the present, is divided into the TERTIARY and QUATERNARY Periods. The Tertiary Period is subdivided into the PALAEOCENE (from about 65 Ma), EOCENE (from about 54 Ma), OLIGOCENE (from about 38 Ma), MIOCENE (from about 26 Ma) and PLIOCENE (from about 7 Ma) Epochs. The Quaternary Period covers the PLEISTOCENE (from about 2–2.5 Ma) and HOLOCENE (present) Epochs. The Holocene began about 10 000 years ago, but there are some who hold this division to be unreal and that at present we live in an INTERGLACIAL period, still in the Pleistocene Epoch.

geomagnetic induction. The induction of the Earth's magnetic field, postulating a dipole at the centre of the Earth and measuring the Earth's magnetic induction with reference to three axes mutually at right angles, directed towards geographic north, geographic east and vertically downwards to the Earth's centre. This dipole field is overlaid by an irregular non-dipole field, which is continually changing.

geometric growth. *See* EXPONENTIAL GROWTH.

geomorphology. The study of the form and development of the Earth, especially of its surface and physical features, and of the relationship between these features and the geological structures beneath.

geophone. A microphone that is lowered beneath the ground surface or towed by a ship to record shock waves in a seismic survey.

geophyte. *See* RAUNKIAER'S LIFE FORMS.

geosere. A series of CLIMAX formations developed through GEOLOGICAL TIME; the total plant succession of the geological past.

geosphere. The mineral, non-living portion of the Earth (i.e. all of the Earth except for the ATMOSPHERE, HYDROSPHERE and BIOSPHERE).

geostrophic wind. The horizontal wind that blows parallel to the ISOBARS, indicating a balance between the horizontal pressure-gradient force and the horizontal components of the CORIOLIS FORCE. *Compare* GRADIENT WIND.

geosyncline. An elongated basin that has been filled with a great thickness of sediment, usually with intercalated volcanic rocks. The strata of most geosynclines have been affected by OROGENY, often with the

intrusion of BATHOLITHS. Geosynclines have been classified into at least 10 named types, of which only the terms EUGEOSYNCLINE and MIOGEOSYNCLINE are in common use.

geotaxis. *See* TAXIS.

geothermal energy. Energy, as heat, derived from anomalies in the GEOTHERMAL GRADIENT. Normally, temperature in the CRUST increases with depth at a constant rate. Locally, however, water, brine or rock may be much hotter than the surrounding rocks. HOT BRINE or water may be tapped and its heat used. HOT DRY ROCKS may be exploited by drilling two boreholes into them, using explosives or HYDROFRACTURING techniques to shatter the rock between the boreholes and so to render the rock permeable. Cold water under pressure is pumped down one borehole and recovered heated from the other. Geothermal energy is delivered as hot water and can be used only close to its source, to heat buildings or (if necessary with further heating) for electricity generation or other industrial use. The extraction of heat from the local anomaly causes cooling and eventually the subsurface temperature is reduced to that of the surrounding rocks. The water extracted from below ground may contain corrosive and toxic substances dissolved from the rocks, and must be contained within a closed system, its heat being removed by HEAT EXCHANGERS to avoid polluting the environment.

geothermal gradient. The change of temperature with depth in the Earth's crust, usually expressed in degrees per unit depth. The geothermal gradient is usually lowest in SHIELD areas and highest in mid-oceanic RIDGE zones.

geotropism. A growth response of plants in which the stimulus is gravity. Main roots are positively geotropic (i.e. they grow downwards). When placed horizontally, elongation of the cells on the upper side of the growing region increases, causing a downward curvature. Main stems are negatively geotropic. Diageotropic organs (e.g., some RHIZOMES) grow at right angles to the gravitational force. A plagiotropic organ (e.g., a root branch) makes an angle other than a right angle with the line of gravitational force. In a more general sense, plagiogeotropism is used of organs which make any constant angle with the vertical, diageotropism then being a special type of plagiogeotropism. Growth curves are under the control of AUXINS. (2) Formerly a synonym for geotaxis (*see* TAXIS).

germ cell. *See* GAMETE.

germ layers. The layers that can be distinguished during the GASTRULA stage of an animal EMBRYO. In DIPLOBLASTIC species these are the endoderm and ectoderm. In TRIPLOBLASTIC species there is a third layer – the mesoderm. Each layer gives rise to different organs. EPIDERMIS, nervous tissue and nephridia (*see* NEPHRIDIUM) develop from ectoderm. The gut lining and associated GLANDS develop from endoderm. The intervening tissues (e.g., MUSCLE, blood system, kidneys, CONNECTIVE TISSUE) develop from the mesoderm.

germ plasm. The contents of a GAMETE (*see* WEISMANNISM).

ghavial. *See* CROCODILIA.

Ghyben–Herzberg principle. Where fresh water is held in a reservoir rock that is open to the sea, the principle determining the amount of fresh water that can be abstracted from a well before salt contamination occurs. Being more dense, salt water forms a lens beneath fresh water, supporting a column about 2.5 percent higher than itself. If the level of fresh water is lowered by pumping, salt water will rise 40 units for each unit the freshwater level falls.

gibberellins. A group of AUXINS that control growth and development in plants. They cause a marked increase in stem elongation in some plants, and control such processes as flower and fruit formation and dormancy. *See also* CYTOKININS.

gibbons. *See* ANTHROPOIDEA.

gibbsite. ($Al(OH)_3$). One of the major ore minerals of aluminium. Gibbsite is one of the main constituents of BAUXITE and LATERITE.

gibleh. *See* DUST STORM.

giga- (G). The prefix used in conjunction with SI units to denote the unit $\times 10^9$.

gill fungi. *See* AGARICACEAE.

gill pouches. *See* GILLS.

gills. (1) Respiratory organs of aquatic animals. In vertebrates (fish, amphibian tadpoles) the gills are associated with gill slits. These develop as outpushings (gill pouches) of the pharynx which break through to the exterior, meeting inpushings of the EPIDERMIS. Gill slits (or gill pouches and the corresponding epidermal grooves) develop in all chordate EMBRYOS, even those of reptiles, birds and mammals, none of which subsequently develop gills. In invertebrates (e.g., many MOLLUSCA, CRUSTACEA, ANNELIDA and aquatic insect larvae) gills are borne on various parts of the body. (2) Lamellae on which SPORES are formed in agaric fungi (*see* AGARICACEAE).

gill slits. *See* GILLS.

Gingkoales. An order of GYMNOSPERMAE that were abundant during the MESOZOIC, but of which only one representative survives today, the maidenhair tree (*Gingko biloba*), which was found growing in eastern China in 1758. This is a DECIDUOUS tree with fan-shaped leaves, CATKIN-like male CONES and female cones, each of which contains a pair of OVULES.

Givetian. The fifth oldest stage of the DEVONIAN System in Europe.

glabrous. Smooth, not hairy.

glacial drift. The sediments deposited directly by glaciers or indirectly in meltwater streams, lakes or the sea. *See also* BOULDER CLAY, DRIFT.

glacial striation. Scratches on rocks that have been smoothed by ice made by masses of ice containing rocks or grit. Striations are used to reconstruct the local movement of an ice mass.

glacier. A body of ice that originates on land by the compaction and recrystallization of snow and shows evidence of present or past movement. Glaciers occur where winter snowfall exceeds summer melting. Altogether they occupy about 10 percent of the Earth's land surface and contain about 98 percent of the planet's fresh water. There are several types of glacier: ICE SHEETS; valley glaciers, which are ice streams flowing down mountain valleys; piedmont glaciers, intermediate between valley glaciers and ice sheets and comparatively rare, are valley glaciers that spread out across lowland at the foot of a mountain range.

gland. An ORGAN or CELL that makes and pours out (i.e. secretes) one or more specific substances (secretions) (e.g., nectaries in plants, sweat, digestive and mammary glands in mammals). Some excretory organs are also called glands (e.g., hydathodes; *see* GUTTATION). Endocrine glands secrete HORMONES directly into the blood stream.

glass. A hard, amorphous mixture, often transparent, made by fusing oxides of silicon, boron or phosphorus, followed by rapid cooling. Glass is a general term that includes many types of mixture having the same typical composition and physical characteristics.

glass fibre. *See* FIBREGLASS.

glass fibre-reinforced cement. *See* CEMENT.

glazed frost. Clear, glass-like ice deposited on objects by the impact of supercooled (*see* SUPERCOOLING) water droplets in a cloud or fog. The water is spread over the surface of the object before the droplet freezes. Glazed frost forms when the rate of deposition is high, so that the LATENT HEAT of freezing raises the temperature of the capturing body to near 0°C.

glazing. The process of imparting a smooth, lustrous surface to pottery by means of firing a powdered glass on to the surface.

gley. A sticky, organic-rich soil layer which develops on ground that is continuously or frequently saturated with water. *See also* BOG SOIL.

Global 2000 Report to the President. A report drawn up by the COUNCIL ON ENVIRONMENTAL QUALITY and the Department of State at the request of US President Jimmy Carter and presented to him in 1980. It maintained that severe problems would arise from pollution, resource depletion and the extinction of species unless steps were taken urgently to prevent them.

Global Environmental Monitoring System (GEMS). The organization established by the UNITED NATIONS ENVIRONMENT PROGRAMME as part of EARTHWATCH to acquire through monitoring data needed for the rational management of the environment. GEMS monitors changes in climate, RENEWABLE RESOURCES, human health, the long-range transport of pollutants and the oceans.

Global Resource Information Database (GRID). A Geneva-based international organization, established in 1985 by the UNITED NATIONS ENVIRONMENT PROGRAMME and the Swiss Government, that uses computers and software developed by NASA to analyze data on environmental matters. It will integrate data from the GLOBAL ENVIRONMENTAL MONITORING SYSTEM with data from the WORLD HEALTH ORGANIZATION, THE FOOD AND AGRICULTURE ORGANIZATION and others.

***Globigerina* ooze.** *See* RHIZOPODA.

globulins. A group of widely distributed PROTEINS that includes ANTIBODIES and many plant seed proteins.

Gloger's rule. As the mean temperature of the environment decreases (e.g., with increasing latitude) the pigmentation in warm-blooded animals tends to decrease.

glory. A system of coloured, rainbow-like rings surrounding a shadow that is cast on the surface of a cloud. On mountain peaks, when the Sun is low, an apparently hugely magnified shadow of the observer may be cast on to thin cloud, the magnification being due to the observer's assumption that the image is at a much greater distance than is the case because of other objects that can be seen dimly through the mist. The shadow of an aircraft on cloud is often similarly surrounded by coloured rings.

Glossina. See TRYPANOSOMIDAE.

glucagon. A HORMONE secreted by the pancreas of vertebrates that promotes the breakdown of glycogen to glucose (*see* CARBOHYDRATES).

glucose. *See* CARBOHYDRATES.

glume. The basal bracts in a grass spikelet.

glutamic acid. An AMINO ACID with the formula $HOOC(CH_2)_2CH(NH_2)COOH$ and a molecular weight of 147.1.

glutamine. An AMINO ACID with a formula $H_2NCH(CH_2)_2(CONH_2)COOH$ and a molecular weight of 146.1.

glycine. An AMINO ACID with the formula $CH_2(NH_2)COOH$ and a molecular weight of 75.1.

Glycine max. See SOYA.

glycogen. *See* CARBOHYDRATES.

glycolysis. The ANAEROBIC first stage in the liberation of energy from food during RESPIRATION. Glucose is broken down to LACTIC ACID or pyruvic acid by a series of ENZYME-catalyzed reactions, and a small amount of ATP is built up. Glycolysis occurs in all types of organism and is an important source of energy during short bursts of intense muscular activity which outrun the available oxygen supply.

glycosides. Complex CARBOHYDRATE substances which, in the presence of hydrolytic ENZYMES (*see* HYDOLYSIS) yield one or more simple sugars and a non-sugar product – an aglycon – which may be toxic. Plant glycosides are water-soluble, bitter, often produce an odour and may be coloured or colourless. They fall into three groups: cyanogenic glycosides, found in species of *Sorghum, Prunus* and *Linum*, in which the poison is hydrogen cyanide; saponin glycosides, found in species of *Agrostemma, Digitalis* and *Actinea*; solanin glycosides, found in some species of SOLANACEAE. *See also* ANTHOCYANINS.

glyphosphate. A translocated ORGANO-PHOSPHORUS PESTICIDE that is used widely for the control both of broadleaved weeds and of grasses. Its mammalian toxicity is relatively low.

Gnathostomata. Vertebrates that possess jaws. *Compare* AGNATHA.

gneiss. A coarse-grained, banded, metamorphic (*see* METAMORPHISM) rock with alternating layers of dissimilar minerals.

Gnetales. An order of GYMNOSPERMAE that show some resemblance to ANGIOSPERMAE (e.g., in the possession of vessels in the wood). Gnetales comprise only three genera: *Ephedra*, shrubs of warm temperate regions with scale-like leaves; *Gnetum*, tropical, evergreen, mostly climbing shrubs; *Welwitschia*, a long-lived woody plant of the deserts of south-western Africa with two oblong leaves which grow throughout the plant's life.

Gnetum. See GNETALES.

GNP. *See* GROSS NATIONAL PRODUCT.

goats. *See* BOVIDAE.

gob. *See* CULM AND GOB BANKS.

goethite ($FeO.OH$). A hydrated iron oxide-bearing mineral which is a WEATHERING product of iron-bearing minerals and a major component of LIMONITE.

gold (Au). A bright yellow, rather soft metallic element that is not corroded by air or water and is not attacked by most acids, but dissolves in aqua regia (a mixture of one part of nitric acid to four parts of hydrochloric acid by volume). The metal is found native in HYDROTHERMAL veins and in PLACER deposits, and it is also concentrated in some GOSSANS. A_r=196.967; Z=79; SG 19.3; mp 1063°C.

Golden Triangle. (1) The roughly triangular island that accommodates the business centre of Pittsburgh, Pennsylvania. (2) *See* NORTH WEST EUROPEAN MEGALOPOLIS.

Golgi apparatus (Golgi body). A structure present in the CYTOPLASM of plant and animal cells and thought to be concerned with secretion. Electron microscopy has shown the Golgi apparatus to consist of a group of flattened, membrane-bounded sacs and associated VESICLES, often continuous with the ENDOPLASMIC RETICULUM.

gonad. In animals, an organ (ovary, testis) that produces GAMETES.

gonad hormone. A HORMONE (*see* ANDROGENS, OESTROGENS, PROGESTERONE) produced by a GONAD.

gonadotropic hormones (gonadotrophic hormones, gonadotrophins). In vertebrates, HORMONES secreted by the PITUITARY GLAND that control the activity of the GONADS, including their production of hormones. *See also* FOLLICLE-STIMULATING HORMONE, LACTOGENIC HORMONE, LUTEINIZING HORMONE.

GOR. *See* GAS/OIL RATIO.

Gordiacea. *See* NEMATOMORPHA.

Gordon and Franklin. Two rivers in Tasmania that supply a temperate rain forest that is designated a WORLD HERITAGE SITE. Plans by the State Government for a hydroelectric scheme that would have involved damming both rivers and flooding 140 square kilo-

metres (54 square miles) of the forest were overruled by the Federal Government in March 1983, following worldwide protests.

gorilla. *See* ANTHROPOIDEA.

Gorleben. A site in Lower Saxony, West Germany, proposed for a nuclear waste reprocessing and storage plant. The site was the focus of many popular demonstrations, and work was suspended in 1979, following a public inquiry in Hanover and the subsequent decision by the Lower Saxony government to await further research on long-term waste disposal before proceeding further.

gorse. *See* LEGUMINOSAE.

gossan. A cellular mass of hydrated iron oxides (essentially LIMONITE), often with QUARTZ and other GANGUE minerals, from which sulphide minerals have been oxidized and leached out (*see* LEACHING) by downward-percolating waters. The presence of a gossan at the surface usually indicates primary sulphides, as well as SECONDARY ENRICHMENT of the ore vein, at depth.

Gossypium. Genus of tropical and subtropical plants (family: Malvaceae) whose seeds yield cotton, comprising long cellulose fibres that cover the seed coat. The seeds also yield oil when crushed and the residue from crushing (oil cake) is fed to livestock.

graafian follicle. *See* OVARIAN FOLLICLE.

graben. A downthrown block between two normal FAULTS.

grab sampling. (1) Obtaining a sample of an atmosphere in a very short time, such that the time in sampling is insignificant compared with the process or period being sampled. (2) A technique for obtaining BENTHIC material from a lake, river or shallow sea in which a box-like sampler is used, operated from the surface.

grade. (1) A group of things all of which have the same quality or value. (2) *See* GRADED SLOPE. (3) *See* GRADED RIVER. (4)

Alternative word for gradient.

graded aggregates. *See* SORTING.

graded bedding. A sedimentary structure in which the coarsest material is concentrated at the bottom of a BED, and the average grain size decreases towards the top of the bed. Waning currents often produce graded bedding, which is a feature of TURBIDITES.

graded river. A river whose slope and channel have developed to provide the exact velocity needed to transport the sediment load it carries. Grade is established first downstream and gradually extends upstream. Theoretically, the profile is smooth and hyperbolically curved, being steep at the source and becoming more nearly horizontal at base level.

graded slope. A slope that is dynamically stable and will maintain itself in the most efficient configuration.

gradient, geothermal. *See* GEOTHERMAL GRADIENT.

gradient wind. A generalization of the GEOSTROPHIC WIND which disregards friction and assumes the wind to flow parallel to the ISOBARS. This gives a truer representation of the actual wind, especially at high wind speeds and along very curved trajectories. Whereas the geostrophic wind can be calculated from the distribution of pressure along a level surface or pressure surface, the gradient wind must also take account of the curvature of the trajectory.

GRAEL. *See* GREENS.

graft hybrid. *See* CHIMAERA.

Gramineae (grasses). A very large and widespread family of MONOCOTYLEDONEAE, with more than 10 000 species, most of which are herbaceous, but a few are woody (e.g., BAMBUSACEAE). The stems are jointed, the long, narrow leaves originating at the nodes. The flowers are inconspicuous, with a much reduced perianth (*see* FLOWER), and

are wind-pollinated or cleistogamous (*see* CLEISTOGAMY). The fruit is single-seeded, usually a CARYOPSIS. Grasses are the most important of all plants for food. They provide cereal crops (e.g., rice, *Oryza sativa*; maize, *Zea mays*; millet, *Sorghum vulgare*; *Setaria italica*; *Pennisetum typhoideum* and *Panicum* species; wheat, *Triticum* species, *see* EINKORN, EMMER; oat, *Avena sativa*; barley, *Hordeum vulgare*; rye, *Secale cereale*). The sugar-cane (*Saccharum officinarum*) yields sugar from the soft central stem tissues. Valuable pasture grasses of temperate regions include cocksfoot (*Dactylis glomerata*), fescues (*Festuca* species), rye-grasses (*Lolium* species) and timothy (*Phleum pratense*). Grasses also yield fibres, paper, adhesives, plastics and materials for thatching and building.

Gram reaction. A bacteriological staining technique in which Gram's stain (named after Christian Gram) is used to distinguish between Gram-positive BACTERIA (e.g., *Streptococcus, Staphylococcus*), which retain the stain, and Gram-negative bacteria (e.g., *Gonococcus*, which causes gonorrhoea), which do not. This contrast reflects marked differences in the biochemistry of the bacteria.

grana. *See* CHLOROPLAST.

Grandpa's Knob generator. An experimental wind-powered electricity generator, designed to produce 1250 kilowatts, installed in the 1940s at Grandpa's Knob, a hill in Vermont, USA. The rotor failed in a high wind, but during the time it was operational it provided sufficient information to encourage further investigation of the possibilities of large-scale generation of electricity using wind power.

granite. A coarsely crystalline ACID IGNEOUS rock with QUARTZ (at least 10 percent) and ALKALI FELDSPAR as the essential minerals. MICA is commonly present, as is sodium-rich plagioclase feldspar. Granite is used chiefly as AGGREGATE and as polished facing for buildings.

granivore. An animal that eats mainly grain.

granodiorite. A coarsely crystalline acid IGNEOUS rock with QUARTZ and both plagioclase and orthoclase FELDSPAR, the plagioclase predominating. MICA is commonly present. Granodiorite is probably the most voluminous of the PLUTONIC igneous rocks and predominates in most of the BATHOLITHS in the world.

granophyre. A MICROGRANITE with a graphic texture of intergrown QUARTZ and FELDSPAR.

granophyric. *See* GRANOPHYRE.

granulite. A high-grade metamorphic (*see* METAMORPHISM) rock with a granular texture.

granulocyte. *See* BLOOD CORPUSCLES.

graphic. Applied to a rock texture that consists of intergrown QUARTZ and ALKALI FELDSPAR crystals; it is seen in certain PEGMATITES, GRANITES and MICROGRANITES (e.g., GRANOPHYRE).

graphite. A crystalline form of CARBON, the atoms being joined together by strong carbonñcarbon bonds in two dimensions, but only by weak van der Waals' forces in the third dimension. It is found in nature, but is also manufactured from COKE and PITCH to form electrodes or blocks of nuclear reactor MODERATOR graphite. Graphite is also a component of the 'lead' in pencils, being soft owing to the characteristic structure of the crystal. It can be machined, but health precautions are taken in graphite machine shops.

graptolites. A class of HEMICHORDATA whose fossils are used to date rocks from the Lower PALAEOZOIC Era. True graptolites appeared first in the ORDOVICIAN Period, and they became extinct in Britain in the SILURIAN and elsewhere in Europe in the early DEVONIAN. They consisted of one or more branches along which cup-like 'thecae' were arranged, cylindrical in the Lower Ordovician, but showing a variety of shapes in the Upper Ordovician. The name is derived from the resemblance to writing of the carbonaceous film left by their remains, from the Greek *grapho* (write) and *lithos* (stone).

grasses. *See* GRAMINEAE.

grasshoppers. *See* ACRIDIDAE, ORTHOPTERA.

grassland. Herbaceous vegetation that is dominated by grasses. Well over half of the British Isles is grassland. Grassland above the tree limit on mountains or subject to winds and spray on coastlines is natural in origin, but most grassland has been created over the last 2000 years by forest clearance and grazing, by domestic animals and by rabbits. Cessation of grassland management (e.g., cutting, grazing, firing) leads to the formation of scrub and the re-establishment of woodland (*see* DOWNLAND). The vegetation of permanent hay meadows (i.e. fields which are mowed, but not grazed) is of great botanical interest, and such meadows are of great conservation value. They contain tall perennial herbs including dog daisy (*Chrysanthemum leucanthemum*), meadow buttercup (*Ranunculus acris*) and yarrow (*Achillea millefolium*). Permanent pasture (i.e. fields that are both grazed and mowed) is distinguished by the abundance of rosette plants such as daisy (*Bellis perennis*), dandelion (*Taraxacum officinale*) and ribwort (*Plantago lanceolata*). Calcareous grassland is rich in attractive herbs such as orchids, small scabious (*Scabiosa columbaria*), rock-rose (*Helianthemum chamaecistus*), felworts (*Gentianella* species) and horseshoe vetch (*Hippocrepis comosa*). Ley (temporary grassland), sown as a crop and containing only a few species, is replacing permanent grassland in many areas. Grasslands form one of the world's major BIOMES, characteristic of the mid-latitude interiors of continents in both hemispheres (called PRAIRIES in North America, pampas in South America, steppes in Asia and SAVANNAH in Africa). These grasslands may have been formed partly by the clearance of forests by humans and maintained subsequently by grazing, which destroys tree seedlings.

graupel. Soft hail.

grauwacke. *See* GREYWACKE.

gravimeter. An instrument for measuring variations in the Earth's gravitational field.

gravitational water. *See* VADOSE WATER.

gravity, crude oil. *See* API GRAVITY.

gravity anomaly. *See* ISOSTASY.

gray (Gy). The SI unit of the dose of IONIZING RADIATION absorbed by living tissues, being equal to 1 joule of energy imparted to 1 kilogram of mass. The gray replaces the RAD.

grayling zone. *See* RIVER ZONES.

graywacke. *See* GREYWACKE.

Great Barrier Reef. A large CORAL REEF developed on an extensive area of CONTINENTAL SHELF and more or less following the 550 metres contour from the Torres Strait south to a point opposite Rockhampton, Queensland, a total length of a little over 2000 kilometres (1200 miles), and with a width varying from 30 to 300 kilometres. It lies about 30–50 kilometres (20ñ30 miles) from the nearest shore in the north, and about 320 kilometres (200 miles) in the south. The Reef is composed of the calcareous skeletal remains of coral, Mollusca and other marine organisms and has a large area of living coral with its associated fauna, making it probably the world's richest ecosystem in terms of the range of species it supports. Parts of the Reef are exposed at low tide, but most lies below the surface. There are fears that the Reef may be damaged by the CROWN OF THORNS STARFISH and by offshore oil exploration.

grebes. *See* PODICIPEDIDAE.

Green Alternative European Link. *See* GREENS.

green bans. Work bans, often imposed officially by trades unions, on projects that are considered likely to cause environmental damage. A number of such bans have been effective in Australia, where they were first attempted, and in the UK a similar ban by the National Union of Seamen on the dumping of radioactive wastes at sea forced the abandonment of this means of waste disposal.

green belt. An area of land, not necessarily continuous, near to and sometimes surrounding a large built-up area. The area is kept open by permanent and severe restriction on building. *See also* GARDEN CITY.

Green Data Book. A list, prepared and published by the INTERNATIONAL UNION FOR CONSERVATION OF NATURE AND NATURAL RESOURCES, of plants that are rare, endangered or threatened with imminent extinction throughout the world. It is a companion work to the RED DATA BOOK.

greenflies. *See* APHIDIDAE.

greenhouse effect. Worldwide changes in climate and sea levels caused by a warming of the atmosphere due to the release of gases, principally CARBON DIOXIDE, which are transparent to short-wave radiation, but absorb radiation at certain long wavelengths. Incoming short-wave solar radiation (including visible light) and heat are absorbed at the ground surface by objects which then behave as BLACK BODIES, radiating heat (i.e. long-wave radiation) back into space. Certain 'greenhouse' gases, (e.g., carbon dioxide, water vapour and CHLOROFLUOROCARBONS) absorb part of this radiation, then reradiate it in all directions, some downwards and some to the side where it may encounter other molecules of these gases and continue the process. Thus the gases form a 'blanket' trapping outgoing heat, much as the glass or plastic does in a greenhouse. The burning of FOSSIL FUELS and clearance of forests, releasing carbon dioxide held in forest soils, has increased the atmospheric concentration of carbon dioxide. The greenhouse effect was predicted early in the 20th century, and by the late 1980s it had been detected in rising sea levels, general atmospheric warming, melting of ice, and changes in vegetation.

Various unusual patterns of weather throughout the world were attributed to it.

greenhouse gases. *See* GREENHOUSE EFFECT.

green manuring. The agricultural and horticultural practice of growing a plant crop specifically for the purpose of ploughing or digging it into the soil in order to improve soil structure and supply nutrients as it decomposes.

greenockite. *See* CADMIUM.

Greenpeace. An international non-governmental environmentalist organization that campaigns to prevent whaling, to halt the discharge of pollutants into the sea and to oppose the nuclear power industry. It owns sea-going vessels and rubber dinghies with which it shadows, obstructs and sometimes attempts to board other ships whose operations it opposes. The group first came to prominence when members sailed in to an area of the Pacific to prevent the atmospheric testing of nuclear weapons; during preparations for a similar protest in 1985 the Greenpeace converted trawler *Rainbow Warrior* was sunk by French agents in Auckland Harbour, New Zealand.

green revolution. The informal name for a complex group of agricultural development programmes devised by the FOOD AND AGRICULTURE ORGANIZATION OF THE UNITED NATIONS and centred on the introduction of new, high-yielding crop varieties. The overall strategy was contained in the INDICATIVE WORLD PLAN FOR AGRICULTURAL DEVELOPMENT. As a result of its implementation food production has been increased in many countries, especially in Asia, and some (e.g., India), which had formerly relied on imports, have become net exporters of food.

Greens. The name adopted by political parties in the UK, Canada, France, Germany and other countries to indicate their primary concern that people must live in harmony with nature within the limitations of the Earth's finite supply of resources. In general, they oppose the nuclear generation of electricity and economic growth as this is usually measured, demand stringent controls on industry to curb pollution and favour self-sufficiency based on the redistribution of wealth, the introduction of APPROPRIATE TECHNOLOGIES and the autonomy of local communities. In the European Parliament a number of national Green parties, together with other parties sharing many of the same objectives, have formed a Rainbow Group (implying a refusal to accept any position on the conventional political spectrum), officially called the Green Alternative European Link (GRAEL).

greenschists. *See* SCHIST.

greenstone. A general name for a range of slightly altered BASIC IGNEOUS rocks.

gregale. *See* MISTRAL

greisening. *See* DEUTERIC.

grey–brown podzolic soils. Acidic soils that are less leached (*see* LEACHING) than a PODZOL. They are forest soils with a layer of litter at the surface and a thin organic layer overlying a greyish–brown leached layer (the A horizon, *see* SOIL HORIZONS). Below this is a darker brown depositional layer (B horizon). *See also* BROWN PODZOLIC SOIL, SOIL CLASSIFICATION.

grey matter. The part of the vertebrate CENTRAL NERVOUS SYSTEM which contains the nerve cell bodies (NEURONS). It lies mainly inside the WHITE MATTER, but forms a superficial layer in the CEREBRAL HEMISPHERES and CEREBELLUM of higher vertebrates.

greywacke (grauwacke, graywacke). A poorly sorted (*see* SORTING) SANDSTONE characteristic of geosynclinal sequences (*see* GEOSYNCLINE) and composed mainly of angular or subangular rock fragments in an ARGILLACEOUS matrix.

greywethers. A popular name for SARSENS, reflecting their fancied resemblance to grazing sheep.

GRID. *See* GLOBAL RESOURCE INFORMATION DATABASE.

grid pattern. An urban planning design that is based on streets spaced at regular intervals and intersecting at right angles. Grid patterns were adopted in many North American cities, but the concept is very ancient.

Grignard reagents. Organometallic reagents of considerable importance in synthesis of organic compounds. They consist of an ether-soluble organomagnesium halide with the general formula RMgX, where R is an alkyl group (i.e. C_nH_{2n+1}) and X is a halide (*see* HALOGENS). They are named after Francois A.V. Grignard (1871–1935).

grike. An enlarged FISSURE in the surface of LIMESTONE caused by chemical WEATHERING. *See also* CLINT.

grit. (1) Solid particles larger than about 76 micrometres, that are released into the atmosphere, usually as a result of industrial activity. Particles smaller than grit are called dust. Particles larger than 5 micrometres cannot penetrate the ALVEOLI, and so grit presents no great hazard to human health, although it may be a nuisance. (2) An imprecise name for an arenaceous (*see* ARENITE) rock that feels gritty. LIMESTONES, SANDSTONES, ARKOSES and GREYWACKES can be grits. Often the gritty feel is due to angular or subangular grains.

gross domestic product (GDP). The value of all the goods and services produced within a nation in a period of time (usually one year) charged at market prices and including taxes on expenditure, with subsidies treated as negative taxes. Essentially, it is a measure of national income. Divided by the number of the population it yields the per caput GDP or the average national income per person. *Compare* GROSS NATIONAL PRODUCT.

gross national product (GNP). The total monetary value of all goods and services produced within a nation during a period of time (usually one year), making no allowance for depreciation of stock or other consumption of capital, but including investment. GNP is a convenient indicator of the level of economic activity and of changes in that activity from one period to another, but beyond that its high degree of aggregation of data makes it a crude tool. *Compare* GROSS DOMESTIC PRODUCT.

gross primary production. *See* PRODUCTION.

gross production rate. *See* PRODUCTION.

ground layer. *See* LAYERS.

ground moraine. *See* MORAINE.

grounds. Solids that are deposited from suspension when the liquid holding them is left to stand, as in brewing.

ground water. Water that occupies pores and crevices in rock and soil, below the surface and above a layer of impermeable material. It is free to move gravitationally, either downwards towards the impermeable layer or by following a gradient. The upper limit of the ground water is the WATER TABLE, the level below which the ground is saturated. Ground water is distinguished from surface water, which remains at or is close to the land surface, above ground that is not saturated. *See also* AQUIFER, METEORIC WATER, VADOSE WATER.

group heating. *See* DISTRICT HEATING.

growing point. *See* MERISTEM.

growth. A permanent increase in a quantity, such as in the mass and volume of an organism, as a result of it having taken in chemical substances unlike itself and converted them into its own substance. In higher plants growth is localized (*see* MERISTEM) and is more or less continuous, involving the division, enlargement and differentiation of cells. In higher animals there are no specialized growing regions, and growth is confined to an early phase in the life cycle.

growth hormone (somatotrophic hormone,

STH). A HORMONE that causes the growth of the entire body and is produced by the PITUITARY GLAND of vertebrates.

growth rings. *See* ANNUAL RINGS.

groyne. A barrier built on a beach at right angles to the water's edge and entering the sea. Groynes are often built as a series, and they serve to reduce erosion by holding back the LONGSHORE CURRENTS which tend to carry sediments along the beach.

Gruiformes (cranes, bustards, crakes, rails, coots, moorhens). An order of birds, many of which are waders that live in marshy country.

grumosols. Soils composed mainly of CLAYS, which swell in wet weather and crack in dry weather. The name is replaced in the US Soil Taxonomy by the order Vertisols. *See* SOIL CLASSIFICATION.

guanine. *See* DNA, RNA.

guano. (1) Deposits of bird droppings that are used as a fertilizer. Guano is found almost exclusively on islands or near coasts, and especially on the west coast of South America. (2) Fertilizers, especially those made from fishes.

guanotrophy. Enrichment with plant nutrients (e.g., PHOSPHATES) derived from bird droppings. Flocks of roosting gulls can cause guanotrophy in inland waters.

guard cells. *See* STOMA.

guayule (*Parthenium argentatum*). A shrub, native to arid regions of Mexico and the south-western states of the USA. The shrub's cells, and especially those of the roots and stems, contain a rubber that is indistinguishable from traditional natural rubber (*see* HEVEA BRASILIENSIS) once it has been processed.

guillemots. *See* ALCIDAE.

Guinea worm. *See* FILARIA

gulls. *See* LARIDAE.

gully. *See* EROSION.

gully reclamation. The restoration of land on which water erosion has formed channels (gullies) along which soil is carried from higher levels at times of heavy rain or the melting of snow. The gullied land may be ploughed at right angles to the gullies and levelled or terraced, and shrubs or trees may be planted to stabilize the soil. Where gullies are very extensive dams may be built across them to collect soil as it is eroded, until they fill and can be levelled.

gumbo. CLAY-rich ground which forms wet and sticky mud.

gums. *See* EUCALYPTUS.

guncotton. A nitrocellulose made by treating cotton with a mixture of nitric and sulphuric acids (nitrating). It is a very safe and convenient explosive.

gunmetal. An ALLOY of copper, zinc and tin, sometimes with the addition of lead and nickel, that has good resistance to corrosion and wear.

gunpowder. An explosive; a mixture of potassium or sodium nitrate, charcoal and sulphur in varying proportions. It is said to be the first manmade explosive.

Gunung. *See* MOUNT AGUNG.

gustiness. TURBULENCE close to the ground that is caused by buildings or other obstacles that prevent the direct flow of air.

Gutenberg discontinuity. *See* MANTLE.

guttation. The excretion of drops of excess water from glands (hydathodes) on the leaves of many plants (e.g., at the tips of grass leaves).

guyot. A flat-topped SEA-MOUNT; its truncated shape is thought to be due to wave action.

Gy. *See* GRAY.

Gymnophiona. *See* APODA.

Gymnospermae. The primitive woody seed plants (SPERMOPHYTA), whose ovules are not protected by ovaries and are usually borne in CONES. The Gymnospermae includes the orders Bennetitales, Cordaitales, and Cycadofilicales (extinct forms), and the extant CONIFERALES, CYCADALES, GINKGOALES, GNETALES and TAXALES. *Compare* ANGIOSPERMAE.

gynandromorphism. An abnormality found in insects, birds and mammals in which one part of the body is male and the rest female. *Compare* INTERSEX.

gynoecium. *See* FLOWER.

gypsum ($CaSO_4.2H_2O$). An EVAPORITE mineral, hydrated calcium sulphate that is used extensively in making wallboard and plaster, in paint and paper filters, in the production of SULPHURIC ACID, etc. Different forms of gypsum are known as ALABASTER, DESERT ROSE, SATIN SPAR and SELENITE.

gyroscope. A heavy symmetrical disc that is free to rotate about an axis which itself is confined in a framework that allows it to take on any orientation in space. A spinning gyroscope maintains fairly constantly any orientation to which it is set. Gyroscopes are used as stabilizers. In ship and aircraft instruments they are used to show changes in the orientation of the vehicle in which they are contained.

gyttja. A dark-coloured sediment found at the bottom of EUTROPHIC lakes. It has a high organic content and provides a ready source of plant nutrients.

H

H. (1) *See* HENRY. (2) *See* HYDROGEN.

h. *See* HECTO-.

h. *See* PLANCK'S CONSTANT.

ha. *See* HECTARE.

Haber process. An industrial process for synthesizing ammonia (NH_3) from atmospheric nitrogen (N_2) and hydrogen (H_2) by passing the gases at high temperature and pressure through a bed containing a CATALYST (osmium). The gases combine according to the equation $N_2 + 3H_2 \rightarrow 2NH_3$. The process is named after Fritz Haber (1868–1934), who developed it in Germany under the pressure of an acute ammonia shortage during World War I.

habit. (1) The general appearance of a plant (e.g., creeping, erect). (2) The characteristic form of the CRYSTALS that comprise a particular mineral.

habitat. The dwelling place of a species or COMMUNITY, providing a particular set of environmental conditions (e.g., forest floor, sea shore etc.).

habitat diversification. *See* BETA-DIVERSITY.

habitat loss. A principal cause of the disappearance of flora and fauna, the development of areas of natural habitat for agricultural or other human use (e.g., DEFORESTATION, the removal of hedgerows to enlarge fields, the filling in of small ponds no longer needed to water livestock, the over-zealous clearance of wild plants from roadside verges, etc.).

habitat type. A group of COMMUNITIES that resemble one another because of similarities in the habitats they produce (e.g., woodland, grassland).

habituation. A diminishing response to repeated stimulation.

haboob. *See* DUST STORM.

hackle. (1) One of the long erectile feathers on the necks of some birds. (2) A metal comb, made from teeth mounted vertically on a wooden base, used in the preparation of certain vegetable fibres (e.g., flax) to remove the outer sheath of the plant stem that has been partially decomposed by RETTING, leaving behind the fibres.

hackly. Applied to a rock fracture that produces a surface covered with sharp, jagged projections.

hade. The angle between a FAULT PLANE, or the plane of a mineral vein and the vertical. Hade is the complement of DIP.

Haeckel, Ernst Heinrich (1834–1919). A German zoologist who is credited with having first used the word ecology, defining it as 'the study of the economy, of the household, of animal organisms'. Haeckel supported enthusiastically the views of DARWIN, CHARLES ROBERT, who believed that Haeckel was instrumental in advancing them in Germany. Later, Haeckel attempted to expand darwinian evolutionary theory into a philosophical and religious system, eventually denying the existence of a personal God, freedom of will and the immortality of the soul. He based this on his belief that higher forms of life evolved from simpler forms and that every cell has psychic properties, so that psychic and psychological processes are merely extensions of the physical.

haematite (hematite, Fe$_2$O$_3$). An iron oxide mineral; the most important ore of iron. It occurs as a widespread ACCESSORY MINERAL in IGNEOUS rocks, in HYDROTHERMAL veins and as ooliths (*see* OOLITE), a replacement mineral and CEMENT in SEDIMENTARY ROCKS. The major exploited occurrence is in the PRECAMBRIAN BANDED IRONSTONE formations. Haematite has a cherry-red STREAK. It is used as a pigment and in anticorrosion paints. Varieties of haematite include kidney iron ore, specularite and micaceous haematite.

haemocoel. *See* COELOM.

haemocyanin. A blue–green, copper-containing RESPIRATORY PIGMENT found in the blood of some MOLLUSCA and ARTHROPODA.

haemoglobin (hemoglobin). An iron-containing RESPIRATORY PIGMENT found in the blood of vertebrates and a few invertebrates. The oxygenated form is scarlet, and the deoxygenated form is bluish–red. Substances that are able to form stable compounds with haemoglobin (e.g., CARBON MONOXIDE) reduce the amount of oxygen reaching tissues and can cause ANOXIA. *See also* METHAEMOGLOBINAEMIA.

hagfish. *See* AGNATHA, CYCLOSTOMATA.

hair follicle. An inpushing of the EPIDERMIS in mammals that surrounds the hair root and produces the hair. *See* OVARIAN FOLLICLE.

halarch succession. *See* HALOSERE.

half-bog soil. *See* BOG SOIL.

half-life. The time required for the decay or disappearance of half of a substance that decays in a regular, exponential way, such that if the amount of the substance starts at 100, after the first half-life period 50 will remain, after the second half-life period 25, then 12.5, 6.25, etc. The term is most commonly used in respect of the decay of radioactive substances.

half-value layer. The thickness of a given material that will reduce the intensity of a beam of radiation to one-half of its original value.

halinokinesis. The tendency of a HALITE to flow under the pressure of OVERBURDEN. It is important in creating oil traps on the flanks and over the crests of the resulting SALT DOMES.

halite (rock salt, NaCl). A widely distributed EVAPORITE mineral that occurs together with other water-soluble minerals (e.g., SYLVITE) and other minerals (e.g., GYPSUM). It may be colourless, white, red, yellow or blue. *See also* HALINOKINESIS.

halo. A misty circle, or series of circles, around the Sun or Moon, preceding the arrival of rain.

halo-. A prefix, from the Greek *halos*, meaning salt.

halocline. The boundary between two masses of water whose SALINITIES differ.

halogenation. The incorporation of one of the HALOGENS, most commonly CHLORINE or BROMINE, into a chemical compound.

halogens. The very reactive elements FLUORINE, CHLORINE, BROMINE, IODINE and astatine (in descending order of reactivity), which together form Group VII of the periodic table. The name is derived from the Greek *halos* (salt) because halogens react with metals to produce salts, called halides (e.g., chlorine reacts with sodium to produce sodium chloride, which is common salt). They also react readily with organic compounds (e.g., organochlorines, which are chlorinated hydrocarbons, and chlorofluorocarbons, which are compounds of carbon with chlorine and fluorine).

halomorphic soil. A soil that contains an excess of salt or of an alkali.

halophyte. A plant that grows in soil containing a high concentration of salt (e.g., samphire, *Salicornia*, which grows in salt marshes).

halosere (halarch succession). The stages in a plant SUCCESSION that begin under saline conditions.

hamada (hammada). A desert region where the surface is BEDROCK. *Compare* ERG, REG. *See also* DESERT PAVEMENT.

hammer mill. A crushing machine in which swinging hammers are pivoted to a revolving arm and crush material against a grid of steel bars. Hammer mills are used in the preliminary treatment of ores and to reduce the bulk of waste materials prior to their disposal.

hamra. A red, sandy soil that also contains CLAY.

handy man. *See HOMO.*

hanger. A wood, often a beechwood, situated on a hillside.

hanging valley. A valley formed by a tributary GLACIER, which flowed into a main glacier so that the surfaces of the two ice masses were level, but the bed of the tributary was at a higher level than that of the main glacier. When the glaciers disappeared, a main valley remained with a tributary valley entering at some distance up its side. Where such valleys carry rivers, they end in a waterfall.

hanging wall. The higher side of an inclined FAULT or VEIN, or the ore limit on the upper side of an inclined ORE BODY.

haploid. *See CHROMOSOMES.*

haplont. The haploid (*see* CHROMOSOMES) stage in the life history of an organism. In animals this is the GAMETE. In BRYOPHYTA, and many ALGAE and FUNGI, the haplont is the dominant stage (gametophyte). In ferns and seed-bearing plants, the haplont is reduced. *Compare* DIPLONT. *See also* ALTERNATION OF GENERATIONS.

haptotropism (thigmotropism, stereotropism). A growth response of plants in which the stimulus is localized contact (e.g., the contact with a support which makes a tendril coil around it).

hardness. (1) Resistance to deformation. In metals this is usually measured by pressing a hardened steel ball or diamond pyramid into the metal for a given time under a given load. The hardness of minerals is usually measured by a scratch test invented by Friedrich Mohs in 1812. He arranged 10 minerals in order of their hardness, so that each will scratch all those below it on the scale. Diamond, at 10 on MOHS'S HARDNESS SCALE, is the hardest and talc, at 1, the softest. (2) Of water, a measure of the amount of dissolved mineral salts (especially calcium), so that the more salts there are the harder the water is said to be. Hardness increases the amount of soap or other detergent required in washing, and it causes scaling in boilers and pipes and furring in kettles.

hardpan. A strongly compacted subsurface soil layer. When cemented (*see* CEMENT) with SILICA it is called a duripan or silcrete, with calcium carbonate a caliche or petrocalcic horizon, with CLAY a claypan and with ferric oxide a LATERITE or plinthite. A thin layer of ferric oxide may be called an iron pan. A fragipan is an acid cemented SOIL HORIZON with a platey structure that occurs between the depositional horizon of the soil and the parent material.

hardwood. A BROADLEAVED TREE or its timber; hardwood species have vessels (*see* TRACHEA) in their wood. *Compare* SOFTWOOD.

hares. *See LAGOMORPHA.*

harmattan. A hot, dry wind that blows from the north-east over the southern Sahara, usually in winter. It is accompanied by blowing sand and dust, which can be seen from afar and which reduces visibility considerably within the affected area. The harmattan is known locally as 'the doctor' for the relief it brings from the damp heat that precedes it.

harmonic. A pure tone, a sinusoidal component in a complex periodic wave, of a FREQUENCY that is an integral multiple of the FUNDAMENTAL FREQUENCY of the wave. If a component (an overtone) in a sound has a frequency twice that of the fundamental it is called the second harmonic.

harvestmen. *See* PHALANGIDA.

haustorium. An organ of parasitic plants (e.g., dodder, *Cuscuta*; some fungi) that withdraws food material from the tissues of the host plant. *See also* PARASITISM.

Hauterivian. *See* NEOCOMIAN.

Hawaiian Floral Region. The part of the PALAEOTROPIC REALM that comprises the Hawaiian islands.

Hawaiian goose. *See* ANATIDAE.

Hawaiian-type volcano. *See* SHIELD VOLCANO.

hawks. *See* FALCONIFORMES.

hawthorn. *See* CRATAEGUS.

hay meadows. *See* GRASSLAND.

hazardous waste. A waste that contains any substance harmful to life. Such a substance may be toxic (e.g., pesticides, compounds of arsenic, cyanides), flammable (e.g., hydrocarbons), corrosive (e.g., strong acids or alkalis) or oxidizing (e.g., nitrates or chromates), and some may be hazardous on more than one count.

hazel. *See* CORYLLUS AVELLANA.

HCH. *See* LINDANE.

He. *See* HELIUM.

head. *See* COMBE ROCK.

headstream. *See* RIVER ZONES.

Health and Safety Executive (HSE). In the UK, the statutory body established to administer the Health and Safety at Work Act, 1974. It is appointed by and is responsible to the Health and Safety Commission, which interprets the Act, issuing and promulgating regulations and sponsoring research and training, and which reports to the Government minister responsible for each industry or activity with which it deals. The HSE operates through inspectorates (e.g., Factories Inspectorate and NUCLEAR INSTALLATIONS INSPECTORATE). *See also* MAJOR HAZARD INCIDENT DATA SERVICE.

hearing loss. The amount, in DECIBELS, for a specified ear and FREQUENCY by which the threshold of audibility for that ear exceeds the normal threshold.

heart rot. *See* TRACE ELEMENTS.

heart urchins. *See* ECHINOIDEA.

heartwood (duramen, truewood). The central part of the wood, which, in the living tree, no longer contains live cells. It functions only for support and not for the conduction of water. The heartwood is generally darker than the SAPWOOD and is more resistant to decay.

heat. *See* OESTROUS CYCLE.

heat exchanger. A device for transferring heat from one fluid or body to another. An example is a coiled pipe carrying hot water through a tank of cooler water in order to warm one, cool the other, or both.

heath. (1) A shrub of the genus *Erica* (*see* ERICACEAE). (2) An area with poor acid soil, typically dominated by ling (*Calluna*) or heaths (*Erica*). Although heathland is usually associated with sandy or gravelly soils, it may develop in CHALK or LIMESTONE areas where the topsoil has become acid through LEACHING.

heat of formation. The heat evolved when 1 gram-molecule of a compound is formed from its constituent elements.

heat of fusion. The heat that is needed to convert a given weight of a solid to its liquid phase.

heat of solution. The amount of heat taken in or given out when a substance is dissolved in a large amount of solvent.

heat pump. A device that transfers heat from a cool region to an adjacent one that is warmer, normally for the purpose of space heating or cooling (i.e. air conditioning), by exploiting the temperature gradient between the two regions. A chemical substance (e.g., a CHLOROFLUOROCARBON) whose boiling point is close to the ambient temperature is circulated in a closed system between the two regions and is compressed in the warmer of them. In the cooler region the fluid vaporizes, absorbing energy in order to do so; in the warmer region it is compressed and condenses, giving up energy as heat. A refrigerator or deep freezer uses the same system, but to cool a small space rather than to heat a large one.

heat recovery. The transfer of heat from a substance that has been heated during an industrial process to a substance about to be treated in order to avoid wasting energy (e.g., the recovery of heat from furnace or incinerator waste gases). The term also refers to the use of waste heat for another purpose (e.g., the use of hot water for DISTRICT HEATING).

heat transfer. The exchange of heat between one body and another, which may proceed by: (a) conduction, where heat diffuses through solid materials or stagnant fluids; (b) CONVECTION; (c) radiation, where heat moves as electromagnetic waves.

heave. The horizontal displacement between the upthrown and downthrown sides of a FAULT.

heavenly cross. A SUN PILLAR crossed by a horizontal bar.

heavy liquids. A group of dense liquids that

are used to separate out ACCESSORY MINERALS (i.e. the HEAVY MINERALS). Commonly used heavy liquids include bromoform (SG 2.87), methylene iodide solution (SG 3.2) and Clerici's solution (SG 4.25). The specific gravity of QUARTZ IS 2.65, of CALCITE 2.71 and of FELDSPAR 2.55–2.76.

heavy minerals. The detrital ACCESSORY MINERALS of a sediment or sedimentary rock that are of high specific gravity (usually set arbitrarily at more than 2.87, the specific gravity of bromoform). Heavy minerals are concentrated in PLACER deposits (e.g., BLACK SANDS). *See also* HEAVY LIQUID.

heavy water (deuterium oxide, D_2O). Water in which the HYDROGEN has been replaced by DEUTERIUM. *See also* NUCLEAR REACTOR.

Hebb–Williams maze. A standard device used to measure intelligent behaviour in experimental animals. It consists of an enclosure with start and goal boxes at opposite ends. The floor is marked out into 36 5-inch squares and barriers are arranged along the boundaries of the square so that there are many blind alleys, but only one correct path to the goal. The number of times a hungry rat must be released from the start box in order to learn the correct path is measured, and the total number of errors it makes in learning its way through the maze is recorded.

hectare (ha). A metric unit of area, equal to 10 000 square metres (2.471) acres).

hecto- (h). The prefix used in conjunction with SI units to denote the unit x 10^2.

hekistotherm. A plant that thrives with very little heat (e.g., arctic mosses and lichens, alpine plants).

heliophobe (sciophyte, shade plant, skiophyte). A plant that grows best in shady places (e.g., dog's mercury, *Mercurialis perennis*). *Compare* HELIOPHYTE.

heliophyll. A plant whose leaves are similar

in structure on both sides and are arranged more or less vertically. *Compare* SKIOPHYLL.

heliophyte (sun plant). A plant that grows best in full sunlight (e.g., cactus). *Compare* HELIOPHOBE.

heliosis. Discoloration on leaves caused by high intensities of sunlight. *See also* SOLARIZATION

heliotaxis. *See* TAXIS.

heliotropism. *See* PHOTOTROPISM.

helium (He). An inert gaseous element that occurs in some NATURAL GASES in the USA, occluded in some radioactive ores (e.g., MONAZITE, PITCHBLENDE) and in the atmosphere at about one part in 200 000. It is used to fill balloons because of its properties as a lifting agent, in some metal-working processes and as a tracer in determining the migration of oil and gas in geological structures. It also has medical, commercial and scientific uses. An ALPHA-PARTICLE is a helium nucleus (two protons and two neutrons) with no surrounding electrons, emitted during certain types of radioactive decay. $A_r = 4.0026$; $Z = 2$; mp $-272.2°C$; bp $-268.934°C$.

helminths. Parasitic (*see* PARASITISM) worms especially members of the PLATYHELMINTHES or NEMATODA.

helm wind. A steady strong wind that blows down the westward (i.e. LEE) slopes and for some distance across the lowlands in the northern Pennines, England, when the prevailing wind over the wider area is from the east. It is caused by wave-like disturbances in the air stream as the wind flows over the ridges of low hills, which accelerate the wind blown down the lee slope.

helophyte. *See* RAUNKIAER'S LIFE FORMS.

Helsinki Convention for the Protection of the Marine Environment of the Baltic Sea. An agreement among seven states bordering the Baltic to reduce pollution, allotting specific tasks to individual states (e.g., Finland is responsible for dealing with oil spills). The Convention was signed in 1973 and came into force in May 1980.

hematite. *See* HAEMATITE.

Hemichordata (Enteropneusta). A small subphylum of the CHORDATA that comprises marine animals, very different from other chordates, which may be worm-like and burrowing (e.g., *Balanoglossus*) or sedentary (e.g., *Cephalodiscus*).

hemicryptophyte. *See* RAUNKIAER'S LIFE FORMS.

Hemimetabola. *See* EXOPTERYGOTA.

hemimetabolous. *See* METAMORPHOSIS.

Hemiptera (Rhynchota, bugs). A large order of insects (division: EXOPTERYGOTA), with piercing and sucking mouth-parts, most of which feed on plant juices, a few on animals. *See also* HETEROPTERA, HOMOPTERA.

hemoglobin. *See* HAEMOGLOBIN.

hemp. *See* CANNABIACEAE.

henry (H). The derived SI unit of self-induction and mutual induction, being the inductance in a closed circuit such that a rate of change of current of 1 ampere per second produces an induced electromotive force (EMF) of 1 volt. It is named after Joseph Henry (1797–1878).

Hepaticae (liverworts). A class of BRYOPHYTA whose members live in damp places or in water. A simple liverwort has a small, flat, green, repeatedly forked, ribbon-like body, lying close to the ground, to which it is attached by unicellular, hair-like RHIZOIDS. Some liverworts resemble mosses in possessing leaf-like expansions, but unlike mosses they have no strands of conducting tissue.

heptachlor. An ORGANOCHLORINE insecticide (a CYCLODIENE INSECTICIDE) which breaks down in the soil to heptachlor epoxide, a stable and more poisonous substance than heptachlor itself. Heptachlor is no longer approved for agricultural use in the UK.

herb (herbaceous plant). A non-woody VASCULAR PLANT having no parts that persist above the ground.

herbaceous plant. *See* HERB.

herbicide. A chemical that is used to kill weeds. *See also* CARBAMATES, DINITRO PESTI-CIDES, HORMONE WEEDKILLERS, SOIL-ACTING HERBICIDES, TRANSLOCATED HERBICIDES, UREA HERBICIDES.

herbivores. Plant-eating animals; primary consumers in a FOOD CHAIN.

herb layer. *See* LAYERS.

herbosa. Vegetation that is made up of non-woody plants (i.e. HERBS).

Hercynian. The processes and products of the OROGENY that affected large parts of Europe in the Upper PALAEOZOIC (DEVONIAN to PERMIAN). Hercynian massifs are exposed in South-West England, Ireland, Brittany, central France, the Iberian Peninsula, Bohemia and in the Ardennes–EifelñRhenish Schiefergebirge belt. Fragments of Hercynian massifs occur within the massifs of the ALPINE OROGENY. *See also* AMORICAN, VARISCAN.

heritability. The estimate of the extent to which a characteristic is inherited; the percentage variation in the characteristic in a population that can be accounted for by genetic variation within that population (e.g., in humans, height has a heritability of 90–95 percent). This estimate contributes nothing to the NATURE–NURTURE controversy (one cannot say that 90 percent of height is determined genetically and 10 percent by environmental factors) and use of the heritability factor in descriptions of behaviour has been criticized.

hermaphrodite (bisexual). Applied to an individual that contains both male and female functional reproductive organs (e.g., earthworm, buttercup). *Compare* DIOECIOUS, MONOECIOUS, UNISEXUAL.

heroin (diacetylmorphine). A drug obtained by the alkylation of morphine. It is a narcotic ALKALOID with an effect resembling that of morphine and is classed as a dangerously addictive drug.

herons. *See* CICONIIFORMES.

herpetology. The scientific study of reptiles.

hertz. *See* FREQUENCY.

hesperidium. *See* CITRUS.

heterocaryon. *See* HETEROKARYON.

heterochrosis. Abnormal coloration (e.g., pale plumage in dark-feathered birds).

heterocyclic compound. Cyclic carbon compound in which other atoms, typically nitrogen, oxygen or sulphur, form part of the ring structure.

heterodont. Possessing different types of teeth (e.g., incisors, canines and molars in a mammal). *Compare* HOMODONT.

heteroecious parasite. *See* PARASITISM.

heterogametic sex. The sex in which the SEX CHROMOSOMES are dissimilar (X and Y) or in which there is an unpaired X chromosome. In LEPIDOPTERA and many vertebrates (birds, reptiles, some amphibians, fish) the female is heterogametic. In most other organisms, including humans, the male is heterogametic. *Compare* HOMOGAMETIC SEX.

heterogamy. (1) (anisogamy) The production of unlike GAMETES, which differ in size and/or form. *See also* OOGAMY. (2) The production of unlike gametes, which differ as to the CHROMOSOMES they contain. (3) The production of flowers of more than one type in the same inflorescence. *Compare* HOMO-GAMY. (4) The alternation during the life cycle of an animal of two types of reproduction involving gametes (e.g., PARTHENOGENE-SIS and SYNGAMY in APHIDIDAE).

heterokaryon (heterocaryon). A cell that contains two or more nuclei of differing genetic constitution. *Compare* DIKARYON, HOMOKARYON.

Heterokontae. *See* XANTHOPHYTA.

Heterometabola. *See* EXOPTERYGOTA.

heterophyte. A parasite (*see* PARASITISM) or SAPROPHYTE.

Heteroptera. A suborder of insects (order: HEMIPTERA). Members typically have forewings that are horny. The suborder includes many plant bugs, the water bugs (e.g., water boatmen) and the wingless bed bug (*Cimex lectularius*). *See also* CAPSIDAE.

heterosis. *See* HYBRID VIGOUR.

heterospory. The production of two kinds of SPORES, as in seed-producing plants and some clubmosses. Microspores give rise to male gametophytes, whereas megaspores grow into female gametophytes. *Compare* HOMOSPORY. *See also* EMBRYO SAC, POLLEN.

heterostyly. Variation in the length of the style in different flowers of the same species (e.g., thrum-eye and pin-eye in the primrose). This ensures cross-pollination by insects, as the anthers of one type of flower are at the same height as the stigmas of the other.

heterothallism. The condition in which the THALLUS of an alga or fungus is not self-fertile. *Compare* HOMOTHALLISM.

heterotrophic. Applied to organisms that need ready-made organic food material from which to produce most of their own constituents and (except in the case of a few organisms such as PHOTOTROPHIC PROTOZOA) to ob-

tain all their energy. Animals, FUNGI, most BACTERIA and a few flowering plants are heterotrophs. Whether they are herbivores, carnivores, parasites or SAPROPHYTES, they depend on AUTOTROPHIC organisms for their supply of food.

heteroxenous parasite. *See* PARASITISM.

heterozygote. An organism that has dissimilar ALLELOMORPHS in respect of a particular character. *Compare* HOMOZYGOTE.

Hettangian. A stage of the JURASSIC System.

Hevea brasiliensis. The tree (family: EUPHORBIACEAE) that is the source of the best natural rubber. It was introduced from Brazil to the Far East in the late 19th century. *See also FICUS, MANIHOT,* MORACEAE.

hexacanth. Six-hooked.

Hexapoda. *See* INSECTA.

hexoses. *See* CARBOHYDRATES.

Hg. *See* MERCURY.

hibernation. The state of DORMANCY in certain animals in winter, during which their metabolic rate is much reduced and, in warm-blooded species, the body temperatures drops until it is very close to that of the surroundings. In the UK, reptiles, amphibians, some mammals and many invertebrates hibernate. Some mammals (e.g., dormouse) do not stir during the hibernation period; others (e.g., bats, hedgehogs) become active and feed during warm spells. *Compare* AESTIVATION. *See also* DIAPAUSE.

hiemilignosa. Monsoon forest and bush, in which the woody plants have small leaves that are shed during the hot, dry season.

high-. In metallurgy and other fields, a prefix that denotes a large content of the substance to which it is attached (e.g., high-chromium steel is steel containing a large amount of chromium; high-alumina cement is cement with a high content of alumina).

high-alumina cement. *See* ALUMINOUS CEMENT.

high forest. *See* FOREST.

highland brook. *See* RIVER ZONES.

high-level inversion. An atmospheric temperature INVERSION formed high (300 metres or more) above the ground surface by the descent of air in anticyclonic (*see* ANTICYCLONE) conditions, and its warming as it is compressed by the higher pressure at a lower level. The inversion may continue provided wind speeds below it are sufficient to sustain the downward movement of air.

high-level wastes. RADIOACTIVE WASTES produced by the nuclear industry. In the UK, the term refers only to the wastes from fuel reprocessing; in other countries, it refers to spent fuel. The quantity involved is small, but it is hot, intensely radioactive and contains ACTINIDES with very long HALF-LIVES. No decision has been taken on a method for final disposal of high-level waste; at present it is stored in tanks of water and kept cool. After storage for up to 10 years much of the heat has been lost, making the waste easier to handle. Probably it will then be processed into cylinders of a glass-like or rock-like substance, held for several years in a temporary store to cool further and then moved for final disposal at a site where it must remain isolated from the environment for up to 1000 years.

high-pressure area. *See* ANTICYCLONE.

high-sulphur crude. *See* CRUDE OIL.

high-temperature gas-cooled reactor (HTGR). A NUCLEAR REACTOR in which enriched uranium (*see* URANIUM ENRICHMENT) is the fuel, made into particles coated with carbon and silicon dioxide (silica). The coolant is helium, the MODERATOR graphite. One HTGR has been built in Britain (the Dragon reactor), and one is working at Fort St Vrain, Colorado, owned by Gulf Central Atomic. Operating temperatures (1000°C) are much higher than those in the MAGNOX REACTOR or ADVANCED GAS-COOLED REACTOR, and only ceramic materials can be used. All orders for HTGRs have been cancelled, and no new ones are being built.

high-yielding varieties (HYVs). Varieties of hybrid wheat and maize, developed mainly in Mexico at the Rockefeller Institute, and of rice, developed mainly in the Philippines at the International Rice Research Institute (IRRI), which form the basis of the GREEN REVOLUTION. They are designed to grow well in low latitudes. Provided the water supply is adequate they respond well to fertilizer applications, growing and maturing rapidly, so in favourable conditions more than one crop can be raised in a year on the same land. The varieties are short-strawed, which makes them less prone to LODGING than traditional varieties.

Hirudinea (leeches). A class of mainly aquatic ANNELIDA whose members have a sucker at each end of the body. Many leeches prey on invertebrates, but some are able to pierce the skin of vertebrates and suck blood.

histic. Applied to soil surface layers that are high in organic carbon and seasonally saturated with water. *See also* SOIL CLASSIFICATION.

histidine. An AMINO ACID with the formula $C_3H_3N_2CH_2CH(NH_2)COOH$ and a molecular weight of 155.2.

histogens. Clearly defined zones of primordial tissue found in apical MERISTEMS and elsewhere in plants, from which new, differentiated tissue develops.

histolysis. The breakdown of tissue.

Histosols. *See* SOIL CLASSIFICATION.

HIV. Human immunodeficiency virus (*see* ACQUIRED IMMUNE DEFICIENCY SYNDROME).

hoar frost. A white, crystalline deposit of ice on the surface of solid objects that is caused by the direct SUBLIMATION of water vapour from the air. *See also* DEW.

Holarctica. The ZOOGEOGRAPHICAL REGION consisting of the PALAEARCTIC REGION and NEARCTIC REGION together.

Holarctic Realm. The FLORAL REALM covering the northern hemisphere as far south as the Tropic of Cancer and consisting of eight regions: ARCTIC FLORAL REGION, EURO-SIBERIAN FLORAL REGION, IRANO-TURANIAN FLORAL REGION, SINO-JAPANESE FLORAL REGION, MEDITERRANEAN FLORAL REGION, HUDSONIAN FLORAL REGION, PACIFIC NORTH AMERICAN FLORAL REGION, ATLANTIC NORTH AMERICAN FLORAL REGION.

holism. The belief that complex systems may be understood only when viewed in their entirety. *Compare* REDUCTIONISM, VITALISM.

Holocene (Recent). The younger subdivision of the QUATERNARY in which we are living at present, the previous subdivision being the PLEISTOCENE. The Holocene is approximately the time since the end of the last glaciation, about the last 10 000 years. Holocene also refers to rocks deposited during this time. However, some climatologists propose that the present is merely another INTERGLACIAL of the Pleistocene and that the use of the term Holocene is premature.

holocoen. The whole environment, comprising the BIOCOEN and ABIOCOEN.

Holometabola. *See* ENDOPTERYGOTA.

holometabolous. *See* METAMORPHOSIS.

holophytic nutrition. The characteristic mode of nutrition of a green plant, that is the synthesis of organic compounds from carbon dioxide, water and mineral salts, by means of light absorbed by chlorophyll (*see* PHOTOSYNTHESIS). *Compare* AUTOTROPH, HOLOZOIC NUTRITION.

Holothuria. *See* HOLOTHUROIDEA.

Holothuroidea (Holothuria, sea-cucumbers). A class of ECHINODERMATA whose members have sausage-shaped bodies and no arms, but tube feet around the mouth that are enlarged to form tentacles. In some species food is extracted from mud and shovelled into the mouth by the tentacles; in others the tentacles are sticky and used to catch small organisms.

holotype. *See* TYPE SPECIMEN.

holozoic nutrition. The characteristic mode of nutrition of animals (i.e. the eating of plants, animals or their solid products). *Compare* HETEROTROPHIC, HOLOPHYTIC NUTRITION.

holy fire. *See* ERGOT.

homeothermy. *See* HOMOIOTHERMY.

home range. The total area occupied over the years by a group of animals or an individual animal. That part of the home range in which an animal spends most of its time, and where it sleeps and gives birth to and raises its young, is the core area.

Hominoidea. The superfamily of PRIMATES which includes the family Hominidae (humans), Pongidae (great apes) and Hylobatidae (gibbons).

Homo. The genus (family: Hominidae) which includes modern humans (*H. sapiens sapiens*). *H. habilis* (handy man), whose remains were found in Olduvai Gorge, Tanzania, was a very early species, alive more than 1.5 Ma, whose members could make stone tools and stand erect. The cranial capacity was small, and some authorities dispute the inclusion of this species in the genus. *H. erectus* includes Java man (formerly *Pithecanthropus erectus*), who lived about 0.5 Ma, walked erect and had prominent ape-like brow ridges. Fossil remains from other parts of the world are now attributed to this species, including China or Peking man (formerly known as *Sinanthropus pekinensis*), who made tools and used fire. *H. heidelbergensis* was a robust species with powerful jaws, thought by some to belong to *H. erectus*. *H. sapiens neander-*

thalensis (Neanderthal man) had a large cranial capacity, well-developed brow ridges and no chin prominence, and became extinct probably about 50 000 years ago. It is associated with the Mousterian culture in which the dead were buried ritually. *H. sapiens rhodesiensis* (Rhodesia man), found at Broken Hill, Zambia, had massive brow ridges combined with some modern features. *H. sapiens fossilis* (Cro-Magnon man) had a high cranium and broad face without prominent brow ridges. His remains, found throughout Europe, are associated with sculpture and paintings. Some Cro-Magnon people were contemporaneous with the last of the Neanderthalers. The successor of Cro-Magnon man was modern man, *H. sapiens sapiens*. *See also* MESOLITHIC, NEOLITHIC, PALAEOLITHIC.

homocaryon. *See* HOMOKARYON.

homocyclic compounds. Cyclic chemical compounds that contain a ring consisting wholly of atoms of the same element.

homodont. Possessing teeth that are all similar, as in most reptiles, amphibians and fish. *Compare* HETERODONT.

homoeostasis. The maintenance of constancy within a biological system, either in terms of interaction between the organisms of a community or as regards the internal environment of an individual.

homogametic sex. The sex in which the nucleus of each cell contains a pair of X chromosomes (*see* SEX CHROMOSOMES). In Lepidoptera and many vertebrates (birds, reptiles, some amphibians, some fish) the male is homogametic. In most other organisms the female is homogametic. *Compare* HETEROGAMETIC SEX.

homogamy. (1) Inbreeding brought about by isolation. (2) The production of GAMETES that are all alike with regard to the CHROMOSOMES they contain. (3) The production of flowers all of the same sexual type (male, female, or HERMAPHRODITE). *Compare* HETEROGAMY. (4) The condition in which the anthers and stigmas of a flower mature simultaneously. *Compare* DICHOGAMY.

homogeneity. The even distribution of species or characteristics, so that a sample taken at any one point resembles one taken at any other.

homoiosmotic. Applied to animals (e.g., vertebrates) that maintain their body fluids at a constant OSMOTIC PRESSURE. *Compare* POIKILOSMOTIC.

homoiothermy (homeothermy, warmbloodedness). The maintenance of a constant body temperature regardless of variations in the external temperature. This occurs in birds and mammals. *Compare* POIKILOTHERMY.

homokaryon (homocaryon). A cell that contains two or more nuclei of identical genetic constitution. *Compare* DIKARYON, HETEROKARYON.

homologous. (1) Applied to an organ in one species that has a fundamentally similar structure, development or origin to an organ in another species. Homologies are clear during embryonic development, but organs may become much modified later. Homologous organs may have very different functions (e.g., the wings of a bat, the paddles of a whale and the forelegs of a horse are homologous; the auditory ossicles of the mammal EAR are homologous with certain bones involved in the articulation of the jaw in fishes). *Compare* ANALOGOUS. (2) (homologous chromosomes) *See* CHROMOSOMES, MEIOSIS.

Homoptera. A suborder of insects (order: HEMIPTERA) whose members have forewings of uniform consistency. The suborder includes aphids (APHIDIDAE), cicads, froghoppers, mealy-bugs, and scale insects. These plant bugs often do considerable damage to crops by sucking the sap, discharging HONEYDEW which blocks the STOMATA, and transmitting diseases.

homospory. The production of only one

kind of SPORE, as in many PTERIDOPHYTA. This gives rise to a gametophyte which generally produces both male and female reproductive organs. Occasionally, environmental conditions may lead to male and female organs being borne on different gametophytes. *Compare* HETEROSPORY.

homothallism The condition in which the THALLUS of a fungus or alga is self-fertile. *Compare* HETEROTHALLISM.

homozygote. An organism that has identical ALLELOMORPHS in respect of a particular character. *Compare* HETEROZYGOTE.

honeydew. A sweet, sticky substance produced by certain sap-eating insects which are compelled to ingest excessive amounts of sap in order to obtain adequate amounts of the proteins and minerals in which it is deficient, excreting the surplus as sugars. In hot, dry weather some species of aphids (*see* APHIDIDAE) and SCALE INSECTS excrete such large amounts that it coats the leaves, especially of maples, lindens and roses, and may fall from the tree as a fine mist. Several insect species, including some ants, feed on honeydew.

hoodoo. An unusually shaped erosion remnant of rock.

hookworms. Parasitic (*see* PARASITISM) NEMATODA that are responsible for much debilitation in humans throughout tropical and subtropical regions. Heavy infestation causes anaemia and the retardation of mental and physical development. *Ancylostoma duodenale* and *Necator americanus* are common in humans; other species parasitize other mammals. Adult hookworms feed on blood and tissue from the wall of the intestine. Eggs pass out in faeces, and the larvae subsequently enter a new host by burrowing through the skin. Wearing shoes and practising the sanitary disposal of human faeces prevents infection.

hop. *See* CANNABIACEAE.

horizon. An approximately horizontal layer within a soil or sediment that is clearly distinct and distinguishable from the layers above and below it. *See also* SOIL HORIZONS.

hormone. A chemical produced by one part of an organism and transported to other parts, where a minute quantity exerts control over specific metabolic functions. *See also* ANDROGENS, AUXINS, CORPUS LUTEUM, ENDOCRINE ORGAN, GONADOTROPIC HORMONES, INTERMEDIN, OESTROGENS, PITUITARY GLAND, PROGESTERONE, THYROXIN.

hormone weedkillers (auxin-type growth regulators). A group of herbicides (e.g., 2,4,-D, MCPA) that are synthetically produced organic compounds whose effects are similar to those of the natural growth-regulating substances of plants (AUXINS). They are absorbed by roots or leaves and translocated (*see* TRANSLOCATION) to the growing points, where they inhibit growth or cause deformed growth which results in the death of the plant. Some are toxic to Dicotyledoneae and non-toxic to Monocotyledoneae and have great importance in the selective control of weeds in cereals and grass. *See also* DICHLORPROP, MECOPROP, 2,4,5,-T, TRANSLOCATED HERBICIDES.

horn. A pyramidal mountain peak formed between three or more CIRQUES.

hornbeam. *See* CARPINUS BETULUS.

hornblende. A calcium magnesium aluminosilicate mineral that belongs to the AMPHIBOLE group. It has a range of compositions which include the typical $Ca_2(Mg,Fe)_5Si_8O_{22}(OH)_2$.

hornfels. A hard, splintery rock formed by the thermal METAMORPHISM of ARGILLACEOUS rock adjacent to a large igneous INTRUSION.

horn silver. *See* SILVER.

hornwort. *See* BRYOPHYTA.

horseflies. *See* BRACHYCERA.

horsehair worms. *See* NEMATOMORPHA.

horseshoe crab. *See* XIPHOSURA.

horsetails. *See* EQUISETALES.

horst. An uplifted block between two normal FAULTS.

host. (1) An organism that harbours a parasite (*see* PARASITISM). A definitive (primary) host harbours the mature stage of a parasite, whereas an intermediate (secondary) host harbours only immature stages. The sheep is the definitive host for the liver fluke (*Fasciola hepatica*) and the snail (*Lymnea truncatula*) is its intermediate host. (2) In COMMENSALISM, the partner that receives no benefit from the relationship. (3) An organism into which an experimental graft is transplanted.

hot blast. A technique, devised in 1828 by James Neilson, for reducing fuel consumption in iron foundries. A blast of air is required in a furnace to increase the temperature at which the fuel burns and to assist the purification of the metal (*see* BESSEMER PROCESS). The use of hot air was found to reduce fuel consumption by about 40 percent. In modern furnaces the hot gas leaving the furnace is used to preheat the incoming blast. *See also* BLAST FURNACE, CUPOLA.

hot brine. Brackish subterranean water whose temperature is markedly higher than that dictated by the GEOTHERMAL GRADIENT and therefore may be used as a source of GEOTHERMAL ENERGY. In some areas (e.g., New Zealand), hot brines have a low SALINITY, but elsewhere the mineral content may be high, making the water highly corrosive and so difficult to handle and to dispose of once its heat has been extracted.

hot dry rock. A body of rock that, because of an anomaly in the GEOTHERMAL GRADIENT, is substantially hotter than the rock surrounding it. GEOTHERMAL ENERGY may be extracted from hot dry rocks.

HSE. *See* HEALTH AND SAFETY EXECUTIVE.

HTGR. *See* HIGH-TEMPERATURE GAS-COOLED REACTOR.

Hudsonian Floral Region. The part of the HOLARCTIC REALM that comprises North America from southern Alaska and the northern shores of the Great Lakes north to northern Alaska and Labrador.

human immunodeficiency virus (HIV). *See* ACQUIRED IMMUNE DEFICIENCY SYNDROME.

humate. A SALT or ESTER of a humic acid derived from HUMUS during the decomposition of organic material in the soil.

humic coal. *See* COAL.

humidification. Increasing artificially the water content of air or other gases (e.g., in air conditioning or central heating systems) to reduce the accumulation of STATIC ELECTRICITY. *See also* HUMIDIFIER.

humidifier. A device for HUMIDIFICATION that brings the air into contact with water at the same temperature (heating or cooling the air if necessary), the operation of the device often being controlled automatically by a sensor that monitors atmospheric humidity (a humidistat). If the water is very cold it can be used to cool and dehumidify the air simultaneously, using a DEHUMIDIFIER. Air may also be dried by passing it over a bed of HYGROSCOPIC crystals (e.g., lithium chloride).

humidistat. *See* HUMIDIFIER.

humidity. The amount of water vapour present in the air, measured as the ABSOLUTE HUMIDITY (i.e. the weight of water in a unit volume of air), relative humidity (i.e. the amount of water vapour expressed as the percentage of the amount that would be present were the air saturated) or specific humidity (i.e. the mass of water vapour in a unit volume of air).

humidity mixing rate. *See* ABSOLUTE HUMIDITY.

humification. The microbial breakdown of

organic matter in the soil to form HUMUS.

Humulus. *See* CANNABIACEAE.

humus. The more or less decomposed organic matter in the soil. Besides being the source of most of the mineral salts needed by plants, humus improves the texture of the soil and holds water, so reducing the loss of nutrients by LEACHING. Mild humus (mull) is produced in soil containing abundant earthworms, where decay is rapid. *See also* MOR.

hurricane. The strongest wind (force 12) on the BEAUFORT SCALE, with a speed of more than 122 kilometres per hour (more than 76 mph), which causes total devastation. Hurricanes are uncommon over land, although not at sea, but a hurricane passing through a populated area brings major disaster. *See also* BATH PLUG VORTEX, TROPICAL CYCLONE.

Huygens' construction. Each point on a WAVE FRONT may be regarded as a new source of secondary wavelets. From this construction, if the position of the wave front at any given time is known, its position at any subsequent time may be determined. The construction is named after Christian Huygens (1629–95).

hybrid. An organism produced by crossing parents of different taxa (e.g., different species, subspecies or varieties) and thus different GENOTYPES. Hybrids are often sterile. *See also* ALLOTETRAPLOID.

hybrid swarm. A large, often very varied population that results from hybridization with subsequent crossing and backcrossing (*see* BACKCROSS).

hybrid vigour (heterosis). Increased vigour (e.g., in terms of fertility or growth) that results from the crossing of two genetically different lines.

hydathode. *See* GUTTATION.

hydragyrum. *See* MERCURY.

hydranth. *See* POLYP.

hydrarch succession. *See* HYDROSERE.

hydrates. SALTS that contain water of crystallization (i.e. water retained as the salt crystallizes from aqueous solution).

hydration. The chemical addition of water to a compound.

hydraulic. Pertaining to water (or by extension to other fluids), in motion or at rest, that is contained (e.g., in pipes or stream channels) and that can exert pressure or be made to do work.

hydraulic civilization. A civilization that depends on sophisticated water management (e.g., drainage systems, irrigation) to sustain its agriculture. *See also* CHINAMPAS, FERTILE CRESCENT, INDUS VALLEY.

hydraulic geometry. The study of the shape of stream channels, based on measurements of the width (the shortest distance from bank to bank), depth at a particular point, vertical cross-section at right angles to the direction of flow and slope or gradient.

hydrazine (N_2H_4). A reactive chemical intermediate in the production of explosives, photographic chemicals and antioxidants. It is also used as a rocket fuel.

hydrocarbon cracking. The decomposition by heat, with or without CATALYSTS, of petroleum or heavy petroleum fractions to give materials of lower boiling points which can be used as motor fuels, domestic heating oils, etc. The process is usually performed in an oil refinery complex. *See also* CRACKING.

hydrocarbons. Strictly, chemical compounds composed only of hydrogen and carbon; more loosely, many organic (i.e. carbon-based) compounds that contain other elements (e.g., fossil fuels are sometimes described as hydrocarbon fuels).

Hydrocharitaceae. A family of tropical and temperate freshwater and marine

MONOCOTYLEDONEAE, some of which (e.g., frogbit, *Hydrocharis morsus-ranae*) have floating leaves, but most have ribbon-like submerged leaves. The female plant of Canadian pondweed (*Elodea canadensis*) was introduced into Europe in the mid-19th century and has spread rapidly through inland waters. Water soldier (*Stratiotes aloïdes*) floats at the surface in summer and sinks in the autumn.

hydrochore. A plant whose seeds, TURIONS or other reproductive structures are dispersed by water (e.g., alder, *Alnus* species, coconut, *Cocos*).

hydrochory. Dissemination by water.

Hydrocorallinae. *See* CORAL.

hydroelectric power. The generation of electricity by turbines turned by the flow of water, on a small (including domestic) or large industrial scale. Where a hydroelectric scheme requires the building of a LARGE DAM and lake there may be an increase in earthquake activity.

hydrofluoric acid (HF). One of the strongest and most corrosive acids known. It reacts with a wide variety of materials. A solution of hydrogen fluoride in water, it is used for etching glass, in the manufacture of microprocessors, in metal processing and for cleaning stonework.

hydrofoil. An arrangement of flat struts beneath the hull of a high-powered boat that pushes the boat out of the water at speed due to the upward pressure on the struts. This reduces drag, as less of the hull is in contact with the water and so permits greater speeds. It is used mainly in rivers and lakes where waters are calm.

hydroforming. A catalytic (*see* CATALYST) process for the dehydrogenation (removal of hydrogen) of paraffins with their conversion to cyclic and AROMATIC hydrocarbons.

hydrofracturing. A technique for making large cracks in subterranean rocks by injecting water under high pressure. It is used for the extraction of GEOTHERMAL ENERGY from HOT DRY ROCK.

hydrogen (H). A gas; the lightest element. It is flammable and has a wide range of uses in the synthesis of, for example, AMMONIA, hydrochloric acid (HC1), METHANOL and in the hydrogenation of coal. It has been proposed as an alternative liquid fuel (called LH2) for aircraft and road vehicles. It is also widely used in laboratories, especially as a carrier gas in GAS CHROMATOGRAPHY. $A_r=1.00797; Z=1$.

hydrogen cyanide. A poisonous gas that has been implicated in industrial accidents. It is used as a fumigant and as a starting material in the production of nylon and other polymers, pharmaceuticals and dyestuffs. *See also* CYANIDE, GLYCOSIDES.

hydrogen ion concentration. *See* pH.

hydrogen peroxide (H_2O_2). An unstable, highly reactive, colourless liquid that can be produced as a by-product of using hydrogen as a fuel. It is used in aqueous solution as a bleaching agent, antiseptic, oxidizing agent and as an oxidant for such rocket fuels as HYDRAZINE.

hydrogen sulphide (H_2S). A poisonous, evil-smelling gas (the smell is of rotten eggs) that is used in chemical industries, rayon manufacture and analytical laboratories. It is manufactured, but also produced naturally by the anaerobic decomposition of sulphur-containing organic materials. At low concentrations it can cause headaches, at high concentrations it is lethal to humans.

hydrograph. A graphic representation of the level of water in AQUIFERS or water courses, displayed as the rate of flow against time.

hydroid. *See* HYDROZOA.

hydrological cycle. The movement of water between the oceans, ground surface and atmosphere by EVAPORATION, PRECIPITATION and the activity of living organisms, as one of the major BIOGEOCHEMICAL CYCLES. Of the Earth's water, 97 percent is in the oceans; of the remainder 98 percent at present is frozen in the polar ice caps and glaciers. Each day about 875.3 km^3 of water evaporates from the oceans and about 160.5 km^3 is lost from the land surface, by evaporation and TRANSPIRATION. About 775.3 km^3 falls on the seas as precipitation and 100 km^3 is carried in the air from the sea over the land, which receives about 260.5 km^3 of precipitation (equal to the 100 km^3 carried from the sea plus the 160.5 km^3 lost from the land surface). About 100 km^3 returns from the land to the sea through rivers, thus completing the cycle. Small absolute losses to the cycle, from the top of the atmosphere into space, are made good by JUVENILE WATER.

hydrologic sequence. A series of samples, taken as vertical sections from soils derived from the same parent material, which show increasingly poor drainage down a slope, the lowest sample often being waterlogged. *See also* CATENA.

hydrology. The scientific study of water on the land surface and beneath it, including its chemical composition and movement, with particular reference to irrigation, drainage, erosion, flood control, etc.

hydrolysis. (1) The formation of an acid and a base from a SALT by the ionic dissociation of water. (2) The decomposition of organic compounds by interaction with water (e.g., the formation of alcohols and acids from ESTERS).

hydrophilous. *See* HYGROPHILOUS.

hydrophyte. *See* RAUNKIAER'S LIFE FORMS.

hydromorphic soil. A soil that contains excess water.

hydroponics. The growing of plants without soil by suspending them with their roots immersed in water that is enriched with essential nutrients or by rooting them in an inert material (e.g., quartz sand) and supplying them with a nutrient solution. The technique can produce large yields in a small space, but preparing the correct balanced mixture of nutrients is usually complex.

hydrosere (hydrarch succession). The stages in a plant SUCCESSION that begins in water or a wet habitat and progresses towards drier conditions. *Compare* XEROSERE.

hydrosphere. That part of the Earth's surface which is covered by water (i.e. oceans, seas, ice caps, lakes, rivers, etc.).

hydrostatic pressure. The pressure exerted by water that is at rest, equally at any point in the water body.

hydrothermal. Applied to any geological process that involves heat or superheated water. Such processes fall into two main types: alteration (e.g., of FELDSPAR to KAOLIN, or of OLIVINE and PYROXENE to SERPENTINE) and deposition. Many metalliferous ore deposits are thought to be of hydrothermal nature, originating as a concentration of VOLATILES in a MAGMA. The major subdivisions of hydrothermal deposits are HYPOTHERMAL, MESOTHERMAL and EPITHERMAL, according to temperature and depth.

hydrotropism. A growth response (i.e. a TROPISM) in a plant in which the stimulus is water.

Hydrozoa. A large class of CNIDARIA whose members usually have an alternation of sessile POLYP colonies and free-swimming medusae (*see* MEDUSA). The group includes hydroids, some of which (e.g., *Hydra*) are solitary and have no medusa stage, the stinging corals and pelagic polyp colonies such as the Portuguese man-of-war. *See also* ACTINOZOA, SCYPHOZOA.

hygrometer. An instrument used to measure atmospheric moisture.

hygropetric. Inhabiting wet rock surfaces.

hygrophilous (hydrophilous). Inhabiting wet places.

hygrophyte. A plant that is found only in a moist habitat and is very sensitive to dry conditions.

hygroscopic. Applied to substances that absorb water readily from the atmosphere.

hygroscopic moisture. Water that is present in the soil, but held by surface tension forces that are too strong for it to be accessible to plants.

Hymenoptera. A large order of insects (division: ENDOPTERYGOTA) whose winged members have two pairs of membranous wings coupled together in flight by small hooks. The Symphyta (e.g., saw-flies and wood-wasps) are unwaisted and have saw-like or drill-like ovipositors, which enable them to insert eggs into leaves, stems or wood. The Apocrita, in which there is a narrow waist between the thorax and abdomen, includes bees, wasps, ants and ichneumons. *See also* ACULEATA, PARA-SITISM.

hypabyssal. Applied to a rock or igneous INTRUSION that crystallized near the surface. In general, such a rock is medium-grained, and the intrusion is typically a SILL or DYKE. Hypabyssal is intermediate between PLU-TONIC and VOLCANIC.

hyperkinesis. Excessive motility of an organism or muscle.

hyperparasite. *See* PARASITISM.

hyperphagia. Voracious eating, often in-discriminately of edible or inedible substances, and proceeding past the point of normal satiation. *Compare* APHAGIA.

hyperplasia (hyperplasy). An abnormal increase in the number of cells in part of an organism (e.g., GALLS, animal tumours).

hyperplasy. *See* HYPERPLASIA.

hypertonic solution. *See* OSMOSIS.

hypertrophic. Applied to bodies of water that are grossly enriched with plant nutrients. *See also* EUTROPHICATION.

hypertrophy. An increase in size that is caused by HYPERPLASIA or by enlargement of cells or fibres (e.g., in well-exercised muscles). *Compare* INVOLUTION.

hypha. In FUNGI, one of the mass of filaments that make up a MYCELIUM. Hyphae are tubular and branched, and may have cross-walls.

hypocentre. The point on the Earth's surface that is directly below the centre of a nuclear bomb explosion.

hypogeal. Applied to germination in which the COTYLEDONS remain below ground, as occurs in the pea. *Compare* EPIGEAL.

hypogene. Applied to mineral deposits that were formed by ascending aqueous solutions. *Compare* SUPERGENE.

hypolimnion. The colder, non-circulating layer of water in a lake, lying below the THERMOCLINE. *Compare* EPILIMNION. *See also* THERMAL STRATIFICATION.

hypophysis cerebri. *See* PITUITARY GLAND.

hypostatic gene. A GENE whose expression is prevented by the presence of another, non-allelic (*see* ALLELOMORPH) gene. *Compare* EPISTATIC GENE.

hypothalamus. Part of the brain floor in vertebrates. In mammals it controls body temperature and produces hormones that influence the PITUITARY GLAND lying below it.

hypothermal. Applied to ore deposits that were formed from an ascending, essentially aqueous solution at high temperatures, generally ranging from 300 to 500°C, and at great depths. *Compare* EPITHERMAL, MESO-THERMAL.

hypotonic solution. *See* OSMOSIS.

hypoxia. *See* ANOXIA.

hypsodont. Applied to high-crowned teeth, which are typical of herbivorous mammals (e.g., ungulates). The crowns must be high because they are worn away continually by grazing. Some hypsodont teeth grow continually from open roots (e.g., in rodents).

Hyracoidea (hyraxes). An order of small, herbivorous, placental mammals (subclass EUTHERIA), resembling short-tailed squirrels, that have four toes on the front feet and three on the hind feet. They show similarities to both rodents and elephants.

hyraxes. *See* HYRACOIDEA.

HYVs. *See* HIGH-YIELDING VARIETIES.

Hz. *See* FREQUENCY.

I

I. *See* IODINE.

IAA. *See* INDOLE-3-ACETIC ACID.

IAEA. *See* INTERNATIONAL ATOMIC ENERGY AGENCY.

IBP. *See* INTERNATIONAL BIOLOGICAL PROGRAMME.

IBPGR. *See* INTERNATIONAL BOARD FOR PLANT GENETIC RESOURCES.

ice accretion. *See* ACCRETION.

ice age. A prolonged period of cold climatic conditions, characterized by snow and ice which persist throughout the year in regions that are free from summer snow and ice in non-glacial periods, and by major extensions of GLACIERS. The cause of ice ages is uncertain, but M. Milankovich (a Yugoslav geophysicist) proposed in the 1950s that there is a correlation between climatic change and variations in the Earth's solar orbit and in the inclination of its axis. Alternative hypotheses are based on changes in solar activity and on changes in the atmospheric content of carbon dioxide (*see* GREENHOUSE EFFECT); such changes have been observed, but it is unclear whether they are the cause or the consequence of changes in temperature. There have been many ice ages in Earth history, dating as far back as the PRECAMBRIAN in which at least 15 major groups of ice ages (periods of glaciation interrupted by periods of remission) occurred in the Laurasian and Gondwana continents. Further groups occurred during the CARBONIFEROUS, with glaciation confined to the southern hemisphere. Throughout history, central and southern Africa have experienced glaciation more frequently than any other land area.

The third main group of ice ages occurred during the QUATERNARY, the most recent being the Devensian (or Weichselian), which receded about 10 000 years ago.

ice anvil. The anvil-shaped top of a CUMULONIMBUS cloud, which is formed from minute ice crystals.

ice cap. *See* ICE SHEET.

ice evaporation level. That level in the atmosphere at which the temperature is sufficiently low for water to change between the gaseous and solid phases with, at most, a very transitory liquid phase. The formation of ice directly from saturated air occurs at any temperature below –40°C, and if ice crystals in air at or below this temperature encounter very dry air they will change to the gaseous phase by SUBLIMATION.

ice nuclei. *See* FREEZING NUCLEI.

ice sheet (ice cap). The largest form of GLACIER, being a layer of ice which covers an extensive area and which may be thick enough to bury all but the highest peaks of entire mountain ranges. Almost all of Antarctica is covered by ice that is 2500 metres thick locally; the Greenland ice sheet is 3000 metres thick. Smaller ice sheets occur in Iceland, Spitzbergen and other arctic islands, and there are still smaller ones in the highlands of Norway.

ichneumons. *See* PARASITOID.

ICRP. *See* INTERNATIONAL COMMISSION ON RADIOLOGICAL PROTECTION.

ICSU. *See* INTERNATIONAL COUNCIL OF SCIENTIFIC UNIONS.

IDA. International Development Association. (*see* INTERNATIONAL BANK FOR RECONSTRUCTION AND DEVELOPMENT).

identical twins. See MONOZYGOTIC TWINS.

idiobiology. The study of the biology of individual organisms.

IFC. International Finance Corporation (*see* INTERNATIONAL BANK FOR RECONSTRUCTION AND DEVELOPMENT).

igneous From the Latin *ignis* (fire), applied to rocks that were once molten and have formed by the solidification of MAGMA. Igneous rocks are contrasted with SEDIMENTARY ROCKS and metamorphic rocks (*see* METAMORPHISM), the two other fundamental groups of rocks. Igneous rocks are classified according to their SILICA content as ACID, INTERMEDIATE and BASIC, by the minerals present as ULTRABASIC, ALKALINE and CALC-ALKALINE, by the grain size of the groundmass (i.e. ignoring PHENOCRYSTS), and by the texture of the rock. *See also* ANDESITE, BASALT, DIORITE, DOLERITE, GABBRO, GRANITE, GRANODIORITE, MICROGRANITE, PERIDOTITE, RHYOLITE, SYENITE.

ignimbrite. A PYROCLASTIC rock composed of unsorted (*see* SORTING) PUMICE and other material, characteristically with glass shards, which is formed by the explosive disintegration of pumice, which has been flattened. Ignimbrites often show FLOW STRUCTURES and range from loose, granular deposits to completely glassy, OBSIDIAN-like rocks; many RHYOLITES have been reinterpreted as ignimbrites. No volcanic eruption producing ignimbrite has been recorded, but it is believed that such an eruption would be related to one producing NUÉE ARDENTES.

IGY. See INTERNATIONAL GEOPHYSICAL YEAR.

IIED. See INTERNATIONAL INSTITUTE FOR ENVIRONMENT AND DEVELOPMENT.

illuviation. The deposition and precipitation of material leached (*see* LEACHING) and/or eluviated (*see* ELUVIATION) from the A

horizon into the B horizon (*see* SOIL HORIZONS).

ilmenite ($FeTiO_3$). The main ore mineral of titanium. Ilmenite is widely distributed in rocks, but major ores result from MAGMATIC DIFFERENTIATION in ANORTHOSITES and in PLACER deposits. Nearly all the titanium produced is used as the oxide, a white pigment, in paints. Titanium alloys are used in the aerospace industry.

imaginetic centre. A term coined by Alvin Toffler in *Future Shock* (1970) to describe places where people noted for their creative ability might be provided with technological assistance and encouraged to examine present or anticipated crises and to speculate freely about the future.

imago. The adult, sexually mature stage of an insect.

imbricate. Applied to a structure in which tabular masses (e.g., flat pebbles, THRUST sheets) overlap one another.

IMCO. *See* INTERGOVERNMENTAL MARITIME CONSULTATIVE ORGANIZATION.

IMF. International Monetary Fund (*see* INTERNATIONAL BANK FOR RECONSTRUCTION AND DEVELOPMENT).

imhoff tank. A tank in which sedimentation treatment for sewage is combined with anaerobic biological treatment, sewage entering into an upper chamber, solids settling through slots into a lower digestion chamber and sludge removal being automatic. Imhoff tanks are used extensively in Australia.

immission. The reception of a substance (e.g., a pollutant) from a remote source of emission; the opposite of emission.

immunity. The ability of an organism to combat infection by parasites. *See also* ANTIBODY, AUTOIMMUNITY, INTERFERON, LYSOZYME, PARASITISM, PHAGOCYTE, PHYTOALEXINS.

impedence. (1) The quantity that determines the amplitude of a current of given voltage in an alternating current expressed as

$$Z= \{R^2 + [L\ \omega - (1/C\ \omega)]^2\ \}^{1/2}$$

Where Z is the impedance, R the resistance, L the self-inductance, C the capacitance and ω the angular frequency (a constant equal to $2f$, where f is the frequency of the alternating current). (2) In acoustics, a measure of the complex ratio of force or pressure to velocity. *See also* CHARACTERISTIC IMPEDANCE.

impermeable. The opposite of permeable (*see* PERMEABILITY).

impervious. The opposite of PERVIOUS.

impingement. The bringing into contact of two or more objects or substances (e.g., when dust is made to impinge on a dust collector).

implantation (nidation). The initial attachment of the mammalian EMBRYO to the uterus of the mother.

impounding reservoir. *See* RESERVOIR.

imprinting. Rapid and stable learning in a young animal, which results in reaction to a particular object (e.g., a duckling becomes imprinted on its mother and subsequently follows her).

improductive forest. Forest that is incapable of yielding products other than fuel because of adverse conditions; forest that grows slowly or whose trees are stunted.

impulse turbine. One of the two principal types of TURBINE, in which the whole of the available head of water is transformed into kinetic energy before reaching the wheel.

inbreeding. *See* ENDOGAMY.

INCAP. Institution of Nutrition for Central America and Panama (*see* INCAPARINA).

incaparina. A food, developed by the Institution of Nutrition for Central America and Panama (INCAP), for use in developing countries. It is made by adding protein concentrate in the form of oilseed (maize and cottonseed oil enriched with vitamins A and B) to a staple cereal food.

Inceptisols. *See* SOIL CLASSIFICATION.

incinerator. (1) A device in which solid, semi-solid, liquid or gaseous combustible material is burnt as a means of disposal. If the material does not support combustion, auxiliary fuel is added. Many types of industrial and domestic wastes are incinerated, and there are many types of incinerator to deal with different wastes (e.g., some wastes must be burnt at specified temperatures to avoid the emission of toxic by-products). *Compare* OPEN BURNING. (2) *See* BREEDER REACTOR.

incised meander. A MEANDER where the lateral curve in the river channel remains the same, whereas the channel itself cuts deeper vertically. *Compare* INGROWN MEANDER.

income elasticity of demand. The mathematical expression of the relationship between changes in income and demand for particular commodities. Starting with surveys of actual consumer behaviour, this can be expressed as dEY/dYE, where E is the expenditure on the commodity, Y is income, dY is the change in income and dE is the change in expenditure. If the result is less than unity, increases in income will not result in proportional increases in expenditure. The technique is based on the first study of family budgets, published in 1857 by the German civil servant and statistician Ernst Engel (1821–96). Engel's law states that the poorer a family is the greater is the proportion of its income spent on food.

incompetent bed. A layer of rock that accommodates itself to the shape of the interbedded COMPETENT BEDS during folding (*see* FOLD).

indehiscent. Applied to fruits that do not open spontaneously to release seeds. *See also* ACHENE, BERRY, CARYOPSIS, DRUPE, NUT.

independent assortment (independent segregation, Mendel's second law). The law of MENDEL, GREGOR, which asserts the chance assortment of ALLELOMORPHS to gametes. Independent assortment does not apply to genes lying on the same chromosome. *See also* LINKAGE, SEGREGATION.

independent segregation. *See* INDEPENDENT ASSORTMENT.

index fossil. The fossil by whose name a particular unit of rock strata (zone), often containing a characteristic assemblage of fossils (a biostratigraphic zone) is known. A zone represents a period of time during which these organisms had reached a particular evolutionary stage, and zone fossils are used for the chronological correlation of rocks of different facies. A good zone fossil should be common and distinctive, have had a wide geographical range, have tolerated varying environments, and have had a relatively rapid rate of evolution. Important examples include TRILOBITES (CAMBRIAN System), GRAPTOLITES (ORDOVICIAN and SILURIAN Systems) and AMMONITES (JURASSIC and CRETACEOUS Systems).

index mineral. A mineral whose first appearance in passing from lower to higher grades of METAMORPHISM indicates that the zone in question has been reached.

index species. An organism that is adapted to only a narrow range of environmental conditions and is used to characterize those conditions.

Indian Floral Region. The part of the PALAEOTROPIC REALM that comprises all of the Indian subcontinent except for an area in the north-west.

Indian rice. *See* WILD RICE.

indiarubber tree. *See* FICUS.

Indicative World Plan for Agricultural Development (IWP). A comprehensive strategy, drawn up in the 1960s and published in 1970 by the FOOD AND AGRICULTURE ORGANIZATION OF THE UNITED NATIONS, with the aim of increasing the production and availability of food in countries where hunger and malnutrition present serious problems. It includes the introduction of HIGH-YIELDING VARIETIES of cereals, where appropriate, together with the technology to grow them, the improvement of distribution systems and the reform of financial institutions, etc. It was nicknamed the green revolution, and has achieved spectacular successes.

indicator species. Species whose presence indicates certain environmental conditions. *See also* BIOTIC INDEX, *CLEMATIS VITALBA*, COLIFORM BACTERIA, FAECAL *STREPTOCOCCUS*, SAPROBIC CLASSIFICATION.

indifferent species. Species that have no marked affinities for any COMMUNITY.

indigenous. Native or original to an area.

indole-3-acetic acid (IAA). A growth-regulating HORMONE (*see* AUXINS) produced by many plants.

Indore. A small-scale composting technique devised by Sir Albert Howard, while working at an agricultural research station at Indore, India. Most garden compost is made today according to his method or some variant of it.

induration. The hardening of sediments through the action of cementation, pressure or heat.

industrialization. The economic process whereby an increasing proportion of the income of a nation or region is derived from industrial manufacturing, associated with increasing investment in manufacturing industry. It is the phase of economic development in which economic activity becomes increasingly concerned with adding value to primary products by processing them into finished goods, rather than with primary production (agriculture, forestry, mining, etc.). *See also* PRIMARY ECONOMY, QUATERNARY ECONOMY, SECONDARY ECONOMY, TERTIARY ECONOMY.

industrial melanism. The occurrence of dark (melanistic) forms of animals in industrial areas, the dark form increasing through natural selection at the expense of the alternative, pale variety because it is better camouflaged in blackened surroundings. Industrial melanism has been recorded in the peppered moth (*Biston (Pachys) betularia*) and in spiders in the north of England.

Industrial Pollution Inspectorate for Scotland. *See* ALKALI INSPECTORATE.

industrial waste. Solid material that is discarded from trading, commercial and industrial premises, and requires disposal. It can be divided roughly into five categories: (a) general factory rubbish, uncontaminated by factory process waste; (b) relatively inert process waste; (c) flammable process waste; (d) acid or caustic wastes; (e) indisputably toxic wastes.

Indus Valley. The site, in Pakistan, of an early advanced civilization whose agriculture depended on sophisticated irrigation. *Compare* FERTILE CRESCENT.

INFCE. *See* INTERNATIONAL NUCLEAR FUEL CYCLE EVALUATION.

infiltration. The penetration of a permeable (*see* PERMEABILITY) solid body or mass by a fluid that fills spaces within it. It is the process by which water seeps into the soil. *See also* INFILTRATION RATE.

infiltration rate. The speed with which water penetrates the soil. It is governed by the texture of the soil, the amount and type of vegetation cover, and the slope of the ground. *See also* FIELD CAPACITY.

inflorescence. A flowering shoot, comprising stems, stalks and bracts, as well as flowers.

infraneustronic. *See* NEUSTRON.

infrared photography. Photography based on exposure to radiation in the infrared waveband, outside the visible light spectrum. It is used increasingly for environmental surveying from the air since it reveals details of heat distribution, waste discharges and atmospheric conditions that cannot be recorded by conventional photography.

infrasonic. Applied to sounds whose FREQUENCY is below the range of audio frequency.

infusoria. An archaic collective name for all the minute organisms found in infusions of organic substances, especially hay. Most of such organisms belong to the CILIOPHORA.

ingrown meander. A MEANDER where the curve in the river migrates sideways as the channel cuts downward, so enlarging the loop. *Compare* INCISED MEANDER.

initially complex model. *See* MODEL.

initially simple model. *See* MODEL.

inlier. An OUTCROP of older rock that is surrounded by younger rock. Inliers may be the result of faulting (*see* FAULT), folding (*see* FOLD), deep erosion or a combination of some or all of these processes. The opposite, an outcrop of younger rock surrounded by older rock, produced by similar processes, is called an outlier.

inositol (phytic acid). A fat-soluble alcohol, widely distributed in plant and animal products. It is essential to the health of most animals and a necessary component of the diet (a VITAMIN) for some (e.g., mice). Other animals, including humans, synthesize it in the body.

inquiline. An animal that lives in the home of another animal of a different species and shares its food. *See also* COMMENSALISM.

Insecta (Hexapoda, insects). The largest class of animals, comprising about 700 000 known species of ARTHROPODA. Most are terrestrial, at least as adults. Typically they have three pairs of legs, two pairs of wings, one pair of antennae, compound eyes and a body divided into head, thorax and abdomen. Included within this class are Ac-ULEATA (ants, bees, wasps), ANOPLURA (suck-ing lice), APHANIPTERA (fleas), APHIDIDAE (aphids), APTERYGOTA (springtails, etc.), COLEOPTERA (beetles), DERMAPTERA (earwigs), DICTYOPTERA (cockroaches and mantids), DIPTERA (true flies), EPHEMEROP-TERA (mayflies), HEMIPTERA (bugs), HYMEN-OPTERA (ants, bees, wasps, saw-flies, etc.), ISOPTERA (termites), LEPIDOPTERA (butterflies and moths), MALLOPHAGA (biting lice), ME-COPTERA (scorpion flies), NEUROPTERA (lace-wings, etc.), ODONATA (dragonflies), ORTHOPTERA (grasshoppers, etc.), PHASMIDA (leaf insects and stick insects), PLECOPTERA (stoneflies), PSOCOPTERA (booklice, etc.), STREPSIPTERA (stylopids), THYSANOPTERA (thrips, etc.) and TRICHOPTERA (caddis flies).

insecticide. A chemical used to kill insects. *See also* CARBAMATES, DINITRO PESTICIDES, DERRIS, NICOTINE, ORGANOCHLORINES, ORGANO-PHOSPHORUS PESTICIDES, PYRETHRUM, QUASSIA, SYSTEMIC INSECTICIDES.

Insectivora. *See* INSECTIVORE.

insectivore. (1) An insect-eating (insectivo-rous) animal or plant (e.g., sundew). (2) A member of the order Insectivora, a group of small, primitive, placental mammals (subclass: EUTHERIA) that includes shrews, moles and hedgehogs.

inselberg. From the German meaning is-land mountain; an isolated, steep-sided hill produced by EXFOLIATION processes.

in situ. In its place of deposition, growth or formation; a term used of minerals, fossils, rocks, etc.

insolation. The reception of solar radiation at a surface.

instar. Any stage of development between moults of the EXOSKELETON in an insect larva or NYMPH.

instinct. A general name, little used by scientists because of its imprecision, for the genetically acquired motivation for innate (unlearned) behaviour. Instinctive behav-iour is based on an elaborate system of re-flexes which, when activated in response to internal or external stimuli, produce a fixed pattern of actions.

Institution of Nutrition for Central Amer-ica and Panama. *See* INCAPARINA.

intensity. The strength of an energy field, especially an electromagnetic field (e.g., the brightness of visible light) measured as the rate of energy flow per unit area. The term is also extended to sound, sound intensity being expressed in watts per square metre and, for plane or spherical free progressive waves, is equal to $p^2/\rho c$, where p is the sound pressure and ρc the CHARACTERISTIC IMPED-ANCE of the system.

interbedded. Laid down in sequence, be-tween one layer and another, or applied to a repetitive sequence of two or more rock types.

intercalary meristem. *See* MERISTEM.

interception. The catching of rainfall by the leaves, branches, twigs, etc. of plants and the holding of it there until it evaporates, thus preventing it from reaching the ground.

interferon. A PROTEIN made by animal cells when they are invaded by VIRUSES. Interferon is non-specific and inhibits the reproduction of viruses. Produced by biotechnological methods using recombinant DNA, inter-feron may prove effective as a drug to com-bat viral infections, including certain forms of cancer.

interfluve. The area of land between two adjacent streams.

interglacial. A period of ice retreat and warmer conditions between two ICE AGES.

Intergovernmental Maritime Consulta-tive Organization (IMCO). The interna-tional body which regulates many aspects of the operation of ships on the high seas and of the pollution of the seas. In 1972, for example, IMCO drew up the International Convention for the Prevention of Pollution from Ships (the London Convention) forbidding the discharge of certain substances from ships at sea. This was later applied (and came to be known as the London Dumping Convention) to regulate the deliberate disposal of wastes by dumping them at sea from ships. The earlier Oslo Convention deals with the discharge of substances resulting from seabed operations.

intermediate. Applied to IGNEOUS rocks that contain intermediate amounts (commonly set at 50–60 percent) of SILICA in their chemical composition. Most of the silica is in the form of silicate minerals (e.g., FELDSPARS, MICAS, AMPHIBOLES, PYROXENES and FELDSPATHOIDS) and there is less than 10 percent QUARTZ. SYENITE, ANDESITE, DIORITE and TRACHYTE are intermediate rocks. In petrology, intermediate is contrasted with ACIDIC, BASIC and ULTRABASIC.

intermediate host. *See* HOST.

intermediate-level wastes. RADIOACTIVE WASTES produced by the nuclear industry and consisting of substances used to clean gases and liquids before their discharge from nuclear installations, sludges from cooling ponds where spent fuel is stored while awaiting reprocessing, and materials contaminated with plutonium. Intermediate-level wastes are bulky and contain ACTINIDES with long HALF-LIVES, but they are not hot. They cannot be disposed of until they have been incorporated in concrete, bitumen or some other solid, after which they were formerly dumped in the deep ocean. Since the moratorium on the dumping of radioactive wastes at sea they have been stored on land.

Intermediate Technology Development Group (ITDG). A non-governmental organization, founded by E.F. Schumacher and based in London, that collects and disseminates information relating to machines and devices that are easily transportable, reliable and that can be maintained, repaired and eventually replaced by locally made versions, making them suitable for use in remote rural ideas. Being well-made, but using unsophisticated technologies, they comprise an intermediate technology, which is more advanced than traditional technologies, but simpler than technologies used in industrial countries.

intermedin. A HORMONE secreted by the PITUITARY GLAND that stimulates the expansion of pigment cells, causing darkening of the skin in some fish, amphibians and reptiles.

intermittent sampling. Sampling successively for limited periods of time throughout an operation or for a predetermined period of time.

internal combustion engine. An engine powered by the combustion of fuel in an enclosed space, the power being used to produce mechanical motion. The high temperatures and pressures produced lead to the formation of such pollutants as NITROGEN OXIDES, and inefficient combustion may lead to emissions of CARBON MONOXIDE and HYDROCARBON residues.

internal environment. The physical and chemical conditions inside the body of a plant or animal. *See* HOMEOSTASIS.,

internal respiration. *See* RESPIRATION.

International Atomic Energy Agency (IAEA). An agency formed initially to exchange experience and information privately among those working with radioactive materials. Later, with the advent of civil nuclear power, it was adopted by the United Nations. It is concerned with all aspects of atomic energy and the commercial and scientific uses of radioisotopes. It is a partner with the FOOD AND AGRICULTURE ORGANIZATION OF THE UNITED NATIONS in the Division of Atomic Energy in Food and Agriculture. Its headquarters are in Vienna.

International Bank for Reconstruction and Development (World Bank). An international institution formed as a result of the

1945 United Nations Monetary and Financial Conference at Bretton Woods, New Hampshire, USA to facilitate trade and development in war-ravaged Europe. It now has 117 members and works in association with other agencies, including development banks for Africa, Asia and Latin America, the International Finance Corporation (IFC) and the International Development Association (IDA). The International Monetary Fund (IMF) was also founded at Bretton Woods, with the aim of reforming and stabilizing currencies.

International Biological Programme (IBP). A world study of biological productivity and human adaptability, initiated by the INTERNATIONAL COUNCIL OF SCIENTIFIC UNIONS, to run from 1964 to 1974. The study was divided into seven sections: productivity of terrestrial communities; production processes on land and in water; conservation of terrestrial communities; productivity of freshwater communities; productivity of marine communities; human adaptability; the use and management of biological resources. More than 40 countries participated, many of them undertaking special studies to complement and extend the general investigation.

International Board for Plant Genetic Resources (IBPGR). International organization that is in overall charge of seed banks throughout the world. It is a member of the CONSULTATIVE GROUP FOR INTERNATIONAL AGRICULTURAL RESEARCH.

International Commission on Radiological Protection (ICRP). A non-governmental scientific organization, established in 1928 that recommends limits for human exposure to ionizing radiation. Most governments base their national permitted doses on ICRP recommendations.

International Conference on Population. An international meeting held under United Nations auspices in Bucharest (1974) and in Mexico City (1984) to discuss the growth of world population, and especially of populations in many Third World countries, and strategies to contain it.

International Convention for the Prevention of Pollution from Ships. See INTERNATIONAL MARINE CONSULTATIVE ORGANIZATION.

International Council of Scientific Unions (ICSU). A non-governmental organization, based in Paris, that encourages the international exchange of scientific information, initiates programmes which require international cooperation (e.g., the INTERNATIONAL BIOLOGICAL PROGRAMME) and publishes the findings of studies and reports on matters relating to the social and political responsibilities, and treatment, of scientists. See also INTERNATIONAL GEOPHYSICAL YEAR, INTERNATIONAL GEOSPHERE–BIOSPHERE PROGRAMME, INTERNATIONAL YEARS OF THE QUIET SUN.

International Development Association. See INTERNATIONAL BANK FOR RECONSTRUCTION AND DEVELOPMENT.

International Finance Corporation. See INTERNATIONAL BANK FOR RECONSTRUCTION AND DEVELOPMENT.

International Geophysical Cooperation. See INTERNATIONAL GEOPHYSICAL YEAR.

International Geophysical Year (IGY). An international programme of cooperative research, involving more than 70 nations, conducted during the 30-month period from July 1957 to December 1959 after the 18-month programme planned originally had been extended by a further 12-month programme called International Geophysical Cooperation. The programme was organized by the INTERNATIONAL COUNCIL OF SCIENTIFIC UNIONS and was directed toward a systematic study of the Earth and its environment, covering 11 fields: aurora and air glow; cosmic rays; geomagnetism; glaciology; gravity; ionospheric physics; longitude and latitude determinations; meteorology; oceanography; seismology; and solar activity. The IGY is recognized as having provided the basis for the ANTARCTIC TREATY, dedicating the Antarctic to peaceful uses and signed by 12 nations active in that region (more joining later) and, through the ICSU, IGY generated new international organizations to continue the work it had begun.

International GeosphereñBiosphere Programme. A 10-year scientific programme launched in 1986 by the INTERNATIONAL COUNCIL OF SCIENTIFIC UNIONS to collect sufficient information to permit adverse environmental events to be predicted a century in advance.

International Hydrological Decade. A 10-year programme of international scientific cooperation in research on water problems, planned and directed by the UNITED NATIONS EDUCATIONAL, SCIENTIFIC AND CULTURAL ORGANIZATION (UNESCO), that ran from January 1965. National committees were formed in 96 countries, and among the Decade's scientific achievements were: a preliminary survey of sediment transport to the oceans, providing data concerning more than 120 river basins; a study of the water budget, energy balance and circulation of one of the Great Lakes; a study of the Chad Basin in Africa. Postgraduate training courses in hydrology were started in a number of countries as part of a general acceleration in the training of hydrologists generated by the programme.

International Institute for Environment and Development (IIED). An international organization, founded in 1971, and with offices in Buenos Aires, London and Washington, that plans and initiates development projects that can be sustained since they neither degrade the environment in which they occur nor deplete the resources on which they are based. IIED is privately funded, and its field projects, mainly in Latin America, Africa and South-East Asia, are implemented with the help of local nongovernmental organizations.

International Monetary Fund. *See* INTERNATIONAL BANK FOR RECONSTRUCTION AND DEVELOPMENT.

International Nuclear Fuel Cycle Evalu-

ation (INFCE). A study, proposed by US President Carter in 1977, and completed in February 1980, of the implications for the nuclear industry of the proliferation of nuclear weapons.

International Planned Parenthood Federation (IPPF). An international nongovernmental organization, based in London, that aims to study and publicize matters pertaining to family planning and population policies. National family planning associations are affiliated to the IPPF, which is recognized officially by the United Nations.

International Referral System (IRS). A programme for the exchange of information on environmental problems, started as part of the Action Plan agreed at the 1972 UNITED NATIONS CONFERENCE ON THE HUMAN ENVIRONMENT.

International Seabed Authority. The organization, based in Kingston, Jamaica, established under the LAW OF THE SEA CONVENTION to regulate exploration for and exploitation of the mineral resources of the oceans by administering a licensing system. A condition of licences is that details of the technologies employed shall be made available to all signatories to the Convention and profits shall be shared.

International Tropical Timber Agreement. The first international commodity agreement to include a clause on conservation, adopted in 1983 and ratified in 1985, to promote international trade in tropical timbers and to help producing and consuming nations to collaborate in improving the management of tropical forests. It also encourages research and development in forest management, reforestation and wood processing, utilization and marketing.

International Union for Conservation of Nature and Natural Resources (IUCN). An independent international body, founded in 1948, and based at Morges, Switzerland, that promotes and initiates scientifically based conservation measures. It cooperates with the United Nations and other intergovernmental agencies, and with its sister organization the World Wildlife Fund, which exists primarily to raise funds, mainly by national appeals, and to allocate them. *See also* GREEN DATA BOOK, RED DATA BOOK.

International Whaling Commission. The international body that regulates commercial whaling by allocating annual quotas for each whaling country and each species. It also seeks to implement a global moratorium on whaling to allow time for proper assessment of the effects of the industry on the species of commercial value. *See also* PACKWOOD–MAGNUSSON AMENDMENT, PELLY AMENDMENT.

International Years of the Quiet Sun (IQSY). An international cooperative programme for the study of the Earth's environment, based on observations made during 1964 and 1965, when solar activity (the number of sunspots) was at a minimum. IQSY (the initials were transposed because it was believed that in this form the acronym would be easier to pronounce) covered meteorology, geomagnetism, aurora, air glow, the ionosphere, solar activity, cosmic rays, aeronomy and space research, with emphasis on space measurements. The IQSY programme was planned under the auspices of a body formed by the INTERNATIONAL COUNCIL OF SCIENTIFIC UNIONS.

interoceptor. A sense organ (e.g., a receptor in the wall of the gut) that detects stimuli originating inside an animal's body or resulting from substances introduced into the body. *Compare* PROPRIOCEPTOR.

intersex. An animal that is intermediate between male and female because of an abnormality in the SEX CHROMOSOMES or hormones (*see* FREEMARTIN). Its cells are all genetically identical, unlike those of a gynandromorph (*see* GYNANDROMORPHISM).

interstadial. A relatively brief and minor period of remission during a major ICE AGE. *Compare* INTERGLACIAL.

intraclinal. *See* TOPOTYPE.

intractable waste. *See* SPECIAL WASTE.

intrinsic rate of increase. The rate at which a population is increasing in number, measured by deducting the instantaneous death rate from the instantaneous birth rate.

introgression (introgressive hybridization). The infiltration of GENES from one species or population into another, closely related one. This is accomplished by hybridization (*see* HYBRID) and subsequent backcrossing (*see* BACKCROSS).

intrusion. (1) A body of IGNEOUS rock that has invaded pre-existing rock. (2) The process of formation of an intrusive body (intrusion). Intrusive bodies are classified according to their size, shape and relationship to the country rock (i.e. the rocks they invade). *See also* BATHOLITH, BOSS, DYKE, LACCOLITH, LOPOLITH, PLUG, SILL, STOCK.

intrusive. Applied to rocks forming intrusions. *Compare* EXTRUSIVE.

inverse density dependence. A proportionate decrease in mortality, or increase in fecundity, as population density increases.

inversion. A condition in which a dense substance lies over a less dense substance (e.g., in the formation of a DIAPIR). In an atmospheric temperature inversion the air temperature increases, and therefore density decreases, with height. Such inversions occur locally in very still air and tend to be stable because rising air, warmed at the surface, loses its BUOYANCY and is trapped when it meets air at the same temperature and density as itself, so tending to reinforce the inversion. Pollutants entering the air close to ground level are similarly trapped, and so temperature inversions are sometimes associated with severe pollution incidents (e.g., the DONORA SMOG INCIDENT).

Invertebrata (Acraniata). The invertebrate animals. *Compare* VERTEBRATA.

inverted relief. A landscape in which SYN-CLINES form the high ground and ANTICLINES the low ground.

invisible hand. A concept proposed by Adam Smith (1723ñ90) in *The Wealth of Nations* according to which, by seeking to promote his or her own economic advantage an individual contributes unconsciously to the public good by producing as much value as he or she can and by supporting domestic rather than foreign industry out of a concern for personal security. Thus the individual is 'led by an invisible hand to promote an end which was no part of his intention'.

involucre. A number of free or united BRACTS.

involuntary muscle. *See* MUSCLE.

involution. (1) A decrease in the size of part of an organism. *Compare* HYPERTROPHY. (2) The development of abnormal forms in an old culture of microorganisms. (3) Rolling in at the edges in a plant organ (e.g., a leaf).

iodine (I). A blackish–grey crystalline solid element; one of the HALOGENS. Iodine is very VOLATILE, giving off a violet vapour. It is slightly soluble in water and readily soluble in ethanol (to give tincture of iodine). Compounds are found in seaweeds, and sodium iodate ($NaIO_3$) occurs in Chile saltpetre. Iodine is used in medicine, chemical analysis and photography. The radioisotope iodine-131 (HALF-LIFE 8.6 days) is used to diagnose and treat disorders of the thyroid gland. A_r=126.90; Z=53; SG 4.95; mp 114°C; bp 184°C.

ion. An atom that has gained (anion) or lost (cation) one or more (negatively charged) electrons and thus carries a negative (anion) or positive (cation) electrical charge, respectively.

ionization. The process whereby atoms acquire a net electrical charge through the gain or loss of electrons.

ionizing radiation. Radiation that is ca-pable of energizing atoms sufficiently to remove electrons from them. In this state atoms become more reactive, so that ionizing radiation increases chemical activity and in this way produces biological effects, including effects that involve alterations induced in DNA. X-rays and gamma-rays are the only electromagnetic waves that cause ionization in biological material. They are deeply penetrative (although not to the same extent, X-rays being the more penetrative), producing ionization sparsely, but uniformly along their track. ALPHA-PARTICLES and BETA-PARTICLES cause intense ionization, alpha-particles being the more strongly ionizing. Both particles, as well as neutrons and protons, produce increased ionization toward the end of their tracks, since ionization increases as the speed of the particle decreases. Beta-particles penetrate living tissue for a few millimetres (i.e. barely through the skin), whereas alpha-particles penetrate only for a fraction of a millimetre (i.e. not through the skin).

ionosphere. The layer of the upper atmosphere, extending upwards from about 80 kilometres above the surface, in which atoms tend to be ionized (*see* IONIZATION) by incoming solar radiation.

IPPF. *See* INTERNATIONAL PLANNED PARENTHOOD FEDERATION.

IQSY. *See* INTERNATIONAL YEARS OF THE QUIET SUN.

Irano-Turanian Floral Region. The region of the HOLARCTIC REALM that comprises central Asia north of the Himalayas from the western edge of the Black Sea to central China.

Irene Serenade. A Greek oil tanker that sank in the Bay of Pylos, Greece, on 23 February 1980. It was carrying a cargo of 100 000 tonnes of crude oil and caused severe pollution to beaches in the bay and for some distance further along the coast.

iron. (1) (Fe) A metallic element; an essential MICRONUTRIENT. Its magnetic properties are modified in natural iron ores. A_r=55.847; Z=26; SG 7.86; mp 1535°C. *See also* HAEMATITE, LODESTONE, SMELTING, STEEL. (2) A METEORITE composed mainly of nickelñiron alloys.

iron bark. *See* EUCALYPTUS.

iron ore. *See* HAEMATITE.

iron pan. *See* HARDPAN.

iron pyrites. *See* PYRITES.

irradiation. Exposure to electromagnetic waves, and especially to IONIZING RADIATION.

irritability. The characteristic ability of organisms to respond to stimuli.

IRS. *See* INTERNATIONAL REFERRAL SYSTEM.

isallobaric winds. *See* WIND CLASSIFICATION.

island arc. A belt of andesitic (*see* ANDESITE) and basaltic (*see* BASALT) volcanic islands arranged approximately in an arc. Island arcs lie parallel to OCEANIC TRENCHES and are associated with the deep-focus earthquakes of a BENIOFF ZONE and with intense magnetic and gravitational anomalies.

island biogeography. The study of the fauna and flora of islands, based on two concepts: (a) the biological diversity and character of an island depends on its size and distance from the mainland (e.g., large mammals are unlikely to be found on remote islands because they find it more difficult to cross the sea); (b) the number of species on an island will attain an equilibrium determined by the rates at which new species arrive and earlier species become extinct (i.e. new arrivals compete for resources, so reducing the populations of their competitors, and species are liable to become extinct when their populations fall below certain thresholds). An understanding of island biogeography is especially valuable in the planning of nature reserves.

isobar. A line drawn on a map to connect points of equal atmospheric pressure, usually, but not always, measured at ground level and corrected to sea level.

isobilateral. Applied to leaves in which the blade lies more or less vertically and has the same structure on both sides. *Compare* DORSIVENTRAL.

isocline. A FOLD in which the adjacent limbs are parallel.

isogamete. *See* GAMETE.

isogamy. The production of GAMETES that are all alike. This occurs uncommonly in ALGAE, FUNGI and PROTOZOA. *Compare* HETEROGAMY.

isogeneic (isogenic, syngeneic). Having the same GENES. *Compare* ALLOGENEIC.

isogenic. *See* ISOGENEIC.

isohaline. Having the same level of SALINITY.

isohel. A line drawn on a map to connect points of equal average hours of sunshine.

isohyet. A line drawn on a map to connect points of equal average rainfall.

isolating mechanisms. Factors (e.g., geographical isolation, sterility of HYBRIDS) that tend to prevent the exchange of GENES between related populations, so facilitating their evolutionary divergence.

isoleucine. An AMINO ACID with the formula $(CH_3)CH_2CH(CH_3)CH(NH_2)COOH$ and a molecular weight of 131.2.

isoline. *See* ISOPLETH.

isomer. (1) One of two or more chemical compounds that have the same formula, but possess different properties due to the different arrangement of atoms within the molecule. (2) Two or more atomic nuclei that have the same atomic number and mass number, but different energy states.

isophytochrone. A line drawn on a map to connect points where the growing season for plants is the same length.

isopleth (isoline). A line drawn on a map to connect points where the value for the phenomenon being plotted is equal (e.g., ISO-BAR).

Isopoda. *See* MALACOSTRACA.

Isoptera (termites). An order of primitive, mostly tropical insects (division: EXOPTERY-GOTA) that live in colonies and have a highly developed CASTE system. Termites inhabit tunnels in the ground, inside wood or in large mounds, which they construct from wood or earth. They are equipped with gut flora, which enable them to digest cellulose. Termites cause serious economic damage by feeding on and destroying wooden buildings, living trees, paper, etc. Termites excrete METHANE and are the main source of this gas in the atmosphere.

isoseismic line. An imaginary line that joins points of equal magnitude of earthquake shock; it normally forms a closed curve around the epicentre. Isoseismic lines can be deduced from questioning witnesses, or from a field survey, or from closely spaced SEISMOGRAPHS.

isostasy. Equal standing; the concept that an overall constancy of mass (m) of the Earth's CRUST is maintained above a theoretical level of compensation within the Earth, so that in principle any column of identical cross-sectional area above that level will have the same mass. Thus, as mass is the product of volume and density (i.e. $m = vd$), variations in volume (v) on the surface, such as a mountain range, are compensated for by variations in density (d) beneath these surface features. The mountain range will be less dense and have a less dense root, than the thinner, denser crust underlying the ocean floor. This relates to the uniformity of the Earth's gravitational attraction (g) over the Earth's surface, except in localized areas of gravity anomaly where geological processes have raised or lowered the crust out of isostatic balance. For example, areas depressed by ice sheets in Scandinavia acquired a negative gravity anomaly and, over the last 10 000 years, have been and still are rising to regain isostatic equilibrium. Land may also rise to compensate for loss of load when material is eroded from the tops of mountains and transported to the sea. This process is known as isostatic readjustment. *See also* RAISED BEACH PLATFORM.

isostatic readjustment. *See* ISOSTASY.

isotherm. A line drawn on a map to connect points of equal temperature.

isotonic. Having the same OSMOTIC PRESSURE.

isotope. Different forms of an element; the nucleus of an isotope of a particular element is identical to that of another isotope of the same element except that it has a different number of neutrons.

isotropic. Having uniform properties throughout, in all directions.

itai-itai. A disease, caused by cadmium poisoning, that leads to bone deterioration. It was first noted in Japan, and is known by its Japanese nickname, which means 'ouch-ouch'.

ITDG. *See* INTERMEDIATE TECHNOLOGY DEVELOPMENT GROUP.

IUCN. *See* INTERNATIONAL UNION FOR CONSERVATION OF NATURE AND NATURAL RESOURCES.

IWC. *See* INTERNATIONAL WHALING COMMISSION.

IWP. *See* INDICATIVE WORLD PLAN FOR AGRICULTURAL DEVELOPMENT.

Ixtoc 1. An oil-well off the coast of Yucatán, Mexico, which blew out on 3 June 1979, releasing oil into the Gulf of Mexico at a rate of 30 000 barrels per day, creating an oil pollution incident greater than the previous largest, caused by the grounding of the *Amoco Cadiz*. The well was partially capped in October, but leaking continued until relief wells were drilled to release the pressure. The well was finally sealed in March 1980, by which time it had released about three million barrels.

J

J. *See* JOULE.

jasper. A form of CHALCEDONY.

Java man. *See* HOMO.

Jaymoytius. *See* ACRANIA.

Jeanneret, Charles Edouard (1887-1965) (Le Corbusier). One of the most influential architects and urban planners of modern times, who was Swiss by birth, but French by training. He studied and later worked in Paris after a short stay in Berlin. He regarded the city as a machine and produced an ideal design for the CONTEMPORARY CITY to accommodate three million persons. He advocated dwellings with flat roofs and roof gardens. His concepts influenced the design of many buildings and cities, and he produced plans for Algiers, São Paulo, Rio de Janeiro, Buenos Aires, Barcelona, Geneva, Stockholm, Antwerp and Moscow.

JET. *See* JOINT EUROPEAN TORUS.

jet propulsion. Locomotion powered by the discharge under pressure of a fluid through an orifice, the reaction to this producing a thrust in the direction opposite to that of the discharge. This is based on Newton's law which states that to a force acting in one direction there must be an equal force acting in the opposite direction. The principle is exploited by some animals (e.g., squid) and in aircraft where the fluid is the heated product of the combustion of kerosene in air.

jetstream. A core of fast-moving (50–100 metres per second, 100–200 knots) air occurring near the TROPOPAUSE in mid latitudes. Jetstreams are a few hundred metres deep, some tens of kilometres wide and move horizontally, but often in an irregular wavy pattern, from west to east in both hemispheres. A sharp difference in temperature occurs across them, the warmer air lying to the side of the jetstream closer to the equator.

joint. A fracture in rock, often across BEDDING PLANES, along which little or no movement has taken place. *See also* COLUMNAR JOINTS.

Joint European Torus (JET). A project, sponsored by the European Economic Community and based in the UK, to develop a FUSION REACTOR based on the torus (i.e. doughnut-shaped) design.

jojoba *(Simmondsia chinensis).* A hardy shrub, native to arid regions in Mexico and the south-west of the USA, whose seeds yield about 50 percent liquid wax by weight. The wax is perhaps the only substance suitable as a replacement for sperm whale oil. Jojoba wax does not oxidize or go rancid readily and can be used in detergents, linoleum, cosmetics, pharmaceuticals and other commercial products.

joule (J). The derived SI unit of energy, being the work done when a force of 1 newton displaces a point 1 metre, and the work done per second by a current of 1 ampere flowing through a resistance of 1 ohm. It is named after James Prescott Joule (1818–89).

Juncaceae. A family of MONOCOTYLEDONEAE that live mainly in wet or cold situations. Rushes (*Juncus* species), the chief members of the family, provide valuable fodder for sheep in hill pastures where grasses are poor and are also used to make baskets, mats, etc. The flowers of the Juncaceae are more complete than those of grasses, having a perianth of six brownish or green segments.

Jurassic. The middle period of the MESO-ZOIC Era, usually dated as beginning 190–205 Ma and lasting about 60 million years. Jurassic also refers to the rocks deposited during the Jurassic Period; these are called the Jurassic System, which is divided into Lower (or Lias), Middle (or Dogger) and Upper (or Malm). The Lower Jurassic is divided into four stages (Hettangian, Sinemurian, Pliensbachian and Toarcian), the Middle into two (Bajocian and Bathonian) and the Upper into five (Callovian, Oxfordian, Kimmeridgian, Portlandian (or Volgian) and Purbeckian).

juvenile water. Water that arises from underground magmatic sources and so enters the HYDROLOGICAL CYCLE for the first time.

K

4K. *See* MCPA.

K. (1) *See* KELVIN. (2) *See* POTASSIUM.

k. *See* KILO-.

kame. (1) A general name for a wide range of depositional forms of GLACIAL DRIFT, including ESKERS, MORAINES, steep-sided hummocks, alluvial cones, valley-side terraces and crevasse fillings. Most kames, but not all, are composed of stratified sand and gravel. (2) In Scotland, a long, steep-sided ridge.

kame terrace. A terrace-like deposit of OUTWASH DEPOSITS along the side of a valley, formed by streams flowing between a GLACIER and the sides of its trough.

kaolin (china clay). A decomposition product of FELDSPAR that is essentially hydrated aluminium silicate (kaolinite, the mineral from which kaolin is extracted, is $Al_2Si_2O_5(OH)_4$). Feldspar itself occurs in GRANITE, from which matrix the clay is extracted. Globally, deposits are very limited. China clay mining at the site where kaolin forms (there are also secondary deposits, accumulated away from the site of formation) is an open-cast process (*see* OPEN-CAST MINING), often with deep workings, since the decomposition producing the clay proceeds from the lowest levels upwards, so the quality of the clay often improves at greater depths. The white spoil heaps, made from sand, flat areas of white micaceous wastes and turquoise lakes, where water is stored, form spectacular landscapes, being popular among makers of science fiction films. Areas affected have been reduced since changes in the way spoil heaps are built and successes in seeding them with grasses now allow the heaps to be landscaped and reclaimed as they are completed. The wastes can be used for construction. Some metals, including tin, are extracted as by-products of china clay working. The clay itself is used principally in paper making, as a filler in rubber and plastics, in medicines, paints and in the manufacture of fine ceramics (i.e. china).

kaolinite. *See* KAOLIN.

kaolinization. *See* DEUTERIC.

Kara-Bogaz-Gol. A gulf 160 kilometres (100 miles) long, 140 kilometres (87 miles) wide, and 2–3 metres (6ñ10 feet) deep linked by a narrow channel to the eastern side of the Caspian Sea. Water from the Caspian filled the gulf and evaporated, thus draining the Caspian whose level fell 2.6 metres (8.5 feet) in 50 years due to increasing demands for water and dry weather. In 1983, the channel was dammed to stem the loss and the gulf dried out, but there were fears that its accumulated salts might blow on to and contaminate adjacent farm land.

Karnian. *See* TRIASSIC.

karst. Limestone topography, where solution channels have formed deep caverns. Depressions occur where the roofs of these caves have collapsed, giving a rough, uneven terrain.

karyogamy. The fusion of two nuclei in SYNGAMY.

karyokinesis. *See* MITOSIS.

karyotype. The chromosomal (*see* CHROMOSOMES) characteristics (i.e. number, shape, size) of the body cells of an individual or species.

katabatic wind. A wind caused by cold air flowing downhill, which occurs most commonly when a hillside cools at night. The air in contact with the surface is cooled and descends into the valley (*see* FROST HOLLOW). *Compare* ANABATIC WIND.

katabolism. *See* CATABOLISM.

katadromy. *See* CATADROMY.

kata-front. A FRONT at which cold air is descending significantly. *Compare* ANA-FRONT. *See also* COLD FRONT.

kelvin (K). The SI unit of thermodynamic temperature; that temperature, pressure and volume of water at which the solid, liquid and gaseous phases are in balance (the triple point) contains 273.16 kelvins. A temperature expressed in K is equal to the temperature in degrees Celsius plus 273.15°C, the intervals between kelvins and degrees Celsius being identical. The name degrees kelvin and symbol °K have been discontinued, the correct usage now being simply kelvin or K.

Kemeny Commission. The body established by US President Carter, under the chairmanship of John Kemeny and consisting of 12 members, to inquire into the events surrounding the nuclear accident at THREE MILE ISLAND. The Commission reported in October 1979 and was critical of many aspects of the US nuclear industry and of the NUCLEAR REGULATORY COMMISSION.

kenaf (ambari hemp). A tropical plant (*Hibiscus cannabinus*) that can also grow in temperate climates. It is cultivated for its fibre. In some countries (e.g., New Zealand, USSR, USA) it is being considered as a crop for making pulp for the paper industry. Like a number of other fibre plants it is commonly called hemp, although it is unrelated to the true hemp plant (*Cannabis sativa*).

keratin. A sulphur-containing PROTEIN present in epidermal (*see* EPIDERMIS) products such as nails, hoofs, claws, hair, beaks, feathers, the outermost layer of skin and horn in vertebrates. Keratin also occurs in the skeletons of some sponges (*see* PORIFERA) and CNIDARIA.

keratinized. Applied to cells that are dead and have been transformed into scales consisting largely of KERATIN.

kerogen. An insoluble organic material found in SEDIMENTARY ROCKS. Microscopically, kerogen appears to be macerated plant remains, and chemically it differs from CRUDE OIL in its high content of oxygen and nitrogen. Kerogen is the HYDROCARBON contained in OIL SHALES.

kerosene (kerosine, paraffin oil). A mixture of hydrocarbons which are obtained during the distillation of petroleum. It is used as a fuel for lighting, heating and in some (e.g. jet) engines.

ketones. A series of organic compounds with the general formula RCOR' where R and R' are univalent hydrocarbon RADICALS (groups). The ketones include ACETONE (CH_3COCH_3) and ethyl methyl ketone ($CH_3CH_2COCH_3$).

ketose. A monosaccharide (*see* CARBO-HYDRATES) that contains a keto ($>C=O$) group.

kettle-hole. A depression in GLACIAL DRIFT caused by the melting of ice which once formed part of the deposit.

kettle-lake. A lake that fills a KETTLE-HOLE.

Keuper. *See* TRIASSIC.

keys. *See* FRAXINUS EXCELSIOR.

keystone species. A species whose removal from the ecosystem of which it forms part leads to a series of extinctions in that system.

kg. *See* KILOGRAM.

khamsin. *See* DUST STORM, SIROCCO.

kidney iron ore. A variety of HAEMATITE occurring in a kidney-shaped form.

kieselguhr. *See* DIATOMACEOUS EARTH.

killas. In South-West England, a miners' name for SLATES and PHYLLITES.

kiln. A furnace used for drying, CALCINATION and for the baking of bricks, pottery and refractory (heat-resistant) products.

kilo- (k). The prefix used in conjunction with SI units to denote the unit x 10^3.

kilocalorie (Calorie). One thousand CALORIES; a unit of heat now replaced by the JOULE, but still used in parallel with the joule in denoting the CALORIFIC VALUE of foods.

kilogram (kg). The SI unit of weight, equal to 1000 grams, and defined in terms of the international prototype which is in the custody of the Bureau Internationale des Poids et Mesures at Sèvres, near Paris.

kimberlite. An altered PERIDOTITE containing MICA and PYROXENES, which is found in BLUE GROUND. Some kimberlites contain diamonds.

Kimmeridgian. A stage of the JURASSIC System.

kinaesthetic sense. The ability to detect movement by means of PROPRIOCEPTORS.

kinase. An ENZYME that activates another enzyme.

kinesis. A random locomotory movement of an organism in response to a stimulus, the direction of movement bearing no relation to the position of the stimulus. *See also* TAXIS.

kinetin. One of the CYTOKININ group of plant hormones.

kinetosome. *See* CILIA.

king crab. *See* XIPHOSURA.

kingdom. *See* CLASSIFICATION.

kingfishers. *See* ALCEDINIDAE.

klendusity. The ability of an organism to avoid disease because of the way it grows (e.g., when the susceptible stage in the life history does not coincide with the seasonal occurrence of a parasite).

klepto-parasite. *See* PARASITISM.

knot. A unit of speed equal to 1 nautical mile per hour (0.515 metres per second). It is still used in many countries as a measure of the speed of winds, ocean currents, ships and aircraft.

kopje. The South African name for a BUTTE. Many kopjes are capped by DOLERITE SILLS.

Krakatoa (Krakatau). A volcanic islet in the Sunda Strait, Indonesia, between Java and Sumatra. It erupted very violently on 20 May 1883, emitting very large amounts of particulate matter and gases into the atmosphere and causing widespread destruction. Studies of the subsequent recolonization of the neighbouring islands have provided valuable insights into ISLAND BIOGEOGRAPHY. *See also* MOUNT AGUNG, NATURAL POLLUTANT, SULPHATE, SULPHUR DIOXIDE.

Krebs, Hans Adolf (1900ñ81). The German-born, naturalized British biochemist who won the Nobel Prize for medicine and pharmacology in 1953 jointly with Fritz Lipmann. He was knighted, for his contributions to biochemistry and in particular for his discovery of the CITRIC ACID CYCLE in the metabolism of carbohydrates.

Kreb's cycle. *See* CITRIC ACID CYCLE.

krill. *See* MALACOSTRACA.

kwashiorkor. A deficiency disease caused by protein deprivation. The name means 'the sickness the child develops when another is born', and the disease is caused most frequently in impoverished communities by premature weaning of infants on to an inadequate diet when the birth of a new infant creates a more imperative claim on the mother's milk.

L

Labiatae. A cosmopolitan family of DI-COTYLEDONEAE, most of which are aromatic, square-stemmed herbs (e.g., dead-nettles) or small shrubs. Many (e.g., thyme, sage, mint, oregano) are used as culinary herbs, others (e.g., lavender, rosemary) yield oils and perfume.

laccolith. An IGNEOUS body with a dome-shaped upper surface and a flat base, which is concordant with the strata into which it is intruded.

Lacertilia. *See* SQUAMATA.

lactic acid. An organic acid produced during the metabolism of many types of cell (e.g., in vertebrate muscle fibre and by BACTERIA, causing the souring of milk). *See also* GLYCOLYSIS.

lactoflavin. *See* RIBOFLAVIN.

lactogen. *See* LACTOGENIC HORMONE.

lactogenic hormone (lactogen, prolactin, luteotropic hormone, LTH). A HORMONE produced by the PITUITARY GLAND of vertebrates. It stimulates the secretion of PROGESTERONE from the CORPUS LUTEUM and initiates milk production in mammals. In pigeons it stimulates the production of CROP MILK.

lactose. *See* CARBOHYDRATES.

lacustrine. Pertaining to a lake.

Ladinian. *See* TRIASSIC.

ladybirds. *See* COLEOPTERA.

Lagomorpha (Duplicidentata, rabbits, hares).   An order of very successful herbivorous animals, with chisel-shaped, continually growing incisor teeth used for gnawing. (In contrast to RODENTIA the upper pair are accompanied by a smaller, second pair.) Lagomorphs are serious pests in many parts of the world, including Australia, where rabbits were introduced in the 18th century. *See also* MYXOMATOSIS.

lahar. A mud flow of water-saturated volcanic ash, or the deposit formed by such a mud flow.

laissez faire. An economic doctrine of non-interference by government or other institutions, with the free play of market forces. The underlying theory holds that human affairs, like processes in nature, are subject to general laws which operate for the good of the system as a whole. If each individual is permitted to work for his or her economic advantage, then this will tend toward the general good, and harmony will be achieved. The concept was first enunciated, and the phrase coined (*'laissez faire, laissez passer'*) by the Physiocrats in the 18th century, and it came to be supported by many leading economists, including Adam Smith (1723–90), David Ricardo (1772–1823), Thomas Robert Malthus (1776–1834), John Stuart Mill (1806–73) and John Locke (1632–1704).

Lake Superior-type iron ore. An alternative name for BANDED IRONSTONES, from the region in which they were first studied.

Lamarck, Jean Baptiste de (1744ñ1829). A French biologist who advanced the first clear theory of evolution, based on the inheritance of acquired characters and more especially acquired habits. He maintained that if, in response to the effects of the environment, part of an animal's body was subjected to an unusual degree of use or disuse, this organ would become modified, and the modification would be handed on to the progeny. He postulated, for example, that the long neck of the giraffe has evolved because the shorter-necked ancestor stretched upward for its food, thus lengthening the neck, a development that was inherited by its offspring. Similarly, the poorly developed wings of flightless birds had evolved because they ceased to be used for flight. Lamarck's theory of evolution was superseded by that of Charles Darwin. *Compare* LYSENKO, TROFIM DENISOVICH.

lamarckism. *See* LAMARCK, JEAN BAPTISTE DE.

Lamellibranchiata (Pelecypoda). A class of MOLLUSCA, including oysters, cockles, and mussels, which have a bivalve shell. They are aquatic, feeding and breathing by means of currents of water set up by CILIA. *See also* TEREDO.

laminar motion. One of the two types of motion that occur in fluids in which individual particles of the fluid follow smooth, well-defined paths, the majority of them moving in the same direction. *Compare* TURBULENCE.

lampreys. *See* AGNATHA, CYCLOSTOMATA.

lamp shells. *See* BRACHIOPODA.

lancelets. *See* ACRANIA, AMPHIOXUS.

land breeze. *See* BREEZE.

land drainage. The removal of water from wet or waterlogged land in order to make the land suitable for cultivation, building development, etc. Since many areas of wetland are located in river valleys that are otherwise fertile, or near to coasts popular with tourists, the economic incentive to drain them is compelling, and wetlands are under threat in many countries. Wetlands also provide important habitats for many species of plants and animals (especially birds) and are the subject of international conservation programmes.

land-fill. The disposal of refuse by tipping it on land. Often the refuse is used to fill in old mine workings or low-lying land, to reclaim land from water or to create a feature on flat land. If the refuse is deposited in prepared trenches or holes, over which earth can be heaped at the end of each day, this is called controlled tipping in the UK and SANITARY LAND-FILL in the USA.

landing and take-off cycle (lto cycle). All the operations performed by an aircraft between the time it is at a height of 915 metres on its landing approach and the time it reaches the same height after its next departure. The cycle is used in the measurement of and the setting of limits to the rate of emission and duration of pollutants released into the air, leading to standards that relate to specified pollutants and to particular types and sizes of aircraft.

landnam. A short period of time during which a forest is cleared and the clearing occupied by swidden farmers (*see* SWIDDEN FARMING). After a few years the clearing is abandoned and the forest regenerated.

land reclamation. The treatment of any unusable land (e.g., slag heaps, quarries, gravel pits, etc.) usually by filling with refuse (*see* LAND-FILL) or levelling, until the land can be brought into productive use.

Landsat. A series of US observation satellites, the first of which was launched in January 1975 to replace the earlier EARTH RESOURCES TECHNOLOGY SATELLITE. In 1986 Landsats 4 and 5 were still operational, and 6 and 7 were planned. Landsats 4 and 5 have provided images of the surface with a 30-metre resolution. *Compare* SPOT.

land use. The deployment of land for any use. Competition for limited areas of land requires the establishment of priorities among claims, which is the object of land use planning.

lanugo. The fine covering of hair on a FOETUS which is shed before birth.

lapilli. PYROCLASTIC fragments ejected by a VOLCANO which are 4–32 millimetres in diameter.

lapse rate. The rate at which temperature decreases in the atmosphere as height above the surface increases. Within the TROPOSPHERE the average lapse rate is 6–8°C per kilometre. *See also* ADIABATIC LAPSE RATE, DRY ADIABATIC LAPSE RATE, WET ADIABATIC LAPSE RATE.

large dam. A DAM that is more than 150 metres (492 feet) high or whose volume is more than 15 million cubic metres (19.6 million cubic yards) or whose reservoir has a capacity of more than 14.8 billion cubic metres (12 million acre-feet). Large dams have been linked to increased earthquake frequency in seismically active areas, with chemical and therefore ecological alteration of waters downstream of them, and, in low latitudes, with increases in water-borne diseases due to providing open water throughout the year in areas where a dry season formerly controlled the disease VECTORS..

Laridae (gulls, terns). A family of colonial sea-birds with webbed feet and powerful flight. Most gulls alight on the water to seize food, whereas terns dive from the air.

larva. An immature stage of an animal, usually very different in form from the adult, able to feed itself, but very rarely capable of sexual reproduction. *See also* BLADDERWORM, LEPTOCEPHALUS, METAMORPHOSIS, NAUPLIUS, NYMPH, PAEDOGENESIS, PLUTEUS, TROCHOPHORE, VELIGER.

laser. Acronym for *Light Amplification by Stimulated Emission of Radiation*; a device that produces a powerful, highly directional, monochromatic, coherent (i.e. its waves are in phase) beam of light. Lasers consist of a transparent cylinder with a reflecting surface at one end and a partially reflecting surface at the other. Light waves are reflected back and forth, some of them emerging at the

partially reflecting end. The light source may be a ruby, whose chromium atoms are excited by a flash lamp so that they emit pulses of highly coherent light, or a mixture of inert gases that produce a continuous beam, or a cube of treated gallium arsenide which emits infrared radiation when an electric current passes through it.

lasion. A more or less dense growth of interdependent organisms attached to surfaces submerged in water, above the bottom. *See also* PERIPHYTON.

latent heat. The energy, as heat, that is absorbed when a substance changes phase from solid to liquid (melting) or liquid to gas (vaporization, evaporation) and emitted when it changes phase in the opposite direction. About 334.7 joules per gram are released when water freezes. The emission of latent heat causes large hail to become wet when falling through a cloud of supercooled (*see* SUPERCOOLING) droplets many degrees colder than 0°C because the freezing of some droplets on to the hailstones soon raises their temperature to 0°C.

laterite. Earthy, granular or concretionary mass, chiefly of iron and aluminium oxides and hydroxides, occurring as a layer or as scattered nodules in tropical soils. *See also* HARDPAN, SOIL CLASSIFICATION.

lateritic soil. Soil containing LATERITE. Such soil is often unsuitable for agriculture, being weathered (*see* WEATHERING) until the laterite forms a hard, impermeable layer. *See also* LATERIZATION, SOIL CLASSIFICATION.

laterization. The process of forming LATERITE, resulting from the removal of silica, alkalis and alkaline earths, and the consequent enrichment in the iron and aluminium compounds that form laterite. The process may be accelerated by the removal of surface vegetation, which serves to bind together soil particles and to maintain the cycling of nutrients. Removal leaves the soil exposed to rain. *See also* ELUVIATION, EROSION, LATERITIC SOIL, LEACHING, SOIL CLASSIFICATION, SOIL HORIZONS.

Late Stone Age. *See* NEOLITHIC.

Latimeria. *See* CROSSOPTERYGII.

laurilignosa. Laurel forest and laurel bush, forming subtropical rain forest.

lava. A molten liquid extruded on to the surface of the Earth, or the solidified product subsequently formed. ACIDIC lavas are more viscous than BASIC lavas and give rise to more violent eruptions.

lawn sand. A mixture of mercurous chloride (calomel) and ferrous sulphate used to control moss in turf. It is harmful to fish.

law of minimum. Plants have minimum requirements of certain mineral salts (*see* MACRONUTRIENTS, MICRONUTRIENTS). If a soil does not supply this minimum, the plants requiring those particular salts cannot grow, regardless of the abundance of other nutrients.

Law of the Sea Convention. The international convention based on agreement reached at the final session of the UNITED NATIONS CONFERENCE ON THE LAW OF THE SEA in 1982, under which an INTERNATIONAL SEABED AUTHORITY was established.

lawrencium. *See* ACTINIDES.

layers. (1) In plant ecology, a COMMUNITY exists in three dimensions, with plants of different heights forming strata at various levels above the ground, each stratum supporting a characteristic population of invertebrate animals and often birds. Four such horizontal vegetation layers are recognized: the tree layer; shrub (bush) layer; field (herb) layer; ground (moss) layer. (2) Tissue layers (e.g., ANNUAL RINGS, GERM LAYERS). (3) A branch that strikes root and can grow independently after becoming detached from the parent plant.

LD$_{50}$. The concentration of a substance that causes the death of one-half of a population exposed to it within a given period of time. *See also* APPLICATION FACTOR.

leachate. *See* LEACHING.

leaching. The removal of the soluble constituents of a rock, soil or ORE (that which is leached being known as the leachate) by the action of percolating waters. Leaching is a major process in the development of porosity in LIMESTONES, in the SECONDARY ENRICHMENT of ores and in the formation of soils. *Compare* ELUVIATION.

leaching field. A system of open pipes in covered trenches that permits effluent from a SEPTIC TANK to enter surrounding soil.

lead (Pb). A metallic element, compounds of which are used in many industries and may accumulate in biological systems. The metal itself is used in lead–acid batteries, for roofing and was formerly used in plumbing; lead pipes still carry public water supplies in some cities. In humans, small doses produce behavioural changes, larger doses can result in paralysis, blindness and eventual death. Its addition to petrol (as TETRAETHYL LEAD) has caused its widespread distribution, especially in the vicinity of major roads. It may also be a serious pollutant near heavy metal smelters. Pregnant women are especially vulnerable, and there is evidence that young children can be harmed by very small amounts accumulated in the body mainly from airborne sources (i.e. vehicle exhausts). $A_r 2 = 07.19$; $Z = 82$; SG 11.34; mp 327.4°C.

leaf insect. *See* PHASMIDA.

lean-burn engine. A petrol-powered car engine designed to burn a fuel/air mixture containing a higher proportion of air (i.e. a lean mixture) than would be acceptable in other engines. This reduces both fuel consumption and the production and emission of unburnt hydrocarbons and nitrogen oxides.

learning. Behaviour that enables an animal to modify its actions with changing situations as a result of experience. *See also* CONDITIONED REFLEX, INSTINCT.

Le Chatelier principle. For general physical systems, if the system is subjected to a constraint, whereby the equilibrium is modified, a change takes place which, if possible, partially annuls the constraint.

lecithin. A complex fatty substance containing phosphorus (i.e. a phospholipid). It is a constituent of UNIT MEMBRANES of plant and animal cells and is present in the yolk of most eggs. *See also* FAT, LIPIN.

Le Corbusier. *See* JEANNERET, CHARLES EDOUARD.

lectotype. A specimen chosen from original material as a substitute for a missing TYPE SPECIMEN. *Compare* NEOTYPE. *See also* PARATYPE.

lee. The side away from wind, ice or current. *Compare* STOSS.

leeches. *See* HIRUDINEA.

lee waves. *See* MOUNTAIN WAVES.

legume. (1) The fruit of a member of the LEGUMINOSAE. A pod that releases its seeds by splitting lengthways down both sides, often explosively. (2) Crops of plants belonging to the Leguminosae (e.g., clover, peas, beans).

Leguminosae. A very large, cosmopolitan family of DICOTYLEDONEAE that includes herbs, trees, climbers, water plants, etc. They are of great value as soil-enriching crops because most bear ROOT NODULES containing symbiotic, nitrogen-fixing (*see* NITROGEN FIXATION) bacteria (*Rhizobium* species). Many Leguminosae species are important sources of human food and animal fodder (e.g., peas, beans, lupins, vetches, clover). *Acacia* and other trees yield timber and resin. *Crotalaria* is cultivated for its fibre (Bombay or Madras hemp). *Ulex* (gorse, furze, whin) is a spiny evergreen shrub (a XEROPHYTE) that is common in the UK.

lek. *See* LEK BREEDING.

lek breeding. A social arrangement among animals in which a group of males assembles on a traditional breeding ground (a lek). An order of dominance is established among them, and the dominant male then undertakes most of the breeding with visiting females (*see* POLYGYNY). Lek breeding has been observed in such birds as the sage grouse and in animals such as the Uganda kob.

lemurs, flying. *See* DERMOPTERA.

lenticel. A pore in the corky outside layer of a woody stem through which gases can diffuse.

lentic water. Standing water (i.e. in lakes, ponds, marshes, etc.) *Compare* LOTIC WATER.

Lepidoptera (butterflies, moths). A large order of insects (division: ENDOPTERYGOTA) whose wings and bodies are covered with minute scales. The larvae (caterpillars) are largely herbivorous and almost all the adults feed on nectar, by means of a long, coiled, tubular proboscis. The order includes serious pests such as the cabbage white butterfly (*Pieris*), clothes moth (*Tinea*), flour moth (*Ephestia*) and cotton bollworm (*Platyhedra gossypiella*). Lepidoptera are important agents of pollination, and some control weeds. The moth *Cactoblastis cactorum* was successfully introduced into Australia to control the prickly pear. *Bombyx mori* is the commercial silkworm moth. Butterflies (Papilionoidea) can be distinguished from moths by their clubbed antennae and the absence of a mechanism for hooking fore and hind wings together. The division of Lepidoptera into butterflies and moths is artificial, and these features may be found separately in some moths.

Lepidosauria. *See* SQUAMATA.

Lepidosiren. *See* DIPNOI.

leptocephalus. The transparent pelagic larva of the eel, which migrates across the Atlantic from the breeding grounds off the West Indies (*see* SARGASSO SEA) to the European rivers where it matures.

leste. *See* SIROCCO.

lethal gene. A GENE that kills the individual bearing it by causing some disorganization of metabolism (e.g., the production of a toxin). If the lethal gene is RECESSIVE, it kills only individuals homozygous for it.

Leucaena. A genus of leguminous (*see* LEGUME) shrubs and small trees native to Mexico, but now planted widely in the tropics and subtropics to provide forage for livestock, GREEN MANURE, fuel wood and timber, to halt soil erosion and to improve soil fertility. The species most widely grown is *L. leucocephala*, of which there are several varieties.

leucine. An amino acid with the formula $(CH_3)_2CHCH_2CH(NH_2)COOH$ and a molecular weight of 131.2.

leucism. A type of ALBINISM involving the absence of pigment from fur or feathers, but not from other parts of the body.

leucite. *See* FELDSPATHOIDS.

leucocratic. Consisting mainly of light-coloured (i.e. felsic) minerals. *Compare* MELANOCRATIC.

leucocytes (white blood cells). *See* BLOOD CORPUSCLES.

leucoplasts. *See* PLASTIDS.

leukaemia. A disease of the blood, often fatal, in which white blood cells (leucocytes) proliferate.

levanter. *See* MISTRAL.

leveche. *See* SIROCCO.

levée. A ridge of ALLUVIUM at the side of a river where the coarse suspended load carried by flooding water has been dumped as the current is checked at the bank.

level. The horizon at which an ORE BODY is being worked and often used to cover all the working at one horizon.

ley. *See* GRASSLAND.

LH. *See* LUTEINIZING HORMONE.

Li. *See* LITHIUM.

Lias. (1) Strata deposited during Lower JURASSIC times. (2) An INTERBEDDED sequence of SHALE and LIMESTONES.

lice. *See* ANOPLURA.

lichen. A dual organism formed by a green alga (*see* CHLOROPHYTA) or blue–green alga (*see* CYANOPHYTA) and a fungus (*see* FUNGI) living together symbiotically (*see* SYMBIOSIS). Lichens live attached to rocks, tree trunks, etc. and may be encrusting, upright and branched, or leaf-like and flat. They are important primary colonizers of bare surfaces and are dominant in mountainous or arctic situations where conditions are too harsh for other plants to grow. Reindeer moss is the staple diet of reindeer in Lapland. Many lichens are very sensitive to air pollution.

LIDAR. Acronym for Light Detection And Ranging; a technique for detecting and tracking chimney plumes, often after they have ceased to be visible to the naked eye. The equipment consists of a laser transmitter, which emits brief, high-intensity pulses of coherent light, and a receiver, which measures the energy of that light when backscattered (*see* BACKSCATTER) by atmospheric aerosols, and interprets the range.

life form. (1) The characteristic form of a plant or animal species at maturity (e.g., herb, worm). (2) *See* RAUNKIAER'S LIFE FORMS.

life table. A description of the age-specific

survival of cohorts of individuals in relation to their age or stage of development.

Light Detection And Ranging. *See* LIDAR.

light minerals. In petrology, usually refers to the FELSIC minerals (e.g., QUARTZ, FELDSPAR, FELDSPATHOIDS) present in a rock.

light reactions. *See* PHOTOSYNTHESIS.

light-water reactor (LWR). A NUCLEAR REACTOR that uses light water (as opposed to HEAVY WATER, deuterium oxide) as a coolant and MODERATOR. *See also* BOILING WATER REACTOR, PRESSURIZED WATER REACTOR.

lignin. A complex non-carbohydrate substance deposited in the walls of many plant cells, especially those of woody tissue, imparting strength and rigidity.

lignite. *See* COAL.

lignosa. Vegetation that is made up of woody plants.

lilac. *See* OLEACEAE.

Liliaceae. A large cosmopolitan family of MONOCOTYLEDONEAE, most of which are herbaceous plants with RHIZOMES, BULBS or CORMS, and flower parts arranged in whorls of three. They include ornamental plants (e.g., lilies, croci, daffodils) and food plants (e.g., onions, asparagus).

limb. The part of a FOLD between adjacent hinges.

lime. (1) Slaked lime (*see* CALCIUM HYDROXIDE). (2) Quicklime (*see* CALCIUM OXIDE).

limestone. A SEDIMENTARY ROCK composed largely of the CARBONATE MINERALS, CALCITE or, less frequently, aragonite. Limestones can be divided into: (a) those formed by the evaporation of aqueous solutions (e.g., TUFA, TRAVERTINE and oolitic limestones, *see* OOLITE); (b) those formed of calcite produced biochemically by living organisms (e.g., CHALK, COQUINA, coral limestone, *see* CORAL

REEF, foraminiferal limestone, *see* RHIZOPODA); (c) those formed by the mechanical accumulation of fragments of pre-existing limestone. Many limestones have been formed by a combination of two or more of these methods, usually followed by a complex DIAGENESIS. Diagenesis or METASOMATISM may convert the limestone into a DOLOMITE rock. Limestones are important as AQUIFERS, as RESERVOIR ROCKS for hydrocarbons, as building stone and AGGREGATE and, with clay, for making cement. *See also* CALCIUM CARBONATE.

limestone pavement. An area of LIMESTONE that has been exposed by glacial activity and shaped by subsequent WEATHERING to produce a bare rock surface of CLINTS separated by GRIKES supporting many rare plants. In the UK, limestone pavements were severely damaged by the removal of stone for use as garden ornament. Under the Wildlife and Countryside Act, 1981, provision was made to protect them by allowing a Limestone Pavement Order to be issued where a pavement was of particular value, making it an offence to damage that pavement.

Limestone Pavement Order. *See* LIMESTONE PAVEMENT.

limiting factor. An environmental factor (e.g., temperature) that restricts the distribution or activity of an organism or population.

Limits to Growth. *See* CLUB OF ROME.

limnetic. (1) Living in the open water of a lake or pond. (2) Inhabiting marshes, lakes or ponds.

limnic. Applied to sediments deposited in freshwater lakes. Limnic also refers to the environment of deposition and, in geology, is contrasted with PARALIC.

limnology. The study of the physical, chemical and biological components of fresh water.

limonite. A mixture of amorphous and CRYPTOCRYSTALLINE hydrated iron oxides and hydroxides formed as a weathering product of iron-bearing minerals. Important occurrences of limonite are as bog iron ore, LATERITE, OCHRE, GOETHITE and in GOSSAN.

lindane (HCH, benzene hexachloride, BHC, gamma-BHC). An ORGANOCHLORINE used to control insect soil pests, aphids, grain weevils, mites, etc. It is harmful to bees, fish, livestock, etc. and is persistent, although less so than DDT, and it is somewhat PHYTOTOXIC.

linear growth (simple interest, arithmetic growth). An increase in value based on interest calculated on the original value only, so that the sum added at the end of each period remains constant. *Compare* EXPONENTIAL GROWTH.

line squall. A belt of severe thunderstorms accompanying a COLD FRONT.

linkage. The association of non-allelomorphic (*see* ALLELOMORPH) genes so that they are inherited together and do not show INDEPENDENT ASSORTMENT. Linkage occurs between two genes when they are situated on the same chromosome. *See also* CROSSING-OVER.

linnaeite (Co_3S_4). One of the major ores of cobalt. Linnaeite is found in HYDROTHERMAL veins, typically with nickel and/or copper sulphides. Cobalt is used in various high-temperature alloys.

linnaeon. *See* LINNEON.

Linnaeus, Carolus (1707–78). A Swedish biologist, the author of *Systema Naturae*, in which he evolved the system of binomial nomenclature still used today. Each organism was designated by a generic name, referring to the group to which it belonged, and a specific name for itself (e.g., *Homo sapiens*). Linnaeus did not believe in evolution, and his classification was an artificial one, based on obvious external characteristics and not on relationships between organisms.

linneon (linnaeon). A species according to LINNAEUS, CAROLUS, often a superspecies.

linoleic acid. *See* VITAMIN F.

linolenic acid. *See* VITAMIN F.

linseed oil. *See* LINUM.

Linum. A genus of DICOTYLEDONEAE (family: Linaceae) including flax (*Linum usitatissimum*), which yields linen fibre, obtained by RETTING away the softer tissues in water. Linseed oil and cattle cake are obtained from the seeds. *See also* GLYCOSIDES.

linuron. A soil-acting TRANSLOCATED HERBICIDE of the UREA group. It is used to control corn marigold in cereals and many weeds in root crops. It can be irritating to the skin and eyes, and is harmful to fish.

lipase. An ENZYME that breaks down fat into its constituent fatty acid and alcohol.

lipid. *See* FAT.

lipin (lipine). A complex fatty substance (e.g., LECITHIN, a phospholipid found in UNIT MEMBRANES).

lipoid. *See* FAT.

liquefied petroleum gas (LPG). A gas, composed mainly of butane (C_4H_{10}), propane (C_3H_8) and pentane (C_5H_{12}), that is produced from the gas associated with crude oil. It is sold in pressurized containers as bottled gas.

liquid metal fast breeder reactor (LMFBR). A BREEDER REACTOR that uses plutonium dioxide or uranium dioxide as a fuel and molten sodium and potassium as a coolant. Being a breeder reactor, it uses no MODERATOR.

lithification. The conversion of sediment into consolidated SEDIMENTARY ROCK. Many processes of DIAGENESIS result in lithification, but diagenesis may continue after lithification has taken place.

lithite. *See* STATOCYST.

lithium (Li).   A light, silveryñwhite, alkali metal; the lightest solid known. Chemically it resembles sodium, but is less active. It is used in alloys and may be used as a fuel in FUSION REACTORS. A_r=6.939; Z=3; SG 0.534; mp 179°C.

lithocyst. *See* STATOCYST.

lithosere.   The stages in a plant succession that begins on an exposed rock surface.

Lithosols.   In the USDA soil taxonomy, surface deposits with no SOIL HORIZONS developed. *See also* SOIL CLASSIFICATION.

lithosphere.   The portion of the Earth above the LOW-VELOCITY ZONE, including the CRUST and part of the upper MANTLE.

Little Ice Age. *See* FERNAU GLACIATION.

littoral currents.   Currents that move parallel to the shore within the surf zone (i.e. within the stretch where waves break), produced by waves breaking obliquely on the shore.

littoral drift.   The movement of material in LITTORAL CURRENTS within the surf zone (i.e. within the stretch where waves break). *See also* BEACH DRIFT, LONGSHORE CURRENTS, LONGSHORE DRIFT.

littoral zone.   (1) The part of a lake extending from the shore down to the limit for rooted vegetation. (2) The intertidal zone of a sea. *Compare* SUBLITTORAL ZONE. *See also* SHORE ZONATION.

liverworts.   *See* BRYOPHYTA, GEMMATION, HEPATICAE.

living fossil. *See* FOSSIL.

Llandeilian.   The third oldest series of the ORDOVICIAN System in the UK.

Llandoverian.   The oldest series of the SILURIAN System in Europe.

Llanvirnian.   The second oldest series of the ORDOVICIAN system in the UK.

lm. *See* LUMEN.

LMFBR. *See* LIQUID METAL FAST BREEDER REACTOR.

LNG. *See* NATURAL GAS.

load on top.   A method of carrying oil in tankers, where the cargo floats on top of water in one of the tanks (the slop tank). The slop tank water is used first to wash out the other tanks, and residues from ballast water are added to it. The oil carried on top of this water can be refined. The system is designed as an alternative to the highly polluting process of discharging ballast residues and water that has been used for washing tanks into the sea.

local allocation tax.   A system of block grants made by central government to local governments (e.g., in Japan) that is calculated as a percentage of revenue from certain national taxes (in Japan, 52 percent of national income, corporate business and liquor taxes). *See also* LOCAL TRANSFER TAX.

local forecast.   A weather forecast for a small enough area for the weather to be uniform all over it and usually involving reference to features of weather that are not shared with neighbouring localities.

local nature reserve. *See* NATURE RESERVE.

local transfer tax.   A system of block grants made by central government to local governments (e.g., in Japan) calculated as a percentage of national revenue from certain taxes (in Japan, local road tax, petrol tax and tonnage tax on sea ports). *See also* LOCAL ALLOCATION TAX.

locus.   The position on a chromosome occupied by a particular GENE or its ALLELOMORPH.

locust. *See* ACRIDIDAE, ORTHOPTERA.

lode.   A VEIN, or system of veins, usually

containing a metalliferous mineral, which, with any intervening COUNTRY ROCK, may be mined as one unit.

lodestone. A variety of MAGNETITE that acts as a compass needle when it is free to swing.

lodging. The mechanical collapse of a cereal crop. This can occur in heavy rain or hail, and the crop may be rendered more vulnerable if the plants are very tall and unable to give adequate mechanical support to the weight of the ears. Many traditional varieties of cereals respond to additional applications of fertilizer by growing to a greater height as well as by producing a heavier ear, thus making them liable to lodging. This has been remedied by the breeding of short-stemmed varieties. A lodged crop is difficult to harvest, much of the grain falls from the ear and is lost, and if the grain becomes moist it may begin to germinate, so making it unsuitable for most uses.

loess. A very fine, unconsolidated and unstratified permeable material, generally grey to buff in colour. Loess is composed mainly of angular particles of such minerals as QUARTZ, FELDSPAR and CALCITE, in a CLAY matrix. Most loess seems to be wind-blown glacial ROCK FLOUR, although some appears to be of desert origin. Loess makes a fertile soil.

logarithmic reproduction curve. A reproduction curve with logarithmic axes.

logarithmic wind profile. A wind profile of a type frequently observed in the lower atmosphere, which is dominated by roughness effects, taking the form: $(u/u_*) = (1/k)\log_e(z/z_o)$ where u is the mean speed at height z, u_* is the friction velocity, k is von Karman's constant (about 0.4) and z_o is the roughness length, a parameter approximately equal to one-thirtieth of the average height of the surface obstacles.

logistic curve. An S-shaped curve on a graph, often characteristic of population growth and stabilization. The curve rises slowly at first, then more and more steeply, finally flattening at a new level.

London Dumping Convention. *See* CONVENTION ON THE PREVENTION OF MARINE POLLUTION BY DUMPING OF WASTES AND OTHER MATTER; INTERGOVERNMENTAL MARITIME CONSULTATIVE ORGANIZATION.

London smog incidents. Two episodes, in 1952 and 1962, when descending cold air, trapped under very stable inversions, caused the Thames Valley to become a reservoir of cool, still, moist air. Fog, with chemical and physical pollutants added, remained stationary for four days in 1952 and five days in 1962. In 1952, 4000 people died and in 1962 about 700 died, although most were elderly and suffering from chronic respiratory complaints. *See also* SMOG.

long-day plants. *See* PHOTOPERIODISM.

long-range forecast. A weather forecast of a speculative nature for a period beyond that for which physical and mechanical laws can be the basis of forecasting because of the complexity or uncertainty of starting conditions. Thus a forecast for a month or season is long-range.

longshore currents. Currents that move parallel to the coast and are generated by waves, winds and tides. *See also* BEACH DRIFT, LITTORAL CURRENTS, LITTORAL DRIFT, LONGSHORE DRIFT.

longshore drift. The movement of material in LONGSHORE CURRENTS along the coast generally, not only in the surf zone (i.e. within the region where waves break). *See also* BEACH DRIFT, LITTORAL CURRENTS.

looping. The behaviour of a chimney PLUME, when large-scale thermal eddies bring puffs of concentrated pollutants to the ground for a few seconds before carrying them aloft again.

lophodont. Applied to teeth in which the cusps on the crown are joined up to form ridges. This type of tooth is typical of many hoofed animals (e.g., rhinoceros, which has a simple lophodont pattern; horse, which has a complex pattern) and rodents. *Compare* BUNODONT, SELENODONT.

lopolith. A saucer-shaped IGNEOUS INTRUSION.

Los Angeles smog. *See* SMOG.

lotic water. Flowing water (i.e. rivers, streams). *Compare* LENTIC WATER.

loudness. The INTENSITY of a sound as perceived by the human ear, dependent on sound pressure (*see* SOUND PRESSURE LEVEL) and FREQUENCY. *See also* DECIBEL, PERCEIVED NOISE LEVEL, SOUND LEVEL.

Love Canal. Area in Niagara Falls, New York where industrial wastes were buried in drums in the 1940s and early 1950s. The drums corroded, leaking substances that included suspected CARCINOGENS. On 7 August 1978, Love Canal was declared a disaster area, and about 240 families were evacuated. Medical examinations revealed that 11 people had chromosome damage, but this was not confirmed by further investigation. In 1982, the Center for Disease Control declared part of the evacuated area fit for habitation, and it was not believed that any of the residents would suffer permanent injury.

lower crust. *See* SIMA.

lower shore. *See* SHORE ZONATION.

lowland reach. *See* RIVER ZONES.

low-level wastes. RADIOACTIVE WASTES consisting of clothing and equipment from hospitals and laboratories where radioactive substances have been used, slightly contaminated soil and rubble from demolished buildings in which radioactive substances were stored or used. Low-level waste contains RADIONUCLIDES with short HALF-LIVES together with traces of radionuclides with longer half-lives. It presents no serious hazard to the public and is suitable for disposal

by shallow burial in trenches or dumping at sea beyond the CONTINENTAL SHELF.

low-pressure area. *See* CYCLONE.

low-sulphur crude. *See* CRUDE OIL.

low-velocity zone (LVZ). A zone of the Earth where earthquake SHOCK WAVES travel at much reduced speeds. The top of the zone varies between 70 and 150 kilometres in depth, and the bottom between 200 and 360 kilometres. The LVZ is possibly the zone over which lithospheric plates move.

LPG. *See* LIQUEFIED PETROLEUM GAS.

LSD. *See* LYSERGIC ACID DIETHYLAMIDE.

LTH. *See* LACTOGENIC HORMONE.

lto cycle. *See* LANDING AND TAKE-OFF CYCLE.

Ludlovian. The third oldest series of the SILURIAN System in Europe.

lumen (1m). The derived SI unit of luminous flux, being the amount of light emitted per second in unit solid angle of 1 steradian by a uniform point source of 1 CANDELA intensity.

luminous flux. *See* FLUX.

lung. A highly vascular organ for breathing air. In land vertebrates lungs are derived from branches of the gut. In lungfish (*see* DIPNOI) the air bladder is modified as a lung. In terrestrial MOLLUSCA the lung is a fold of epidermis enclosing an air space. Spiders and scorpions have lung books in depressions in the body wall.

lungfish. *See* DIPNOI.

Lusitanian flora and fauna. Plants and animals that occur in south-western Europe (e.g., Iberian Peninsula) and again in the far west and south-west of the British Isles (parts of Ireland, southern Cornwall), mainly in coastal areas. A typical plant is the pale butterwort (*Pin-*

guicula lusitanica). Such COMMUNITIES are believed to be relics of an earlier INTER-GLACIAL.

lustre. The appearance of a mineral in reflected light, usually described by such terms as metallic, submetallic, adamantine (like diamond), vitreous, resinous, greasy, silky, pearly and earthy.

luteinizing hormone (LH). A HORMONE produced by the PITUITARY GLAND of vertebrates. It initiates ovulation, the production of OESTROGENS by the ovaries and the development of the CORPUS LUTEUM in mammals. In males it initiates the production of ANDROGENS by the testes.

luteotropic hormone. *See* LACTOGENIC HORMONE.

lutite. *See* ARGILLITE.

lux (lx). The derived SI unit of illumination, being the illumination provided by 1 LUMEN per square metre.

LVZ. *See* LOW-VELOCITY ZONE.

L waves. *See* EARTHQUAKE.

LWR. *See* LIGHT-WATER REACTOR.

L$_x$ noise levels. Noise levels that exceed the usual or permitted level for a proportion of a measured period. Thus an L$_{50}$ noise level is one that exceeds the usual or permitted level for 50 percent of the time. *See also* DECIBEL.

lx. *See* LUX.

Lycopodiales (Lycopsida, clubmosses). A group of largely extinct PTERIDOPHYTA that were abundant during the CARBONIFEROUS, when tree-like species (e.g., *Lepidodendron*, common in COAL MEASURES) existed. Present-day species are small plants, most superficially resembling mosses. *Lycopodium* and *Selaginella* grow in boggy ground, and *Isoëtes* (quillwort) is an aquatic plant.

Lycopsida. *See* LYCOPODIALES.

lymphocytes. *See* BLOOD CORPUSCLES.

lynchet. A ridge, from a few centimetres to several metres high, formed after a long period of ploughing on a hillside as disturbed soil migrates downslope. The deep furrow along the upper edge of the ploughed area forms a negative lynchet, and accumulated soil along the lower edge forms a positive lynchet. Lynchets are not the result of deliberate terracing. Often they are of great historical value providing records of past land use.

Lysenko, Trofim Denisovich (1898–1976). A Russian geneticist; the chief proponent of neo-lamarckism, known as lysenkoism, a revival of the theory of the inheritance of acquired characters. *See also* LAMARCK, JEAN BAPTISTE DE.

lysenkoism. *See* LYSENKO, TROFIM DENISOVICH.

lysergic acid diethylamide (LSD). A hallucinogenic drug manufactured from lysergic acid, a crystalline substance obtained from ERGOT.

lysine. An AMINO ACID with the formula $NH_2(CH_2)_4CH(NH_2)COOH$ and a molecular weight of 146.2.

lysogenic bacterium. A bacterium carrying a non-virulent form of virus (a BACTERIOPHAGE). The genetic material of the virus, attached to that of the bacterium and reproduced when the host cell divides, is known as prophage.

lysosomes. Membrane-bounded particles, present in CYTOPLASM, that liberate ENZYMES responsible for the breakdown (AUTOLYSIS) of dead or damaged cells. Lysosomes are probably involved in dissolution of tissue during METAMORPHOSIS.

lysozyme. An ENZYME that kills BACTERIA by destroying their cell walls. Lysozymes occur in mammalian body fluids (e.g., tears) and are also produced by the skin.

M

2M. *See* MCPA.

M. *See* MEGA-.

m. (1) *See* METRE. (2) *See* MILLI-.

Ma. Millions of years.

Maastrichtian. A stage of the CRETACEOUS System.

MAB. *See* MAN AND THE BIOSPHERE PROGRAMME.

MAC. *See* MAXIMUM ALLOWABLE CONCENTRATION.

Macaronesian Floral Region. The part of the PALAEOTROPIC REALM that comprises the islands off the coast of West Africa.

machair. Herb-rich CALCAREOUS grassland that grows on shell sand along the western coast of Scotland, which is of great botanical interest and valuable for grazing.

Mach number. The ratio of the speed of an object to the speed of sound in the undisturbed medium through which the object is travelling. Thus an aircraft travelling supersonically has a Mach number greater than one. Mach numbers are used in preference to more conventional units (miles per hour, knots, etc.) to represent the airspeed of high-flying aircraft because where the air density is very low airspeed indicators, which measure the resistance offered by the air to the moving aircraft, give readings that require major correction before they can be of navigational use.

macrogamete (megagamete). The larger of the types of GAMETE (i.e. the female gamete) (anisogamy) produced during HETEROGAMY. *See also* OOGAMY.

macronutrient. An element or compound that is needed in relatively large quantities by an organism. Crop plants require amounts ranging from a few kilograms to a few hundred kilograms per hectare of carbon, hydrogen and oxygen (supplied from carbon dioxide and water), as well as nitrogen, potassium and phosphorus (supplied from the soil). *Compare* MICRONUTRIENT. *See also* FERTILIZER.

macrophage. One of the large irregularly-shaped cells found in CONNECTIVE TISSUE and body fluids of vertebrates. Macrophages remove dead cells and foreign particles by engulfing them.

macrophagous. Applied to animals that feed at intervals on pieces of food which are large in relation to their body size. *Compare* MICROPHAGOUS.

macrophytes. Large aquatic plants (e.g., crowfoot, water lily), as opposed to phytoplankton (*see* PLANKTON) and other small ALGAE.

Madagascan Floral Region. The part of the PALAEOTROPIC REALM that comprises Madagascar and its offshore islands.

Mad Hatter's disease. Mental derangement caused by the absorption of mercury in small doses over a long period. The disease affected hatters who used mercury in the manufacture of felt.

Madison process. An industrial process for fractionating in a dry state the components of mixed urban wastes on site. It produces material suitable as a feedstock for a BIOPLEX.

Madreporaria. *See* CORAL.

maestrale. *See* MISTRAL.

maestro. *See* MISTRAL.

mafic. Applied to the FERROMAGNESIAN MINERALS or to IGNEOUS rocks relatively rich in such minerals (i.e. synonomous with BASIC and MELANOCRATIC). *Compare* FELSIC.

magma. A molten material composed of silicates and VOLATILES (water and gases) in complex solution; it originates within the lower CRUST or MANTLE. Magmas may contain appreciable amounts of solids and usually have temperatures in the range 700–1100°C. Within the molten mass various processes, broadly called MAGMATIC DIFFERENTIATION, operate to produce fractions of different compositions. These fractions can then solidify adjacently, producing mixed magmas and layered INTRUSIONS, or one or more fractions may migrate independently. The volatiles become concentrated at the top of the magma and can then migrate through the COUNTRY ROCK and/or the Earth's surface producing METASOMATISM, HYDROTHERMAL deposits, FUMAROLES, etc. Magma that reaches the surface and loses most of its volatile substances is known as lava, and the solidified products of magma are called igneous rocks.

magmatic differentiation. Any process that tends to produce two or more separate fractions in a MAGMA. Processes that have been suggested include the separation of early formed crystals from a melt by gravity settling (suggested for some CHROMITE deposits) and the development of immiscible liquids (suggested for some MAGNETITE and some sulphide deposits).

magnesite. *See* CARBONATE MINERALS.

magnesium (Mg). A light, silvery–white metallic element that tarnishes easily in air and burns with an intense white flame to magnesium oxide. It occurs as magnesite (*see* CARBONATE MINERALS), DOLOMITE, CARNALLITE and in many other compounds. It is used in alloys, in photography, signalling and in incendiary bombs, and also has medical uses. It is an essential nutrient (*see* MACRONUTRIENT). $A_r = 24.312$; $Z = 12$; SG 1.74; mp 651°C.

magnetic anomaly. A fluctuation from the normal value for the Earth's magnetic field due to a local concentration or deficiency of magnetic minerals and a change in the internal magnetic field, which is recorded in rocks at the time of their formation (*see* GEOMAGNETIC INDUCTION, PALAEOMAGNETISM). The symmetry of magnetic anomalies shown by rocks on either side of oceanic RIDGES (e.g. mid-Atlantic ridge) is regarded as substantial evidence for the theory of SEA-FLOOR SPREADING.

magnetite (Fe_3O_4). An iron oxide mineral of the SPINEL group; an important source of iron. Magnetite occurs as a widespread ACCESSORY MINERAL of many rocks and is found in economic deposits in BASIC IGNEOUS rocks as a product of high-temperature MAGMATIC DIFFERENTIATION in metasomatized (*see* METASOMATISM) LIMESTONES and in PLACER deposits. Magnetite also occurs with HAEMATITE in the feebly recrystallized BANDED IRONSTONES. Magnetite is also an important source of the metal vanadium, which is used in iron and steel. *See also* LODESTONE.

magnetohydrodynamics (MHD). A technique for generating electricity by passing an ionized (*see* IONIZATION) gas through a magnetic field. Three systems are being considered experimentally: (a) open cycle, which is the most advanced; (b) closed-cycle plasma; (c) closed-cycle liquid metal. Theoretical efficiencies of generation could be in the region of 50–60 percent.

magnetosphere. The space surrounding the Earth or other celestial body in which there is a magnetic field associated with that body.

Magnox reactor. A UK-built NUCLEAR RE-ACTOR, which was the first in the world to operate commercially. Magnox reactors, which use carbon dioxide as a coolant and graphite as a MODERATOR, derive their name from a magnesium oxide alloy used to clad the uranium fuel rods. The fuel is natural uranium. Eight Magnox stations were built, the first in 1956 at Calder Hall, Cumbria. The Magnox operates at 300°C. *Compare* ADVANCED GAS-COOLED REACTOR.

mahogany. *See* MELIACEAE.

maidenhair tree. *See* GINGKOALES.

main sewer. A sewer that collects sewage from a large area.

maize. *See* GRAMINEAE, TEOSINTE.

Major Hazard Incident Data Service. An information service, formed in September 1986 by the UK HEALTH AND SAFETY EXECUTIVE and Safety and Reliability Directorate of the UNITED KINGDOM ATOMIC ENERGY AUTHORITY to record details of major events and to make the accumulated data available worldwide.

major quadrat. *See* QUADRAT, MAJOR.

make-up water. Water that is used to replenish a system that loses water through leakage, evaporation, etc.

malachite ($CuCO_3.Cu(OH)_2$). A copper CARBONATE MINERAL found in the enriched oxidized zone of a copper sulphide vein that has undergone SECONDARY ENRICHMENT and, rarely, as a CEMENT in a SANDSTONE. Malachite is an important ore of copper.

Malacostraca. A large group of CRUSTACEA of great diversity, with compound eyes and, typically, a tail fan, and a carapace covering the thorax. The most important groups are: (a) Isopoda (e.g., woodlice, water and shore slaters, some fish lice), a large group that includes aquatic, terrestrial and parasitic forms and in which the body is flattened dorsoventrally; (b) AMPHIPODA (e.g., sand-hoppers, freshwater shrimps) in which the body is laterally compressed; (c) Decapoda (e.g., crabs, lobster, crayfish, prawns, shrimps), a very diverse group whose members have five pairs of thoracic walking or swimming legs, often with large pincers on the first pair; (d) Euphausiacea (krill) marine, PELAGIC, mostly luminescent animals that are often abundant and form an important part of the food of whales.

malathion. An ORGANOPHOSPHORUS PESTICIDE used to control aphids, leafhoppers, codling moth, mites, etc., but resistant strains of aphids and mites have appeared. It is harmful to fish and bees, but it is one of the less toxic organophosphorus compounds. Toxicity to mammals may be increased by prior exposure to PARATHION or other organophosphorus compounds that inhibit detoxifying mechanisms.

Malaysian–Papuan Floral Region. The part of the PALAEOTROPIC REALM that comprises the southern part of the Malaysian Peninsula, the islands of Indonesia and Papua New Guinea.

maleic hydrazide. A growth regulator (*see* ANTIGIBBERELLIN) used to control grass and other weeds on verges and in amenity areas. It enters the plant mainly through the foliage and inhibits growth by preventing cell division. Its toxicity to mammals is low.

malignant pustule. *See* ANTHRAX.

mallee scrub. A scrub vegetation, consisting mainly of low *EUCALYPTUS* bushes. It is characteristic of the dry subtropical regions of south-east and south-west Australia.

Mallophaga (biting lice). A group comprising several orders of small, wingless, flat-bodied insects (EXOPTERYGOTA), most of them associated with birds, but a few with mammals. Some are important pests of poultry, sheep and cattle, but none affects humans. They do not pierce the skin as do sucking lice (*see* ANOPLURA), but scrape the skin and chew feathers.

malm. Rock strata that were deposited during Upper JURASSIC times.

malpighian tubules. The main excretory organs of insects, named after the Italian physiologist Marcello Malpighi (1628–94), who is regarded as the founder of the microscopic study of anatomy.

Malthus, Thomas Robert (1766–1834). An English economist and clergyman who argued, in his *Essay on the Principle of Population* (1798), that since human populations have the capacity to increase in size exponentially (i.e. each set of parents is able to produce more than the two children needed to replace them), but the resources needed to sustain them (e.g., cultivable land) can be increased, at best, only in a linear fashion (i.e. bringing new land into cultivation does not facilitate the bringing of yet more land into cultivation), the size of a population must be limited either by famine, warfare and disease, or by 'moral restraint'. DARWIN, CHARLES ROBERT, and WALLACE, ALFRED RUSSELL, were influenced by his ideas, although Malthus was concerned only with human populations, and his predictions were shown to be incorrect.

maltose. *See* CARBOHYDRATES.

mamma. A breast-like cloud formation in the base of anvil cloud (*see* CULUMONIMBUS) caused by FALLOUT or instability at the cloud base.

Mammalia (mammals). A class of warm-blooded (*see* HOMOIOTHERMY), air-breathing vertebrates with many distinctive features, including hair, three auditory ossicles (*see* EAR) and HETERODONT teeth. *See also* EUTHERIA, MARSUPIALIA, MONOTREMATA.

mammoth tree. *See* TAXODIACEAE.

Man and the Biosphere Programme (MAB). An international scientific programme under the auspices of the United Nations Educational, Scientific and Cultural Organization that aims to build up a worldwide network of areas for study that together will represent all the 194 BIOGEOGRAPHICAL PROVINCES and investigate such matters as CARRYING CAPACITY and wildlife conservation.

manatee. *See* SIRENIA.

mandioca. *See* MANIHOT.

maneb. A fungicide of the dithiocarbamate group (*see* CARBAMATES) that is used to control diseases such as potato blight and tomato-leaf mould. It can be irritating to the eyes and skin.

manganese (Mn). A reddish-white, hard, brittle metallic element, which occurs as PYROLUSITE and is used in many alloys. It is an essential MICRONUTRIENT for plants. A_r=54.938; Z=25; SG 7.20; mp 1244°C.

manganese nodules. Hydrated manganese oxide (with some iron) concretions, found on the surface and up to a depth of a few centimetres in three major oceans over millions of square kilometres. They were first discovered by the *Challenger* expedition of 1873-76, and their origin and evolution are still not fully understood. The nodules also contain significant concentrations of other metals (e.g., copper, cobalt, nickel).

mange. *See* ACARINA.

mangold-wurzel. SEE *BETA VULGARIS*.

mangroves. Trees and shrubs of the genera *Rhizophora, Brugiera, Sonneratia* and *Avicennia* or, more generally, COMMUNITIES dominated by them. They occur along tidal estuaries, in salt marshes and on muddy coasts in tropical America and Asia, where they form dense thickets. The American mangrove formations consist mainly of the common (or red) mangrove (*A. nitida* or *A. marina*) of the family Verbenaceae. Asian formations also include members of other families (e.g., *Sonneratia*; family: Lythraceae). Some mangroves produce ADVENTITIOUS aerial roots, which descend in an arch, strike at some distance from the parent stem and send up new trunks, so that the forest spreads in a very dense fashion. Other species produce PNEUMATOPHORES. The seeds produce a long embryonic root which grows downwards while the fruit is still attached to the tree and which may take root before the fruit falls, so that the new plant produces shoots almost at once. The roots beneath the mud are aerated by aerial roots which project above the mud and which have minute openings (LENTICELS) into which air diffuses and passes down to the main roots beneath. Wood from some species is hard and durable, the bark yields a substance used in tanning, and the fruit of the common mangrove is sweet and wholesome. *See also* RHIZOPHORACEAE.

Manidae. *See* PHOLIDOTA.

Manihot. A genus (family: EUPHORBIACEAE) of herbs and shrubs that are native to America. *M. esculenta (utilissima)* is cassava (also known as manioc or mandioca), widely cultivated for its starchy tuberous (*see* TUBER) roots from which tapioca is prepared. The juice of some types is poisonous and must be squeezed out or otherwise removed. When evaporated, the juice forms an antiseptic syrup used for preserving meat. Cassava contains only small amounts of proteins with sulphur, and over-reliance on it as food can cause the deficiency disease KWASHIORKOR. *M. glaziovii* and some other species yield rubber when tapped.

manila hemp. *See* ABACÁ.

manioc. *See* MANIHOT.

Mankind at the Turning Point. *See* CLUB OF ROME.

mantids. *See* DICTYOPTERA.

mantle. The upper part of the interior of the Earth, between the CRUST and the CORE. The upper boundary of the mantle is the MOHOROVIČIĆ DISCONTINUITY, at a depth of about 50-70 kilometres below the Earth's surface, and the lower boundary is the Gutenberg discontinuity, at a depth of about 2900 kilometres. The upper mantle is probably composed of ULTRABASIC rocks, which are the source of most BASIC and ultrabasic MAGMAS.

maples. *See* ACER.

maquis. A plant formation found in the Mediterranean area comprising mainly low-growing, XEROPHILOUS evergreen trees and shrubs. It results mainly from the deterioration of the original vegetation by grazing and burning.

mar. *See* MOR.

marasmus. A deficiency disease, found most commonly in babies less than one year old and caused by general undernutrition aggravated by premature weaning. *See also* KWASHIORKOR.

marble. A metamorphosed LIMESTONE, formed under thermal or regional METAMORPHISM, the recrystallization to a SACCHAROIDAL texture destroying all fossils. Nevertheless, many fossiliferous limestones are sold as marbles.

marble gall. *See* GALL.

marijuana. *See* CANNABIACEAE.

marine park. A permanent reservation on the seabed for the conservation of species. Marine parks exist in the USA, Japan, the Philippines, Kenya, Israel, Australia (where two parks have been designated on the GREAT BARRIER REEF) and the UK (where NATURE RESERVES have been designated, e.g., in the Isles of Scilly).

marine regression. The lowering of the sea level relative to the land, or the retreat of the sea by the build-out of TERRIGENOUS sediments. *Compare* MARINE TRANSGRESSION.

marine transgression. The raising of sea level in relation to the level of the land.

marl. A CALCAREOUS MUDSTONE.

marram grass. *See* AMMOPHILA.

marsh. An area of waterlogged ground with a largely mineral basis, in contrast to the peat of BOG and FEN.

marsh gas. *See* METHANE.

Marsupialia (marsupials). An order (infraclass: Metatheria) of mammals whose living representatives are native only to Australasia and America. In marsupials the PLACENTA either does not develop or is less efficient than that of the EUTHERIA, and the young are born in a very underdeveloped state. Most female marsupials have a pouch (marsupium) to which the newborn find their way across the mother's fur and inside which they are suckled and develop further. The American opposums are arboreal, with prehensile tails, and the females usually have no pouch. Australian marsupials show considerable diversity and include herbivores (e.g., koala 'bear'), carnivores (e.g., Tasmanian 'wolf' (probably extinct), Tasmanian devil, marsupial 'cat'), burrowing forms (e.g., wombat, marsupial 'mole'), jumping forms (e.g., kangaroo, wallaby) and gliding forms (e.g., phalangers).

marsupium. The pouch in which many female members of the MARSUPIALIA and echidna (*see* MONOTREMATA) suckle their young.

masonry cement. *See* MORTAR.

mass production. The making of large numbers of identical products in a factory, often using production line techniques.

mass spectrometer. An apparatus that separates electrically charged particles in accordance with their masses. It is commonly used in chemical analysis, especially of organic compounds.

mass wasting. The downslope movement of weathered rock material by gravity alone. The process may be aided or hastened by the presence of ice and/or water. *See also* CREEP, SOLIFLUCTION.

Mastigophora. *See* FLAGELLATA.

mating. *See* ASSORTATIVE BREEDING.

maximum allowable concentration (MAC). The concentration of a pollutant that is considered (in regulations) to be harmless to healthy adults during their working hours, assuming they are not in contact with the pollutant outside working hours.

maximum permissible body burden. The concentration of a RADIOISOTOPE that will deliver not more than the MAXIMUM PERMISSIBLE DOSE to a critical body organ when inhaled or ingested at a normal rate.

maximum permissible dose. The dose of IONIZING RADIATION, accumulated during a specified time, of such magnitude that no injury may be expected to result during the lifetime of the individual exposed, and no intolerable burden is likely to accrue to society through genetic damage to the descendants of the exposed person. *Compare* MAXIMUM PERMISSIBLE BODY BURDEN.

maximum permissible level. A general term describing the greatest degree of contamination that is permitted from any source, but especially of radioactive substances.

mayflies. *See* EPHEMEROPTERA.

mb (millibar). One-thousandth of a BAR; the unit most commonly used in reporting atmospheric pressure.

MCP. *See* MCPA.

MCPA (MC, 4K, 2M). A hormonal TRANSLOCATED HERBICIDE that is used to control many broadleaved weeds in cereal and grass crops. It is not very persistent, breaking down in the soil within a few weeks of application.

MCPP. *See* MECOPROP.

meadow. An area of permanent grassland, and especially one that is cut for hay. Water meadows are similar grass fields that are regularly flooded by river water.

meander. A freely developed curve in a river as it flows over more or less level land with only a shallow seaward gradient. *See also* INCISED MEANDER, INGROWN MEANDER, MEANDER BELT, MEANDER SCROLL, OX-BOW LAKE, POINT BAR.

meander belt. The total area covered by a system of river MEANDERS, usually 15–18 times the width of the river channel when it is full.

meandor scroll. The arc-shaped depression of a former MEANDER that is now partially infilled. *Compare* OX-BOW LAKE.

mean free path. (1) The average distance travelled by a molecule in a fluid between collisions with other molecules. (2) The average distance a sound wave travels in a room between successive reflections.

mechanical turbulence. *See* FORCED CONVECTION.

mecoprop (MCPP). A hormonal TRANSLOCATED HERBICIDE that is used to control many broadleaved weeds in cereals, turf and orchards.

Mecoptera (scorpion-flies). A small ancient order of mainly carnivorous flying insects (division: ENDOPTERYGOTA) some of which have changed little since the PERMIAN. The males often have the abdomen upturned.

median. In statistics, the value of the VARIATE which divides the total frequency into two halves.

Mediterranean Action Plan. An international programme to reduce pollution in the Mediterranean Sea that grew out of a series of conferences sponsored by UNITED NATIONS ENVIRONMENT PROGRAMME and attended by representatives of 18 of the countries bordering the Mediterranean Sea, as part of the REGIONAL SEAS PROGRAMME. By 1982, four protocols and a number of treaties (e.g., the ATHENS TREATY ON LAND-BASED SOURCES OF POLLUTION) had been agreed and signed.

Mediterranean Floral Region. The region of the HOLARCTIC REALM that comprises southern Europe and North Africa, around the shores of the Mediterranean Sea.

medulla. The central part of an organ (e.g., the pith of a stem, the inner part of a kidney). *Compare* CORTEX.

medulla oblongata. In vertebrates, the part of the brain that merges into the spinal cord. Its main functions are the coordination of impulses from sense organs and the regulations of involuntary actions (e.g., heart beating and breathing).

medullary rays. Thin, vertical plates of living cells that run radially through the VASCULAR tissue of the stems and roots of plants. Food materials are stored in and conducted radially through medullary rays.

medusa. A disc- or bell-shaped, free-swimming generation in CNIDARIA that reproduces sexually, usually giving rise to the POLYP generation. The medusa stage is absent in the ACTINOZOA, and is the dominant or only generation in the SCYPHOZOA.

mega- (M). The prefix used in conjunction with SI units to denote the unit x 10^6.

Megachiroptera. *See* CHIROPTERA.

megachromosomes. Giant CHROMOSOMES that occur in the salivary glands and some other tissues of many DIPTERA, including the fruit fly *DROSOPHILA*. Each chromosome is greatly thickened, consisting of many identical parallel CHROMATIDS, and each exhibits a pattern of transverse bands marking the arrangement of GENES along its length. These chromosomes have been studied intensively in *Drosophila*, and by associating changes in the banding pattern with genetic differences, many genes have been located on specific chromosomes.

megagamete. *See* MACROGAMETE.

megalecithal. *See* TELOLECITHAL.

megalopolis. A large urban area produced by the expansion of adjacent conurbations until their boundaries meet (e.g., areas of the eastern seaboard of the USA, MEGALOPOLITAN CORRIDOR, and the NORTH-WEST EUROPEAN MEGALOPOLIS).

megalopolitan corridor. The region between Boston, Massachusetts, and Washington, DC, USA.

megaspore. *See* HETEROSPORY, SPORE.

megasporophyll. *See* FLOWER.

megatherm. A plant that needs a continuously high temperature (i.e. a tropical plant).

meiosis (reduction division). Cell division during which the CHROMOSOME number is halved. It occurs in all organisms that reproduce sexually; in animals during GAMETE formation and in many plants during SPORE formation. The process usually consists of two successive divisions. During the first, the chromosomes arrange themselves in HOMOLOGOUS pairs, then the partners move apart to opposite sides of the cell (*see* CROSSING-OVER). The second division resembles MITOSIS with half the normal number of chromosomes. The final result is the formation of four haploid daughter cells from one diploid cell.

melanins. The dark pigments that are present in many animals, often in special cells (e.g., melanocytes, melanophores). *See also* CHROMATOPHORES, INDUSTRIAL MELANISM.

melanistic. *See* INDUSTRIAL MELANISM.

melanocratic. Consisting of dark minerals (i.e. FERROMAGNESIAN MINERALS). *Compare* LEUCOCRATIC. *See also* MAFIC.

melanocytes. *See* MELANINS.

melanophores. *See* MELANINS.

Meliaceae. A family of DICOTYLEDONEAE, consisting of trees and shrubs of warm regions and including many species that yield valuable timber. *Khaya* is African mahogany, and *Swietenia* is the mahogany of tropical America.

meltdown. In a NUCLEAR REACTOR, the overheating of the core to an extent that causes the FUEL ELEMENTS to melt. In theory this might lead to the melting of all fuel elements and a substantial part of the MODERATOR if this is of a solid material. In practice the process proves to be self-limiting and only a small proportion of the fuel elements melt. *See also* CHINA SYNDROME.

melting point (mp). The constant temperature at which the solid and liquid phases of a substance are in equilibrium at a given pressure. Melting points are usually quoted for standard atmospheric pressure (760 mm Hg).

membrane. *See* PLASMA MEMBRANE, UNIT MEMBRANES.

Mendel, Gregor (1822–84). An Austrian monk who conducted precise experiments in heredity, using the garden pea, and formulated the basic laws of genetics. *See also* INDEPENDENT ASSORTMENT; SEGREGATION, LAW OF.

mendelevium. *See* ACTINIDES.

mendelian population. A population that interbreeds.

Mendel's laws. *See* MENDEL, GREGOR; INDEPENDENT ASSORTMENT; SEGREGATION, LAW OF.

menstrual cycle. *See* OESTROUS CYCLE.

Mercalli scale. *See* MODIFIED MERCALLI SCALE.

mercaptans. Organic compounds with the general formula R–SH, meaning that the thiol group (–SH) is attached to a RADICAL (e.g., CH_3 or C_2H_5). The simpler mercaptans have strong, repulsive odours, but these become less pronounced with increasing molecular weights and higher boiling points. Mercaptans may be produced in oil refinery feed preparation units as a result of incipient CRACKING, where the offensive gases are burnt in plant heaters. Mercaptans that arise in cracking units are removed by scrubbing (*see* SCRUBBER) with caustic soda, removed from the caustic soda by stream stripping and then burnt.

Mercurialis perennis. See EUPHORBIACEAE.

mercury (Hg, quicksilver, hydragyrum). A liquid metallic element which may damage the nervous system (*see* MAD HATTER'S DISEASE) if ingested or inhaled, and whose organic compounds are very poisonous. It is used extensively in the chemical and plastics industries, in scientific instruments, in dentistry (in amalgam), in detonators and formerly in heavy-duty switchgear, where it has been largely replaced by electronic devices. A_r=200.59; Z=80; SG 13.6; mp -39°C; bp 357°C.

Merioneth. The third oldest series in the CAMBRIAN System.

meristem. A region of active cell division in a plant. Many plants (e.g., seaweeds, mosses, ferns) have unicellular meristems, but in flowering plants (*see* ANGIOSPERMAE) meristems consist of groups of cells. Cells formed by meristems become differentiated into various tissues depending on the location of the meristem. Apical meristems (growing points) are at the tips of stems, roots and their branches; cambium lies between the XYLEM and PHLOEM, and produces new xylem and phloem during SECONDARY THICKENING; phellogen (CORK CAMBIUM) is a secondary meristem that arises in the cortex of woody plants and produces cells that become corky; intercalary meristems occur in grasses at the bases of internodes.

merogony. The fertilization of an egg fragment or of an egg that has no nucleus. In certain marine organisms, an egg that is separated into two parts can develop into two larvae, the larva from the egg that contained no nucleus possessing only the paternal set of CHROMOSOMES. An egg fragment treated with a parthenogenetic (*see* PARTHENOGENE-

SIS) agent also develops to the BLASTULA stage.

meromixis. The permanent stratification of water masses in lakes. Sometimes dissolved substances create a gradient of density differences with depth, such that the complete mixing and circulation of water masses is prevented.

mesa. A steep-sided plateau of horizontally bedded rock, topped with resistant cap rock.

mesarch. All the stages in a plant SUCCESSION that begins in a moderately damp habitat.

mesocarp. *See* PERICARP.

mesoclimate. A local climatic effect, occurring over an area several kilometres wide and 100–200 metres high, where the climate differs from the regional climate.

mesoderm. *See* GERM LAYERS.

mesogloea. *See* DIPLOBLASTIC.

mesohaline. Applied to brackish waters in which the salinity is between 5 and 15 parts per thousand. *Compare* EUHALINE, FRESH WATER, OLIGOHALINE, POLYHALINE.

Mesolithic (Middle Stone Age, Protoneolithic, Epipalaeolithic). The transitional period in the development of human societies between the PALAEOLITHIC and NEOLITHIC. The period occupies different spans of time in different places, and the concept is a developmental rather than a chronological one.

mesopelagic. Applied to organisms that live below the EPIPELAGIC zone, in twilight conditions, down to the start of the BATHYPELAGIC zone at about 1000 metres. *Compare* ABYSSOPELAGIC.

mesophilic microorganisms. Microorganisms whose optimum temperature for growth lies between 20 and 45°C. Examples include pathogenic (*see* PATHOGEN) bacteria that infect mammals and birds. *Compare* PSYCHROPHILIC MICROORGANISMS, THERMOPHILIC MICROORGANISMS.

mesophilous. Applied to plants or plant communities that are associated with neutral soil conditions. *Compare* CALCICOLE, CALCIFUGE.

mesophyll. The tissue inside the blade of a leaf. In DORSIVENTRAL leaves, the upper (palisade) mesophyll consists of elongated, tightly packed cells containing numerous CHLOROPLASTS. The lower (spongy) mesophyll consists of cells that contain fewer chloroplasts and are loosely packed, the air spaces between them communicating with the STOMATA.

mesophyte. A plant that grows in conditions that are neither very wet nor very dry. *Compare* XEROPHYTE. *See also* RAUNKIAER'S LIFE FORMS.

mesosaprobic. Applied to a body of water in which organic matter is decomposing rapidly and in which the oxygen level is considerably reduced. *Compare* CATAROBIC, OLIGOSAPROBIC, POLYSAPROBIC. *See also* SAPROBIC CLASSIFICATION.

mesosphere. (1) The part of the atmosphere that extends from the IONOSPHERE to the EXOSPHERE (i.e. from about 400 to 1000 kilometres above the surface of the Earth). It is sometimes considered to be part of the exosphere. (2) The part of the atmosphere that lies between the STRATOSPHERE and the THERMOSPHERE (i.e. from about 40 to 80 kilometres above the surface of the Earth).

mesothelioma. A major type of lung cancer. It may be caused by the repeated inhalation of ASBESTOS fibres.

mesotherm. A plant of warm temperate regions. An example of a mesotherm is maize.

mesothermal. Applied to ORE deposits that have formed from an ascending, essentially aqueous solution at fairly high temperatures (200–300°C). Mesothermal deposits are HYDROTHERMAL deposits formed in conditions intermediate between those that produced HYPOTHERMAL and EPITHERMAL deposits.

mesotrophic. Applied to freshwater bodies that contain moderate amounts of plant nutrients and are therefore moderately productive. The term is also supplied to MIRES that are moderately rich in plant nutrients. *Compare* DYSTROPHIC, EUTROPHIC, OLIGOTROPHIC. *See also* EUTROPHICATION.

Mesozoic. One of the eras of geological time, which occurred between the PALAEOZOIC and CENOZOIC, and which comprises the TRIASSIC, JURASSIC and CRETACEOUS Periods. Mesozoic also refers to the rocks formed during this time.

messenger RNA. *See* RNA.

meta-. (1) A prefix meaning changed or altered. (2) In geology, an abbreviation for metamorphosed (*see* METAMORPHISM) (e.g., a metabasalt is a metamorphosed basalt). (3) In chemistry, a prefix meaning that an organic compound contains a benzene ring substituted in the 1 and 3 positions. (4) In chemistry, a prefix meaning that an acid or a SALT is the least hydrated (*see* HYDRATION) of several compounds of the same name.

metabiosis. The beneficial exchange of growth factors among species.

Metabola. *See* PTERYGOTA.

metabolism. All the chemical reactions that take place in a living organism, comprising both ANABOLISM and CATABOLISM. Basal metabolism is the energy exchange of an animal at rest.

metabolite. A substance that is involved in METABOLISM (i.e. an essential nutrient). *Compare* ANTIMETABOLITE.

metagenesis. The alternation of sexual and asexual generations in the life cycle of an animal (e.g., many CNIDARIA). *See also* ALTERNATION OF GENERATIONS.

metal. A substance that possesses certain qualities: metallic lustre, malleability, ductility, high specific gravity and good conductivity of heat and electricity. Metals are generally electropositive and combine with oxygen to give bases. Substances possessing certain of these qualities, but also other non-metallic properties (e.g., arsenic) are called metalloids.

metaldehyde. A molluscicide that is used to kill slugs and snails (i.e. MOLLUSCA).

metalloid. *See* METAL.

metamere. *See* SEGMENTATION.

metameric segmentation. *See* SEGMENTATION.

metamerism. *See* SEGMENTATION.

metamorphic aureole. The zone of COUNTRY ROCK around an IGNEOUS INTRUSION in which new textures and minerals have developed in response to the heat dissipated by the intrusion. The size of an aureole depends on the nature of the country rocks and on the size, temperature and volatile (*see* VOLATILES) content of the intrusion.

metamorphism. The alteration of the texture and/or composition of a rock by the action of heat and/or pressure and, in some cases, by the addition of extra material (*see* METASOMATISM). Metamorphism is considered to occur in the solid state. It can involve heat alone (producing a METAMORPHIC AUREOLE around an igneous rock), pressure alone (e.g., along a THRUST, producing a MYLONITE) or heat and pressure together, producing metamorphic rocks over a large area (called regional metamorphism). The increase in heat and pressure alters the texture of a rock. Thus as temperature and pressure increase, a MUDSTONE can change into a SLATE, PHYLLITE, SCHIST or possibly a GNEISS; at the same time the original minerals become unstable and new ones form. Experiment has revealed the stability field of many minerals, and so the conditions under which metamorphic rocks formed within a metamorphic zone (characterized by an INDEX MINERAL) can be postulated.

metamorphosis. The transformation of a larval into an adult stage which occurs in AMPHIBIA, some fishes (*see* LEPTOCEPHALUS) and many groups of invertebrates. In insects, incomplete (hemimetabolous) metamorphosis (e.g., in cockroaches) is a gradual process. Complete (holometabolous) metamorphosis (e.g., in the housefly) is a rapid and drastic change, involving the extensive breakdown of larval tissue and the rebuilding of adult organs, and it occurs during the pupal stage (*see* PUPA). *See also* LYSOSOMES.

metam sodium. *See* METHAM SODIUM.

Metaphyta. The kingdom that includes all multicellular plants.

metaplasm. The non-living constituents of PROTOPLASM.

metaquartzite. A metamorphic (*see* METAMORPHISM) rock that consists of QUARTZ grains that have been welded together by pressure. In thin section the quartz can be seen to have been stained, and the contacts between grains to be sutured. Metaquartzite is sometimes called quartzite, but this name also covers ORTHOQUARTZITE.

metasediment. Metamorphosed (*see* METAMORPHISM) SEDIMENTARY ROCK.

metasomatism. The process in which pre-existing rock is partly or wholly altered by the addition of new material. Metasomatism accompanied by raised temperatures is sometimes called contact metamorphism.

metaxenia. The influence of pollen from different parents on the development of the fruit of a plant causing, for example, variations in maturation time.

Metazoa. The kingdom that includes all the animals.

meteoric water. Water that has fallen as rain and has percolated down through the VADOSE zone to the WATER TABLE. Chemical reactions within the zone introduce sulphate, carbonate and bicarbonate ions into the water.

meteorite. A solid extraterrestrial body that has fallen to the surface of the Earth. According to their chemical compositions meteorites are classified in three major groups: irons; stony irons; stones. Irons are alloys of nickel and iron. Stony irons are similar, but also contain SILICATE minerals. Stones consist primarily of PYROXENES, with variable amounts of OLIVINE, nickel–iron alloys and other minerals. Stones are further subdivided into chondrites, which contain small, spherical mineral bodies (chondrules) and achondrites, which contain no chondrules. Some chondrites are carbonaceous and contain HYDROCARBONS, FATTY ACIDS and AMINO ACIDS. The study of meteorites, which are four to five billion years old, provides information about the composition of the Earth's MANTLE and CORE.

meteorology. The study of the atmosphere, its structure, composition and phenomena (e.g., weather).

methaemoglobinaemia. An illness caused by the presence in the blood of methaemoglobin, a compound of HAEMOGLOBIN and oxygen which is more stable than oxyhaemoglobin and so does not yield up its oxygen. In extreme cases methaemoglobinaemia may lead to asphyxiation (in infants, the 'blue baby' syndrome). It can occur in infants and ruminants (*see* RUMINANTIA) when nitrates ingested from food or water are reduced to nitrites by bacteria in the digestive system, nitrites being able to pass through the gut wall and into the bloodstream to form stable compounds with blood haemoglobin. Hydroxylamine, another product of the bacterial reduction of nitrate to nitrite, can also pass through the gut wall, and it plays a principal role in the destruction of VITAMIN A, as well as having a haemolytic (i.e. haemoglobin-destroying) action which can cause anaemia. The risk to infants of methaemoglobinaemia has led to the imposition of limits (50 mg/1 in the EEC) to the amount of nitrate (arriving principally through the LEACHING of nitrogen-based fertilizer from farms) in waters abstracted for the public supply.

metham sodium (metam sodium). A soil fumigant of the dithiocarbamate group of pesticides (*see* CARBAMATES) that is used to control soil fungi, to kill earthworms and other soil fauna, and to check the growth of weed seedlings. It is used to sterilize potting composts and greenhouse soils. It and its decomposition products are irritating to the skin and eyes of humans.

methanal. *See* FORMALDEHYDE.

methane (CH_4). The simplest hydrocarbon; a product of natural or artificial anaerobic decomposition. In its natural form it is also known as marsh gas. It is combustible and is used as a fuel. Methane is the principal constituent of NATURAL GAS. In the atmosphere it is oxidized slowly (to carbon dioxide and water vapour), but while it remains in the air it is a greenhouse gas (*see* GREENHOUSE EFFECT), the main sources being termites and ruminants (in both cases as a by-product of the bacterial breakdown of cellulose in their guts).

methanol (methyl alcohol, wood alcohol, CH_3OH). An alcohol produced by the oxidation of METHANE using a catalyst, but formerly it was made from wood by a dry distillation process. It is used as a solvent, a fuel and as a raw material in the manufacture of synthetic resins.

methionine. An AMINO ACID with the formula $CH_3S(CH_2)_2CH(NH_2)COOH$ and a molecular weight of 149.2.

methyl alcohol. *See* METHANOL.

methylated spirits. *See* DENATURING.

methyl chloride. *See* CHLOROMETHANE.

methylene blue stability test. A test to

measure the ability of an effluent to remain in an oxidized condition when it is incubated anaerobically. The sample is mixed with the dye methylene blue and the test continues for five days at 20°C. If the blue colour formed at the start remains at the end of the period the effluent is considered stable.

methyl isocyanate. A highly toxic compound with an LD_{50} of 5 ppm in rats that is used in the manufacture of some CARBAMATES (e.g., aldicarb, CARBARYL). It is a gas at normal temperatures, heavier than air and attacks proteins, including blood proteins and lung tissue, causing respiratory difficulty or failure if inhaled. It is also intensely irritating to the skin and eyes, causing ulceration. Severe ulceration of the cornea may lead to permanent blindness. It is the gas that escaped at BHOPAL.

metoecious parasite. *See* PARASITISM.

metoxenous parasite. *See* PARASITISM.

metre (m). The SI unit of length, defined in 1960 as the length that is equal to 1 650 763.73 wavelengths *in vacuo* of the radiation corresponding to the transitions between the levels $2p_{10}$ and $5d_5$ of the isotope krypton-86. One metre is equal to 39.3701 inches.

metropolis. The chief city, often the capital city, of a country or region.

Meuse Valley incident. An air pollution incident which occurred in the Meuse Valley, France, in December, 1930, when cold weather and KATABATIC WINDS, together with a temperature INVERSION, caused fog contaminated with industrial pollutants to persist for several days. Several hundred people became ill, about 60 died, and many cattle had to be slaughtered.

MFTF. *See* MIRROR FUSION TEST FACILITY.

Mg. *See* MAGNESIUM.

MHD. *See* MAGNETOHYDRODYNAMICS.

mica. A group of SILICATE MINERALS with a single, perfect cleavage which produces characteristic flexible cleavage flakes. Micas are composed of rings of SILICA tetrahedra forming layers joined by CATIONS (e.g., potassium, magnesium, iron, lithium, aluminium, calcium). Biotite, a potassium-iron mica, and muscovite $(K_2Al(Si_6Al_2)O_{20}(OH,F)_4$ are the commonest micas. Micas occur in IGNEOUS rocks, and muscovite is also common in sediments. Mica is used as a thermal and electrical insulator. Inhalation of fragments of it can cause SILICOSIS.

micaceous haematite. *See* HAEMATITE.

micro- (m). A prefix used in conjunction with SI units to denote the unit x 10^{-6}.

microbial metallurgy. A form of biotechnology in which bacteria are used in metallurgical processes. Bacterial action is useful in separating valuable metals (e.g., uranium) from some ores, and also in producing sulphur from PYRITES and GYPSUM, and SULPHURIC ACID from sulphur.

Microchiroptera. *See* CHIROPTERA.

microclimate. The climate within a very local area.

microcrystalline. Composed of crystals that can be resolved, but only with the aid of a microscope. *Compare* CRYPTOCRYSTALLINE.

microgabbro. A medium grain size rock of gabbroic (*see* GABBRO) composition, which is more commonly called dolerite in the UK or diabase in North America.

microgamete. The smaller of the two types of GAMETE which are produced during HETEROGAMY; the male gamete. *See also* OOGAMY.

microgranite. A medium grain size, acidic IGNEOUS rock whose mineralogical and chemical composition is similar to that of GRANITE. PORPHYRITIC microgranite is usually called quartz porphyry, and microgranite showing a small-scale GRAPHIC texture (called micrographic or granophyric) is called granophyre. Microgranites grade with descending grain size into RHYOLITES. Microgranites form SILLS, DYKES and PLUGS.

micrographic. *See* GRAPHIC, MICROGRANITE.

microhabitat. A very local habitat (e.g., a decaying tree stump).

microlecithal. Applied to eggs (e.g., human egg) that contain little yolk, *Compare* TELOLECITHAL.

micrometre. *See* MICRON.

micron (micrometre, μm). One-thousandth of a millimetre (i.e. one-millionth of a metre); a unit used in measuring microscopically small objects. *See also* NANOMETRE.

micronutrient. An element or compound that is required by living organisms, but in very small amounts. Plants require a few grams to a few hundred grams per hectare of iron, manganese, zinc, born, copper, molybdenum and cobalt. *Compare* MACRONUTRIENT. *See also* TRACE ELEMENT, VITAMIN.

microorganism. Any living organism that is too small to be seen with the naked eye.

microphagous. Applied to animals (e.g., baleen whales, BALAENOIDEA) that feed, usually more or less continuously, on particles (e.g., PLANKTON) that are minute in relation to their own body size. *Compare* MACROPHAGOUS.

Micropodidae. *See* APODIDAE.

Micropodiformes. *See* APODIFORMES.

microrelief. Small-scale TOPOGRAPHY, which is measured in centimetres.

microsere. All the stages in a plant SUCCESSION that occur within a MICROHABITAT.

microspecies. A small species or subspecies that shows a very slight variability from the main species of which it is part (e.g.,

small variants among dandelions, roses or blackberries).

microspore. *See* HETEROSPORY, SPORE.

microsporophyll. *See* FLOWER.

microtherm. A plant that grows in cool temperate regions (e.g., oats).

mictium. A mixture of species, such as that occurring in a transitional zone between two distinct habitats. *See also* ECOTONE.

middle shore. *See* SHORE ZONATION.

Middle Stone Age. *See* MESOLITHIC.

Midland hawthorn. *See* CRATAEGUS.

mid-oceanic ridge. *See* RIDGE.

Mie scattering. The scattering of sunlight, predominantly in a forward (i.e. downward) direction by atmospheric particles of more than about 0.1 micrometres radius. The mechanism was first proposed by G. Mie in 1908. *See also* AEROSOL, BACKSCATTER, RAYLEIGH SCATTERING.

migmatite. *See* MIGMATIZATION.

migmatization. The process of alteration of a high-grade metamorphic (*see* METAMORPHISM) rock to GRANITE. Migmatization involves alteration of the composition of the metamorphic rock, principally by increasing the sodium and potassium content. Migmatites are rocks consisting of thin alternating layers or lenses of granite and GNEISS in which ghosts of the pre-existing banding can often be seen in the granitic parts.

migration. (1) The establishment of a plant species in a new area. (2) Movements that particular animals carry out regularly, often between breeding places and winter feeding grounds. *See also* ANADROMY, CATADROMY.

migrule. The unit, or agent, of plant migration (i.e. the part that is transported to a new location where it gives rise to newly colonizing plants; a DIASPORE.

mildews. A type of fungus (*see* FUNGI) that causes mildew diseases in plants (i.e. any diseases manifested by a visible MOULD) or similar damage to plant products. Powdery mildews ASCOMYCETES) grow on the surface of host plants. Downy mildews (PHYCOMYCETES) penetrate deeply.

mild humus. *See* MULL.

milk sugar. *See* CARBOHYDRATES.

milk teeth. *See* DECIDUOUS.

millet. *See* GRAMINEAE.

milli-. A prefix used in conjunction with SI units to denote the unit x $10^{-3.}$

millimicron. *See* NANOMETRE.

milling. *See* PULVERIZATION.

millipedes. *See* MYRIAPODA.

mimicry. *See* BATESIAN MIMICRY, MÜLLERIAN MIMICRY.

Minamata. A bay and town in Japan that gave its name to a disease of the central nervous system caused by MERCURY poisoning and affecting the people consuming the fish and shellfish caught in Minamata Bay. The pollution was traced in 1959 to the discharge of effluents from the Chisso factory, which was manufacturing ACETALDEHYDE and VINYL CHLORIDE. The effluents accumulated in the sediment in the Bay as dimethylmercury, a toxic organic mercury compound ingested by marine organisms. Between 1953 and 1960, 43 people died, and many more were incapacitated by Minamata disease. The damage was twofold: (a) direct poisoning of the nervous system; (b) similar teratogenic (*see* TERATOGEN) effects due to the ability of the toxin to cross the placental barrier following ingestion of the dimethylmercury by the mother. The tragedy was particularly severe because

fishing was an important industry and because the fish and shellfish were a staple protein food for the community.

mineral. A natural solid and homogeneous inorganic substance whose chemical composition is either fixed or falls within a defined range.

minerotrophic. *See* SOLIGENOUS.

minimal area. (1) The smallest area that is certain to contain a particular species. (2) The smallest area that contains a representative sample of a COMMUNITY.

minimum factor. In forestry, the factor which limits the distribution of forest by the intensity of its occurrence. The limit may be imposed by a high intensity or a low one.

minnow reach. *See* RIVER ZONES.

Miocene. A subdivision of the CENOZOIC Era, usually ranked as an EPOCH, which follows the OLIGOCENE. The Miocene is thought to have lasted from 26 to 7 Ma. Miocene also refers to rocks deposited during this time, called the Miocene Series.

miogeosyncline. A GEOSYNCLINE in which the sequence is relatively thin and there are no volcanic rocks lying close to a CRATON.

miracidium. *See* TREMATODA.

mire Usually, a synonym for peatland (*see* PEAT) that includes both BOG and FEN. The term can also be used to include mineral as well as peaty soils if the WATER TABLE or throughput of base-poor to base-rich water is high.

Mirror Fusion Test Facility (MFTF). An experimental device for research into nuclear fusion energy (*see* FUSION REACTOR) which uses a magnetic container of an alternative design to the more advanced TOKAMAK FUSION TEST REACTOR or torus (*see* JOINT EUROPEAN TORUS).

mispickel. *See* ARSENOPYRITE.

Mississippian. In North American stratigraphy, a system that is equivalent to the Lower CARBONIFEROUS. *See also* GEOLOGICAL TIME.

mist. (1) Liquid droplets, less than 10 micrometres in diameter, that are generated by condensation. (2) In meteorology, a suspension of water droplets in the atmosphere which reduces visibility to 1–2 kilometres. *Compare* FOG, SMOG.

mistral. A dry, cold northerly wind which blows down the Rhône valley and into the Gulf of Lyons. Similar winds may occur wherever the movement of a cold air mass is obstructed by mountains. Such winds blow through gaps in the mountains, sometimes for days on end, and may be associated with very deep air currents and with falling pressure if they are strong. The cold northerly wind that blows over northern Italy in winter is called the maestrale; the similar cool westerly wind of Corsica and Sardinia is called the maestro. Other names include: gregale of Euroclydon, Greece; tramontana; etesian winds; pampero, in Argentina; and levanter, which blows through the Strait of Gibralter. *See also* BORA, FÖHN WIND, SOUTHERLY BUSTER.

mites. *See* ACARINA.

mitochondria (chondriosomes). Membrane-bounded particles that are present in the CYTOPLASM of all EUKARYOTIC cells. Mitochondria contain ENZYME systems which are responsible for providing energy in the form of ATP.

mitosis (karyokinesis). The normal process by which a cell nucleus divides into two daughter nuclei, each of which has an identical complement of CHROMOSOMES. The following stages are recognized: (a) the prophase, during which the chromosomes (each consisting of two CHROMATIDS) condense and become visible; (b) the metaphase, in which the nuclear membrane usually disappears and a gelatinous ellipsoid structure (the spindle), consisting of a number of protoplasmic (*see* PROTOPLASM)

threads, is formed, the chromosomes becoming attached to the spindle threads while lying in the equatorial plane; (c) the anaphase, during which the chromatids separate and move towards opposite poles of the spindle; (d) the telephase, in which the two sets of chromatids form new nuclei, each surrounded by a membrane. *See also* MEIOSIS.

mixed economy. An economic system in which some components operate according to free-market forces (*see* LAISSEZ FAIRE), whereas other components are subject to central control as in a PLANNED ECONOMY. These days the free-market elements are usually subject to a variable degree of central regulation.

mixed forest. *See* FOREST.

mixing length theory. A theory that is used to calculate the behaviour of molecules in a fluid that are involved in a turbulent exchange (*see* TURBULENCE). The mixing length is the average distance, perpendicular to the main flow, covered by a moving particle, a fluid in turbulent exchange being displaced in a direction perpendicular to its direction of flow. *See also* MEAN FREE PATH.

mixotroph. An organism that is both HETEROTROPHIC and AUTOTROPHIC (e.g., an insectivorous plant).

MM. Million; used in connection with BARRELS or cubic feet in figures for fossil fuel reserves.

Mn. *See* MANGANESE.

Mo. *See* MOLYBDENUM.

mobile belt. An elongated region of the Earth's CRUST that is undergoing uplift and subsidence, earthquakes and volcanic activity, and probably folding (*see* FOLD) and faulting (*see* FAULT) as well, to produce a mountain chain.

mock sun (sun dog). A bright spot seen in the sky at an angle of 22° to one or both sides of the Sun, and sometimes faintly discoloured. Mock suns are caused by the refraction of light by ice crystals in the lower atmosphere.

mode. In statistics, the value of the VARIATE that is possessed by the greatest number of individuals.

model. A simplified description of a SYSTEM, used as an aid to understanding the system. Models may exist as mental constructs (i.e. the ideas we have of the world and which we use to interpret the information we receive through our senses) or more formally as verbal or mathematical descriptions. Mathematical models are constructed from numerical values given to the components of the system and the relationships among components. If the model is sufficiently accurate to be approximately true it can be used to predict the behaviour of the system when selected values are changed. Models may be constructed as initially complex models, which start with a replica of the system and discard non-essential details, or as initially simple models, which begin with the most essential elements of the system and add others as they are needed. Apart from mathematical models, models may also be physical (i.e. models of buildings, working models of machines, prototypes, etc.). *See also* ANALOGUE, ANALOGUE COMPUTER, DIGITAL COMPUTER.

moderator. In a NUCLEAR REACTOR, a substance surrounding the fuel elements that has the property of slowing down the high-energy neutrons emitted by the decay of the fuel. This increases the likelihood that neutrons will encounter other atoms of fissile uranium-235, rather than being absorbed by atoms of uranium-238, so tending to sustain the CHAIN REACTION.

modified Mercalli scale. A scale used to measure the intensity of earthquake shock, based on the structural damage caused by the earthquake, and ranging from instrumental (I) to catastrophic (XII). Because the extent of structural damage depends on factors other than the intensity of the earthquake

(e.g., the design of buildings) the modified Mercalli scale has been largely replaced by the more objective RICHTER SCALE.

modular building. *See* MODULARISM.

modularism. The technique of making whole structures more permanent by making their component substructures less permanent. Thus, a modular building consists of a permanent framework into which interchangeable modules can be fitted and, when necessary, removed and replaced. The pen fitted with a disposable ink cartridge is based on this principle: the framework (the pen) has a long life, the substructure (the cartridge) a short life.

Mohorovicic discontinuity (Moho). A seismic discontinuity within the Earth below which shock waves from the focus of an EARTHQUAKE travel with increased velocities. This was first proposed in the early 20th century by the Serbian seismologist A. Mohorovicic and is now accepted as a major structural divide in the Earth, the region above the discontinuity being the CRUST and the region below, down to the CORE of the Earth, the MANTLE.

Mohs's hardness scale. A system, devised by Friedrich Mohs (1772–1839), for classifying the hardness of minerals on the basis of which will scratch and mark which. The softest mineral is talc (hardness 1), corundum has a hardness of 9, and the hardest of all minerals is diamond (hardness 10).

mol. *See* MOLE.

molasse. An association of predominantly non-marine CLASTIC sediments, including BRECCIA conglomerate, ARKOSE, SANDSTONE and SHALE, that formed during the MIOCENE as a result of the rapid erosion of the Alps. *Compare* FLYSCH.

mole (mol). The SI unit of amount of substance, being that amount which contains as many elementary units as there are atoms in 0.012 kilograms of carbon-12. The elementary units must be specified. One mole of a compound has a mass equal to its molecular weight in grams.

molecular diffusion. The spontaneous intermixing of different substances by molecular movement, giving uniform concentrations. *See also* DIFFUSION, EDDY DIFFUSION.

mole drain. *See* SOIL DRAINAGE.

mollic. Applied to dark, organic-rich surface layers of soil that are rich in calcium and magnesium and have a strong structure, so the soil remains 'soft' when it is dry. *See also* SOIL CLASSIFICATION.

Mollisols. *See* SOIL CLASSIFICATION.

Mollusca. A large phylum of soft-bodied, unsegmented, mainly aquatic animals, most of which have calcareous shells. Locomotion is usually by means of a large muscular 'foot'. *See also* AMPHINEURA, CEPHALOPODA, DECAPODA, GASTROPODA, LAMELLIBRANCHIATA, SCAPHOPODA.

molluscicide. A chemical used to kill MOLLUSCA. Examples include metaldehyde and copper sulphate, which is used to control the water snail *Limnaea trunculata*, the VECTOR of the liver fluke parasite in sheep.

molten salt reactor. A BREEDER REACTOR that uses molten salt as a coolant and thorium as the fertile material (the blanket). The molten salt reactor is a thermal breeder, and although work on it is not so advanced as on the LIQUID METAL FAST BREEDER REACTOR, it is considered to be a possible alternative.

molybdenite (MoS_2). The mineral molybdenum sulphide from which most of the world's MOLYBDENUM is produced as a by-product of mining PORPHYRITIC COPPER deposits, and also from the geologically analogous porphyritic molybdenum deposits.

molybdenum (Mo). A hard, white metallic element that occurs as MOLYBDENITE; it is used in special steels and alloys. It is an essential MICRONUTRIENT. A_r=95.94; Z=42; SG 10.2; mp 2620°C.

momentum. *See* ACCELERATION.

monazite $((Ce,La,Th)PO_4)$. A phosphate mineral of the rare earth metals cerium (Ce), lanthanum (La) and thorium (Th), which occurs as a rare ACCESSORY MINERAL in GRANITES and PEGMATITES, but is concentrated in some PLACER deposits (particularly BLACK SANDS) from which it is obtained as a by-product of mining for titanium minerals and zircon. The rare earth metals are used in ALLOYS and as CATALYSTS, and thorium is used in some NUCLEAR REACTORS. Monazites in pegmatites have been used extensively in determining the RADIOMETRIC AGES of the rocks.

Monera. The kingdom that includes all the PROKARYOTIC organisms.

monitoring programme. The systematic measuring, quantitatively or qualitatively, of a phenomenon or the presence of a substance, over a period of time. Experiments and demonstrations are observed (monitored) in this way. Monitoring programmes are often used to provide information on the distribution in space or time of pollutants, so that effective measures may be developed to limit their harmful effects should the results of the programme indicate that remedial measures are necessary.

monocarpic. Applied to plants that flower only once during their lives.

monocaryon. *See* MONOKARYON.

monoclimax. The theoretical development within a regional climatic type of a single CLIMAX formation. *Compare* POLYCLIMAX.

monocline. A localized zone of steeply dipping beds in a region of horizontal to low DIP.

Monocotyledoneae (monocotyledons). One of the two classes of flowering plants (ANGIOSPERMAE) in which the EMBRYO has only one COTYLEDON. The leaves are usually parallel-veined. The VASCULAR BUNDLES in the stem are scattered and do not contain cambium (*see* MERISTEM). Flower parts are usually in threes or multiples of three. Most monocotyledons are small plants (e.g., grasses, lilies, orchids), but a few are large (e.g., palm trees). The class includes very important food-producing plants. *See also* COCOS NUCIFERA, GRAMINEAE, *MUSA*.

monoculture. The cultivation of a crop of the same type in successive years to the exclusion of all other crop types.

monocytes. *See* BLOOD CORPUSCLES.

monoecious. Applied to organisms in which both male and female reproductive organs are borne in the same individual. In flowering plants, the term is restricted to individuals with separate male and female flowers on the same plant. *Compare* DI-OECIOUS.

monogamy. Permanent pair bonding; among species that form permanent pair bonds both parents are generally involved in the rearing of young. *Compare* POLYGAMY.

monogenetic. *See* PROVENANCE.

monohybrid inheritance. The inheritance of one pair of ALLELOMORPHS. The monohybrid ratio indicates the proportion of different types of individuals in the F_2 generation that are producing by crossing individuals differing in respect of a single pair of allelomorphs. This ratio is 3:1 if one character is dominant.

monohybrid ratio. *See* MONOHYBRID INHERITANCE.

monokaryon (monocaryon). A fungal HYPHA or MYCELIUM made up of cells each of which contains one haploid nucleus. All the nuclei are identical. *Compare* DIKARYON.

monomer. A chemical compound that consists of single molecules. *Compare* POLYMER.

monomolecular layer. A layer that is one molecule thick (e.g., a film of oil on water). Such a layer may form a boundary between two fluids preventing the natural transfer of molecules between them.

monophagous. Applied to an animal that feeds on only one kind of food. An example is the black hairstreak butterfly (*Strymonidia pruni*), which feeds only on blackthorn (*Prunus spinosa*).

monophyletic. Applied to a natural, closely related group of species or other taxa (*see* TAXON), all of which have a common origin. *Compare* POLYPHYLETIC.

monophyodont. Applied to an animal that has only one set of teeth during its lifetime. *Compare* DIPHYODONT, POLYPHYODONT.

monosaccharides. *See* CARBOHYDRATES.

monosome. An unpaired CHROMOSOME. A single X chromosome is normally present in the heterogametic sex of some organisms. Monosomes occur abnormally in many AN-EUPLOIDS.

monotocous (uniparous). Applied to animals that produce young singly. *Compare* DITOCOUS, POLYTOCOUS.

Monotremata. An order of MAMMALIA (subclass: Prototheria) that lay eggs and have many primitive reptilian features. They are represented today only by the duck-billed platypus (*Ornithorhynchus*) of Australia and the spiny anteaters (the echidna (*Tachyglossus*) of Australia and *Zaglossus* of New Guinea).

monovular twins. *See* MONOZYGOTIC TWINS.

monoxenic parasite. *See* PARASITISM.

monozygotic twins (identical twins, monovular twins, uniovular twins). Twins that are derived from a single fertilized egg. They are genetically identical and therefore always of the same sex. *See also* POLYEMBRYONY. *Compare* DIZYGOTIC TWINS.

monsoon forest. *See* FOREST.

Monte Carlo method. In statistics, the construction of an artificial STOCHASTIC model of the process being studied to provide a basis for sampling experiments.

montmorillonite ($Al_4(Si_4O_{10})_2(OH)_4.xH_2O$). A CLAY MINERAL with a layer lattice structure and characteristically a variable water content so that the clay swells as water is taken in between the layers.FULLER'S EARTH consists mainly of montmorillonite, but the purest form is found in BENTONITE, formed by the WEATHERING of volcanic ash.

moor (moorland). An open area of acid PEAT, usually on high ground and occupied by heathers (*Erica, Calluna*), sedges and certain grasses (e.g., purple moor grass, *Molinia, Caerulea*).

mor (mar, raw humus). An acidic, crumbly HUMUS layer of soil that is clearly marked off from the mineral soil beneath. It is poor in animal life, especially in earthworms. *Compare* MULL.

Moraceae. A family of tropical and subtropical DICOTYLEDONEAE, most of which are trees and shrubs containing latex. The economically important genera include: mulberry (*Morus*), cultivated for its fruit and its leaves, which are used as food for silkworms; *Castilloa*, which is a source of rubber; *Brosium*, which yields the breadnut and a milky drink made from the latex; *Artocarpus*, the source of breadfruit; *Broussonetia*, the inner bark of which is used as paper and cloth. *Ficus* species also belong to this family.

moraine. The debris produced by a GLACIER. Ground moraine is an undulating, relatively flat-topped deposit of BOULDER CLAY which is exposed after the retreat of a glacier or ICE SHEET. A terminal (end) moraine is a ridge-like accumulation of GLACIAL DRIFT (usually boulder clay) which is formed at the ice margin during a lull in a glacial retreat.

morbidity rate. The incidence of sickness in a population. *Compare* MORTALITY RATE.

morphactins. A group of synthetic plant growth regulators that cause such effects as dwarfing, bushiness and the inhibition of germination.

morphallaxis. The gradual development of a particular form from a mass of undifferentiated cells, and especially the remodelling of tissue from existing animal cells during the regeneration of a missing part. *See also* AUTOTOMY.

morphology. The study of the form of organisms or of the Earth's physical features (*see* GEOMORPHOLOGY).

mortality rate. The incidence of death in a population. *Compare* MORBIDITY RATE.

mortar. A construction material that is used to hold together bricks or stones. Mortars can be made from a mixture of SAND and CEMENT, or from pure hydrated (slaked) lime (calcium hydroxide), with or without the addition of volcanic ash (*see* POZZOLANA), coal or wood ash, crushed bricks or any fired clay in powder form. Masonry cement, used as mortar, is a mixture of PORTLAND CEMENT, an air-entraining plasticizer and an inert filler.

morula. An early animal EMBRYO, before the BLASTULA stage, which consists of a solid ball of cells.

mosaic. (1) In genetics, a CHIMAERA. (2) Applied to an egg in which various cytoplasmic areas give rise to definite structures in the EMBRYO, so that if an area of egg is damaged the corresponding structure will not develop correctly. (3) In a plant, a patchiness in the green colour caused by a viral disease. (4) A vegetation pattern in which two or more COMMUNITY types are interspersed (e.g., grassland interspersed with patches of scrub).

mosquitoes. *See* CULICIDAE.

mosses. *See* MUSCI.

moss animals. *See* Polyzoa.

moss layer. *See* Layers.

most seriously affected countries (MSA countries). A list prepared each year by the Food and Agriculture Organization of the United Nations of the countries suffering most seriously from food shortages and therefore in most urgent need of assistance.

mother-of-pearl clouds. *See* Nacreous Clouds.

motor pattern. Those aspects of animal behaviour which can most easily be observed (i.e. what an animal does) and which involve the movement of muscles (which includes the making of sounds). Motor patterns provide the information used in the compilation of Ethograms.

mould. (1) Any fungus (*see* Fungi) that produces a superficial growth of Mycelium (e.g., *Penicillium notatum,* the source of the antibiotic penicillin). (2) A soil rich in Humus (e.g., leaf mould).

moult. (1) The periodic shedding of hair (e.g., the spring moult in mammals when the thick winter coat is lost) or feathers. Most birds moult in the autumn. Ducks also moult in the summer and become flightless for a time owing to the loss of their large wing feathers; the drab moult plumage of the drake is called eclipse plumage. (2) *See* Ecdysis.

Mount Agung (Gunung). A volcano in Indonesia that erupted violently in March 1963, emitting large amounts of particulate matter and gases, whose movement and effects were monitored. Large quantities of sulphur dioxide (or possibly sulphates) reached a height of about 18 kilometres, spread over the world within about six months, lasted for several years and produced twilight displays of colours. The temperature of the lower equatorial stratosphere increased by about 6–7°C shortly after the eruption, probably because of the presence of particles which may have increased the usual stratospheric particle content about 10-fold. It is assumed that the Mount Agung eruption was typical of tropical volcanic eruptions (e.g., those of Mount Tambora and Krakatoa). *See also* Natural Pollutant.

mountain waves (lee waves). Waves, usually a few miles in length, that are set in motion when stable air (*see* Static stability) moves over a mountain ridge. The formation of such waves is more likely if the air is more stable at lower than at higher levels. The vertical velocity of the air within the waves may be 300 metres per minute or more, making the downdraughts a hazard to aircraft. Clouds formed in the wave crest indicate the vertical extent of the waves, which may be several times the height of the mountains. With waves of large amplitude, an eddy with reverse motion may form, called rotor flow.

Mount St Helens. A volcano in Washington State, USA, that erupted in May 1980, releasing large amounts of ash, fragments of stone and pumice. It caused landslides, mud flows and floods, which devastated a large area.

Mount Tambora. A volcano in Indonesia that, in 1815, produced what is believed to have been the most violent eruption in historical times, reducing the height of the mountain from 4000 to 2800 metres and emitting large quantities of gases and particles into the atmosphere. *Compare* Mount Agung, Mount St Helens. *See also* Natural Pollutant, sulphate, sulphur dioxide.

mp. *See* Melting point.

MSA countries. *See* Most seriously affected countries.

mucoid feeding. Feeding by means of mucus which is extruded from the mouth, then ingested together with such small particles as have stuck to it. It is a method employed by some Mollusca.

mudbank. *See* Flocculation.

mudstone. A lithified, ARGILLACEOUS SEDIMENTARY ROCK which is not FISSILE. *Compare* SHALE.

mulberry. *See* MORACEAE.

mulch. A material (e.g., straw, paper, plastic sheeting) that is used to protect the surface of the soil from erosion. Where the mulch is composed of biodegradable material it becomes incorporated into the topsoil to become HUMUS.

mull (mild humus). A loose, crumbly HUMUS layer, mingling with the mineral soil beneath, and produced where decay is rapid. Its acidity is low, and it contains a rich fauna, including abundant earthworms. *Compare* MOR.

müllerian mimicry. The resemblance between one species of animal equipped with warning coloration and a defence (e.g., distastefulness) against predators and another similarly protected species. Both gain from the resemblance because, after attacking a member of one species, a predator learns to avoid all animals of their appearance and so both species are protected. *Compare* BATESIAN MIMICRY.

multiple factors (polygenes). A group of GENES that combine to control the development of a single character (e.g., stature in humans). A wide range of variation is produced by numerous combinations of these genes.

multivoltine. Applied to organisms that produce several generations of young in a year. *Compare* BIVOLTINE, UNIVOLTINE.

municipal waste. Substances discarded as unusable by private households, offices, shops, etc., but not the waste products of industrial processing or manufacturing. Municipal waste is composed typically of paper, organic matter, plastics, metals and non-metallic minerals (e.g., ash). Historically, the content of plastics has increased rapidly, and the ash content has decreased. Such waste is generally collected for disposal by incineration, LAND-FILL, COMPOSTING or DESTRUCTIVE DISTILLATION, and in some areas it is used as fuel to generate heat or power.

Musa. A genus of huge tropical monocotyledonous (*see* MONOCOTYLEDONEAE) herbs which have large, oval leaves. The bananas (e.g., *M. sapientum*) and the plantain (*M. paradisiaca*) are widely cultivated and produce a very high yield of food per acre of land. ABACÁ (Manila hemp), a valuable fibre, is obtained from the leaf stalks of *M. textilis*.

Muschelkalk. *See* TRIASSIC.

Musci (mosses). A class of BRYOPHYTA that have prostrate or erect stems containing conducting tissue. Their leaves are usually only one cell thick, except in the midrib region. Mosses occur in both moist and dry terrestrial habitats, and in water. *Sphagnum* species are an important constituent of BOG.

muscle. Highly contractile animal tissue, which is classified as two main types: striped and smooth. Striped (striated, voluntary, skeletal) muscle consists of elongated, multinucleate fibres whose CYTOPLASM is made up of longitudinal fibrils, with alternating bands of different composition. Striped muscle contracts rapidly and is mainly concerned with moving skeletal parts in vertebrates and in ARTHROPODA. Smooth (plain, involuntary) muscle consists of individual spindle-shaped cells. It performs slow, often sustained contractions and is found in the walls of such organs as the gut and blood vessels. Similar muscle is found in many invertebrates (e.g., MOLLUSCA, ANNELIDA).

muscovite. *See* MICA.

mushroom rocks. Sandblasted rocks, which have been sculptured into shapes reminiscent of a longitudinal cross-section through a mushroom by sand carried by the wind at a height of less than 2 metres. *See also* CORRASION.

mushrooms. *See* AGARICACEAE.

mussels. *See* LAMELLIBRANCHIATA.

mustard gas (dichlorodiethyl sulphide, (ClCH$_2$CH$_2$)$_2$S). One of the first chemical warfare agents to be used in World War I. Stockpiles of the gas were disposed of when they were no longer required, and occasionally they raise fears of local pollution.

mutagen. An agent (e.g., X-rays, gamma-rays, mustard gas) that can induce MUTATION.

mutation. A sudden change in the CHROMOSOMES of a cell, most mutations being due to changes in the DNA of individual GENES, others to alterations in the structure or number of chromosomes (*see* ANEUPLOID, POLYPLOID). Mutations may occur during GAMETE formation, when they produce a heritable change, or in body cells (somatic mutation). Most mutations are harmful, but evolution proceeds through the operation of natural selection on mutations, harmless or the occasional beneficial mutations surviving. *See also* MUTAGEN.

mutualism. Formerly a synonym for SYMBIOSIS, but now used variously to mean: (a) a type of symbiosis in which neither partner is essential for the well-being or life of the other; (b) both symbiosis and COMMENSALISM; or (c) association between organisms of different species, including PARASITISM.

my. An abbreviation meaning millions of years.

mycelium. The vegetative part of a fungus that consists of a mass of minute filaments (*see* HYPHAE).

mycetocytes. Specialized cells in the body of an insect that contain symbiotic (*see* SYMBIOSIS) microorganisms. Sometimes mycetocytes are localized in special organs (mycetomes).

mycetomes. *See* MYCETOCYTES.

mycetophagous. Applied to an animal that feeds on FUNGI.

Mycetozoa. *See* MYXOMYCOPHYTA.

mycoplasmas (pleuropneumonia-like organisms, PPLO). A group of PROKARYOTIC organisms, without cell walls, that are usually regarded as very small BACTERIA. Some are SAPROPHYTES, others cause disease in plants or animals.

mycorrhiza. The symbiotic (*see* SYMBIOSIS) association of the root of a higher plant with a fungus. In an ectotrophic mycorrhiza (e.g., heath, pine trees) the fungal MYCELIUM covers the outside of the roots; in an endotrophic mycorrhiza (e.g., orchids) the fungus grows inside the cells of the root CORTEX.

mycotrophic. Applied to plants that possess a MYCORRHIZA.

mycotrophy. The nutritional relationship between certain FUNGI and VASCULAR PLANTS and BRYOPHYTA. *See also* MYCORRHIZA.

myiasis. Infestation by the parasitic larvae of certain flies (DIPTERA) (e.g., the warble-fly).

mylonite. A very compact, welded ROCK FLOUR produced at the bases of THRUSTS.

Myocastor coypus. *See* COYPU.

myoglobin. A PROTEIN that is found in MUSCLE and whose function is the temporary storage of oxygen. A molecule of HAEMOGLOBIN consists of four myoglobin-like molecules.

Myriapoda (centipedes and millipedes). Terrestrial ARTHROPODA with many similar leg-bearing segments. Centipedes (Chilopoda) are carnivorous, with the first pair of appendages modified as poison claws and with the body flattened dorsoventrally. Millipedes (Diplopoda) are mainly vegetarian, with cylindrical bodies and two pairs of legs to each apparent segment.

myrmecochore. A plant whose reproductive structures are dispersed by ants (e.g., violets, whose seeds bear oil bodies – elaiosomes – which are attractive to ants).

myrmecophily. (1) Living with ants. Some animals live inside ants' nests (e.g., larvae of the large blue butterfly, *Maculinea arion*), and some plants (*see* MYRMECOPHYTE) are inhabited by ants. (2) A plant pollination by ants.

myrmecophyte. A plant which, by means of special structures, shelters ants and may provide food for them, possibly in return for protection against insect predators. For example, *Acacia sphaerocephala* has thorns that are inhabited by ants, the ants feeding from nectaries and food bodies on the leaves.

Mysticeti. *See* BALAENOIDEA.

myxobacteria. A small group of BACTERIA, formed by the gliding together of rod-shaped vegetative cells, often to produce fruiting bodies.

myxomatosis. A viral disease affecting rabbits and, occasionally, hares. It is endemic in South America, where it is transmitted by mosquitoes. Myxomatosis was introduced deliberately into Australia in 1950 and somehow reached the UK in 1953. By 1955 it had virtually wiped out the rabbit population in some areas, and nationally had reduced the rabbit population by about 90 percent. This led to startling changes in vegetation (*see* DOWNLAND VEGETATION). Rabbit populations in both the UK and Australia have recovered gradually and many more rabbits now survive each fresh outbreak. In the UK the VECTOR of the disease is the rabbit flea.

Myxomycetes. *See* MYXOMYCOPHYTA.

Myxomycophyta (slime moulds, slime fungi). A group of organisms that includes the Mycetozoa (Myxomycetes, Protomyxa), intermediate between FUNGI and PROTOZOA. They commonly occur as SAPROPHYTES on decaying vegetable matter. The vegetative stage usually consists of a multinucleate, amoeboid mass of PROTOPLASM (a plasmodium) which is mobile and can ingest food. Reproduction is by means of SPORES produced in plant-like structures (sporangia). A common species is the bright yellow flowers of tan (*Fuligo septica*).

Myxophyceae. *See* CYANOPHYTA.

Myxophyta. *See* CYANOPHYTA.

N

N. (1) *See* NEWTON. (2) *See* NITROGEN.

n. *See* NANO-.

Na. *See* SODIUM.

nacreous clouds (mother-of-pearl clouds). Thin clouds that appear at heights of 20–30 kilometres above the surface and are seen most often in Scandinavia in winter, when the Sun is several degrees below the horizon. A temperature of about -80°C is believed to be required for their formation. Their lifetimes are not known because conditions for good visibility do not last long. The low stratospheric temperatures associated with them and their seasonal and geographical variations support the hypothesis that they are produced when local saturation occurs. The required temperature may be reached partly as a result of MOUNTAIN WAVES. It has been suggested that the injection of water vapour (e.g., from aircraft exhausts) in the stratosphere might modify high-level cloud formation by raising the FROST POINT temperature and so permitting nacreous clouds to form over a wider area.

Namurian. The third oldest STAGE of the CARBONIFEROUS System in Europe.

nanism. A stunting of plant growth that occurs, for example, in trees at the limit of their range on mountains.

nano- (n). A prefix used in conjunction with SI units to denote the unit x 10^{-9}.

nanometre (nm). One-thousandth of a MICRON, or 10 ångströms.

nanoplankton. The smallest of the phytoplankton (*see* PLANKTON).

β-naphthylamine. *See* BETA-NAPHTHYLAMINE.

nappe. (1) A faulted (*see* FAULT), overturned FOLD. (2) A large body of rock that has moved forward by recumbent folding (*see* RECUMBENT FOLD) and/or thrusting (*see* THRUST).

NASA. *See* NATIONAL AERONAUTICS AND SPACE ADMINISTRATION.

nastic movement. A plant response that is independent of the direction of the stimulus (e.g., the drooping of the leaves of *Mimosa pudica* when they are touched (seismonasty); the daily opening and closing of flowers, the 'sleep movements' of leaves). Nastic response to illumination is called photonasty and to temperature thermonasty. Response to the alternation of day and night, related to changes in both illumination and temperature, is called nyctinasty.

National Aeronautics and Space Administration (NASA). The US government agency that is responsible for space exploration and for the launching and management of satellites (e.g., LANDSAT).

National Audubon Society. A US-based society, with branches in each state, named after the American artist and naturalist John James Audubon (1785-1851) and founded around the turn of this century. It is concerned with wildlife conservation, especially with the protection of birds.

National Environmental Policy Act. *See*

COUNCIL ON ENVIRONMENTAL QUALITY.

national nature reserve. *See* NATURE RESERVE.

national park. As defined in 1975 by the INTERNATIONAL UNION FOR CONSERVATION OF NATURE AND NATURAL RESOURCES, an extensive area of land that has not been significantly altered by human activities, that contains landscapes, species or ECOSYSTEMS of scientific, educational or aesthetic value and is set aside in perpetuity to be managed by the state for conservation. In UK planning, an extensive tract of country, designated by the COUNTRYSIDE COMMISSION, which, by reason of its natural beauty and the opportunities it affords for open-air recreation, shall be preserved and enhanced for the purpose of promoting its enjoyment by the public. The UK definition differs slightly from the international definition because there are few parts of the UK which have not been altered by human activity, and because the UK national parks were defined and the first of them designated before the international definition was produced. There are no national parks in Scotland.

National Parks and Access to the Countryside Act, 1949. *See* AREA OF OUTSTANDING NATURAL BEAUTY.

National Parks Commission. *See* COUNTRYSIDE COMMISSION.

National Park Service. The branch of the US Department of the Interior that is responsible for the management of sites of historic interest, areas set aside for public recreation and about 180 national parks.

National Radiological Protection Board (NRPB). In the UK, an organization established under the Radiological Protection Act, 1970, to study ways of improving the protection of the public from exposure to ionizing radiation and to provide advisory and technical services, for which it is permitted to charge.

National Society for Clean Air. A UK voluntary organization that seeks to reduce air pollution by creating a body of informed public opinion that will exert political pressure.

native element. An element that occurs in the uncombined state as a mineral. The 20 native elements can be divided into those in which the native element is the major, or only, ORE MINERAL: CARBON (graphite and diamond), sulphur, gold and the platinoid metals (platinum, palladium, rhodium, iridium, ruthenium, osmium); and the other elements whose major source is from compounds: IRON, nickel, COPPER, ARSENIC, SELENIUM, SILVER, ANTIMONY, tellurium, MERCURY, LEAD and bismuth.

natric. Applied to a layer of soil that is enriched with SODIUM.

natural frequency. The FREQUENCY at which a system oscillates freely after excitation.

natural gas. A mixture of gases (HYDROCARBONS and non-hydrocarbons) that is found in nature, often associated with deposits of petroleum. The principal component of natural gas is usually METHANE, and the term is often restricted to such a gas, although natural gases composed predominantly of CARBON DIOXIDE and HYDROGEN SULPHIDE are known. Under subsurface RESERVOIR conditions, some natural gases are liquids. Commercially, natural gas is transported by pipelines or, refrigerated, in a liquid state (as liquefied natural gas, LNG) in LNG carriers.

natural increase. The rate of population growth, calculated by subtracting the number of deaths from the number of births in a given period (or vice versa, if deaths exceed births and the population is decreasing).

naturalists' trusts (county trusts for nature conservation). In Britain, bodies of naturalists organized roughly on a county basis in England and Wales, with the Scottish Wildlife Trust serving the whole of Scotland. The trusts seek to conserve natural habitats by acquiring and managing NATURE RESERVES and by offering advice and education regarding the preservation of wildlife in Britain. The Norfolk Naturalists' Trust was the first to be formed in 1926. The trusts are sponsored by the Royal Society for Nature Conservation (formerly the Society for the Promotion of Nature Conservation and before that the Society for the Promotion of Nature Reserves).

natural pollutant. A substance of natural origin that may be regarded as a pollutant when present in excess. Examples include volcanic dust, particles of sea salt, OZONE formed photochemically or by lightning and products of forest fires. *See* KRAKATOA, MOUNT AGUNG, MOUNT ST HELENS, MOUNT TAMBORA, SULPHATE, SULPHUR DIOXIDE.

natural resource ecosystem. A natural ECOSYSTEM in which one part is of use to humans.

Natural Resources Defense Council. A US non-governmental environmental organization that monitors resource use and environmental pollution, especially to ensure the strict observance of existing legislation.

natural selection. *See* DARWIN, CHARLES ROBERT.

Nature Conservancy Council (NCC). The official body, established in Britain by Act of Parliament in 1973 to be responsible for the conservation of flora, fauna and geological and physiographical features throughout Great Britain. It establishes and maintains national NATURE RESERVES, gives advice and education on nature conservation, carries out research into related topics and advises the Secretary of State for the Environment on the designation of SITES OF SPECIAL SCIENTIFIC INTEREST. The Council is financed by the DEPARTMENT OF THE ENVIRONMENT and replaced the Nature Conservancy, formed in 1949.

nature–nurture. The controversy over the categorization of behaviour into that which

is innate (instinctive), and so attributable to nature, and that which is learned, and so attributable to nurture. *See also* EUGENICS, HERITABILITY, INSTINCT, LARMACK, JEAN BAPTISTE DE.

nature reserve. In British planning, an area of land and/or water that is managed primarily to safeguard the flora, fauna and physical features it contains. Reserves that are declared and managed by the NATURE CONSERVANCY COUNCIL on behalf of the DEPARTMENT OF THE ENVIRONMENT are called national nature reserves (NNR). Those designated by local planning authorities (in consultation with the Nature Conservancy Council) are called local nature reserves. County NATURALISTS' TRUSTS and other voluntary bodies are often associated with the management of local nature reserves and, in addition, have established a large number of reserves of their own.

nauplius. A free-swimming larva, typical of many CRUSTACEA, that has an egg-shaped unsegmented body, three pairs of appendages and a single eye.

nautical mile. Defined originally as the average length of one minute of latitude (which varies slightly with longitude), and now defined as 6080 feet. The international nautical mile is 1852 metres (6076.12 feet).

nautilus. *See* CEPHALOPODA.

Neanderthal man. *See* HOMO.

Nearctic Region. The ZOOGEOGRAPHICAL REGION, forming part of ARCTOGEA, and consisting of North America as far south as Mexico.

NEF. *See* NOISE EXPOSURE FORECAST.

negative feedback. *See* FEEDBACK.

negative lynchet. *See* LYNCHET.

negative price. An economic concept which imputes a price to substances that have no value but may cause damage to, or necessitate expenditure by, others (e.g., industrial effluents).

neighbourhood noise. Noise from any source that may cause disturbance and annoyance to the general public in their homes or going about their business, but not industrial noise experienced by workers.

nekton. The strongly swimming animals of the PELAGIC zone of a lake or sea (e.g., fish).

Nemathelminthes. A phylum used by some taxonomists, comprising the NEMATODA and NEMATOMORPHA.

nematoblast (cnidoblast, thread cell). A type of cell, found in large numbers in most CNIDARIA, that holds a bladder (nematocyst) containing a hollow coiled thread which is shot out when the cell is stimulated and becomes attached to the animal's prey. Some threads can penetrate skin and contain poison, so exerting a paralyzing effect.

Nematocera (craneflies, mosquitoes). A suborder of DIPTERA with long, many-jointed antennae.

nematocide. A chemical used to control nematode worms (NEMATODA) (e.g., certain DINITRO PESTICIDES, ORGANOCHLORINES and ORGANOPHOSPHORUS PESTICIDES used to kill eelworms infesting potatoes, sugar-beet, etc.).

nematocyst. See NEMATOBLAST.

Nematoda (roundworms). A very large group, usually regarded as a phylum, of unsegmented worms, without a COELOM, usually pointed at both ends. Some are free-living in the soil, water and even in liquids such as vinegar. Many are parasites (see PARASITISM), (e.g., potato root eelworm, sugar-beet eelworm, pinworms of vertebrates). See also FILARIA, HOOKWORMS, NEMATHELMINTHES.

Nematomorpha (Gordiacea, horsehair worms, threadworms). A group of extremely long (up to about 90 centi-

metres), slender freshwater and marine worms that resemble NEMATODA. The larvae parasitize (see PARASITISM) insects or CRUSTACEA. The Nematomorpha are regarded by some taxonomists as a separate phylum. See also ASCHELMINTHES, NEMATHELMINTHES.

Nemertea (Nemertini, proboscis worms, ribbon worms). A small phylum of flattened, unsegmented mainly marine animals with a large, reversible proboscis, used for catching prey. Nemertea resemble PLATYHELMINTHES in their general body organization.

Nemertini. See NEMERTEA.

nene. See ANATIDAE.

Neoceratodus. See DIPNOI.

Neocomian. A subdivision (usually ranking as a STAGE) of the CRETACEOUS System, and comprising the Berriasian or Ryazanian, and the Valanginian, Hauterivian and Barremian.

neo-Darwinism. The current theory of evolution, which combines Darwin's theory of evolution by natural selection (see DARWIN, CHARLES ROBERT) with modern concepts of genetics.

Neogea. The ZOOGEOGRAPHICAL REGION comprising Central and South America.

Neogene. The later of the two periods of the CENOZOIC Era, including the MIOCENE and PLIOCENE Epochs; Neogene also refers to the rocks formed during this time. In some usages, Neogene also includes the PLEISTOCENE and HOLOCENE (Recent). The terms PALAEOGENE and Neogene are restricted mainly to European usage.

Neolithic (Late Stone Age). The stage of development of early agricultural societies which obtained some of their food by the cultivation of domesticated crop plants and/or the husbanding of domesticated livestock. Neolithic peoples used polished stone tools, although today many peoples whose agriculture is at a Neolithic stage of development use metals, usually imported in ready-made form. *Compare* MESOLITHIC, PALAEOLITHIC.

Neolithic revolution. *See* AGRICULTURAL REVOLUTION.

neoplasm. A tumour, often malignant.

neossoptiles. The feathers of a nestling bird. *Compare* TELEOPTILES.

neotony. The temporary or permanent persistence of pre-adult stages or structures in animals. In extreme examples PAEDOGENESIS occurs, as in the axolotl, a salamander that breeds in the larval form. Neoteny has probably played a part in human evolution as many adult features (e.g., large head, reduced amount of hair) resemble those of foetal apes.

Neotropical Realm. The FLORAL REALM that includes the southern tips of California and Florida, Central America and the islands of the Caribbean, and South America south to Paraguay and the northern half of Chile. It is divided into three regions: CENTRAL AMERICAN FLORAL REGION; PACIFIC SOUTH AMERICAN FLORAL REGION; PARANO-AMAZONIAN FLORAL REGION.

neotype. A specimen chosen to replace a TYPE SPECIMEN when all the original material is missing. *Compare* LECTOTYPE, PARATYPE.

NEPA. National Environmental Policy Act (*see* COUNCIL ON ENVIRONMENTAL QUALITY).

nepheline. *See* FELDSPATHOIDS.

nephridium. An osmoregulatory (*see* OSMOREGULATION) and excretory organ, typical of many animals (e.g., ACRANIA, ANNELIDA, PLATYHELMINTHES), that consists of a tubule open to the exterior at one end. The other end is often closed, ending in hollow flame cells (solenocytes) containing CILIA.

neptunium. *See* ACTINIDES.

neritic. Applied to the parts of the sea lying over the CONTINENTAL SHELF, usually less than 200 metres deep. *See also* LITTORAL ZONE, OCEANIC, SUBLITTORAL ZONE.

nerve cell. *See* NEURON.

nerve net. A network of NEURONS which make up a simple nervous system in CNIDARIA. In animals with central nervous systems nerve nets may exist in some parts of the body.

ness. A promontory or headland.

net primary aerial production. *See* PRODUCTION.

net primary production. *See* PRODUCTION.

net reproduction rate. The average number of female babies that will be born to a newly born female during her lifetime, if current rates of birth and death continue. If the net reproduction rate is larger than unity the population will grow, and it will decline if the rate is less than unity, although the age structure of the population may postpone the increase or decrease.

neuron (neurone, nerve cell). The fundamental unit of a nervous system. Each neuron consists of a nucleated nerve cell body from which fine processes (AXONS) along which nerve impulses pass. The places where the processes make contact with those of other neurons are called synapses. *See also* CENTRAL NERVOUS SYSTEM, GANGLION, NERVE NET.

neurone. *See* NEURON.

Neuroptera. An order of predatory insects (division: ENDOPTERYGOTA) whose members possess two similar pairs of wings covered in a delicate network of veins. They include lacewings, alder-flies, ant-lions, and snake-flies.

neurula. A vertebrate EMBRYO at the stage where the brain and spinal cord are beginning to form.

neustron. Organisms associated with the surface film of water. Infraneustronic animals (e.g., mosquito larvae) suspend themselves from the underside of the film; supraneustronic species (e.g., pond-skaters) live on the upper side of the film.

neutral fat. *See* FAT.

névé (firn). Snow that has survived a summer without melting and that has become granular. It is an intermediate stage in the conversion of snow to GLACIER ice.

New Caledonian Floral Region. The part of the PALAEOTROPIC REALM that comprises the islands of New Caledonia (Vanuatu).

New Red Sandstone. The RED BEDS of PERMO-TRIASSIC age in Europe. The vast majority of red beds in Devon are Permo-Triassic. *Compare* OLD RED SANDSTONE.

newton (N). The derived SI unit of force, being the force required to give a mass of 1 kilogram an acceleration of 1 metre per second per second. It is named after Sir Isaac Newton (1642–1727). The unit of pressure is the newton per square metre (N/m^2 or $N\,m^{-2}$), sometimes called the pascal. *See also* ATMOSPHERE, BAR.

newts. *See* AMPHIBIA.

New Zealand Floral Region. The part of the AUSTRAL REALM that comprises New Zealand and its offshore islands.

NGO. *See* NON-GOVERNMENTAL ORGANIZATION.

niacin. *See* NICOTINIC ACID.

niche. (1) The specific part of a habitat occupied by an organism. (2) The role an organism plays in an ECOSYSTEM. In different communities very different organisms may play comparable roles (e.g., on coral islands the earthworm niche is filled by land crabs).

niche diversification. *See* ALPHA DIVERSITY.

nickel carbonyl ($Ni(CO)_4$). An intermediary compound in the Mond process whereby nickel is extracted from the impure metal by the direct action of carbon monoxide on the finely divided metal. Nickel carbonyl is formed as a volatile substance (b.p. 43°C), and at 200°C the gas decomposes into pure nickel and carbon monoxide (which can then be used again). Nickel carbonyl was the first metal carbonyl to be discovered (by Ludwig Mond, after whom the process is named, in 1890). It is very toxic, explosive, a respiratory irritant and possibly a CARCINOGEN.

nick point. A sudden steepening of the gradient of a river bed, usually where it flows over resistant rock. It is produced by an intersection of new and old graded profiles of a river when a previous base level has been altered.

nicotine. An ALKALOID extracted from tobacco leaves and used as a CONTACT INSECTICIDE to control aphids and other insects on horticultural and fruit crops. It may be sprayed or used as a fumigant (*see* FUMIGATION), and although poisonous to mammals it loses its toxicity rapidly after application.

nicotinic acid (niacin, *p-p* factor). An important VITAMIN of the B group, found in malt extract, yeast and liver, and synthesized by many microorganisms. It plays a vital role in respiration. In humans, a deficiency causes pellagra.

nidation. *See* IMPLANTATION.

nidicolous. Applied to birds (e.g., blackbird) that are helpless and often naked when hatched and are fed for some time in the nest. *Compare* NIDIFUGOUS.

nidifugous. Applied to birds (e.g., mallard) that are well developed when hatched, with a downy covering and a store of yolk remaining inside the body. They are able to leave the nest at once. *Compare* NIDICOLOUS.

nife. An acronym for nickel and iron (NiFe), which are presumed to be the major constituents of the core of the Earth.

night jars. *See* CAPRIMULGIDAE.

night soil. Human excrement, which used to be collected by night before sewerage systems provided for continuous removal.

NII. *See* NUCLEAR INSTALLATIONS INSPECTORATE.

NIMBY. An acronym for not in my back yard (*see* BACKYARDISM).

NIREX (Nuclear Industry Radioactive Waste Executive). The UK Government agency established in 1982 to consider the relative merits of alternative methods for the disposal of INTERMEDIATE-LEVEL RADIOACTIVE WASTE and LOW-LEVEL RADIOACTIVE WASTES, and to find and develop the facilities needed. NIREX identified a need for shallow land disposal sites to take low-level wastes, and deep sites to take intermediate-level wastes.

nitrate. A salt (NO_3^-) or ester of nitric acid (HNO_3). Nitrates are the chief source of the nitrogen used by plants in their synthesis of proteins (*see* NITRIFICATION). They are highly soluble in water and so prone to LEACHING from land supporting vegetation. Where nitrogen is supplied as a soluble fertilizer the rate of leaching may increase, leading to NITRATE POLLUTION.

nitrate bacteria. *See* NITRIFICATION.

nitrate pollution. The contamination of fresh water by NITRATE, often caused by the LEACHING of nitrogen fertilizers from farm land. Since such contamination can occur suddenly, following heavy rain, and since nitrates are not routinely removed from water before it enters the public supply, nitrate pollution can affect humans. It causes no known harm in adults, but in young infants it can cause METHAEMOGLOBINAEMIA. In the EEC the maximum permitted limit for nitrates in water is 50 milligrams per litre, and in years to come this may be enforced by restricting the use of agricultural fertilizers in especially susceptible areas.

nitrating. *See* GUNCOTTON.

nitrification. The conversion by aerobic soil bacteria (nitrifying bacteria) of organic nitrogen compounds into NITRATES (NO_3^-), which can be absorbed by green plants. Dead organic matter is broken down into substances such as ammonia, which reacts with calcium carbonate forming ammonium carbonate. Ammonium carbonate is oxidized to nitrite (NO_2^-) by nitrite bacteria (*Nitrosomonas*), then nitrites are converted to nitrates by nitrate bacteria (*Nitrobacter*). *See also* NITROGEN CYCLE.

nitrifying bacteria. *See* NITRIFICATION.

nitrite bacteria. *See* NITRIFICATION.

nitrites. *See* NITRIFICATION.

Nitrobacter. *See* NITRIFICATION.

nitrogen (N). An element; an odourless, colourless, chemically inactive gas that forms about 80 percent of the atmosphere. It is an essential MACRONUTRIENT for plants and is used in fertilizers. The main natural source of nitrogen for industrial use is Chile saltpetre ($NaNO_3$), but atmospheric nitrogen is also fixed industrially (e.g., by the HABER PROCESS) to form ammonia (NH_3) and ammonium (NH_4^+) compounds. $A_r = 14.0067$; $Z=7$. *See also* NITRATE, NITRATE POLLUTION, NITRIFICATION, NITROGEN CYCLE, NITROGEN FIXATION.

nitrogen cycle. The circulation of nitrogen atoms through the soil, water and air, brought about mainly by living organisms. Inorganic nitrogen compounds (mainly NITRATES) are synthesized into organic compounds by AUTOTROPHIC plants. These die or are eaten by animals, and the organic nitrogen compounds in their bodies and excretory products eventually return to the soil or water. NITROGEN FIXATION augments the supply of organic nitrogen by taking nitrogen from the air. Nitrifying bacteria convert organic nitrogen into nitrates (*see* NITRIFICATION), and denitrifying bacteria convert some of the nitrates into gaseous nitrogen, which returns to the atmosphere (*see* DENITRIFICATION). *See also* ROOT NODULES.

nitrogen dioxide. *See* NITROGEN OXIDES.

nitrogen fixation. Any reaction as a result of which gaseous nitrogen forms a soluble compound that is available as a plant nutrient either directly or after it has engaged in further reactions. In soils, certain nitrogen-fixing bacteria and CYANOPHYTA utilize gaseous nitrogen and release it for the use of other organisms as metabolic waste products or as a decay product following their death. Some nitrogen-fixing bacteria (e.g., *Azotobacter* and *Clostridium pasteurianum*) are free-living in the soil. Others live symbiotically in the ROOT NODULES of LEGUMINOSAE. Nitrogen is also fixed under high-energy conditions by lightning, producing NITROGEN OXIDES, and industrially in the manufacture of nitrogen-based fertilizers (e.g., by the HABER PROCESS). *See also* NITRIFICATION, NITROGEN CYCLE.

nitrogen monoxide. *See* NITROGEN OXIDES.

nitrogen oxides (NO_x). A range of compounds formed by the oxidation of atmospheric nitrogen under high-energy conditions, (e.g., by COSMIC RAY bombardment of the uppermost region of the atmosphere, by lightning, in some industrial furnaces and in high-compression INTERNAL COMBUSTION ENGINES). The range of oxides includes nitrous oxide (dinitrogen oxide, N_2O), nitric oxide (nitrogen monoxide, NO), and nitrogen dioxide (dinitrogen tetroxide, N_2O_4). NO_x are involved in the formation of photochemical SMOG and ACID RAIN, and those formed in the STRATOSPHERE or migrating upwards across the TROPOPAUSE engage in reactions involving OZONE, depending on circumstances and the quantities involved either increasing or depleting the OZONE LAYER.

nitrosamines. Yellow oils that form naturally in the environment by the reaction of aqueous nitrous acid (HNO_2) with secondary AMINES (RR'NH). They can be formed in the human stomach from nitrites derived from food and secondary amines. Following animal experiments with dimethylnitrosamine, certain nitrosamine and nitrosamide compounds have been found to be active carcinogens, producing tumours in more sites than most known carcinogens. However, there is no evidence that they cause cancer in humans, there being no correlation between appropriate forms of cancer and dietary or diet-derived nitrosamines in studies of human populations.

Nitrosomonas. *See* NITRIFICATION.

nitrous oxide. *See* NITROGEN OXIDES.

nm. *See* NANOMETRE.

NNI. *See* NOISE AND NUMBER INDEX.

NNR. National nature reserve (*see* NATURE RESERVE).

Noah principle. The view that all species on Earth have the right to exist and should be preserved, so that any conservation effort must take account of all species, not simply certain ones.

nobelium. *See* ACTINIDES.

node. *See* ANTINODE.

nodum. A unit of vegetation, regardless of its rank (e.g., ASSOCIATION, SOCIETY, CONSOCIATION).

noise. Random sound or electromagnetic radiation that does not form part of, and tends to obscure, the sound or radiation carrying information (the signal) to an observer. The efficiency of communication may be measured as a signal-to-noise ratio.

noise and number index (NNI). An index for the measurement of disturbance by aircraft noise, developed in the UK and now used widely. The index is based on a social survey carried out in the vicinity of Heathrow Airport, London.

noise criteria. Sets of graph curves that relate sound levels in octave bands to speech interference, and thus to their acceptability for particular applications, often in a working environment.

noise exposure forecast (NEF). A technique for predicting the subjective effect of aircraft noise on the average person, expressed in NEF units, and sometimes drawn on maps as lines connecting points of equal NEF value. The estimate is based on values assigned to a range of factors (e.g., frequency of aircraft movements, distribution of movements by day and night, aircraft weights, flight profiles, etc.).

noise-rating curves. Sets of graph curves that relate sound levels in octave bands to their acceptability for particular applications. When an octave-band analysis is plotted on a graph of noise-rating curves, the number of the curve which is reached by the level in one or more bands is the noise-rating number (NRN) of the noise. DECIBEL units are used more commonly than noise-rating numbers.

noise-rating numbers. *See* NOISE-RATING CURVES.

noise reduction coefficient (NRC). The average ABSORPTION COEFFICIENTS of a surface or material at 250 Hz, 500 Hz, 1 kHz, and 2 kHz.

noise zoning. The classification of areas according to their use, taking particular account of noise levels, so that land use planning may contain noisy activities within areas where they will cause minimum disturbance to residents.

nomad. A wanderer; a person, usually a member of an extended family group or tribe, who moves from place to place and has no settled residence. Most nomadic tribes are pastoralist, cultivating no plant crops and subsisting on the products of their herds and flocks of animals, and moving from one area of pasture to another according to the season or to need.

non-abstractive use. Of water, a use that returns the water almost immediately as surface water (e.g., hydroelectric power generation). *Compare* ABSTRACTIVE USE.

non-governmental organization (NGO). A voluntary body, usually with an international membership, that is accorded official status by the United Nations (UN), enabling it to send delegates to attend certain meetings in an observer or consultant capacity, and to provide information and views to UN committees.

non-identical twins. *See* DIZYGOTIC TWINS.

non-parametric. Applied to a statistical test that is independent of any specific form of distribution.

non-porous. *See* POROUS.

non-renewable resource. A natural resource which, in terms of human time scales, is contained within the Earth in a fixed quantity and therefore can be used once only in the foreseeable future (although it may be recycled after its first use). This includes the FOSSIL FUELS and is extended to include mineral resources and sometimes GROUND WATER, although water and many minerals are renewed eventually (*see* HYDROLOGICAL CYCLE). The distinction is drawn according to time scales. If a resource will be renewed, but only over a geological time span counted in millions of years, it is considered non-renewable. *Compare* RENEWABLE RESOURCE.

Norian. *See* TRIASSIC.

normal distribution. *See* GAUSSIAN DISTRIBUTION.

normal fault. A FAULT in which the FAULT-PLANE dips to the downthrown side.

northern lights. *See* AURORA BOREALIS.

North-West European Megalopolis (Golden Triangle). An area of north-western Europe that extends from the Ruhr in West Germany westwards as far as Paris, northwards to include most of The Netherlands, Luxembourg and Belgium, and reaches an apex in the Manchester–Liverpool CONURBATION in England. Although severely affected by industrial decline during the 1980s, especially in England, within this roughly triangular area population densities and industrial activity are much greater than they are outside the area. Prior to the decline, there was a tendency for the Megalopolis (so termed by the EEC) to exert a strong economic attraction, drawing more industry and population into itself, so causing developmental problems outside the area. The majority of the EEC's most serious pollution problems occur within the region, which includes or is traversed by the Danube, Rhine, Elbe, Meuse, Moselle, Seine, Weser, Thames, Severn, Mersey, etc. The region also experiences the most severe urban deprivation.

notochord. A longitudinal, elastic supporting rod, consisting of vacuolated (*see* VACUOLE) cells surrounded by a sheath, which lies beneath the CENTRAL NERVOUS SYSTEM and is present at some stage of development in all CHORDATA. A notochord is present in adult ACRANIA and AGNATHA, but in most vertebrates it is more or less obliterated by the development of the vertebral column.

Notogea (Australasian Region, Australian Region). The ZOOGEOGRAPHICAL REGION comprising Australia, New Zealand, Tasmania, New Guinea and the islands south and east of WALLACE'S LINE.

nouméite. *See* GARNIERITE.

novel protein foods. High-protein foods that are not derived from livestock products or are animal in origin, but are converted into food suitable for humans or livestock by novel processes. For human consumption, most novel protein foods are derived from SOYA, other leguminous crops (*see* LEGUMINOSAE) or cereals, but for livestock feeds the wastes from food processing, abbatoirs, livestock excrement or from the cultivation of unicellular organisms on hydrocarbon SUBSTRATES are also immediately or potentially available. Proteins derived from microorganisms (e.g., YEASTS) are called SINGLE-CELL PROTEINS. Novel proteins may be used directly, in powdered form, as meat extenders to increase the protein content of

traditional meat products (e.g., in meat pies, sausages, etc.) or may be spun to produce long fibres or extruded through a perforated surface, then textured, coloured and flavoured to produce meat substitutes (e.g., textured vegetable protein, TVP).

NO$_x$. *See* NITROGEN OXIDES.

noy. A unit of noisiness, related to the PERCEIVED NOISE LEVEL, and measured in perceived noise DECIBELS (PNdB).

NPK. Nitrogen (N), phosphorus (P) and potassium (K); three major MACRONUTRIENTS, the principal constituents of most industrially manufactured fertilizers. The initials are sometimes used as a synonym for artificial fertilizer.

NRC. (1) *See* NOISE REDUCTION COEFFICIENT. (2) *See* NUCLEAR REGULATORY COMMISSION.

NRN. Noise-rating number (*see* NOISE-RATING CURVES).

NRPB. *See* NATIONAL RADIOLOGICAL PROTECTION BOARD.

nuclear fission. The splitting of the nucleus of a heavy atom into two, with the emission of neutrons and energy. *Compare* NUCLEAR FUSION. *See also* CRITICAL MASS, NUCLEAR REACTOR.

nuclear fuel cycle. The sequence of operations by which fuel is obtained, used and the waste products disposed of in the production of explosives for use in nuclear or thermonuclear weapons and in the generation of electrical power by NUCLEAR REACTORS. For civil use the cycle begins with the mining of the ore and the extraction of uranium dioxide (yellowcake). URANIUM ENRICHMENT of the fuel enhances the content of uranium-235 for those reactors that require it. The fuel is then made into FUEL ELEMENTS, after which it is ready for use in the reactor. When the uranium-235 content of the fuel falls below a threshold value the fuel is spent. It is removed either for disposal as waste or for reprocessing (e.g., in a THERMAL OXIDE RE-

PROCESSING PLANT). Reprocessing involves removing the fuel from its containers and extracting from it such usable material as remains to produce fuel that can be returned to a reactor and waste requiring final disposal (*see* HIGH-LEVEL WASTES, INTERMEDIATE-LEVEL WASTES, LOW-LEVEL WASTES, RADIOACTIVE WASTE).

nuclear fusion. A nuclear reaction in which the nuclei of light atoms are fused to form heavier atoms with the release of energy. Fusion reactions occur in the explosion of thermonuclear (hydrogen) bombs and supply the energy of stars. The same principle underlies the FUSION REACTOR. *Compare* NUCLEAR FISSION.

Nuclear Industry Radioactive Waste Executive. *See* NIREX.

Nuclear Installations Inspectorate (NII). In the UK, an organization established under the Nuclear Installations Acts, 1965 and 1969, to administer the licensing and operation of civil nuclear installations of all kinds, and the storage, handling and transport of all radioactive materials.

Nuclear Non-Proliferation Act. US legislation, signed by President Carter in 1978, that aims to limit the spread of nuclear weapon capability by restricting the supply of nuclear power facilities and fuels by the USA to other countries. It forbids the supply of nuclear fuels, reactors or reactor components to countries that do not possess nuclear weapons unless those countries agree not to explode nuclear devices or otherwise use nuclear materials in ways inconsistent with guidelines laid down by the INTERNATIONAL ATOMIC ENERGY AGENCY. The Act provided the basis for the Nuclear Non-Proliferation Treaty, which commits signatory governments to broadly similar objectives.

Nuclear Non-Proliferation Treaty. *See* NUCLEAR NON-PROLIFERATION ACT.

nuclear reactor. A device for generating heat by the controlled NUCLEAR FISSION of atoms of uranium-235, or by the nuclear fusion of light atoms (*see* FUSION REACTOR). The centre of a thermal fission reactor is the core, consisting of a MODERATOR into which a CRITICAL MASS of fuel elements and regulatory controls rods are inserted. The moderator (often water, HEAVY WATER or GRAPHITE) is used to slow down escaping neutrons emitted by the unstable uranium-235 atomic nucleus, without absorbing them, so reducing them to energies at which they are more likely to bombard other uranium atoms, causing further fissions and releasing more neutrons. Control rods (of cadmium or boron) absorb neutrons and so slow the reaction, and are raised or lowered into or out of the core to regulate the output of energy. The energy is released as heat and other radiation. The heat is transferred to a coolant (usually water or carbon dioxide) contained in a sealed system of pipes which pass through or around the core, and from the coolant to a separate supply of water which it converts into steam to operate the turbine generators. Individual reactor systems differ in the coolants, moderators and types of fuel they use. A fission reactor will continue to function until its fuel is depleted to a level at which its power output falls off to an unacceptable level, although the life of the fuel is extended, in all fission reactors, by the production of some plutonium, which is also consumed. Some reactors are dedicated to the production of plutonium for weapons production, and their fuel is removed before the plutonium can be consumed. A breeder reactor uses no moderator. Its core is surrounded by a 'blanket' of uranium-238 (the more common natural ISOTOPE of uranium and of no value as a nuclear fuel) which is bombarded by fast neutrons and so converted into plutonium, which can be used as a fuel in conventional fission reactors. Breeder reactors are sometimes called fast because they utilize fast neutrons. *See also* ADVANCED GAS-COOLED REACTOR, BOILING WATER REACTOR, CANADIAN DEUTERIUM–URANIUM REACTOR, DRAGON REACTOR, MAGNOX REACTOR, PRESSURIZED WATER REACTOR, RMBK REACTOR, STEAM-GENERATING HEAVY-WATER REACTOR.

Nuclear Regulatory Commission (NRC).

The US federal body that is responsible for licensing the construction and operation of nuclear reactors and with supervizing their observance of safety requirements.

nuclear waste. *See* HIGH-LEVEL WASTE, INTERMEDIATE-LEVEL WASTE, LOW-LEVEL WASTE, RADIOACTIVE WASTE.

nuclear winter. The theory, developed over a number of years and summarized in an article in *Science*, December 23 1983, by R. Turco, O. Toon, T. Ackerman, J. Pollack and C. Sagan, that a full-scale thermonuclear war would cause a global climatic cooling of 20–40°C. The theory assumed widespread fires that released black smoke and so shielded the ground surface from sunlight. Subsequent climatic models modified the original scenario, and the nuclear winter theory is not accepted by most climatologists.

nucleic acid. *See* DNA, RNA.

nucleolus. A compact structure inside a cell nucleus, which is concerned with the synthesis of ribosomal RNA.

nucleoprotein. A compound made up of a protein and a nucleic acid. *See also* DNA, RNA.

nucleotides. *See* DNA, RNA.

nucleus. In a cell, a membrane-bound body that contains the CHROMOSOMES and, in EUKARYOTIC cells, one or more nucleoli (*see* NUCLEOLUS).

nuclide. The nucleus of an atom, characterized by the number of protons and neutrons comprising the nucleus, and in some cases by the energy state of the nucleus.

nuée ardente. An incandescent cloud of gas and LAVA, violently erupted from some VOLCANOES that produce ACIDIC MAGMAS. Nuées ardentes flow downhill at great speeds, with the material buoyed up by expanding, heated, volcanic gases and trapped air, and may do great damage (e.g., the destruction of St Pierre, Martinique in 1902).

nullisomic. Applied to cells showing a kind of aneuploidy (*see* ANEUPLOID) in which both members of a homologous pair of CHROMOSOMES are absent from the diploid set.

numerical abundance. The number of individual plants of a particular species that are present in an area.

numerical forecast. A weather forecast that is made using a computer to calculate the expected events and basing the calculations on fundamental laws of physics and mechanics.

nunatak. A hill or mountain peak that is high enough to escape glaciation. There are many such hills in Greenland, projecting above the ice. The word is derived from the Eskimo language in which it means lonely peak. *See also* REFUGIUM.

nut. A dry, INDEHISCENT, single-seeded fruit, usually with a woody wall (e.g., acorn, beech, hazel (cob) nut).

nutation (circumnutation). A slow, spiralling growth movement in a plant caused by the continual alteration in the position of the most actively growing part of the organ.

nutrient budget. A budget drawn up for a living system in terms of nutrients.

nutrient stripping. The removal of nutrients from a substance. This may be the removal of nutrients from sewage to reduce rates of EUTROPHICATION in receiving waters, or it may be the incidental removal of nutrients from food during processing.

nyctinasty. *See* NASTIC MOVEMENT.

nymph. A young stage of an exopterygote (*see* EXOPTERYGOTA) insect (e.g., mayfly, dragonfly, earwig). It resembles the adult in many ways, but its wings are absent or incompletely developed, and it is sexually immature.

O

O. *See* OXYGEN.

OA. *See* OXYGEN ABSORBED.

oak. *See* QUERCUS.

oak-apple. *See* GALL.

oasis. *See* ARID ZONE.

oat. *See* GRAMINEAE.

obligate parasite. *See* PARASITISM

oblique strip fault. *See* FAULT.

obliterative shading (obliterative counter-shading). A very common type of animal coloration in which the dorsal surface is relatively dark and there is a gradual lightening towards the ventral side. This tends to obliterate the solid appearance of the animal, affording very effective camouflage.

obsidian. A glassy, ACIDIC IGNEOUS rock with a CONCHOIDAL fracture and a composition similar to that of RHYOLITE.

occluded front. The boundary of a WARM FRONT when it has been overtaken and lifted by an advancing COLD FRONT. When the air behind the front is included in the description it is called an occlusion.

occlusion. *See* OCCLUDED FRONT.

oceanic. Applied to parts of the oceans that are deeper than 200 metres (i.e. they lie seaward from the CONTINENTAL SHELF). *Compare* NERITIC.

oceanic ridge. *See* RIDGE.

oceanic trench. A narrow, elongated depression at the margin between an ocean basin and continental CRUST, characterized by large negative gravity anomalies (*see* ISOSTASY) and seismic activity. The trench marks the site of a destructive plate margin (*see* PLATE TECTONICS), where an oceanic plate plunges down into the MANTLE beneath an adjacent plate.

ocellus. A simple light receptor, composed of a small number of sensory cells and a single lens, found in many invertebrates. *See also* EYE.

ochre. Natural red and yellow pigments composed of LIMONITE.

octane number. A parameter that defines the quality of petrol (gasoline). High-octane fuel has better antiknock properties than low-octane fuel; knocking is premature combustion of the fuel, making a characteristic sound and causing loss of power and, eventually, damage to the engine. The octane number can be increased by further refining to increase the proportion of isooctane (2,2,4-trimethylpentane) in the fuel or, more cheaply, by the addition of tetraethyl lead.

Octapoda. An order of CEPHALOPODA whose members have eight tentacles and no shell (e.g., octopus, argonaut). *Compare* DECAPODA.

octave. The interval between two sounds, one of which has a FREQUENCY twice that of the other.

octopuses. *See* OCTAPODA.

octroi. A tax or duty raised on goods entering a town or area. It was formerly used in Europe and is still used in India.

Odonata (dragonflies, damselflies). An order of large, often highly coloured insects (division: EXOPTERYGOTA) whose members have two similar pairs of membranous wings. The adults are strong fliers, preying on insects which they seize while on the wing. The NYMPHS are aquatic and predatory with a lower lip (labium) modified as an extensible, prehensile mask used to catch prey.

Odontoceti (toothed whales). A suborder of the CETACEA that includes the sperm whale, killer whale, narwhal, porpoises and dolphins, all of which are predatory mammals that feed on fish and aquatic invertebrates.

OECD. *See* ORGANIZATION FOR ECONOMIC COOPERATION AND DEVELOPMENT.

oenology. *See* ENOLOGY.

oestrogens (estrogens). Natural or synthetic steroid HORMONES, which stimulate the growth and activity of the reproductive system and the development of SECONDARY SEXUAL CHARACTERS in female vertebrates. Oestrogens are secreted by the ovary and, in small amounts, by the adrenal gland, PLACENTA and the testes.

oestrous cycle (estrous cycle). A reproductive cycle that occurs during the breeding season in many female mammals, provided they are not pregnant. One phase of the cycle is oestrus (heat), during which ovulation occurs and there is an urge to mate. OESTROGEN and PROGESTERONE levels vary according to the phase of the cycle, which is controlled by HORMONES secreted by the PITUITARY GLAND. The menstrual cycle of some primates (e.g., humans, apes and Old World monkeys) is a modified oestrous cycle in which there is no well-marked oestrous phase, and destruction of the uterine lining produces bleeding.

offensive industry. Industry that is noxious or offensive by reason of the processes it employs, the materials it uses, its products or the wastes it discharges.

ohmΩ. The derived SI unit of resistance; the resistance between two points of a conductor when a constant difference of potential of 1 volt applied between the points produces in the conductor a current of 1 ampere. It is named after Georg Ohm (1787–1854).

oil, crude. *See* CRUDE OIL.

oil field. An area underlain by one or more oil (or gas, giving gas field) POOLS which are adjacent or overlapping in plan view, and which are related to a single geological feature that is either structural or stratigraphic. The size of an oil field refer to the volume of recoverable oil (the reserves), which is given either in BARRELS or tonnes (depending on the gravity of the oil, 1 tonne = 6.5–7.5 barrels).

oil pool. *See* POOL.

oil shale. A black, brown or green SHALE, more or less impregnated with KEROGEN, from which gaseous and liquid petroleum can be obtained by DESTRUCTIVE DISTILLATION at 300–400°C. With increasing kerogen content, oil shales grade imperceptibly through BOGHEAD COALS into CANNEL COALS.

oil slick. Oil, discharged naturally or by accident or design, that floats on the surface of water as a discrete mass and is moved by wind, currents and tides.

okta. *See* CLOUD AMOUNT.

old man's beard. *See* CLEMATIS VITALBA.

Old Red Sandstone (ORS). The continental FACIES of the DEVONIAN System. *Compare* NEW RED SANDSTONE.

Oleaceae. A family of Dicotyledoneae, some members of which are of economic importance. These include ash (*Fraxinus*), olive (*Olea europea*), which yields oil from its fruits (a DRUPE), and lilac (*Syringia vulgaris*).

Olea europea. *See* OLEACEAE.

Oligocene. A subdivision of the CENOZOIC Era, usually ranked as an EPOCH, that follows the EOCENE and is thought to have lasted from about 38 to 26 Ma. Oligocene also refers to rocks deposited during this time, called the Oligocene Series.

Oligochaeta. An order of terrestrial and freshwater ANNELIDA whose members (e.g., earthworm) have few bristles (chaetae), no muscular, lateral projections (parapodia) along the body and no specialized head appendages. Oligochaetes are HERMAPHRODITE, and the eggs develop inside a cocoon. Earthworms are very important in the maintenance of soil fertility because their tunnels aerate and drain the soil, and some species continually bring to the surface fresh soil containing HUMUS. There may be up to 7.4 million earthworms per hectare of land (3 million per acre). In Western Europe they bring about 6 kilograms of soil per square metre (11 pounds per square yard) to the surface each year.

oligohaline. Applied to slightly BRACKISH waters whose salinity is 0.5–5°/oo. *Compare* EUHALINE, FRESH WATER, MESOHALINE, POLYHALINE.

oligophagous. Applied to an animal that feeds on only a few kinds of food.

oligosaprobic. Applied to a body of water in which organic matter is decomposing very slowly and in which the oxygen content is high. *Compare* CATAROBIC, MESOSAPROBIC, POLYSAPROBIC. *See also* SAPROBIC CLASSIFICATION

oligotrophic. (1) Applied to MIRES (e.g., RAISED BOGS and BLANKET BOGS) and freshwater bodies that are poor in plant nutrients. Oligotrophic water bodies are unproductive, and their waters are clear because planktonic (*see* PLANKTON) organisms are sparse. (2) Applied to any lake whose HYPOLIMNION does not become depleted of oxygen during the summer. Even moderately productive lakes may be classified as oligotrophic if they are deep enough for the hypolimnion never to become deoxygenated. *Compare* DYSTROPHIC, EUTROPHIC, MESOTROPHIC. *See also* EUTROPHICATION.

olive. *See* OLEACEAE.

olivine. A group of rock-forming SILICATE MINERALS which range in composition from Mg_2SiO_4 to Fe_2SiO_4. Olivines occur mostly in BASIC and ULTRABASIC rocks and are easily altered by WEATHERING or HYDROTHERMAL processes to SERPENTINE minerals.

ombrogenous (ombrotrophic, ombrophilous). Originating as a result of wet climates; applied to acid peatlands (i.e., BOGS, wet HEATHS) that receive water from precipitation rather than from the ground. *Compare* SOLIGENOUS.

ommatidia. *See* EYE.

omnivore. An animal that eats food of both plant and animal derivation (e.g., badger, humans).

onager (*Equus hemippus onager*). A wild ass of south-west Asia that was hunted for its meat in PALAEOLITHIC times.

onchocerciasis. A disease, prevalent in equatorial Africa and the American tropics, that causes impaired vision, leading eventually to blindness and fibrous tumours or nodules, most commonly on the head and shoulders. The disease is caused by the parasitic roundworm *Onchocerca volvulus* (*see* NEMATODA), transmitted to humans by several species of blackfly.

onchosphere (oncosphere). An early stage in the life history of a tapeworm (*see* CESTODA), consisting of a spherical, chitinous (*see* CHITIN) shell which contains a six-hooked (hexacanth) EMBRYO. The onchosphere develops from the egg and precedes the cysticercus (*see* BLADDERWORM) stage.

one-crop economy. The dependence of a nation or region largely on a single economic activity, thus making it vulnerable to changes that affect the production or marketing of its limited range of commodities. Some Arab countries are dependent on oil revenues for most of their foreign earnings, Iceland is heavily dependent on fishing, and in Central America a number of countries (especially Honduras, Costa Rica and Panama, the so-called banana republics) derive about one-third of their export earnings from the production and export of bananas.

onion weathering. *See* SPHAEROIDAL WEATHERING.

ontogeny (ontogenesis). The sequence of development during the whole life history of an organism.

Onychophora (walking worms). A small group of terrestrial, worm-like, many-legged animals, usually regarded as ARTHROPODA, but with a thin, unjointed CUTICLE and other worm-like features (*see* ANNELIDA). A single living genus (*Peripatus*) lives beneath stones and dead bark in warm climates.

onyx. *See* CHALCEDONY.

oocyte. A cell that gives rise by MEIOSIS to an animal ovum (egg). During this process a primary oocyte usually gives rise to one large secondary oocyte and a cell with very little CYTOPLASM (i.e. polar body). The second meiotic division produces the ovum and another polar body.

oogamy. The production of two kinds of GAMETES. The female gamete (egg) is large and non-motile, the male one small and motile.

oogenesis (ovogenesis). The cell divisions and changes that result in the formation of an ovum (egg). *See also* OOCYTE.

oogonium. (1) The female sex organ of some FUNGI and ALGAE, inside which large, non-motile female GAMETES (eggs, oospheres) are formed. (2) A cell in the ovary

of an animal that gives rise to OOCYTES.

oolite. A SEDIMENTARY ROCK composed predominantly of concentrically layered sphaeroidal grains (ooliths). Ooliths forming at the present day are CALCAREOUS and are found in tropical regions where currents roll sand-sized particles around in evaporating shallow water. Some ancient ooliths are dolomitic (*see* DOLOMITE) or siliceous (*see* SILICA) or are composed of various ores of iron.

ooliths. *See* OOLITE.

oosphere. The female GAMETE of FUNGI and some plants, produced in an OOGONIUM, which develops into an OOSPORE.

oospore. A resting SPORE with a thick wall, formed from an OOSPHERE after fertilization.

open burning. The outdoor burning of wastes (e.g., lumber, sawdust, scrapped cars, tyres, textiles, etc.) and the burning of open waste dumps (refuse tips). Inevitably, it is a source of atmospheric pollution. *Compare* INCINERATOR.

open-cast mining. The working of coal seams or mineral deposits (e.g., KAOLIN) that lie close to the surface by stripping off the OVERBURDEN rather than by tunnelling. *Compare* STRIP MINING.

open-cut mining. *See* STRIP MINING.

open community. A community that is readily colonized by other organisms because some NICHES remain unoccupied.

open country. In UK planning, as defined in the Countryside Act, 1968, an area consisting wholly or predominantly of mountain, moor, down, cliff, foreshore, beech, sand dunes, woodlands, rivers or canals.

open fold. *See* FOLD.

open-hearth furnace. A REVERBERATORY FURNACE, containing a cup-shaped hearth, that is used for melting and refining pig iron,

iron ore and scrap for steel production. A large amount of dust from ore and other materials, and splashings from SLAG, are carried away in the noxious stream of waste gas.

operon. A group of GENES that control the synthesis of one ENZYME system.

Ophidia. *See* SQUAMATA.

Ophiuroidea (brittle stars). A class of ECHINODERMATA whose members have flattened, star-shaped bodies with arms sharply marked off from the disc. They move mainly by muscular movements of the arms and eat small CRUSTACEA, MOLLUSCA and detritus. *See also* ASTEROIDEA.

Opiliones. *See* PHALANGIDA.

Opisthobranchia. *See* GASTROPODA.

opportunity cost. The real cost of satisfying a want, expressed in terms of the cost of sacrificing an alternative activity. If capital (or labour) might earn a particular sum in some activity other than the one in which it is presently engaged, then that is the opportunity cost of the present activity. It may or may not be a net cost when present actual income is set against it.

optical activity. The power of a substance to rotate the plane of polarized light transmitted through it. All living systems show this property, as do all asymmetric molecules.

optimum population. The number of individuals that can be accommodated within an area to the maximum advantage of each of them. The concept may be interpreted ecologically or economically.

Opuntia. A genus of fleshy-stemmed CACTACEAE, most of which have small fleshy leaves that drop off early. Prickly pear (*O. vulgaris*) has become a troublesome weed since its introduction into Australia. Some species are used for hedging, others as food

for cochineal insects. The fruits of some are edible.

OR. *See* ORIENTATION RESPONSE.

oraches. *See* CHENOPODIACEAE.

orang-utan. *See* ANTHROPOIDEA.

Orchidacea (orchids). A very large cosmopolitan family of MONOCOTYLEDONEAE, all of which are perennial herbs with tuberous roots and vertical stocks or RHIZOMES. Epiphytic (*see* EPIPHYTE) species, often with aerial roots, are an important feature of tropical vegetation. A few species are SAPROPHYTES (e.g. bird's nest orchid, *Neottia nidus-avis*). MYCORRHIZAE occur almost invariably. Pollination is usually by insects, to which pollen masses (pollinia) become attached. The pods of the genus *Vanilla* are used for flavouring. Many orchids are cultivated for their flowers.

order. *See* CLASSIFICATION.

Ordovician. The second oldest PERIOD of the PALAEOZOIC Era, usually taken as beginning some time between 500 and 530 years BP. Ordovician also refers to the rocks formed during this time, called the Ordovician System, and in the UK is divided into five SERIES (Arenigian, Llanvirnian, Llandeilian, Caradocian, Ashgillian). The Ordovician probably lasted 65–70 million years. Geologists in most other countries place the Tremadocian in the Ordovician, but in the UK it is placed in the CAMBRIAN. The Ordovician is zoned (*see* INDEX FOSSIL) using GRAPTOLITES.

ore. A MINERAL or mineral aggregate, more or less mixed with GANGUE, that can be worked and treated at a profit.

ore body. A mass of ORE that is capable of being worked. Metalliferous ore bodies can be classified into: (a) IGNEOUS segregations formed by MAGMATIC DIFFERENTIATION; (b) intrusive (see INTRUSION) bodies (e.g., PEGMATITE, DIAMOND pipes, HYDROTHERMAL veins); (c) metasomatic (*see* METASOMATISM) bodies;

(d) sedimentary ore bodies (e.g., BANDED IRONSTONES, LATERITES, PLACER deposits).

ore mineral. A MINERAL from which a useful metal may be extracted.

organ. A part of a plant or animal that forms a structural unit (e.g., leaf, heart, skin).

organelle. One of the permanent structures with specialized functions within cells.

organic farming (biological husbandry). Farming without the use of industrially made (i.e. 'artificial') FERTILIZERS or PESTICIDES, according to principles laid down by Sir Albert Howard, Lady Eve Balfour and others, and as modernized and interpreted in Britain by the Soil Association.

Organization for Economic Cooperation and Development (OECD). An intergovernmental institution concerned with a wide range of issues that have economic or developmental implications for member states, but with many interests in the world in general. The members, all developed countries, are: Australia, Austria, Belgium, Canada, Denmark, Finland, France, West Germany, Greece, Iceland, Ireland, Italy, Japan, Luxembourg, The Netherlands, New Zealand, Norway, Portugal, Spain, Sweden, Switzerland, Turkey, the UK and the USA. Yugoslavia has a special status entitling it to participate in many OECD activities.

organochlorines (chlorinated hydrocarbons). Organic compounds containing chlorine. Many organochlorines have biocidal properties (*see* BIOCIDE) and are used as the active ingredients for pesticides with a high persistence, due to their chemical stability and low solubility in water. Pesticide organochlorines include DDT, ALDRIN and LINDANE. These have proved very effective in the control of insect pests, especially of insect VECTORS of disease in tropical regions, but some insects have become immune to them. They persist in sublethal doses in animals, where they accumulate in fatty tissue. Low concentrations of residues may be increased as they pass up the FOOD CHAIN, and toxic levels may thus occur in the bodies of predators, causing death or sterility (e.g., in some birds). Because of these side effects, the use of many organochlorine insecticides has been limited in most countries. *See also* CHLORDANE, ENDOSULFAN, ENDRIN, HEPTACHLOR, POLYCHLORINATED BIPHENYLS.

organogeny (organogenesis). The formation of ORGANS during the development of an individual organism.

organomercury fungicides. Poisonous compounds, containing MERCURY compounds, used as sprays or seed treatments to control fungal diseases in crops.

organophosphorus pesticides. A group of non-persistent PESTICIDES of varying vertebrate toxicity that appear to inhibit the action of cholinesterase (*see* ACETYLCHOLINE), an enzyme that cancels chemical messages within the nervous system. Because these pesticides break down rapidly, danger to wildlife is of short duration compared with that caused by ORGANOCHLORINES. *See also* DEMETRON-S-METHYL, DISULFOTON, FENITROTHION, GLYPHOSPHATE, MALATHION, PARATHION.

Oriental Region. The ZOOGEOGRAPHICAL REGION, forming part of ARCTOGEA and consisting of Asia south of the Himalayas and Indonesia apart from Celebes and New Guinea.

orientation response (OR). The physiological response of an animal to sudden changes in its immediate environment (e.g., pricking up its ears at a new sound, dilating its pupils at a new sight, etc.). The response represents the defensive mechanisms of an animal reacting to novelties that might threaten it.

original (prototype). In modelling, that which the MODEL seeks to simulate or describe.

Origin of Species by Means of Natural Selection, The. See DARWIN, CHARLES ROBERT.

ornithophily. Pollination by birds (e.g., humming birds).

orogenesis. *See* DIASTROPHISM.

orogenic belt. An elongated region of intensely deformed rocks, produced during an OROGENY. Young orogenic belts are characterized by fold mountains, but with age erosion strips off the higher levels to expose metamorphic rocks (*see* METAMORPHISM) and BATHOLITHS.

orogeny. A period of mountain building, involving the formation of FOLDS and FAULTS, METAMORPHISM and IGNEOUS INTRUSION. Most orogenies affected GEOSYNCLINES and lasted tens of millions of years, with several maxima. The major orogenies affecting Europe were the Caledonian, Hercynian and Alpine.

orographic lifting. *See* CONVECTION.

orpiment. *See* ARSENIC.

ortet. An organism that gives rise to a CLONE. *Compare* RAMET.

orthoclase. *See* FELDSPAR.

orthogenesis (directive evolution, orthoselection). An evolutionary trend that is thought to be caused by the inherent tendency of a group to develop in a particular way.

orthograde. Applied to animals (e.g., higher apes and humans) that walk erect, on two legs.

Orthoptera (Saltatoria, grasshoppers, crickets, locusts). An order of terrestrial insects (division: EXOPTERYGOTA) with biting mouthparts, narrow, thickened forewings covering membranous hind wings and hind legs usually enlarged for jumping. Most are poor fliers, and some are wingless. *See also* ACRIDIDAE.

orthoquartzite (quartzarenite, quartzose sandstone, silicaceous sandstone). A SANDSTONE composed of detrital QUARTZ grains cemented together by SILICA, such that the silica content (i.e. grains plus cement) is at least 90 percent (95 percent in some classifications). Quartzite is used loosely as a synonym, but this term also includes METAQUARZITE.

orthoselection. *See* ORTHOGENESIS.

Orycteropodidae. *See* TUBULIDENTATA.

Oslo Convention. *See* INTERGOVERNMENTAL MARITIME CONSULTATIVE ORGANIZATION.

osmoregulation. Control of the water content of an organism. *See also* HOMOIOSMOTIC, OSMOSIS, POIKILOSMOTIC.

osmosis. The passage of water through a SEMIPERMEABLE MEMBRANE from a solution of low concentration (the hypotonic solution) to one of higher concentration (the hypertonic solution). This tends to continue until the solutions are of equal concentration (isotonic) on either side of the membrane. Because cell membranes (PLASMA MEMBRANES) are more or less semipermeable, osmosis plays a part in the movement of water in living organisms. *See also* OSMOTIC PRESSURE.

osmotic pressure. The pressure that develops when a pure solvent is separated from a solution by a membrane that allows only solvent molecules to pass through it.

ossicles, auditory. *See* EAR.

Osteichthyes (bony fishes). The class of fishes that includes the ACTINOPTERYGII and CHOANICHTHYES. *Compare* CHONDRICHTHYES.

osteolepid. Having an outer skin made of armoured bone-like cosmoid scales. The osteolepid CROSSOPTERYGII, from which the AMPHIBIA are descended, lived in freshwater swamps and seas during the DEVONIAN and CARBONIFEROUS. They had functional lungs, probably assisted by gills, and paired pectoral and pelvic fins with bony elements, similar to amphibian limbs.

Ostracoda. A small subclass of CRUSTACEA whose members have a bivalve (*see* VALVE) carapace enclosing the body and reduced trunk limbs. They live in fresh and salt water and may be PELAGIC or bottom-dwellers. *Cypris* is common in ponds.

otocyst. *See* STATOCYST.

otolith. *See* EAR.

outbreeding. *See* EXOGAMY.

outcrop. (1) The exposure of a rock body or a rock surface (e.g., a FAULT or JOINT). (2) The total area, especially as shown on a geological map, in which a rock body or rock surface lies at the Earth's surface, although possibly concealed by a thin superficial deposit such as soil. The verb is to outcrop or to crop out.

outfall sewer. A pipe or conduit that transports sewage, raw or treated, to a final point of discharge.

outlier. *See* INLIER.

outwash deposit (fluvioglacial deposit). A stratified deposit of material that has been eroded by GLACIERS and reworked by meltwater. Outwash deposits are one component of GLACIAL DRIFT.

outwash plain. A stratified deposit of material initially eroded by GLACIERS and then transported and deposited by meltwaters beyond the glacial margin. *See also* OUTWASH DEPOSITS.

ovarian follicle (graafian follicle). An envelope of cells that surrounds the developing egg in many animals. In vertebrates it secretes oestrogens. In most mammals the follicle becomes the CORPUS LUTEUM after bursting at ovulation.

overburden. The material lying above a mineral deposit that must be removed in order to work the deposit.

overstorey (overwood). The uppermost layer in a woodland, usually the trees that form a canopy.

overtone. *See* HARMONIC.

overturn. The circulation of water and nutrients in a lake when the thermal strata become mixed by the wind during the spring and autumn. *See also* THERMAL STRATIFICATION.

overturned fold. A FOLD in which one limb has been tilted beyond the vertical, so that both limbs dip in the same direction. Such a fold will have an inclined AXIAL PLANE.

overwood. *See* OVERSTOREY.

oviparous. Applied to animals (e.g., frogs) that lay eggs in which the EMBRYOS are at the most only slightly developed. *Compare* OVOVIVIPAROUS, VIVIPAROUS.

ovogenesis. *See* OOGENESIS.

ovotestis. An organ present in some HERMAPHRODITE animals (e.g., snails) that produces both eggs and sperm.

ovoviviparous. Applied to animals (e.g., vipers) that retain their eggs within the maternal body until embryonic development is complete. During all or most of its development the EMBRYO is separated from the mother's body by the EGG MEMBRANES. *Compare* OVIPAROUS, VIVIPAROUS.

ovulation. The release of a ripe egg from an ovary.

ovule. The plant structure that develops into a seed after fertilization. *See also* EMBRYO SAC.

ovum. *See* GAMETE.

owls. *See* STRICHFORMES.

ox-bow lake. A crescent-shaped lake left at the site of a former river MEANDER.

Oxfordian. A STAGE of the JURASSIC System.

oxic. Applied to a soil layer from which much of the silica that was combined with iron and alumina has been leached (*see* LEACHING), leaving SESQUIOXIDES of iron and aluminium and 1:1 lattice silicate clays. *See also* SOIL CLASSIFICATION, SOIL HORIZON.

oxidases. Enzymes (DEHYDROGENASES) that catalyze oxidative reactions in which hydrogen is removed from a substrate and combines with free oxygen.

oxidation. The combination of oxygen with a substance or the removal of hydrogen from it, or, more generally, any reaction in which an atom loses electrons (e.g., the change of a ferrous iron, Fe^{2+}, to a ferric ion, Fe^{3+}). The removal of oxygen from a substance, or the addition of hydrogen to it or, more generally, any reaction in which an atom gains electrons is called reduction.

oxidation pond. A shallow lagoon or basin in which waste water is purified by sedimentation accompanied by AEROBIC and ANAEROBIC treatment.

oxidative phosphorylation. The combination of ADP with phosphate to form ATP. RESPIRATION provides the energy needed for this reaction.

Oxisols. *See* SOIL CLASSIFICATION.

oxygen (O). An element; an odourless gas that forms approximately 21 percent of the atmosphere and is essential to most forms of life. It is the most abundant of all elements in the Earth's crust, occurring in rocks, water and air. Industrially it is mixed with other gases (e.g., acetylene, hydrogen) to produce high-temperature flames for welding and metal cutting. $A_r = 15.9994$; $Z = 8$.

oxygen absorbed (OA). A test to measure the amount of dissolved oxygen in a sample of water by observing the quantity of oxygen absorbed by permanganate over fours hours at 27°C.

oxygen sag curve. A line drawn on a graph to trace the dissolved oxygen content of water over a period of time in order to measure the capacity of the water to purify itself following the discharge into it of an effluent.

oxytocin. A HORMONE secreted by the PITUITARY GLAND that stimulates uterine contractions and the secretion of milk in mammals.

oysters. *See* LAMELLIBRANCHIATA.

ozone (O_3). A form of OXYGEN in which the molecule consists of three atoms rather than two. It is very reactive chemically and is an irritant to eyes and respiratory tissues. It occurs naturally at about 0.01 ppm of air; 0.1 ppm is considered toxic. As a pollutant it is harmful to plants and is involved in the formation of photochemical SMOG. It is formed by the recombination of oxygen following its ionization (e.g., by electrical discharges or ultraviolet radiation). *See also* OZONE LAYER.

ozone layer (ozonosphere). A layer of the atmosphere, about 20ñ50 kilometres above the surface, in which the concentration of OZONE is higher than elsewhere in the atmosphere due to the dissociation and reformation of oxygen molecules exposed to high-frequency ultraviolet radiation. The process absorbs energy from the radiation, so preventing it from penetrating the atmosphere further.

ozonometer. An instrument for measuring the amount of OZONE present in the air.

ozonosphere. *See* OZONE LAYER.

P

P. (1) *See* PHOSPHORUS. (2) *See* PHON.

p. *See* PICO-.

P₁. *See* PARENTAL GENERATION.

Pacific North American Floral Region. The part of the HOLARCTIC REALM that comprises North America west of the Rocky Mountains from southern Alaska south to the Mexican border.

Pacific South American Floral Region (Andean Floral Region). The region of the NEOTROPICAL REALM that comprises the Andes and the coastal strip of South America to their west from Ecuador to central Chile.

package treatment plant. A transportable unit for sewage treatment that is capable of achieving a desired quality of effluent. It is sometimes used as a temporary measure prior to the installation of permanent treatment facilities.

Packwood–Magnusson Amendment. The amendment to the US Fishery Conservation and Management Act, 1979, which allows the President to restrict or forbid fishing in US waters by any country whose activities weaken the authority of the INTERNATIONAL WHALING COMMISSION. *See also* PELLY AMENDMENT.

paedogamy. *See* AUTOGAMY.

paedogenesis. Reproduction in animals that have become sexually mature although they are still in a young stage; it occurs in many salamanders. For example, insufficient secretion of THYROXINE by the thyroid (possibly due to the high iodine content of the water in which it lives) leads to reproduction in the tadpole stage of the axolotl (*Amblystoma mexicanum*). *Compare* NEOTONY.

paedomorphosis. The evolutionary process by which youthful characters are introduced into the line of adults. *See also* NEOTONY, PAEDOGENESIS.

PAH. *See* POLYCYCLIC AROMATIC HYDROCARBONS.

pahoehoe. A Hawaiian word used to describe a basaltic lava with a smooth, billowy or ropy surface, formed by a flow fed from below through lava pipes. If the flow is turbulent pahoehoe may be converted to AA.

palade granules. *See* RIBOSOMES.

Palaearctic Region. The ZOOGEOGRAPHICAL REGION, forming part of ARCTOGEA and consisting of Europe, North Africa, Arabia, apart from the south-western corner, and Asia as far south as the Himalayas.

palaeo- (paleo-). A prefix derived, with no change in sense, from the Greek *palaios* meaning ancient.

Palaeocene. The lowest division of the CENOZOIC Era, usually ranked as an EPOCH, which lasted from about 65 to 54 Ma. Palaeocene also refers to rocks deposited during this time, called the Palaeocene Series. Some authors do not recognize the Palaeocene and consider the EOCENE to extend all the way back to the end of the CRETACEOUS.

palaeoecology. The study of past environments, their flora and fauna and the interrelationships among them.

Palaeogene. The earlier of the two PERIODS of the CENOZOIC Era, which includes the PALAEOCENE, EOCENE and OLIGOCENE Epochs. Palaeogene also refers to rocks formed during this time. The terms Palaeogene and Neogene are restricted mainly to European usage, and some authors restrict the Palaeogene to the Eocene and Oligocene only.

palaeogenesis. The persistence of normally embryonic characteristics into a later stage. *See also* NEOTONY.

Palaeolithic (Early Stone Age). The stage in the development of a human society when its members engage in no settled agriculture, obtaining their food by hunting, fishing and the gathering of wild plants. The domestication of crop plants and animals began probably about 11 000 years BP, and so this date marks the beginning of the transition from Palaeolithic to NEOLITHIC culture in the regions affected, although other peoples continued to live by hunting and gathering for much longer, and some survive to the present day. Palaeolithic peoples used unpolished stone tools. *Compare* MESOLITHIC.

palaeomagnetism. GEOMAGNETIC INDUCTION in the geological past of the Earth. It is studied by determining the magnitude and direction of magnetic fields among iron compounds and in rocks at the time of their formation. When molten, ions within rocks are free to move and align themselves with the prevailing magnetic field. When the rock hardens they are no longer free to move, and so retain the alignment they have acquired. Thus information can be deduced from the examination of them regarding variations in the positions of the Earth's magnetic poles at times in the past which are determined by the RADIOMETRIC AGES of the rocks. Such studies have shown that the Earth's magnetic field reverses its polarity (i.e. the North pole becomes the South, and vice versa) periodically.

palaeontology. The study of past life forms.

Palaeotropic Realm. The FLORAL REALM covering Africa, apart from the area bordering the Mediterranean Sea and the tip of southern Africa, but including Madagascar, Arabia and Asia, and the Pacific and Indian Ocean islands from the Himalayas south to New Guinea, but excluding Australia and the whole of Central and South America. The Realm is divided into 14 regions: Macaronesian, Saharo-Arabian, Sudanian-Sindian, Ethiopian, West African, East African, Madagascan, Indian, South-East Asian, Malaysian–Papuan, New Caledonian, Fijian, Polynesian and Hawaiian.

Palaeozoic. One of the ERAS of GEOLOGICAL TIME, occurring between the PRECAMBRIAN and the MESOZOIC, and comprising the CAMBRIAN, ORDOVICIAN, SILURIAN, DEVONIAN, CARBONIFEROUS (Mississippian and Pennsylvanian in North America) and PERMIAN Periods. Palaeozoic also refers to the rocks formed during this time. The Lower Palaeozoic comprises the Cambrian, Ordovician and Silurian; the Upper Palaeozoic comprises the Devonian, Carboniferous and Permian.

paleo-. *See* PALAEO-.

paleosere. The EOSERE of the PALAEOZOIC Period.

palingenesis. The recapitulation of adult stages of ancestors during the early development of descendants (e.g., the formation of gill pouches in the embryos of mammals). The recapitulation theory, or biogenetic law advanced by HAECKEL, ERNEST HEINRICH, (encapsulated in the saying 'ontogeny repeats phylogeny') is now largely discredited, although the embryonic stages of related organisms do resemble each other.

palisade mesophyll. *See* MESOPHYLL.

Palmae (Arecaceae, palms). A tropical and subtropical family of MONOCOTYLEDONEAE, most of which have very large fan-shaped or feathery leaves. Some are rhizomatous (*see* RHIZOME), a few are climbers with thin reed-like stems (e.g., those that yield rattan), whereas others have tall stems, reaching 60 metres in some species, capped by a crown of leaves. The fruit is a BERRY (e.g., date) or DRUPE (e.g., coconut). The palms are of great economic importance, yielding food, fibres, oil, timber, etc. *Phoenix dactylifera* is the date palm, *COCOS NUCIFERA* the coconut palm, species of *Metroxylon, Caryota* and *Arenga* yield sago from their pith and palm sugar from their sap. *Areca* is the source of the betel nut, and the fruit of *Elaeis* yields palm oil, one of the most important of all vegetable oils, which is edible, rich in VITAMINS A and B, and used for cooking and in the manufacture of soaps, and the leaves of *Copernicia* yield a valuable wax.

palms. *See* PALMAE.

palynology (pollen analysis). The study of POLLEN and SPORES. Pollen grains have walls of a material that is highly resistant to all forms of decomposition and have shapes characteristic of the families, genera or, in some cases, even species of the plants from which they came. The identification of fossil pollen contributes to the correlation of strata, particularly those bearing coal or oil, and knowledge of the composition of past plant communities allows deductions to be made regarding the environment and especially the climate in which they lived.

pampas. *See* GRASSLAND.

pampero. *See* MISTRAL.

PAN. *See* PEROXYACETYL NITRATES.

panclimax (panformation). Two or more related CLIMAXES for FORMATIONS that exist under similar climatic conditions and comprise similar life forms and genera, or DOMINANTS. Panclimaxes may arise from common origins in an EOCLIMAX.

Pandanaceae. A family of tropical MONOCOTYLEDONEAE, mainly confined to coasts and marshes, whose members have long, narrow leaves and tall, usually twisted stems, supported by aerial roots. Some screwpines (*Pandanus* species) yield edible fruits (e.g., Nicobar breadfruit), and the leaves are used for weaving and thatching.

Pandionidae. *See* FALCONIFORMES.

panemone. A vertical axis windmill that consists of two or more concave or flat surfaces designed to rotate about the axis. The spinning signs used for advertising purposes outside shops and filling stations are panemones, as are cup ANEMOMETERS and SAVONIUS ROTORS.

panformation. *See* PANCLIMAX.

Panhonlib group. A group of those ships that are registered in and fly the flag of Panama, Honduras or Liberia, the so-called flags of convenience. There are many such ships, but most of their owners are nationals of other countries. The *TORREY CANYON* and *AMOCO CADIZ* belonged to the group, both being registered in Liberia. The *Torrey Canyon* was owned and chartered by US nationals, and manned by an Italian captain and crew.

panicle. A branched INFLORESCENCE.

Panos Institute. An international policy studies and information institute founded in 1986 by a group of individuals who formerly worked for EARTHSCAN. Panos undertakes information research on a wide range of subjects relating economic development to environmental conservation.

Pantopoda (Pycnogonida, sea-spiders). A small class of marine ARACHNIDA with much elongated legs and an extra (fifth) pair of legs in front, used by males for carrying the eggs. Many species are ectoparasites (*see* PARASITISM) of sea-anemones.

pantothenic acid. A VITAMIN of the B group that is essential for the formation of a COENZYME in many organisms, including vertebrates and some BACTERIA.

Pantotheria. A group of JURASSIC mammals that probably gave rise to both the MARSUPIALIA and EUTHERIA.

Papaveraceae. A family of mainly herbaceous DICOTYLEDONEAE, most of which have conspicuous nectarless flowers which are visited by insects for their pollen. Opium is obtained from the poppy (*Papaver somniferum*) by cutting notches in the half-ripened capsules and collecting the hardened exudation of latex.

paper chromatography. *See* CHROMATOGRAPHY.

papilionaceous. Applied to flowers that are shaped and coloured in ways reminiscent of butterflies. Examples include those characteristic of the pea family. *See also* LEGUMINOSAE.

Papilionoidea. *See* LEPIDOPTERA.

paraffin oil. See KEROSENE.

paralic. Applied to sediments formed on the landward side of a coast in an area subject to marine incursion. Paralic is also applied to the environment of such an area. *Compare* LIMNIC.

parameter. A value that is constant and characteristic of a particular set of data (e.g., the mean height of all English women).

paramorph. A variant within a species.

Parano-Amazonian Floral Region. The region of the NEOTROPICAL REALM that comprises Brazil and Bolivia.

paraquat. An organic herbicide that kills plants on contact and is used to control many broadleaved weeds and grasses. It is poisonous to humans, and the accidental ingestion of large amounts, usually by children, has caused some deaths. Used correctly it is rapidly adsorbed on soil particles and so is inactivated.

parasematic coloration. Coloration that protects an animal by diverting the attack of a predator towards a non-vital part of the body (e.g., eye patterns on the wings of some butterflies).

parasite. *See* PARASITISM.

parasitism. A close association between living organisms of two different species during which one partner – the parasite – obtains food at the expense of the other – the host. Many animals and a few flowering plants (e.g., dodder) are parasites. Pathogenic (*see* PATHOGEN) BACTERIA and FUNGI are parasitic on plants and animals, and VIRUSES parasitize plants, animals and bacteria. Parasitism is sometimes used in a wider sense to include COMMENSALISM, MUTUALISM and SYMBIOSIS. Ecoparasites are each adapted to a specific host or closely allied group of hosts. Ectoparasites live for long or short periods on the outside of their hosts. Endoparasites live inside their hosts. Hemiparasites (e.g., yellow rattle) obtain only part of their food supply from their hosts. Kleptoparasites (cleptoparasites) (e.g., skuas) habitually steal food caught by animals of another species. Brood parasites (e.g., cuckoos) entrust the rearing of their young to animals of a different species. Hyperparasites (superparasites) parasitize other parasites. Obligate parasites can live only as parasites. Facultative parasites can live as SAPROPHYTES, but live parasitically under certain circumstances. Xenoparasites either infest hosts not normal to them or are capable of invading only an injured organism. Autoecious parasites spend their whole life cycle on an individual host. Heteroecious (heteroxenous) parasites need more than one host to complete their life cycle or are not host-specific. Metoecious (metoxenous) parasites are not host-specific. Ametoecious (monoxenic) parasites are host-specific. Tropoparasites are obligate parasites during some part of their life cycle, but live non-parasitically for the rest of the time (e.g., the swan mussel, *Anodonta cygnea*, is parasitic as a larva in freshwater fishes).

parasitoid. An animal whose mode of life is intermediate between PARASITISM and predation (*see* PREDATOR) because the parasitoid ultimately kills its host. An example is ichneumons (ACULEATA) which lay eggs inside the larvae or eggs of other insects).

parasympathetic nervous system. *See* AUTONOMIC NERVOUS SYSTEM.

parathion (thiophos). The first ORGANO-PHOSPHORUS PESTICIDE to be used widely in UK agriculture, having been introduced in 1944. It is used to control aphids and other insects, mites, eelworms and woodlice. It is very poisonous to mammals and birds, but breaks down rapidly after application.

paratonic (aitogenic). Applied to plant movements (e.g., TROPISM, NASTIC MOVEMENT) that occur in response to external stimuli. *Compare* AUTONOMIC.

paratype. (1) Any specimen that is cited along with the TYPE SPECIMEN(s) in the original description of a species. (2) All the external factors that affect the manifestation of a genetic character. (3) An abnormal type within a species used, for example, of bacterial colonies.

Paraxonia. *See* ARTIODACTYLA.

Parazoa. Multicellular animals whose bodies are at a very low level of organization and which are not closely related to other members of the METAZOA. Sponges (*see* PORIFERA) are the only living examples.

parenchyma. (1) Loose, soft packing tissue, consisting of living cells. It occurs extensively in plants (e.g., in pith, cortex) and in animals that lack a COELOM (e.g., flatworms). (2) The cells specific to an animal organ, as distinct from its other components (e.g., blood vessels and connective tissue).

parental generation. The parents of the FIRST-FILIAL GENERATION (F_1) in a genetical experiment. For the ALLELOMORPHS being observed, all the males of the P_1 generation are identical to each other, and so are the females.

Paris Convention. *See* CONVENTION FOR THE PREVENTION OF MARINE POLLUTION FROM LAND-BASED SOURCES.

Parker Morris. A set of standards to which all public and much private house building in the UK is required to conform. The standards were first proposed in *Homes for Today and Tomorrow* (The Parker Morris Report), commissioned by the then Ministry of Housing and Local Government and published in 1961. The standards govern such matters as floor space (including storage space), heating, basic fittings, outside appearance and gardens.

parrots. *See* PSITTACIFORMES.

Parthenium argentatum. *See* GUAYULE.

parthenocarpy. The development of seedless fruit without fertilization (e.g., in banana).

parthenogenesis. 'The production of offspring by the development of unfertilized ova. Parthogenesis occurs naturally in some plants (e.g., dandelion) and animals (e.g., aphids), the egg nucleus usually being diploid. Parthenogenesis can sometimes be induced artificially in other animals by cooling and other techniques applied to the haploid ova. *See also* APOMIXIS, CHROMOSOMES, MEROGONY.

partial population curve. The POPULATION DENSITY of a given developmental stage plotted on a graph against time, used for example to show how the number of people in a given age group changes in a particular place.

particle. A small, discrete mass of solid or liquid matter (e.g., AEROSOL, DUST, FUME, MIST, SMOKE and SPRAY, each of which has different properties).

partition chromatography. *See* CHROMATOGRAPHY.

pascal. *See* NEWTON.

Passeriformes (passerines, perching birds, song birds). A very large order of birds, comprising about one-half of all the known species and occupying many land habitats. They have three toes in front and one long toe behind. Often the display behaviour is complicated, the nests elaborate, and the males have a well-developed song. The order includes the larks, swallows, wagtails, pipits, shrikes, orioles, starlings, waxwings, crows, wrens, accentors, warblers, flycatchers, thrushes, tits (chickadees), treecreepers, finches and buntings.

passerines. *See* PASSERIFORMES.

pasteurization. A form of partial sterilization that is widely used to destroy many pathogenic BACTERIA (*see* PATHOGEN) in food without drastically altering its flavour. Milk is pasteurized by being heated for half an hour at 62°C, then being cooled rapidly. The process is named after its originator, Louis Pasteur (1822–95).

pastoralist. *See* NOMAD.

patabiont. An animal that spends all its life in forest-floor litter. *Compare* PATACOLE, PATOXENE.

patacole. An animal that lives temporarily in forest-floor litter. *Compare* PATABIONT, PATOXENE.

patch reef. *See* CORAL REEF.

paternoster lakes. Strings of lakes, forming patterns reminiscent of the beads on a rosary, which are dammed by MORAINES or rock bars in glacially eroded valleys.

pathogen (pathogenic organism). An organism that causes a communicable disease (e.g., influenza virus; typhoid bacillus; malarial parasite, *Plasmodium*; filarial worms).

pathogenic organism. *See* PATHOGEN.

pathological waste. Waste that may contain PATHOGENS and whose disposal might therefore threaten public health (e.g., hospital wastes, some laboratory wastes and any other possibly infected materials).

patina. *See* DESERT VARNISH.

patoxene. An animal that occurs by accident in forest-floor litter. *Compare* PATABIONT, PATACOLE

pattern intensity. The extent to which the density of a pattern varies from place to place. If intensity is high, differences are pronounced and densely populated zones alternate with zones that are sparsely populated.

pavement. All the layers that together comprise a road. Generally there are four: (a) the subbase of unbound AGGREGATE for drainage; (b) the road base, which bears most of the weight, made from aggregates that may be bonded (bound together, usually with BITUMEN) or unbonded; (c) the base course, generally made from ASPHALT, aggregate and bitumen, to provide a shaped surface; (d) the wearing course, made from asphalt, aggregate and bitumen, to take the wear from vehicles and protect the lower layers. Together, the base course and wearing course comprise the road surface, which may be rigid (rigid pavement) or have a slight elasticity under traffic movement (elastic pavement).

pawpaw. *See* CARICA PAPAYA.

P/B. The ratio of primary PRODUCTION to total BIOMASS.

Pb. *See* LEAD.

PBB. *See* POLYBROMINATED BIPHENYLS.

PCB. *See* POLYCHLORINATED BIPHENYLS.

peak sound pressure level. The value, in DECIBELS, of the maximum SOUND PRESSURE, as opposed to the ROOT-MEAN-SQUARE (i.e. effective) sound pressure.

pearly. *See* LUSTRE.

pea soup fog. A particularly opaque and dirty-greenish FOG, characteristic of London in the decades before the 1950s. Such fog is caused largely by domestic smoke. Sometimes there was a layer of relatively clear air close to the ground, giving the impression of nocturnal darkness in daytime. *See also* LONDON SMOG INCIDENTS.

peat. Organic soil, often many metres deep, composed of partly decomposed plant material. It forms under anaerobic conditions in waterlogged areas such as FENS and BOGS.

peatland. *See* MIRE, PEAT.

peck order. The social hierarchy in a group of gregarious animals of one species, based originally on observations of the behaviour of domestic chickens in which the hierarchy was demonstrated in the order in which individuals deferred to those more dominant than themselves when feeding. In many species what was once thought to be a simple social arrangement of this type is now recognized as a much more complex and flexible pattern.

pectin. *See* CARBOHYDRATES.

ped. An individual, natural aggregate of soil particles.

pedalfer. A leached (*see* LEACHING) soil formed where moisture and drainage are sufficient to remove soluble constituents. *Compare* PEDOCAL.

pediment. An eroded wash slope at the foot of a mountain in an arid or semiarid region.

pedocal. A non-leached (*see* LEACHING) soil where moisture and drainage have been insufficient to remove soluble nutrients. It is an alkaline soil, containing a horizon (*see* SOIL HORIZONS) of accumulated carbonates. *Compare* PEDALFER.

pedogenesis. The origin and development of soils.

pedogenic. Applied to effects caused by EDAPHIC FACTORS.

pedology. The study of the morphology and distribution of soils.

pedon. The smallest vertical column of soil that contains all the SOIL HORIZONS present at that point. A collection of pedons is called a polypedon.

pegmatites. IGNEOUS bodies of coarse grain usually found as DYKES associated with a larger body of finer-grained, PLUTONIC rock. The grain sizes are relative: crystals 10 metres long are found in some pegmatites. Pegmatites crystallized at a late stage from the MAGMA, when it was enriched in VOLATILES, and some are sources of minerals of rare elements (e.g., tantalum, lithium, niobium and the rare earths).

Pekilo process. A process for producing protein from starch and sulphite wastes from the food industry, thus disposing of the wastes usefully. The material is placed in fermentation vessels under aseptic conditions and stirred, while fungi are grown in it by continuous cultivation, harvested and processed further to extract the protein. *See also* NOVEL PROTEIN FOODS.

Peking man. *See* HOMO.

pelagic. Applied to organisms of the PLANKTON and NEKTON that inhabit the open water of a sea or lake. *Compare* BENTHOS, DEMERSAL.

Peléan volcano. *See* VOLCANO.

Pelecaniformes. An order of large, aquatic fish-eating birds that have all four toes webbed. They include the gannets, which feed by swimming under water or diving from great heights, the pelicans, which scoop up fish in their large bills while swimming, and the cormorants, which dive from the surface. Most species nest in colonies and all are extremely vulnerable to oil pollution, which clogs their feathers and poisons them when they swallow the oil while trying to clean themselves.

Pelecypoda. *See* LAMELLIBRANCHIATA.

pelicans. *See* PELECANIFORMES.

pelitic. Argillaceous (clayey), applied

mainly to metamorphic (*see* METAMORPHISM) rocks formed from argillaceous rocks.

pellagra. *See* NICOTINIC ACID.

Pelly Amendment. The amendment to the US Fishermen's Protection Act, administered by the Department of Commerce, which allows the President to ban all fish imports from any country whose actions tend to weaken the authority of the INTERNATIONAL WHALING COMMISSION. *See also* PACKWOOD–MAGNUSSON AMENDMENT.

Peltier effect. An electrical effect associated with differences in temperature produced at junctions between dissimilar metals when an electrical current is passed through them. It is named after J.C.A. Peltier (1785–1845).

pene-. A prefix meaning almost, as in penecontemporaneous, PENEPLAIN, etc.

peneplain. A low-lying surface, approaching base level, which is the logical end product of river erosion and the evolution of a landscape.

penetrative convection. *See* CONVECTION.

penguins. *See* SPHENISCIDAE.

penicillin. *See* ANTIBIOTICS, *PENICILLIUM NOTATUM*.

Penicillium notatum. The MOULD from which the ANTIBIOTIC penicillin is derived.

Pennsylvanian. The system that, in North American stratigraphy, is equivalent to the Upper CARBONIFEROUS.

pentadactyl limb. The type of limb that evolved in vertebrates as an adaptation to terrestrial life. The basic skeletal pattern, often much modified by the loss or fusion of parts, consists of a large upper bone, the humerus (forelimb or arm) or femur (hind limb or leg) articulating with two parallel bones, the radius and ulna (forelimb) or tibia and fibula (hind limb). The distal end of the limb consists of several small wrist or ankle bones (carpals and tarsals, respectively) articulating with five palm or sole bones (metacarpals and metatarsals, respectively), which in turn articulate with five sets of finger or toe bones (phalanges).

Pentastomida (tongue worms). A phylum of elongated worm-like parasites (*see* PARASITISM) that are found in the nasal passages of predatory mammals. The larvae, which live mainly in herbivorous mammals, resemble parasitic mites, so the tongue worms are assumed to be related to the ARACHNIDA.

pentlandite. A sulphide mineral, composed of iron and nickel combined in varying proportions ($Ni,Fe)_9S_8$) and formed by the separation of an immiscible sulphide liquid from BASIC MAGMA. Pentlandite is one of the two major ores of nickel (the other being GARNIERITE).

pentoses. *See* CARBOHYDRATES.

peptidases. ENZYMES that break down PEPTIDES and PROTEINS. An example is pepsin, which is secreted by the stomach lining in vertebrates.

peptides. Compounds composed of two or more AMINO ACIDS formed by the condensation of the $-NH_2$ group of one of the acids and the carboxyl (–COOH) group of another, resulting in the –NH–CO– peptide linkage. *See also* POLYPEPTIDES.

peptones. The breakdown products of PROTEINS, consisting of POLYPEPTIDES

perceived noise level. The SOUND PRESSURE LEVEL between one-third of an OCTAVE and one octave of random NOISE at 1 kHz which normal people consider equally noisy to a sound of interest. It is measured as perceived noise decibels (PNdB)

$$PNdB = 40 + 10 \log_2 noy.$$

See DECIBEL, NOY.

perceived risk. The subjective appreciation

of a danger by individuals, which may bear little relation to the statistical probability of injury. Risks that appear to be under voluntary control are perceived as less alarming than those over which the individual seems to have no control. Thus the danger of injury in a road accident seems less threatening than the danger of injury resulting from an accident at a nuclear power plant, although the actual risk of the former is thousands of times greater than that of the latter. *See also* RISK ASSESSMENT.

percentage area method. A method for determining the distribution and density of a species. A square grid divided by wires into 10 or 5 centimetre squares is placed on the ground, and the percentage of the area covered by each species and the percentage of bare ground are estimated. *See also* DOMIN SCALE.

perching birds. *See* PASSERIFORMES.

percolating filter (trickling filter). A bed of inert material through which water is trickled in order to purify it. The bed offers a very large surface area which becomes covered with aerobic microorganisms fed by nutrients carried in the water.

perdominant. A dominant species that is present in all, or nearly all, the associations of a FORMATION.

perennial. Recurring year after year, and applied particularly to a plant that continues to live from year to year. In herbaceous perennials the aerial parts die down in the autumn, leaving an underground structure (e.g., BULK, CORM, RHIZOME or TUBER) to overwinter. In woody perennials (e.g., trees) there are permanent, woody, aerial stems from which the plant makes new growth each year, thus allowing many of these plants to attain great size. *Compare* ANNUAL, BIENNIAL, EPHEMERAL.

perfect gas. A theoretical concept of an ideal gas that would obey the gas laws exactly and would consist of perfectly elastic molecules, the volume occupied by the

molecules and the forces of attraction between them being zero or negligible.

perianth. *See* FLOWER.

pericarp. The part of a fruit formed from the ovary wall (e.g., the flesh of a BERRY, the shell of a hazel NUT). It comprises the skin (epicarp), the middle, often fleshy mesocarp and the stony or membranous inner layer (endocarp).

pericline. An ANTICLINE that plunges in both directions.

peridotite. A group of coarse-grained, ULTRABASIC IGNEOUS rocks that consist essentially of OLIVINE, with or without other FERRO-MAGNESIAN MINERALS, and with no FELDSPAR.

periglacial. Applied to the area surrounding the limit of glaciation and subject to intense frost action and to the living organisms typical of such areas.

period. A major worldwide unit of GEOLOGICAL TIME, corresponding to a stratigraphical system. The geological periods since the PRECAMBRIAN are usually grouped into the PALAEOZOIC, MESOZOIC, and CENOZOIC Eras or, collectively, the PHANEROZOIC Eon. Periods are subdivided into epochs, which are subdivided into ages.

periodic. Applied to phenomena that repeat in identical form at regular intervals of time or space.

periodic table of elements
Table of elements in alphabetical order (International Atomic Weights 1961)

Element	Symbol	Atomic number (Z)	Relative atomic mass (A_r)
Actinium	Ac	89	(227)
Aluminium	Al	13	26.9815
Americium	Am	95	(243)
Antimony	Sb	51	121.75
Argon	Ar	18	39.948

Table of elements in alphabetical order
(International Atomic Weights 1961)

Element	Symbol	Atomic number	Relative atomic mass
Arsenic	As	33	74.9216
Astatine	At	85	(210)
Barium	Ba	56	137.34
Berkelium	Bk	97	(247)
Beryllium	Be	4	9.01218
Bismuth	Bi	83	208.9806
Boron	B	5	10.81
Bromine	Br	35	79.904
Cadmium	Cd	48	112.40
Caesium	Cs	55	132.9055
Calcium	Ca	20	40.08
Californium	Cf	98	(251)
Carbon	C	6	12.011
Cerium	Ce	58	140.12
Chlorine	Cl	17	35.453
Chromium	Cr	24	51.996
Cobalt	Co	27	58.9332
Copper	Cu	29	63.546
Curium	Cm	96	(247)
Dysprosium	Dy	66	162.50
Einsteinium	Es	99	(254)
Erbium	Er	68	167.26
Europium	Eu	63	151.96
Fermium	Fm	100	(253)
Fluorine	F	9	18.9984
Francium	Fr	87	(223)
Gadolinium	Gd	64	157.25
Gallium	Ga	31	69.72
Germanium	Ge	32	72.59
Gold	Au	79	196.9665
Hafnium	Hf	72	178.49
Helium	He	2	4.0026
Holmium	Ho	67	164.9303
Hydrogen	H	1	1.008
Indium	In	49	114.82
Iodine	I	53	126.9045
Iridium	Ir	77	192.2
Iron	Fe	26	55.847
Krypton	Kr	36	83.80
Lanthanum	La	57	138.9055
Lawrencium	Lw	103	(257)
Lead	Pb	82	207.2
Lithium	Li	3	6.941
Lutetium	Lu	71	174.97
Magnesium	Mg	12	24.305
Manganese	Mn	25	54.9380
Mendelevium	Md	101	(256)
Mercury	Hg	80	200.59
Molybdenum	Mo	42	95.94
Neodymium	Nd	60	144.24
Neon	Ne	10	20.179
Neptunium	Np	93	237.0482
Nickel	Ni	28	58.71
Niobium	Nb	41	92.906
Nitrogen	N	7	14.0067
Nobelium	No	102	(254)
Osmium	Os	76	190.2
Oxygen	O	8	15.994
Palladium	Pd	46	106.4
Phosphorus	P	15	30.9738
Platinum	Pt	78	195.09
Plutonium	Pu	94	(242)
Polonium	Po	84	(210)
Potassium	K	19	39.102
Praseodymium	Pr	59	140.0977
Promethium	Pm	61	(147)
Protactinium	Pa	91	231.0359
Radium	Ra	88	226.0254
Radon	Rn	86	(222)
Rhenium	Re	75	186.2
Rhodium	Rh	45	102.905
Rubidium	Rb	37	85.47
Ruthenium	Ru	44	101.07
Samarium	Sm	62	150.4
Scandium	Sc	21	44.9559
Selenium	Se	34	78.96
Silicon	Si	14	28.086
Silver	Ag	47	107.868
Sodium	Na	11	22.9898
Strontium	Sr	38	87.62
Sulphur	S	16	32.06
Tantalum	Ta	73	180.9479
Technetium	Tc	43	98.9062
Tellurium	Te	52	127.60
Terbium	Tb	65	158.9254
Thallium	Tl	81	204.37
Thorium	Th	90	232.0381
Thulium	Tm	69	168.9342
Tin	Sn	50	118.69
Titanium	Ti	22	47.90
Tungsten	W	74	183.85
Uranium	U	92	238.029
Vanadium	V	23	50.9414
Xenon	Xe	54	131.30
Ytterbium	Yb	70	173.04
Yttrium	Y	39	88.9059
Zinc	Zn	30	65.37
Zirconium	Zr	40	91.22

peripheral nervous system. The nervous system other than the CENTRAL NERVOUS SYSTEM, consisting mainly of nerves, each of which is made up largely of a bundle of nerve fibres (long fine processes of NEURONS).

periphyton. The organisms attached to underwater rooted plants.

Perissodactyla. The order of mammals that includes horses, rhinoceroses and tapirs, odd-toed hoofed animals (ungulates) in which the weight-bearing axis of the foot passes through the third toe. In horses the first and fifth digits are absent, and the second and fourth are incomplete (the splint bones). In the rhinoceros all the feet have three digits, and in the tapir the front feet have four digits and the hind feet three. *Compare* ARTIODACTYLA.

perknite. *See* ULTRABASIC.

permafrost. Ground that is permanently frozen. The water in the near-surface layers may melt in summer, with resulting SOLIFLUCTION on slopes. Such features as POLYGONS, stone STRIPES and PINGOS are associated with permafrost.

permanent meadow. *See* GRASSLAND.

permeability. The capacity of a structure (e.g., rock or soil) to permit the movement of a fluid through it. The degree of permeability depends on the size and shapes of the pores in the rock or soil and on the extent, size and shape of the connections between them. Permeability is usually measured in darcys or millidarcys. RESERVOIR ROCKS usually have permeabilities ranging from a few millidarcys to several darcys. The opposite of permeable is impermeable. *See also* PERVIOUS, SEMIPERMEABLE MEMBRANE.

Permian. The youngest PERIOD of the PALAEOZOIC Era, usually taken as beginning some time between 275 and 290 Ma and lasting about 55 millions years. Permian also refers to the rocks formed during this time, called the Permian System, which in Eastern Europe is divided into two series, the Lower and Upper Permian. The Lower Permian is divided in four stages: Asselian, Sakmarian, Artinskian and Kungurian; the Upper Permian is divided into two stages: Kazanian and Tartarian. In north-western Europe the EVAPORITE FACIES of the Permian is divided historically into the Rotliegendes and Zechstein.

Permo-Trias. The PERMIAN and TRIASSIC Systems considered together.

peroxisomes. Membrane-bound particles found in the CYTOPLASM of plant cells. Peroxisomes contain OXIDASES and CATALASE, and are probably the site of PHOTORESPIRATION.

peroxyacetyl nitrates (PAN). A group of compounds, forming a component of photochemical SMOG, that cause irritation to the eyes and respiratory passages of humans and are the primary cause of smog injuries to plants, many of which are sensitive to concentrations of 0.05 ppm or less.

persistence. The capacity of a substance to remain chemically stable. This is an important factor in estimating the environmental effects of substances discharged into the environment. Certain toxic substances (e.g., CYANIDES) have a low persistence, whereas other less immediately toxic substances (e.g., many ORGANOCHLORINES) have a high persistence and may therefore produce more serious effects.

pervious. Permeable (*see* PERMEABILITY) by virtue of mechanical discontinuities (e.g., JOINTS, FISSURES and BEDDING planes in the rock). In US usages, the term is synonymous with permeable. *Compare* IMPERVIOUS.

pesticides. A general term for chemical agents that are used in order to kill unwanted plants (i.e. weeds), animal pests or disease-causing fungi, and embracing insecticides, herbicides, fungicides, nematicides, etc. Some pesticides have had widespread disruptive effects among non-target species.

petiole. Leaf stalk.

petrocalcic horizon. *See* HARDPAN.

petrographic microscope. *See* THIN SECTION.

petroleum. A naturally occurring material composed predominantly of HYDROCARBONS in solid, liquid or gaseous phase or in some combination of these phases depending on the nature of the compounds and RESERVOIR conditions. Petroleum can be divided into NATURAL GAS, CONDENSATE, CRUDE OIL and BITUMEN.

petrological microscope. *See* THIN SECTION.

petrology. The study of the origins, alteration, present condition and decay of rocks and minerals.

PFR. *See* PROTOTYPE FAST REACTOR.

pH. The measure of the acidity or alkalinity of a substance, based on the number of hydrogen ions (H^+) in a litre of the solution, and expressed in terms of $pH = \log_{10}(1/[H^+])$, where $[H^+]$ is the hydrogen ion concentration. Pure water dissociates slightly into hydrogen and hydroxyl ions (H^+ and OH^-, respectively), the concentration of each type of ion being 10^{-7} mole per litre, so the pH of water is $\log_{10}(1/10^{-7}) = 7$, and this is taken to represent neutrality on the pH scale. Smaller values are acidic, larger values alkaline.

phacolith. A lens-shaped IGNEOUS INTRUSION in the crest or rough of a FOLD.

Phaeophyta (Phaeophycaeae, brown algae). A division of ALGAE, mostly marine and BENTHIC, which are common in the intertidal zone (*see* SHORE ZONATION). Phaeophyta contain a brown pigment (fucoxanthin), which masks the CHLOROPHYLL and other pigments present, giving the plants a brown to olive green colour. Some species are filamentous and minute. Others (e.g., bladderwracks and kelps) are very large and differentiated into a basal holdfast and a stem-like portion bearing ribbon-like or leaf-like fronds. Brown seaweeds are used as manure and fodder in some coastal districts, and a few species are edible for humans.

phages. *See* BACTERIOPHAGES.

phagocyte (phagocytic cell). A cell that ingests small particles (*see* PHAGOCYTOSIS).

phagocytic cell. *See* PHAGOCYTE.

phagocytosis. The process, involving flowing movements of the CYTOPLASM, by which a cell ingests particles from its surroundings. In simple forms of life (e.g., some PROTOZOA) phagocytosis is a method of feeding. In higher animals it counters infection from invading organisms (e.g., some white blood corpuscles are phagocytes).

phalanges. *See* PENTADACTYL LIMB.

Phalangida (Opiliones, harvestmen). Long-legged predaceous ARACHNIDA. They are distinguished from spiders by the absence of silk glands and by their undivided bodies.

Phanerogamia. *See* SPERMATOPHYTA.

phanerophyte. *See* RAUNKIAER'S LIFE FORMS.

Phanerozoic. The period of time from the beginning of the CAMBRIAN Period to the present day. *See also* GEOLOGICAL TIME.

phase. (1) A state of matter (i.e. gaseous, liquid or solid); some authorities count plasma (in which most atoms are ionized under conditions of very high energy) as a fourth phase. (2) A measure of whether a periodic phenomenon is in time with another, usually measured in radians (rad; 1 rad = 180°). If one SINE WAVE is at its maximum when another is at its minimum, the two are said to be 1 rad (180°) out of phase.

Phasmida (stick-insects, leaf-insects). An order of mainly tropical insects (division: EXOPTERYGOTA), many of which are wingless, that closely resemble twigs or leaves. Some have no known male sex and reproduce by PARTHENOGENESIS. Phasmida were formerly included in the ORTHOPTERA.

phellem. *See* CORK.

Phellinus pomaceus. *See* CHLOROMETHANE.

phellogen. *See* CORK, MERISTEM.

Phénix. The French experimental BREEDER REACTOR, with 250 megawatt capacity, sited at Marcoule. Phénix was connected to the supply grid at the end of 1973 and operated at full capacity from March 1974. *See also* SUPERPHÉNIX.

phenocopy. A non-inherited variation, caused by environmental influence, that produces an effect like that of a known GENE MUTATION.

phenocryst. A relatively large crystal set in a finer groundmass in an IGNEOUS rock, forming a PORPHYRITIC texture.

phenology. The study of the timing of recurring natural phenomena (e.g., the flowering of particular plants, the arrival or departure of migrating birds, etc.) with particular reference to climatological observations.

phenotype. The outward expression of the GENES (i.e. the characteristics displayed by an organism as determined by the interaction of its genetic constitution and its environment). *Compare* GENOTYPE.

phenylalanine. An AMINO ACID with the formula $C_6H_5CH_2CH(NH_2)COOH$ and a molecular weight of 165.2.

pheromone. A substance released by an animal into its surroundings that influences the behaviour or development of others of the same species (e.g., sexual attractants released by female moths, queen bee substance which prevents the development of other queens).

philoprogenetive. Producing many offspring; prolific.

phloem. VASCULAR tissue through which synthesized food products are transported in PTERIDOPHYTA and SPERMATOPHYTA (seed plants). The characteristic element of phloem is the sieve tube, a longitudinal row of elongated living cells, which communicate by means of perforations in their connecting walls.

Phoca sibirica. *See* BAIKAL.

Pholidota (Manidae, pangolins). Scaly anteaters of Africa and Asia, formerly included in the EDENTATA, but now classified as a separate order. Some are burrowing, others are able to climb trees. Like other anteaters they have an elongated snout, a long tongue, no teeth and large claws, but their skin is covered with horny scales interspersed with hairs.

phon (P). A unit of sound volume based on the SOUND PRESSURE LEVEL compared with a reference level of a pure tone of 1 kHz. The noise of a jet aircraft is about 140 phon. *Compare* SONE.

phoresy. The transport of one animal by another of a different species (e.g., certain nest-dwelling pseudoscorpions temporarily attach themselves, by means of their claws, to birds or insects which then transport them to new feeding grounds).

Phoronidea. A small phylum of colonial, marine, worm-like animals that superficially resemble POLYZOA. Unlike Polyzoa, Phoronidea possess a blood system and HAEMOGLOBIN.

phosgene ($COCl_2$). A very toxic gas, which was used as a weapon in World War I. It interferes with the nervous system and causes inflammation of the lungs. It is given off by many DEGREASING agents (e.g., trichloroethylene) when they are heated, is produced by metal-plating processes and is a combustion product of some organic compounds containing chlorine (e.g., POLYVINYL CHORIDE) if the temperature of combustion is high and carbon monoxide is present.

phosphagens. Organic phosphates (e.g., creatine phosphate in vertebrates) that provide a store of phosphate for building up ATP during muscular contraction.

phosphate. An essential MACRONUTRIENT for plants, supplied naturally from the WEATHERING of rocks and artificially from the mining of phosphate rock, an insoluble EVAPORITE. PHOSPHATE ROCK is processed by treatment with sulphuric acid to produce superphosphate, containing some soluble phosphorus pentoxide (P_2O_5) fertilizer. Superphosphate can be treated further, with phosphoric acid, to increase the P_2O_5 content, giving triple superphosphate. GYPSUM is a byproduct of processing. Phosphates are also used industrially. They can cause EUTROPHICATION when discharged, mainly in sewage, but phosphate is fairly immobile in soils, especially CALCAREOUS ones, and contamination of water from phosphate fertilizer is not serious. *See also* PHOSPHORUS CYCLE.

phosphate rock. A SEDIMENTARY ROCK, rich in the mineral APATITE, which is the only important source of phosphorus. Phosphate rock is treated with dilute acid (*see* PHOSPHATE) to produce a more soluble material containing a higher percentage of such water-soluble compounds as $Ca(H_2PO_4)_2$. *See also* PHOSPHORITE.

phosphorite. Generally, all phosphorus-rich deposits (PHOSPHATE ROCK); more specifically, lithified (*see* LITHIFICATION) phosphate rock as distinct from the usually unlithified pebble phosphates.

phosphorus (P). An element that occurs in several allotropic (*see* ALLOTROPY) forms. The most common form is white phosphorus, a waxy, white, very poisonous, very flammable solid (mp 44°C). Red phosphorus is a dark red, non-poisonous, not very flammable powder. Phosphorus is found in nature only in the combined state, mainly as PHOSPHATE ROCK. It is used in fertilizers, detergents and matches, and is an essential MACRONUTRIENT. $A_r = 30.9738$; $Z = 15$.

phosphorus cycle. The circulation of atoms of phosphorus, brought about mainly by living organisms. The weathering of PHOSPHATE ROCK releases phosphorus into water and thus into soils where it is taken up by plants, for which it is an important macronutrient, being an essential constituent of DNA and RNA and is involved in the ADP–ATP process (phosphorylation) whereby energy is made available to cells. The death and decomposition of living tissues return phosphorus to the soil in a form available to plants, so continuing the cycle. Some of the phosphorus entering the sea finds its way into sedimentary and EVAPORITE deposits, which are eventually raised above sea level as new sources of phosphate rock.

photic zone. The surface waters of a sea or lake, penetrated by sunlight. This zone includes the EUPHOTIC ZONE and DYSPHOTIC ZONE.

photochemical. Applied to chemical reactions whose energy is supplied by sunlight.

photochemical reactions. *See* PHOTOSYNTHESIS.

photochemical smog. *See* SMOG.

photogeology. The geological interpretation of aerial photographs, most commonly taken as overlapping pairs of views taken vertically from cameras with two lenses along a straight line of flight. These can then be seen stereoscopically. Satellite imagery (e.g., from the LANDSAT and SPOT programmes) and other forms of REMOTE SENSING are also used.

photon. A quantum of electromagnetic radiation with zero rest mass and energy equal to the product of the FREQUENCY of the radiation and PLANCK'S CONSTANT. Photons are generated when a particle possessing an electrical charge changes its momentum, when nuclei or electrons collide, and when certain nuclei and particles decay. In some contexts photons are regarded as elementary particles.

photonasty. *See* NASTIC MOVEMENT.

photoperiodicity. *See* PHOTOPERIODISM.

photoperiodism (photoperiodicity). The response of plants or animals to the relative duration of light and dark. The timing of the breeding season in many vertebrates and the flowering period of many plants is regulated by day length. For example, the chrysanthemum is a short-day plant, flowering when the nights are long; lettuce is a long-day plant. *Compare* CIRCADIAN RHYTHM. *See also* PHYTOCHROME.

photophosphorylation. The formation of ATP from ADP and phosphate, utilizing light as the source of energy. *See also* PHOTOSYNTHESIS.

photorespiration. RESPIRATION activated by light that is characteristic of certain plants (e.g., wheat, sugar-beet), but not of maize and sugar cane. The two types of respiration follow different biochemical pathways. *See also* PEROXISOMES.

photosynthesis. The synthesis of living cells of organic compounds from simple inorganic compounds, using light energy. In green plants, absorption of light by CHLOROPHYLL initiates photochemical reactions ('light reactions') in which oxygen is released from water and light energy is converted into chemical energy by the formation of ATP. This is photophosphorylation. Subsequent dark reactions result in the reduction of carbon dioxide (using hydrogen which originated in the water) to CARBOHYDRATES, utilizing the energy stored in ATP

$$CO_2 + 2H_2O \xrightarrow[\text{chlorophyll}]{\text{light energy}} [CH_2O] + O_2$$

CH_2O being the simplest carbohydrate and the oxygen (O_2) being released as a gas. AMINO ACIDS are synthesized by the combination of intermediate products, with the nitrogen being derived from minerals salts. Most of the oxygen in the atmosphere is derived from photosynthesis. Photosynthetic BACTE-RIA do not produce oxygen; they take the hydrogen necessary for the reduction of carbon dioxide from organic compounds or inorganic sources other than water. HETEROTROPHIC organisms depend on the supply of organic material produced by chlorophyll-containing plants during photosynthesis, and so are also dependent on sunlight as their ultimate source of energy.

phototaxis. *See* TAXIS.

phototrophic. Applied to organisms that obtain energy from sunlight. *Compare* CHEMOAUTOTROPHIC, CHEMOTROPHIC. *See also* AUTOTROPHIC, PHOTOSYNTHESIS.

phototropism (heliotropism). (1) Formerly a synonym for phototaxis (*see* TAXIS). (2) In plants, a growth response in which the stimulus is light. Most stems are positively phototropic and will curve towards a light source; some roots (e.g., the aerial roots of ivy) are negatively phototropic. Growth curvatures are under the control of AUXINS.

photovoltaic effect. The production of an electromotive force by radiant energy, commonly light, falling on the junction of two dissimilar materials (e.g., a metal and a semiconductor).

phreaticolous. Applied to organisms that live in subterranean fresh water.

phreatic water. Water that is below the WATER TABLE, in the zone of saturation. *Compare* VADOSE WATER.

phreatophyte. A plant with roots that are long enough to reach the WATER TABLE.

Phthiraptera. A name sometimes given to the biting lice (MALLOPHAGA) and sucking lice (ANOPLURA) together.

phycobilins (biliproteins). Blue and red pigments present in the CYANOPHYTA (blue–green algae) and RHODOPHYTA (red algae). Like the CHLOROPHYLL also present in these organisms, phycobilins absorb light energy.

phycomycetes (downy mildews). FUNGI that usually show a well-marked sexual phase, with the production of thick-walled resting SPORES. The HYPHAE are mostly without cross-walls. The group includes parasites (*see* PARASITISM) that cause white rust of CRUCIFERAE (*Cystopus*), potato blight (*Phytophthora*) and fungal diseases of fish (e.g., *Saprolegnia*) and SAPROPHYTES (e.g., *Mucor*, the common black pin-mould which grows on food such as stale bread).

phyllite. An ARGILLACEOUS metamorphic rock (*see* METAMORPHISM) with shiny CLEAVAGE planes that are due to the presence of flaky minerals. The individual flaky minerals are too small to be seen with the naked eye; with increasing grain size phyllite grades into MICA–SCHIST. Phyllites are formed under low-temperature regional metamorphism.

phylloclade. See CLADODE.

phyllode. A flattened, expanded petiole (leaf stalk) that takes the place of a reduced leaf blade (e.g., in many species of *Acacia*).

Phylloxera. See VITIS.

phylogeny. The evolutionary history of a group of organisms, as distinct from the development of an individual organism.

phylum. *See* CLASSIFICATION.

physiocracy. The rule of nature as this was understood by the Physiocratic School, which flourished in France in the 18th century and which laid the foundations for subsequent economic theories, exerting considerable influence on ADAM SMITH. The physiocrats believed that the land is the only true source of wealth, since only the natural growth of plants produces goods from nothing, manufacturing industry merely altering the form of pre-existing materials. Thus the products of agriculture should be prized highly, and those of manufacture considered of less worth. Humans, being dependent on plants, are subject to universally applicable natural laws, which human laws should not contradict, so the purpose of government is merely to educate the people regarding the natural law, and to foster and remove hinderances to its observance, a general rule they summarized in the phrase *laissez faire, laissez passer*, the origin of LAISSEZ FAIRE economic theory.

physiographic succession. A SUCCESSION whose character is largely determined by topographical factors and the local climates produced by them. A succession on an exposed hillside may proceed differently from that in the valley and even from that on the sheltered side of the same hill.

physiological specialization. The occurrence of a number of genetically distinguishable forms within a species that differ from one another in their biochemical characters, but not in their structure. In disease-producing organisms (e.g., black stem rust of wheat) this is of great importance because physiological differences cause differences in pathogenicity towards different varieties of the host species, complicating the breeding of resistant crop varieties.

Physopoda. *See* THYSANOPTERA.

phytic acid. *See* INOSITOL.

phytoalexins. Substances poisonous to FUNGI that are produced by plants when infected by fungal diseases.

phytobenthos. *See* BENTHOS.

phytochrome. A plant pigment that becomes activated by red light, and in its activated form initiates growth, germination, flowering, etc. In the absence of red light the phytochrome becomes deactivated or lost. Phytochrome regulates PHOTOPERIODISM.

phytocoenosis. The assemblage of plants that inhabit a particular area.

phytogeocoenosis. A plant community and its physical environment.

phytogeography. The study of the distribution of plant species in relation to climate, geography and history.

phytokinins. *See* CYTOKININS.

phytophagous. Applied to animals that feed on plants (i.e. herbivores).

phytoplankton. *See* PLANKTON.

phytoplankton bloom. *See* BLOOM.

phytosociology. The study of plants in the widest sense, and including the study of all phenomena that affect their lives as social units. It is thus the study of plant COMMUNITIES and so covers much of the field of ECOLOGY, including the composition and classification of communities, their MORPHOLOGY, structure, development and change, and the relationships within them among species and between species and their environment.

phytotoxic. Poisonous to green plants.

phytotron. A building used for growing plants under controlled environmental conditions.

pica. The habit of eating non-food materials. It sometimes occurs in children.

Picidae (woodpeckers). A family of birds (order: Piciformes) that are specialized for climbing and boring wood. The bill is strong and sharply pointed for digging out insects and making nest holes in trees, and the long, sticky protrusible tongue is used for extracting insects from holes. The rigid tail is used to support the bird as it climbs and to steady it as it bores into wood with rapid, percussive blows.

pico- (p). A prefix used in conjunction with SI units to denote the unit x 10^{-12}.

picrite. *See* ULTRABASIC.

piedmont glacier. *See* GLACIER.

piezoelectric effect. The interaction between electrical and mechanical stress–strain variables in a medium, whereby energy is converted from one form into another.

piezometer. *See* PIEZOMETRIC LEVEL.

piezometric level. (1) The level to which water in a confined AQUIFER will rise under its own pressure in a borehole, the confining pressure having been removed. (2) The level to which water will rise in a piezometer (a tube inserted into the ground so that the lower end, with a permeable tip, is at a position from which pre-pressure measurements are required).

pigeons. *See* COLUMBIFORMES.

pigeon's milk. *See* CROP MILK.

piliferous. Having hair; used especially of plants.

pillow lava. LAVA shaped like pillows lying one on top of another. The lava can be of any composition from ACID to BASIC, but all recent pillow lavas appear to have been formed under water.

pilot balloons. *See* BALLOONS.

Pinaceae. A family of coniferous trees and shrubs (order: CONIFERAE) that includes firs (*Abies*), pines (*Pinus*), cedars (*Cedrus*) and the deciduous larches (*Larix*). Members of the Pinaceae yield resins, turpentine and bark for tanning, but are most important for their easily worked timber, for which the pines in particular are cultivated on a vast scale.

pingo. A dome-shaped body in a region of PERMAFROST, caused by lenses of ice accumulating beneath the surface.

pinna. In mammals, the external, visible part of the ear.

pinnate. Applied to compound leaves that consist of more than three leaflets arranged in two rows on a single stalk.

pinnatifid. Applied to leaves that are lobed pinnately (*see* PINNATE), but divided completely into leaflets.

Pinnipedia (seals, sea-lions, walruses). A suborder (order: Carnivora) of carnivorous aquatic mammals with fin-like limbs.

pinocytosis. The ingestion of drops of liquid by cells.

Pinus aristata. See BRISTLE-CONE PINE.

pinworms. *See* NEMATODA.

pisciculture. *See* AQUACULTURE.

pisé de terre. *See* RAMMED EARTH.

pistil. *See* FLOWER.

pistillate. Applied to FLOWERS that have carpels, but not stamens, and so are female. *Compare* STAMINATE.

pitch. (1) An aural assessment of sounds so that they can be ordered in a scale from low to high. Pitch is determined by FREQUENCY, but also by sound pressure and waveform. (2) The angle at which a propeller blade meets the air or water. Most propeller-driven aircraft and advanced AEROGENERATORS have variable-pitch propeller blades. (3) A black, carbonaceous, non-crystalline material that is solid at normal temperatures, but flows on slight warming. It occurs naturally, but rarely. Most pitch is obtained as a residue from oil refining.

pitchblende. A naturally occurring mixture of uranium oxides, including URANINITE, found in massive form.

pitchstone. A predominantly glassy, ACIDIC IGNEOUS rock that tends to have a duller LUSTRE and a flatter FRACTURE than OBSIDIAN.

Pithecanthropus erectus. See HOMO.

pitot tube. An instrument for measuring the velocity or pressure of a gas or liquid. It is a tube with two openings, one into the moving fluid and one away from it. The difference in pressure between the ends of the tube is related to the velocity of the fluid; the pressure can be read directly from the opening facing away from the flow. In an aircraft, the airspeed indicator, pressure altimeter and vertical speed indicator are linked to a pitot tube carried outside the aircraft.

pituitary gland (pituitary body, hypophysis cerebri). A gland situated under the floor of, and whose activity is influenced by, the brain in vertebrates. The pituitary secretes a number of hormones (*see* FOLLICLE-STIMULATING HORMONE, GROWTH HORMONE, INTERMEDIN, LACTOGENIC HORMONE, LUTEINIZING HORMONE, OXYTOCIN), some of which control the activity of other ENDOCRINE ORGANS.

placenta. (1) In a flower, the part of the ovary wall which bears the OVULES. (2) An organ by which the EMBRYO of mammals is attached to the uterus, and through which the exchange of oxygen, food, antibodies and waste products takes place, maternal and foetal blood vessels being in close contact. In placental mammals (subclass: EUTHERIA) the embryonic tissues involved are the CHORION and usually the ALLANTOIS. In most marsupial mammals (order: MARSUPIALIA) there is no placenta, but in a few the yolk sac or allantois develops a weak connection with the uterus. The placenta secretes PROGESTERONE and small amounts of OESTROGENS.

placental mammals. *See* EUTHERIA.

placer. A sediment in which transportation (by wind, water or ice) has caused concentrations of heavy minerals (e.g., GOLD, CASSITERITE, RUTILE, etc.) to accumulate, weathered out (*see* WEATHERING) from the rocks or VEINS.

Placodermi. An extinct group of mainly DEVONIAN fishes whose bodies were heavily armoured with bone.

plaggen. Soils with deep cultivation horizons. *See* SOIL HORIZONS.

plagioclase. *See* FELDSPAR.

plagioclimax. A stable plant COMMUNITY, in equilibrium under existing conditions, but which has not reached the natural CLIMAX, or has regressed from it, because of the operation of BIOTIC FACTORS such as human intervention e.g., constantly grazed or mown grassland, woodland managed as COPPICE). *See also* POSTCLIMAX, PRECLIMAX, PROCLIMAX, SERCLIMAX, SUBCLIMAX.

plagiogeotropism. *See* GEOTROPISM.

plagiosere. A plant SUCCESSION deflected from its normal course by BIOTIC FACTORS. This results in the formation of a PLAGIOCLIMAX.

plain muscle. *See* MUSCLE.

Planck's constant (*h*). The universal constant that relates the frequency of radiation (*v*) with its quantum of energy (*E*), such that $E = hv$. The value of Planck's constant is 6.62559 x 10^{-34} joule seconds (6.62559 x 10^{-27} erg seconds). It is named after Max Planck (1858–1947).

plane wave. A wave in which the WAVE FRONTS are parallel with one another and at right angles to the direction of propagation.

plankton. Animals and plants, many of them microscopically small, that float or swim very feebly in fresh or salt water. They are moved passively by winds, waves, or currents. The animals (zooplankton) are chiefly PROTOZOA, small CRUSTACEA, and larval stages of MOLLUSCA and other invertebrates. The plants (phytoplankton) are almost all ALGAE (e.g. diatoms (BACILLARIOPHYTA), PYRROPHYTA, DESMIDIACEAE, etc.) The smallest organisms (e.g., diatoms) are called nannoplankton. All animal life in the open sea depends ultimately on the phytoplankton, which is the basis of FOODS CHAINS leading to fish, whales, seabirds, etc. *Compare* AERIAL PLANKTON.

planned economy. An economic system that is controlled centrally by an authority responsible for all planning connected with the production and distribution of goods.

Compare LAISSEZ FAIRE, MIXED ECONOMY.

planning permission. The official permission that is required by anyone wishing to carry out a particular development. Under UK planning laws, administered by planning authorities ranging from the lowest tier of local government to the Secretary of State for the Environment, permission must be obtained for any development that would alter the use of a building or area of land, or that would have a major effect on the landscape or environment.

planosol. A soil that has a compact, clayey depositional layer developing into HARDPAN. *See also* SOIL HORIZONS.

plant. An organism belonging to the kingdom METAPHYTA and characterized chiefly by a HOLOPHYTIC NUTRITION dependent on the possession of CHLOROPHYLL. Some plants, however, are parasitic (*see* PARASITISM) or saprophytic (*see* SAPROPHYTE). Most have cellulose cell walls, are sedentary and have branching bodies. The distinction between plants and animals becomes blurred in some groups (e.g., slime moulds, MYXOMYCOPHYTA; unicellular ALGAE; PROTOZOA), which are now usually regarded as belonging to a separate kingdom – PROTISTA.

plantain. *See* MUSA.

plantigrade. Applied to animals (e.g., humans) that walk with all of the lower surface of the foot (metacarpals or metatarsals and digits) on the ground. *Compare* DIGITIGRADE, UNGULIGRADE.

plasma. *(1) See* PHASE. *(2) See* BLOOD.

plasmagenes. Cytoplasmic particles or substances that can reproduce and pass on inherited qualities (cytoplasmic inheritance) to daughter cells. They are not inherited through the CHROMOSOMES of the GAMETES in a mendelian way (*see* INDEPENDENT ASSORTMENT, SEGREGATION). Plasmagenes may be present in bodies such as PLASTIDS (plastogenes) or MITOCHONDRIA. *Compare* GENES.

plasmalemma. The cell membrane; in plant cells the term is applied only to the external membrane (ectoplast) and not to the one surrounding the VACUOLE (which is called the tonoplast).

plasma membrane (cell membrane). A very fine membrane, composed mainly of protein and fat, that bounds the CYTOPLASM of cells. It isolates the cell, allowing its internal composition to differ from that of its surroundings, because it is selectively permeable, allowing water and fat molecules to pass readily across it, but preventing the passage of many substances (e.g., proteins and many ions). In active transport, substances are passed across cell membranes against concentration gradients.

plasmodemata. *See* CELL WALL.

Plasmodium. A genus of protozoan (SPOROZOA) parasites (*see* PARASITISM), found in the blood of birds and mammals, that cause malaria in humans, the intermediate hosts being certain species of mosquitoes.

plasmodium. The vegetative stage of slime moulds (*see* MYXOMYCOPHYTA) consisting of an amoeboid, multinucleate mass of PROTOPLASM. *See also* COENOCYTE, SYNCTIUM.

plasmogamy. *See* PLASTOGAMY.

plasmon. A cell's complement of PLASMAGENES.

plastics. Organic POLYMERS of a wide variety of types and uses (e.g., POLYVINYL CHLORIDE, polyurethane, polythene). Commonly plastics fall into two groups — thermoplastic and thermosetting — depending on their behaviour upon being heated.

plastids. Membrane-bound bodies found in the CYTOPLASM of most plant cells (not in FUNGI, CYANOPHYTA or BACTERIA). Leucoplasts are colourless and store food materials; chromoplasts (*see* CHROMATOPHORES) contain pigments (e.g., CAROTENOIDS and CHLOROPHYLL).

plastogamy (plasmogamy). The fusion of CYTOPLASM, but not the nuclei, when cells come together to form a PLASMODIUM.

plastogene. A PLASTID that functions as a PLASMAGENE.

plateau basalt. An extensive flow, or series of flows, of BASALTIC LAVA which, because of differential erosion, forms a plateau. Most plateau basalts are also FLOOD BASALTS, and the terms are often used synonymously.

plate tectonics. The theory that the LITHOSPHERE is made up of several moving major, rigid plates, plus some minor ones, whose edges are defined by belts of EARTHQUAKES, VOLCANOES, BENIOFF ZONES, OCEANIC TRENCHES, ISLAND ARCS, mid-ocean RIDGES and TRANSFORM FAULTS. Boundaries between plates can be: (a) constructive, where plates are moving apart with the production of new material by SEA-FLOOR SPREADING along ridges; (b) destructive, where one plate descends beneath another along a SUBDUCTION ZONE, most of the sediment being scraped off to form fold mountains until eventually the ocean may close, leading to a continental collision; (c) conservative, where plates slide past one another along transform faults. The mechanism causing the moving of plates is believed to result from convection within the LOW-VELOCITY ZONE. The movements of ancient plates far back into the PRECAMBRIAN have been postulated, with variable amounts of agreement. Although there are still unexplained and apparently contradictory features of the Earth as described by this model, the theory has revolutionized the Earth sciences and has provided new ideas in the search for resources. *Compare* CONTINENTAL DRIFT.

platform. A stable, flat area (e.g., a CONTINENTAL SHELF) on which a thin sequence of sediments may accumulate.

platinoids. The platinum group of metals (platinum, palladium, rhodium, iridium, ruthenium, osmium) which occur NATIVE with nickel and copper sulphide as a result of differentiation in BASIC MAGMA and in PLACER deposits. The metals are used as CATALYSTS, in ALLOYS and electroplating, as well as in applications such as electrical contacts, where corrosion or oxidation would be detrimental.

Platyhelminthes (flatworms). A phylum of flattened invertebrate animals that comprises the flukes (TREMATODA), tapeworms (CESTODA) and free-living flatworms (TURBELLARIA). Platyhelminthes are TRIPLOBLASTIC and have no COELOM or blood system. The alimentary canal, when present, has a single opening, the mouth. The reproductive system is complicated; they are usually HERMAPHRODITE.

playa. A lake that exists temporarily after rainfall, or its dried-up bed. Playas are common features of some deserts and may contain EVAPORITE deposits.

Plecoptera (stoneflies). An order of small primitive insects (division: EXOPTERYGOTA) whose members have two pairs of membranous wings. The larvae are aquatic and predatory. The adults are short-lived and do not feed.

pleiotropic. Applied to a GENE that affects more than one characteristic in the PHENOTYPE.

Pleistocene. The older subdivision of the QUATERNARY, usually ranked as an EPOCH, which was the time of the most recent glaciation and is thought to have lasted from about 2 Ma to about 10 000 years ago. Evidence from ocean drilling suggests that a general climatic cooling occurred during the TERTIARY with an acceleration from the late MIOCENE, about 10 Ma. Some climatologists consider we are still living in an INTERGLACIAL period of the Pleistocene. Pleistocene also refers to the rocks deposited during this time, called the Pleistocene Series.

pleomorphism. *See* POLYMORPHISM.

pleurilignosa. Rain forest and rain bush.

pleurodont. Applied to the condition, common in lizards and snakes, in which the teeth are attached by one side to the inner surface of the jaw bones. *Compare* ACRODONT, THECODONT.

pleuropneumonia-like organisms. *See* MYCOPLASMAS.

pleustron. A single-layered plant community that floats on or in a water body (e.g., duckweed on the surface of a pond).

Pliensbachian. A STAGE of the JURASSIC System.

Plinian volcano. *See* VOLCANO.

plinthite. *See* HARDPAN.

Pliocene. A subdivision of the CENOZOIC Era, usually ranked as an EPOCH, which followed the MIOCENE and is thought to have lasted from seven to two million years BP. Pliocene also refers to rocks deposited during this time, called the Pliocene Series.

pliomorphism. *See* POLYMORPHISM.

plovers. *See* CHARADRIIDAE.

plug. A comparatively small, steep-sided IGNEOUS rock body that is roughly circular in plan. Some plugs are in-filled and solidified feeder channels for VOLCANOES and are composed of LAVA or PYROCLASTIC rock.

plume. Chimney effluent composed of gases alone or of gases and particulates (*see* PARTICLE). The form of the plume depends on turbulence in the atmosphere. Descriptions of plumes use such words as looping, coning, fanning, fumigating and lofting.

plumule. (1) The terminal bud in the EMBRYO of a seed plant (SPERMATOPHYTA). (2) In birds, a down feather.

plunge. *See* FOLD.

pluteus. A planktonic (*see* PLANKTON) larva of ECHINOIDEA (sea urchins) and some OPHIUROIDEA (brittle stars). Plutei metamorphose (*see* METAMORPHOSIS) into adults without attaching themselves to a SUBSTRATE.

pluton. A large IGNEOUS INTRUSION that solidified at depth. The term is less commonly used to refer to any igneous intrusion of unknown configuration or to an intrusion that does not fit any other category.

plutonic. (1) Applied to a rock or to a large IGNEOUS INTRUSION which crystallized at depth. HYPABYSSAL and plutonic rocks are contrasted with VOLCANIC ROCKS (2) Applied loosely to any coarse-grained rock. (3) Applied to any process involving heat deep within the Earth's CRUST (e.g., MIGMATIZATION).

plutonium (Pu). A TRANSURANIC ELEMENT; one of the ACTINIDES, different isotopes of which are produced by nuclear reactions. Thirteen isotopes are known, the most stable being $^{244}_{94}$Pu. $^{239}_{94}$Pu is produced in NUCLEAR REACTORS and has a HALF-LIFE of 24 400 years. It is also used in nuclear weapons, 1 kilogram of ^{239}Pu yielding energy equivalent to about 10^{14} joules. $Z = 94$.

PNdB. Abbreviation for perceived noise level in DECIBELS.

pneumatolysis. The alteration of rock, both the IGNEOUS body and COUNTRY ROCK, by VOLATILES produced during the consolidation of MAGMA. KAOLINIZATION, TOURMALINIZATION, GREISENING AND SERPENTINIZATION are examples of pneumatolysis, which itself is one type of METASOMATISM.

pneumatophores. Erect, aerial root branches produced by some plants (e.g., MANGROVES such as *Avicennia* and *Sonneratia*) that grow in water or tidal swamps. Air spaces inside the roots communicate with the atmosphere by means of pores.

pneumoconiosis. An incapacitating disease of humans caused by the retention of inhaled dusts in the lungs, thus reducing lung function. It occurs among workers in the mining industries, especially in coal mining.

Podicipedidae (grebes). A family of diving birds (order: Podicipediformes or Colymbiformes) whose members have lobed toes, nest on floating vegetation, are clumsy on land and have weak and hurried flight. Their courtship displays are often elaborate.

podzol (podsol). An acid soil in which aluminium and iron oxides and hydroxides have been leached (*see* LEACHING) from the upper layers (A horizon, *see* SOIL HORIZONS), leaving a surface of raw HUMUS over an upper pale horizon and a lower, denser, darker horizon.

podzolization (podsolization). The process by which aluminium and iron oxides and hydroxides are leached (*see* LEACHING) from upper layers of soils and precipitated in the zone of deposition beneath. *See also* PODZOL, SOIL HORIZONS.

poikilosmotic. Applied to animals (e.g., many marine invertebrates) that have body fluids whose OSMOTIC PRESSURE varies according to that of the surrounding water. *Compare* HOMOIOSMOTIC.

poikilothermy (cold-bloodedness). The possession of a body temperature that varies, approximating to that of the surroundings, and can be controlled only by moving to a warmer or cooler place. All animals except birds and mammals are poikilotherms. The rate of metabolism of poikilothermic animals slows down in cold conditions, so many of them have to hibernate (*see* HIBERNATION). *Compare* HOMOIOTHERMY.

point bar. A sand or gravel deposit built up by rivers on the inner curve of a MEANDER.

polar body. *See* OOCYTE.

polar desert soil. *See* TUNDRA SOILS.

pole. *See* ROD.

polje. A feature of a KARST landscape that consists of a closed basin with an irregular outline and, usually, steep walls. Surface water disappears down a SWALLOW HOLE after traversing the flattish floor of the polje.

poll. (1) To remove the branches from the main trunk of a living (unfelled) tree. (2) To remove the horns from cattle.

pollard. (1) A tree from which the branches have been removed, leaving a main trunk from which new growth develops at the top. Pollarding was used to produce small timber in places where grazing animals made coppicing (*see* COPPICE) impracticable. (2) An animal whose horns have been removed.

pollen. Microspores (*see* HETEROSPORY) of seed-producing plants (SPERMATOPHYTA). Each pollen grain contains a much-reduced male gametophyte (*see* ALTERNATION OF GENERATIONS). Pollen grains are transferred by wind, water, birds or other animals to the OVULES (in GYMNOSPERMAE) or STIGMAS (in ANGIOSPERMAE) where pollen tubes containing male nuclei grow out and penetrate the EMBRYO SAC.

pollen analysis. *See* PALYNOLOGY.

pollinia. *See* ORCHIDACEAE.

polluter pays principle. *See* EXTERNALITIES.

pollution. The direct or indirect alteration of the physical, thermal, biological or radioactive properties of any part of the environment in such a way as to create a hazard or potential hazard to health, safety or welfare of any living species. Pollution may occur naturally (*see* NATURAL POLLUTANT), but the term is more commonly applied to changes wrought by the emission of industrial pollutants or by the careless discharge or disposal of human domestic wastes or sewage. The term also includes the production of excessive noise (e.g., by aircraft, road vehicles or factories) and the release of excessive heat (*see* THERMAL POLLUTION). *See also* AIR POLLUTION.

pollution accretion. *See* ACCRETION.

polyandry. A breeding system in which one female mates with several males. It is un-

common, but occurs occasionally in certain primate communities (e.g., howler monkeys). *Compare* POLYGYNY.

polybrominated biphenyls (PBB). A range of compounds used as flame retardants. They are persistent in the environment and harmful to animals, including humans. They are similar in many ways to POLYCHLORINATED BIPHENYLS.

Polychaeta (bristle worms). An order of mainly marine annelid (*see* ANNELIDA) worms whose members have numerous bristles (chaetae) borne on flat outgrowths (parapodia) of the body segments. Their heads bear sense organs and tentacles. Males and females are separate, and fertilization is external. Some (e.g., ragworms) are freeswimming; others (e.g., tubeworms, lugworms) live in tubes or burrows.

polychlorinated biphenyls (PCB). A group of closely related chlorinated hydrocarbon compounds (*see* ORGANOCHLORINES) whose principal use has been as liquid insulators in high-voltage transformers. Their use is being reduced due to evidence of their persistence and toxicity in the environment.

polyclimax. A number of CLIMAXES occurring within a climatic region, probably due to the subordination of climatic factors by EDAPHIC FACTORS. *Compare* MONOCLIMAX.

polycyclic aromatic hydrocarbons (PAH). A group of chemical compounds, including benzo *a* pyrene, dibenzopyrene and dibenzoacridine, that are carcinogenic (*see* CARCINOGEN) to humans.

polyembryony. The development of more than one EMBRYO from a single fertilized egg. In some insects (e.g., parasitic HYMENOPTERA) many hundreds of embryos may develop from one egg. *See also* MONOZYGOTIC TWINS, PARASITISM.

polyenergid. Applied to a cell nucleus containing several sets of CHROMOSOMES which have been produced as the result of repeated division within the intact nuclear membrane (*see* ENDOMITOSIS). *See also* POLYPLOID.

polygamy. (1) The production of male, female and HERMAPHRODITE flowers on the same or different plants. (2) The possession of more than one mate. *Compare* MONOGAMY.

polygenes. *See* MULTIPLE FACTORS.

polygenetic. *See* PROVENANCE.

polygon. In TOPOGRAPHY, a unit of polygonally patterned ground surface which is found in regions of PERMAFROST and also in areas of alternating flooding and desiccation, such as PLAYAS. Polygons vary from a few millimetres to several tens of metres in diameter and have several modes of origin. In regions of permafrost, polygons grade into stone STRIPES on hill tops.

polygoneutic. Applied to an animal that produces several broods or litters of young in one season.

polygyny. An animal breeding system that involves one male and several females. *Compare* POLYANDRY.

polyhaline. Applied to BRACKISH waters whose salinity approaches that of sea water (i.e. 18–30°/oo salt). *Compare* EUHALINE, FRESHWATER, MESOHALINE, OLIGOHALINE.

polymer. A substance formed by the joining together (polymerization) of simple basic chemical units (monomers) in a regular pattern. Polymers are commonly based on the carbon system, or on silicon and oxygen atoms.

polymerization. *See* POLYMER.

polymorphism (pleomorphism, pleiomorphism). (1) The occurrence, usually in the same habitat and within an interbreeding population, of several distinct forms making up one species. The different forms are present in fairly constant proportions. Examples include: some colour varieties in plants; feeding, stinging and reproducing POLYPS in

the Portuguese man-of-war (*Physalia*); colour varieties in certain butterflies (e.g., *Heliconius* species of tropical America); human BLOOD GROUPS; queen, drone and worker castes in the honey bee. (2) The occurrence of different forms in the same individual at different stages in its life history (e.g., MEDUSA and POLYP in some CNIDARIA). (3) The occurrence of separate minerals that have the same chemical composition, but different physical structures and properties.

polymorphonuclear leucocytes. *See* BLOOD CORPUSCLES.

polymorphs. *See* BLOOD CORPUSCLES.

Polynesian Floral Region. The part of the PALAEOTROPIC REALM that comprises the Pacific islands east of Indonesia, but excluding the Fijian and Hawaiian islands.

polyp. In CNIDARIA, a sedentary form which may live singly (e.g., sea-anemone) or as a member of a colony (e.g., CORALS). The typical polyp (hydranth) has a tubular body, attached to the SUBSTRATE at one end and with a mouth surrounded by tentacles at the other. In some colonies there is more than one type of polyp (e.g., some corals have different forms for catching and digesting prey). Some polyps produce MEDUSAE.

polypedon. *See* PEDON.

polypeptide. A chain of three or more AMINO ACIDS joined together by the peptide linkage (–CONH–). Polypeptide chains may consist of hundreds of amino acid units. *See also* PROTEINS.

polyphagous. Applied to an animal that feeds on many kinds of food.

polyphyletic. Applied to a taxonomic group of organisms that do not all share a common origin. The group must therefore be an artificial one, the similarities of its members resulting from CONVERGENT EVOLUTION. The phylum POLYZOA is thought to be polyphyletic. *Compare* MONOPHYLETIC.

polyphyodont. Applied to an animal (e.g., a shark) that has a continuous succession of teeth. *Compare* DIPHYODONT, MONOPHYODONT.

polyploid. Applied to organisms or cells that have three or more times the haploid number of CHROMOSOMES. Polyploid individuals are sterile when crossed with normal (diploid) ones. Polyploidy is rare in animals, but fairly common in flowering plants. *Compare* ALLOPOLYPLOID, ALLOTETRAPLOID, AUTOPOLYPLOID, POLYENERGID, TETRAPLOID, TRIPLOID.

polyribosomes. *See* RIBOSOMES.

polysaccharides. *See* CARBOHYDRATES.

polysaprobic. Applied to a body of water in which organic matter is decomposing rapidly and in which the dissolved oxygen is present in very low concentrations or exhausted. *Compare* CATAROBIC, MESOSAPROBIC, OLIGOSAPROBIC, SAPROBIC CLASSIFICATION.

polysomes. *See* RIBOSOMES.

polytocous. (1) Applied to a plant that fruits repeatedly or is CAULOCARPOUS. (2) Applied to an animal that produces many young at a time. *Compare* DITOCOUS, MONOTOCOUS.

polytopic. Applied to an organism or group of organisms that occurs in more than one area. *See also* DISCONTINUOUS DISTRIBUTION.

polytrophic. *See* EUTROPHIC.

polytypic. Applied to a species that has a variety of forms living in different parts of its range. *See also* CLINE.

polyunsaturated fatty acids. *See* FATTY ACIDS.

polyvinyl chloride (PVC). One of the most common plastics, used in the manufacture of clothing, furniture, gramophone records and containers, and produced by the polymerization of VINYL CHLORIDE.

Polyzoa (Bryozoa; sea-mats, corallines). Small colonial animals, all of which are aquatic, most of them marine. They superficially resemble CNIDARIA, but are far more complex. They feed by means of ciliated (*see* CILIA) tentacles. Each individual has a horny, gelatinous or CALCAREOUS case, and the colony may be encrusting, bushy or very like a seaweed in form. The Polyzoa are often divided into two phyla – Endoprocta and Ectoprocta.

pome. A false fruit (e.g., an apple or pear) in which the fleshy part develops not from the ovary but from the receptacle (axis) of the flower.

pondweeds. Freshwater plants with submerged or floating leaves that belong to several genera (e.g., *Potamogeton, Elodea*).

pool. Geologically, a body of oil (oil pool), natural gas (gas pool) or both that occurs in a RESERVOIR under a single pressure system. The oil or gas occurs with water within the pores or fissures in the RESERVOIR ROCK.

poplars. *See* SALICACEAE.

population. A group of individuals, sharing some feature in common and living in a particular defined area, that is considered without regard to interrelationships among them. They may all belong to the same species (e.g., the human population of a particular country) or to some other taxonomic group (e.g., the bird population of a particular area), and the concept of a population is used when describing phenomena that affect the group as a whole (e.g., changes in numbers).

population density. The size of a population in relation to the area in which it occurs, expressed as the number of individuals per unit area.

population dynamics. The study of changes in POPULATION DENSITIES with time.

population ecology. The study of the factors that affect the number of individuals of a particular population present in a specified area over a period of time.

Porifera (sponges). The only phylum in the subkingdom PARAZOA, which differ from all other multicellular animals. Sponges are aquatic (freshwater, as well as marine) and sessile. Their bodies have a single, usually much-branched body cavity, communicating with the surrounding water by numerous pores. The cavity is lined by collared cells bearing flagella (*see* FLAGELLUM), by means of which currents of water containing food particles are passed through the body. Most sponges have horny, calcareous or siliceous (silicon-based) skeletons.

porosity. The percentage of pore space in a rock. Porosity in SEDIMENTARY ROCKS ranges from less than 1 percent to more than 50 percent and depends on the sorting, angularity and packing of the grains, as well as on the degree of cementation (*see* CEMENT) of the rock. The porosity of an original sediment may be increased by LEACHING during DIAGENESIS, or it may be decreased by compaction and cementation. A porous rock need not necessarily be permeable (*see* PERMEABILITY); both CHALK and CLAY have very high porosities (above 30 percent), but neither is permeable. The opposite of porous is non-porous.

porphyritic. Applied to a texture of IGNEOUS rocks in which large crystals (phenocrysts) are set in a finer ground mass.

porphyritic copper. Large, low-grade, disseminated HYDROTHERMAL deposits of copper ore (chiefly CHALCOPYRITE) occurring in finely broken-up intrusions of PORPHYRITIC, silica-rich IGNEOUS rock. Porphyritic copper deposits appear to have been formed during the last 170 million years in areas of widespread volcanism and are being worked by OPEN-CAST MINING at grades of 0.5 percent copper, providing about half the world's annual copper requirement.

porphyritic molybdenum. *See* MOLYBDENUM.

porpoises. *See* CETACEA.

Portland cement. A mixture of ground LIMESTONE and CLAY, invented in 1756 by John Smeaton, and patented in 1834 by John Aspdin, who considered it as strong as the JURASSIC Portland stone of Dorset; it was used in many buildings at that time. Portland CEMENT is the cement most commonly used today. Its essential ingredients are calcium carbonate, silica and alumina.

Portlandian. A STAGE in the JURASSIC System.

positive feedback. *See* FEEDBACK.

positive lynchet. *See* LYNCHET.

postclimax. A stable plant COMMUNITY whose composition reflects more favourable (e.g., cooler or moister) climatic conditions than the average for a region. *Compare* PRECLIMAX, PROCLIMAX, SERCLIMAX, SUBCLIMAX.

postclisere. The series of FORMATIONS that arises when the climate becomes wetter. *Compare* PRECLISERE. *See also* CLISERE.

post-industrial society. A society in which the principal economic activity, and so the main source of wealth and employment, was formerly manufacturing industry, but which has moved beyond that stage of economic development. *See also* QUATERNARY ECONOMY.

potamobenthos. *See* BENTHOS.

potamon. The organisms living in rivers and streams.

potassium (K). A silvery–white, soft, very reactive metallic element that is widely distributed in the form of salts. It is an essential MACRONUTRIENT and is used in fertilizers. A_r=39.102; Z=19; SG 0.86; mp 62.3°C.

potato. *See* SOLANACEAE.

potential climax. The CLIMAX that will replace the existing climax should the climate change. *See also* POSTCLIMAX, PRECLIMAX.

potential instability. *See* CONDITIONAL IN-STABILITY.

potential temperature gradient. The difference between the ADIABATIC LAPSE RATE and the actual LAPSE RATE in the lower atmosphere. If the gradient is zero (i.e. the two are the same) a mass of rising gas will have a constant BUOYANCY. If the gradient is negative the buoyancy of the gas will increase with height; if the gradient is positive, buoyancy will decrease with height up to a point where it reaches zero.

Poza Rica incident. An air pollution incident that occurred in 1950 at Poza Rica, Mexico, in which 22 people died and some 320 were made ill following the discharge into the atmosphere of large quantities of HYDROGEN SULPHIDE. The discharge resulted from the failure of equipment at an oil refinery sulphur-recovery unit, and the gas was trapped beneath an INVERSION.

pozzolana. Volcanic ash, found near Pozzuoli, Italy, and used since Roman times as an ingredient in some cements and mortars.

ppb. Parts per billion (i.e. parts per 10^9).

p–p factor. *See* NICOTINIC ACID.

PPLO. Pleuropneumonia-like organisms. (*see* MYCOPLASMAS).

ppm. Parts per million.

prairie. In North America the natural GRASSLAND occupying much of the central part of the continent in mid latitudes. It may have been formed by the clearance of forest, due to a combination of climate change and human intervention, and then sustained by grazing by large herds of animals (e.g., bison).

prairie soil (brunizem). A grassland soil that is dark brown and mildly acidic at the surface, lying over a leached (*see* LEACHING) layer. There is a brownish subsoil grading down to the parent material, with little or no calcium carbonate. Prairie soils are found in generally cool to warm temperate climates. *See also* SOIL CLASSIFICATION, SOIL HORIZONS.

pre-adaptation. The possession of characteristics that provide an advantage for an organism when it is exposed to new conditions (e.g., the lobed fins of DEVONIAN CHOANICHTHYES, from which the legs of AMPHIBIA were evolved).

Precambrian (Cryptozoic). All geological time before the beginning of the CAMBRIAN Period, and rocks formed during that time. Precambrian rocks are grouped into many local stratigraphic subdivisions. Precambrian time is usually divided into the Archaean Era followed by the Proterozoic Era.

precipitation. (1) The settling out of water from cloud in the form of dew, rain, hail, snow, etc. (2) The formation of solid particles in a solution; the settling out of small particles.

Precipitous Bluff. A geological feature in south-western Tasmania, threatened by plans to exploit the LIMESTONE deposits it contains. The top 300 metres of the Bluff are composed of DOLERITE overlying SEDIMENTARY ROCKS including beds of limestone.

preclimax. A stable plant COMMUNITY whose composition reflects less favourable (warmer or drier) climatic conditions than the average for the region. *Compare* POSTCLIMAX, PROCLIMAX, SERCLIMAX, SUBCLIMAX.

preclisere. The series of FORMATIONS that arises when the climate becomes drier. *Compare* POSTCLISERE. *See also* CLISERE.

precocial. Applied to birds (e.g., ducks) whose newly hatched young are well formed, down-covered and able to run about. *Compare* ALTRICIAL.

predator. An animal that kills other animals for food; a secondary consumer in a FOOD

CHAIN. A predator preys externally on other animals, usually destroying more than one individual, whereas a parasite (*see* PARASITISM) often lives on or in a single host without killing it.

preferential species. Species that are present in several COMMUNITIES, but that are predominant or more vigorous in one particular community.

presbycousis (presbyacusis). Hearing loss due to advanced age, usually at high FREQUENCIES.

presence indicator. A species whose presence is taken to indicate the existence of a particular environmental factor.

pressurized water reactor (PWR). A lightwater NUCLEAR REACTOR which uses enriched uranium (*see* URANIUM ENRICHMENT) as a fuel, light water (*compare* HEAVY WATER) as a MODERATOR, and as a coolant light water held in the liquid state under a pressure of 150 atmospheres, contained inside a vessel of welded steel.

prevailing wind. The direction from which the wind in a particular area blows more frequently than any other.

prickly pear. *See* OPUNTIA.

Pridolian. The youngest SERIES of the SILURIAN System in Europe. In the UK, this series is often called the Downtonian, and is placed at the bottom of the DEVONIAN System.

primary air. Air admitted to a furnace or incinerator during the first part of the firing cycle (i.e. together with the fuel).

primary commodity. *See* PRIMARY ECONOMY.

primary consumer. *See* FOOD CHAIN.

primary economy. An economic system in which the principal activities, and so the main sources of wealth and employment, are connected with agriculture, mining and the obtaining of primary commodities (e.g., unprocessed agricultural produce, minerals and other raw materials). *Compare* QUATERNARY ECONOMY, SECONDARY ECONOMY, TERTIARY ECONOMY.

primary environmental quality standards. *See* ENVIRONMENT QUALITY STANDARDS.

primary minerals. Minerals formed directly from the cooling MAGMA and persisting as the original mineral even in SEDIMENTARY ROCKS (e.g., quartz). *Compare* SECONDARY MINERALS.

primary production. *See* PRODUCTION.

primary productivity. *See* PRODUCTIVITY.

primary rock. Rock produced directly from the MAGMA. *Compare* SECONDARY ROCK.

primary succession. *See* SUCCESSION.

primary treatment. A process for removing most floating solids and those that can be made to settle from waste water, and so to reduce the concentration of suspended solids. *Compare* SECONDARY TREATMENT.

Primates. A mainly arboreal order of MAMMALIA that comprises the ANTHROPOIDEA (monkeys, apes, humans) and the more primitive Prosimii (lemurs, lorises, tarsiers). Primates have large CEREBRAL HEMISPHERES and eyes directed forward, and their thumbs (and sometimes also big toes) are opposable to the other digits.

prisere. A natural plant SUCCESSION which begins on bare ground and culminates in a CLIMAX.

proactinium. *See* ACTINIDES.

probe. An apparatus for sampling or measuring at a distance from the instruments that receive or analyze the data obtained. Probes are commonly used to obtain information from environments that are inaccessible or hostile (e.g., space probes exploring other planets, probes used for chimney or duct sampling). *See also* PITOT TUBE.

Proboscidea (elephants, mammoths, mastodons). An order of herbivorous placental mammals (EUTHERIA). Modern elephants (two species) are characterized by their great size, stout legs, long tusks, trunks and by having three sets of large, grinding molars, developed in a series so that only two pairs are in use at any one time. The earliest known proboscidean (*Moeritherium*) was only 60 centimetres tall and had more complete dentition, small tusks and probably a short snout.

proboscis worms. *See* NEMERTEA.

procaryotic. *See* PROKARYOTIC.

Procellariiformes (tubenoses). An order of PELAGIC oceanic birds that includes petrels, albatrosses, fulmars and shearwaters. They have external tubular nostrils, hooked beaks and long, narrow wings, and they nest in colonies on remote shores.

proclimax. According to terminology introduced by F.E. Clements, any plant COMMUNITY that resembles the CLIMAX in stability or permanence, but lacks the composition expected in the existing climate. *See also* SUBCLIMAX, SERCLIMAX, PRECLIMAX, POSTCLIMAX, DISCLIMAX.

producer. *See* FOOD CHAIN.

production. (1) Gross production rate; the rate of assimilation shown by organisms of a given trophic level. (2) Gross primary production; the assimilation of organic matter or biocontent by a plant COMMUNITY during a specified period. (3) Net primary production; the BIOMASS or biocontent incorporated into a plant community during a specified period of time. (4) Net aerial production; the biomass or biocontent incorporated into the aerial parts (leaf, stem, leaves and associated organs) of a plant community. (5) Net production rate; the assimilation rate (i.e. gross production rate)

minus losses of matter by predation, RESPIRATION and DECOMPOSITION. (6) Primary production; the total quantity of organic matter newly formed by PHOTOSYNTHESIS. *See also* PRODUCTIVITY.

production ecology. The study of BIOMES in terms of the production and distribution of food, and hence the flow of energy within them.

production line. A manufacturing process that breaks fabrication into simple, discrete elements, each of which is performed by one person or group of people, in order to increase individual productivity.

production rate. The number of organisms formed within an area during a given period of time.

production residues. Wastes that result from production and distribution. *Compare* CONSUMPTION RESIDUES.

productivity. (1) Primary productivity; the amount of organic matter made in a given time by the ATROPHIC organisms in an ecosystem. (2) Net productivity; the amount of organic matter produced in excess of that used up by the producing organisms during RESPIRATION, thus representing potential food for the consumers of the ecosystem (*see* FOOD CHAIN). (3) In forestry, the total timber yield per annum. The productivity rating index is the expected yield per unit area divided by the standard yield for the same area and expressed as a percentage.

productivity rating index. *See* PRODUCTIVITY.

product standard. A standard calculated for products, especially for such consumable products as food and detergents, that aims to ensure a particular quality of effluent discharged from their production or use.

profundal zone. The zone of a lake, comprising the deep water and the lake bottom, that lies below the compensation depth (*see* COMPENSATION POINT).

progesterone. A HORMONE secreted by the CORPUS LUTEUM and the PLACENTA in mammals. It is responsible for preparing the uterus for the implantation of the EMBRYO(S) and inhibits OVULATION during pregnancy.

proglottides. *See* CESTODA.

Project Independence. A proposal by the US government, made in January 1974, whose aim was to develop the energy resources of the USA so that by 1980 the country would be self-sufficient in energy and so invulnerable to pressures from overseas suppliers of oil. The Project required the relaxation of many restrictions imposed in earlier years to improve the quality of the environment. Recently it seems to have had little importance in US energy policy considerations.

prokaryote. *See* PROKARYOTIC.

prokaryotic (procaryotic).   Applied to organisms or cells (i.e. prokaryotes) whose genetic material (filaments of DNA) is not enclosed by a nuclear membrane, and that do not possess MITOCHONDRIA or PLASTIDS. BACTERIA and CYANOPHYTA are the only prokaryotic organisms. *Compare* EUKARYOTIC.

prolactin. *See* LACTOGENIC HORMONE.

proline. An AMINO ACID with the formula $NH(CH_2)CHCOOH$ and a molecular weight of 115.1n

pronatalist. *See* ANTINATALIST.

propagule (propagulum).   Any part of a plant (e.g., seed, cutting) that is capable of forming a new individual when separated from the original plant.

propanone. *See* ACETONE.

prophage. *See* LYSOGENIC BACTERIUM.

prophase. *See* MITOSIS.

propolis. A plant resin used by bees for sealing crevices in the hive.

proprioceptor. A sense organ (e.g., the balancing organ of the inner ear; receptors in joints, muscles or blood vessel walls) by which an animal receives information regarding its position or the movements and other changes going on inside its body. Using this information the animal automatically coordinates its movements. The term does not cover sense organs that detect substances introduced into the alimentary canal or respiratory system (i.e. INTEROCEPTOR).

Prosimii. *See* PRIMATES.

Prosobranchia. *See* GASTROPODA.

protandrous. (1) Applied to flowers (e.g., rosebay willow-herb, *Chamaenerion angustifolium*) whose anthers mature before their carpels. *Compare* PROTOGYNOUS. *See also* DICHOGAMY. (2) Applied to hermaphroditic (*see* HERMAPHRODITE) animals that produce first sperm and then eggs (e.g., some NEMATODA).

proteins. Very complex nitrogenous organic compounds made up of many AMINO ACID molecules linked to form one or more chains. Proteins are the basic constituents of living organisms and form an essential part of the food of HETEROTROPHIC organisms. The possible number of combinations of amino acids is immense, so very large numbers of proteins exist, and each species has proteins peculiar to itself. *See also* ANTIBODY, ANTIGEN, CYTOCHROME, ENZYME, GLOBULINS, HAEMOGLOBIN.

proteoclastic enzymes. *See* PROTEOLYTIC ENZYMES.

proteolytic enzymes (proteoclastic enzymes).   ENZYMES that have the power to decompose or hydrolyze (*see* HYDROLYSIS) PROTEINS.

Proterozoic. *See* PRECAMBRIAN.

prothallus. The gametophyte in ferns and

related plants (division: PTERIDOPHYTA). It is a small, but independent generation bearing the sex organs. It shows no differentiation into stems, roots and leaves, but usually forms a green THALLUS prostrate on the soil and attached by rhizoids. *See also* ALTERNATION OF GENERATIONS.

Protista. The kingdom (in some classifications a superphylum or phylum) that comprises all the simple organisms (i.e. BACTERIA, MYXOMYCOPHYTA, PROTOZOA and some ALGAE). The term was formerly restricted to unicellular organisms, and some authorities now regard it as obsolete, with the PROKARYOTIC organisms forming the kingdom MONERA.

proto-. From the Greek *protos*, meaning first, a prefix denoting first or original.

protochordates. *See* CHORDATA.

protogynous. Applied to flowers (e.g., figwort, *Scrophularia*) whose carpels mature before their POLLEN is shed. *Compare* PROTANDROUS. *See also* DICHOGAMY.

Protomyxa. *See* MYXOMYCOPHYTA.

Protoneolithic. *See* MESOLITHIC.

protoplasm. The living matter of a cell, comprising both the NUCLEUS and the CYTOPLASM.

protoplast. The PROTOPLASM, as distinct from the non-living cell wall of a plant cell.

Protopterus. *See* DIPNOI.

protore. A primary mineral deposit from which an economic ORE may be formed by enrichment (e.g., LEACHING) or SECONDARY ENRICHMENT. (2) A deposit that may become economically workable with technological change or an increase in price.

protosoil. A soil at an early stage of development.

Prototheria. *See* MONOTREMATA.

prototroph. A strain of microorganisms, differing from related organisms in the natural state usually as a result of MUTATION, that has no nutritional requirements other than those necessary to the organism in its natural state. *Compare* AUXOTROPH.

prototype. The original, from which later versions follow, or the first, full-scale working model of a new technological device, used for testing under actual operation conditions. *See also* ORIGINAL.

prototype fast reactor (PFR). The UK 250 megawatt fast BREEDER REACTOR built at Dounreay, Caithness, Scotland.

Protozoa. A subkingdom and phylum comprising animals whose bodies are not divided into cells (i.e. they are unicellular or, according to some authorities, non-cellular). *See also* CILIOPHORA, FLAGELLATA, RHIZOPODA, SPOROZOA.

provenance. The terrane or parent rock from which the clasts (*see* CLASTIC) of a sediment are derived. Sediments derived from one source are termed monogenetic, those from several sources polygenetic.

psammitic. Arenaceous (*see* ARENITE) and applied mainly to metamorphic (*see* METAMORPHISM) rocks formed from arenaceous rocks.

psammon. The organisms that inhabit the water lying between grains of sand.

psammophyte. A plant that grows in sandy soil.

psammosere. The stages in a plant SUCCESSION that begins in sandy soil.

pseudaposematic coloration. *See* BATESIAN MIMICRY.

pseudepisematic characters. Lures (e.g., the filamentous dorsal fin forming the 'fishing line' of the angler fish *Lophius*) which enable certain animals to catch their prey.

pseudogamy. The development of an ovum when it is stimulated by the entry of a male GAMETE whose NUCLEUS does not fuse with the egg nucleus. This occurs in some seed plants (SPERMATOPHYTA) and some nematode worms (*see* NEMATODA).

pseudokarst. KARST-like terrain, not on LIMESTONE, whose rough TOPOGRAPHY is due to causes other than solution.

pseudopodium. A temporary protrusion of CYTOPLASM from the surface of a cell (e.g., *AMOEBA*, some white blood cells), serving for locomotion or the ingestion of particles.

psilomelane. A mixture of manganese oxides, and an important source of manganese, formed at or near the surface as a SECONDARY MINERAL.

Psilophytales. An order of extinct PTERIDO-PHYTA, common in the DEVONIAN Period, which were small evergreen plants with horizontal RHIZOMES and erect forked branches in some species bearing many small leaves. The plants reproduced by SPORES formed in terminal sporangia (*see* SPORANGIUM).

Psilopsida. The group of plants, within the TRACHEOPHYTA, that includes the PSILO-PHYTALES and PSILOTALES.

Psilotales. An order of PTERIDOPHYTA, probably related to the fossil order PSILO-PHYTALES, and now represented by a few small plants (*Psilotum* and *Tmesipteris*), most of which are EPIPHYTES found in tropical and subtropical areas.

Psittaciformes (parrots). An order of tree-dwelling birds found mainly in warm climates. They are mainly vegetarian, and some use their powerful beaks to break open hard nutshells. Most nest in holes.

psocids. *See* PSOCOPTERA.

Psocoptera (psocids, barklice, booklice, dustlice). An order of small soft-bodied insects (division: EXOPTERYGOTA), with long antennae and biting mouthparts, some of which are wingless. They live among vegetation, paper or dried materials, feeding on fungi and other organic matter. Some species harbour sheep tapeworms (CESTODA).

Psophocarpus tetragonolobus. *See* WINGED BEAN.

psychosocial stressors. Stimuli which are suspected of causing diseases and which originate in social relationships, affecting the organism through the higher nervous processes.

psychosomatic. Applied to physical effects, including many diseases, caused or influenced by mental processes.

psychosphere. Human thought and culture as an environmental phenomenon. Some writers have suggested that the psychosphere be used to complement concepts of ATMOSPHERE, BIOSPHERE, HYDROSPHERE and LITHOSPERE. *Compare* TECHNOSPHERE.

psychrometer. An instrument for measuring dry bulb and WET BULB TEMPERATURE.

psychrometry. (1) The measurement of the humidity of the atmosphere. (2) The thermodynamics of air and water vapour mixtures, especially as applied to air conditioning.

psychrophilic microorganisms. Microorganisms whose optimum temperature for growth lies below 20°C. *Compare* MESO-PHILIC MICROORGANISMS, THERMOPHILIC MI-CROORGANISMS.

Pteridium aquilinum. *See* BRACKEN.

Pteridophyta. A division of plants, mainly terrestrial, that includes the present-day ferns (FILICALES), clubmosses (LYCOPODIA-LES), horsetails (EQUISETALES) and PSILO-TALES. Pteridophytes have proper roots, stems and leaves, and a well-developed VAS-CULAR system. Asexual SPORES are produced on SPOROPHYLLS, which often resemble foliage leaves or are grouped to form cones. The gametophyte (*see* ALTERNATION OF GENERATIONS) is a small, green PROTHALLUS. Fossil orders include the PSILOPHYTALES and SPENOPHYLLALES.

Pteropsida. The group of TRACHEOPHYTA that includes the ferns (FILICALES) and seed plants (SPERMATOPHYTA).

pteroylglutamic acid. *See* FOLIC ACID.

Pterygota (Metabola). All insects, except those comprising the APTERYGOTA (primitive wingless insects). Some Pterygota are wingless (e.g., lice, fleas), but these are thought to have evolved from winged forms. *See also* ENDOPTERYGOTA, EXOPTERYGOTA.

Pu. *See* PLUTONIUM.

pubescent. Of a plant, covered with soft short hairs.

Puccinia graminis. *See* BERBERIS VULGARIS, RUST FUNGI.

puddingstone. A popular name for CONGLOMERATE: now usually restricted to particular strata such as the EOCENE Hertfordshire Puddingstone.

puddling. (1) The trampling of wet soil by livestock; it causes the soil to become less permeable (*see* PERMEABILITY). (2) An ancient method for making a pond watertight by lining the bottom with clay which was trampled to make it impermeable. *See also* DEW POND.

puffins. *See* ALCIDAE.

Pulmonata. *See* GASTROPODA.

pulses. Food crops of the family LEGUMINOSAE, including the pea (*Pisum sativum*), lentil (*Lens culinaris*), lablab (*Dolichos lablab*), butter bean (*Phaseolus lunatus*), chick pea (*Cicer arietinum*), black gram (*Phaseolus mungo*), green gram (*Phaseolus aureus*), etc., which form important protein foods in many countries.

pulverization (milling, shredding). An intermediate step in refuse disposal in which refuse is broken into small particles, so reducing its volume.

pumice. A highly VESICULAR, ACID PYROCLASTIC rock. *Compare* SCORIA.

pumped storage. A system for generating peak load electricity by turbines turned by a fall of water from a reservoir filled by pumps that lift water from a lower level. During base load periods, when demand for electrical power is steady or low, water is lifted from the lower to the higher level; at times of peak demand it is released. In certain situations wind power can be used for the pumping operation.

pumping station. An installation for raising sewage to a higher elevation, or for pumping mains water supplies.

pupa (chrysalis). The stage between a LARVA and adult in ENDOPTERYGOTA. The pupa appears from the outside to be quiescent, but inside the insect's body is being remodelled. The term chrysalis is often used only for the pupae of LEPIDOPTERA (butterflies and moths). *See also* METAMORPHOSIS.

Purbeckian. The uppermost STAGE of the European JURASSIC System.

pure line. A succession of generations that are homozygous (*see* HOMOZYGOTE) for all GENES. Pure lines are achieved by intensive inbreeding (*see* ENDOGAMY).

pure tone. A sound whose waveform is sinusoidal (*see* SINE WAVE).

putrefaction. The decomposition of organic matter with incomplete oxidation and the production of noxious gases.

putrescible wastes. Wastes of animal or vegetable origin that are degraded bacteriologically.

puy. A volcanic PLUG, especially one in the Auvergne, France.

PVC. *See* POLYVINYL CHLORIDE.

P wave. The primary wave reaching a SEISMOGRAPH from an earthquake. P waves are compressional and travel through the crust, mantle and core. *Compare* S WAVE.

PWR. *See* PRESSURIZED WATER REACTOR.

pycnium. *See* SPERMOGONIUM.

Pycnogonida. *See* PANTOPODA.

pyrethrum. An INSECTICIDE, prepared from the flowers of *Chrysanthemum cinerariaefolium*, that breaks down rapidly after application, does not harm plants, is not very toxic to vertebrates and therefore has a minimal effect on wildlife.

pyridoxine. A VITAMIN of the B group (B_6); an essential for the formation of a COENZYME in many organisms, including birds, mammals and some BACTERIA.

pyrites (fool's gold, FeS_2). A widespread iron sulphide mineral used in the manufacture of SULPHURIC ACID. Because cobalt substitutes for iron in the crystal lattice, pyrites is also an important source of cobalt.

pyroclastic. Applied to material (TEPHRA) blown out by an explosive volcanic eruption and subsequently deposited. Pyroclastic deposits include TUFFS, IGNIMBRITES and some AGGLOMERATES.

pyroclimax. *See* FIRE CLIMAX.

pyrolusite (MnO_2). One of the major ORE MINERALS of MANGANESE, occurring in SEDIMENTARY ROCKS or as a residual deposit concentrated by LEACHING. Pyrolusite is also one of the manganese minerals in deep-sea MANGANESE NODULES and occurs commonly as dendritic forms on JOINT surfaces. Manganese is an almost ubiquitous alloying metal.

pyrolysis. The DESTRUCTIVE DISTILLATION of combustible material achieved by raising it to a high temperature while restricting severely the supply of oxygen, so it cannot burn. The exhaust gases are highly noxious, but can be contained when pyrolysis is conducted on an industrial scale. On a small scale (e.g., in some wood-burning stoves) the process is highly polluting.

pyrometry. The measurement of high temperatures.

pyroxenes. A group of rock-forming SILICATE MINERALS with a wide range of composition. The general formula approximates to $ABSi_2O_6$, where A is usually magnesium, iron, calcium or sodium, and B is usually magnesium, iron or aluminium, and the silicon can also be partly replaced by aluminium. Pyroxenes occur in IGNEOUS rocks, being characteristic of BASIC and ULTRABASIC varieties, and in some metamorphic (*see* METAMORPHISM) rocks. Augite, basically a calcium magnesium aluminosilicate, is the best known pyroxene.

Pyrrophyta (fire algae). A group of mainly unicellular ALGAE, most of which have flagella (*see* FLAGELLUM). They constitute a large part of the PLANKTON in seas and inland waters. The most important members of the group, the Dynophyceae (Dinoflagellata) are often included in the FLAGELLATA (PROTOZOA), some being without CHLOROPLASTS. When abundant, Dinophyceae can cause luminescence and red water in the sea, and mussel poisoning in humans and some sea birds.

Q

Q_{10} (temperature coefficient). The increase in the rate of a process that is brought about by raising the temperature by 10 degrees celsius. In many living systems, within certain limits, the rate is doubled or more than doubled for each 10-degree increase.

Q_{02}. The oxygen uptake of an organism expressed in microlitres per milligram (dry weight) per hour.

quadrat. A sampling area, often 1 metre square, used in studying the composition of an area of vegetation. The area is usually defined by a frame, sometimes subdivided by fine wires, laid on the ground. *See also* DOMIN SCALE; QUADRAT, MAJOR; QUADRAT METHOD.

quadrat, major. A QUADRAT that includes all the more important species, as well as a number of the less important ones.

quadrat method. A method for the intensive study of an environment within a circumscribed area in order to gain a comprehensive knowledge of the wider area. It involves the laying down of QUADRATS and examining closely the species within them. *See also* DOMIN SCALE, VEGETATION STUDY.

quantum. Any observable quantity is quantized when its magnitude, in some or all of its range, is restricted to a discrete set of values. If the magnitude of the quantity is always a multiple of a definite unit, then that unit is called the quantum. The concept can be applied in many fields.

quartz (SiO_2). A crystalline mineral characteristic of ACID IGNEOUS rocks. Because of its resistance to chemical WEATHERING and its lack of CLEAVAGE, it is very abundant in many SEDIMENTARY ROCKS and metamorphic (*see* METAMORPHISM) rocks. Pure quartz (rock crystal) is water-clear, but slight impurities cause the well-known varieties which include amethyst (purple), rose quartz (pink), citrine (yellow) and cairngorm (brown).

quartzarenite. *See* ORTHOQUARTZITE.

quartzite. A rock composed almost entirely of SILICA, which is either a metamorphic METAQUARTZITE or a sedimentary ORTHOQUARTZITE. Quartzites are quarried extensively for AGGREGATE.

quartzose sandstone. *See* ORTHOQUARTZITE.

quartz porphyry. *See* MICROGRANITE.

quassia. An insecticide, extracted from the roots and timber of trees and shrubs of the tropical family Simaroubaceae, that has been used since 1825 as a fly-killer and later for aphid control. It is still used to control plum sawfly and as a bird and mammal repellant.

Quaternary. GEOLOGICAL TIME since the end of the PLIOCENE (i.e. the PLEISTOCENE and HOLOCENE) and ranked either as an ERA, following the CENOZOIC Era, or a PERIOD of the Cenozoic Era, following the TERTIARY Period. Quaternary also refers to rocks formed during this time.

quaternary economy. An economic system in which the principal activities, and so the main sources of wealth and employment, are no longer related to such primary activities as agriculture or mining nor to industrial manufacturing or the service industries.

Thus in theory quaternary economies follow the kinds of economies found in the world at present and might be based, for example, on the acquisition, processing and dissemination of information. *Compare* PRIMARY ECONOMY, SECONDARY ECONOMY, TERTIARY ECONOMY.

Queensland arrowroot. *See* ACHIRA.

Quercus (oaks). A genus of EVERGREEN or DECIDUOUS trees or shrubs (family: Fagaceae) that produce nuts (acorns) borne in cups. Many species yield bark used for tanning, as well as valuable timber. The bark of *Q. suber* is cork. Oaks are the dominant trees in natural woodland throughout much of the British Isles. The two native species are *Q. robur* — the common oak — which is widespread and has long-stalked acorns, and *Q. petraea* — the durmast oak — characteristic of northern and western areas, which has short-stalked acorns.

quicklime. *See* CALCIUM OXIDE.

quicksilver. *See* MERCURY.

R

Ra. (1) *See* RADIUM. (2) *See* RAYLEIGH NUMBER.

rabbits. *See* LAGOMORPHA.

race. (1) A loose term used to describe MICROSPECIES, permanent varieties or particular breeds of fungi, plants or animals. (2) *See* RHIZOME.

raceme. An unbranched INFLORESCENCE whose individual flowers are stalked.

rachion. The line on a lake shore where wave action causes the most disturbance.

rachis. The main axis of an INFLORESCENCE.

rad. (1) A unit formerly used in measuring the amount of IONIZING RADIATION absorbed by living tissues, 1 rad being equal to 100 ergs of energy per gram of tissue. It has been replaced by the GRAY (Gy); 1 rad = 10^{-2} Gy. (2) *See* RADIAN.

radar. *See* ECHO.

radial drainage. River systems that form a radial pattern. Radial drainage is typical of high mountain areas or systems on volcanic cones.

radially symmetrical. *See* ACTINOMORPHIC.

radian (rad). The supplementary SI unit of plane angle, being the angle subtended at the centre of a circle by an arc equal in length to the radius of the circle. Thus $2 (\pi)$ rad = 360°; 1 rad = 57.296°.

radiation. *See* ADAPTIVE RADIATION, ALPHA-PARTICLES, BETA-PARTICLES, GAMMA-RAYS, ION-IZING RADIATION, NUCLEAR FISSION, NUCLEAR FUSION, NUCLEAR REACTOR.

radiation dose equivalent (dose equivalent). A measure of the effect of ionizing radiation on the substance or tissue that absorbs it. Since the effect of radiation in one kind of biological tissue or organ is different from the effect in another tissue or organ, and some types of radiation are more harmful than others, the absorbed dose, which can be measured, is multiplied by weighting factors stipulated by the International Commission on Radiological Protection to give a dose equivalent, for which the SI unit is the SIEVERT. The value in sieverts represents the risk to health from that amount of radiation had it been absorbed uniformly throughout the body.

radiation fog. *See* FOG.

radiation window. A band in the radiation spectrum extending from 8.5 to 11.0 micrometres in wavelength in which little absorption by water vapour occurs. In clear skies radiation from the ground in this band can escape to space, whereas the remainder is absorbed.

radical. A group of atoms present in a series of compounds that maintains its identity through chemical changes which affect the rest of the molecule. A radical is generally incapable of independent existence.

radicle. The part of an EMBRYO in a seed that develops into a root during germination.

Radioactive Substances Act, 1960. *See* RADIOACTIVE WASTE.

radioactive waste. Waste that consists of, or is contaminated with, RADIONUCLIDES. According to their radioactive content, wastes are divided into three categories: LOW-LEVEL, INTERMEDIATE-LEVEL AND HIGH-LEVEL. In the UK, regulations under the Radioactive Substances Act, 1960, require all radioactive wastes to be isolated so they cannot come into direct or indirect contact with humans and then to be disposed of by approved methods. NIREX is the agency concerned with development of facilities for the disposal of low- and intermediate-level wastes. No decision has yet been taken on the final disposal of high-level wastes.

radioactivity. A property exhibited by unstable isotopes of elements that decay, emitting radiation, principally as ALPHA-PARTICLES, BETA-PARTICLES and GAMMA-RAYS.

Radiolaria. *See* RHIZOPODA.

radiolarian ooze. A deep-sea deposit composed chiefly of the siliceous skeletons of Radiolaria (*see* RHIZOPODA).

radiometric age. The time, measured in years before the present, that it has taken for a particular ratio of 'daughter' to 'parent' atoms to be formed by the radioactive decay of the parent atom (*see* RADIOACTIVITY). This assumes a closed system and the absence of native daughter atoms in the original material, and depends on the unique HALF-LIFE ($t_{1/2}$) of particular radioactive nuclides. Of the many radioactive nuclides in nature, four have been found most suitable for determining the radiometric age of rocks: uranium-238 ($t_{1/2}$ = 4510 my) decaying to lead-206; uranium-235 ($t_{1/2}$ = 713 my) decaying to lead-207; potassium-40 ($t_{1/2}$ = 12 700 my) decaying to argon-40; rubidium-87 (t = 470 000 my) decaying to strontium-87. Carbon-14 is a rare isotope occurring in living plants and animals and is continually renewed in the atmosphere by the bombardment of nitrogen-14 by COSMIC RAYS. Biological material up to about 40 000 years ± 5000 years old can be carbon-14 dated by determining the ratio of carbon-14 to carbon-12 in the sample. When a living organism dies it ceases to assimilate carbon-14 and that present decays relatively rapidly. Carbon-14 dating has provided a very useful tool for archaeologists and students of recent Earth history, but has proved unreliable for dating material younger than 1000 BC. *Compare* DENDROCHRONOLOGY.

radionuclide. An unstable NUCLIDE that undergoes spontaneous radioactive decay, emitting radiation as it does so, and changing eventually from one element into another.

radiosonde. *See* BALLOON.

radium (Ra). A very rare, naturally-occurring radioactive, metallic element, whose most stable isotope ($^{226}_{88}$Ra) has a HALF-LIFE of 1620 years. A_r=88; SG 5; mp 700° C.

radon (Rn). An element occurring naturally as a colourless, odourless noble gas, chemically almost inert, that is the immediate breakdown product of radium-226. Of its more than 20 isotopes the most stable is radon-222, with a HALF-LIFE of 3.8 days. Since there is a small amount of RADIUM in most rocks, radon is present close to ground level almost everywhere, although concentrations vary widely from place to place. Its decay products (radon daughters), known as radium A, B, C, C' and C", are very short-lived, but adhere to surfaces. Radon is suspected of causing lung cancer, and some countries have monitored the interior of buildings where levels are likely to be high and have recommended exposure limits.

radon daughters. *See* RADON.

rainbow. A circular arc of coloured light that displays the colours of the spectrum (violet, indigo, blue, green, yellow, orange, red) with red on the outside. The centre of the arc is opposite to the Sun, the rainbow never describes more than a half circle, and the higher the Sun the smaller the arc. Rainbows are caused by the reflection and refraction of light in water droplets in such a way that the light emerges split into its spectrum colours at an angle of 42° to the direction of the Sun's rays, so producing an arc with a radius of

42°. The formation of a rainbow requires water droplets to be falling and the Sun to be shining simultaneously, so that rainbows occur in showery weather.

Rainbow Group. *See* GREENS.

Rainbow Warrior. A converted trawler, owned and operated by GREENPEACE, that took part in many protests against whaling, nuclear weapons testing and marine pollution. It was sunk by French secret agents in July 1985, in Auckland Harbour, New Zealand, subsequently refloated and then scuttled in December 1987.

rain forest. *See* FOREST.

rainmaking. *See* ARTIFICIAL RAIN.

rainout. The removal of particulate matter from the atmosphere by formation of water droplets on the particles which act as CONDENSATION NUCLEI, followed by rain. This is generally a more effective removal mechanism than WASHOUT.

raise (rise). In mining, a small tunnel excavated upwards from a DRIVE or LEVEL.

raised beach. A former beach, made from material left above the present high-water mark when the sea level was higher relative to the land than at present. *See also* RAISED BEACH PLATFORM.

raised beach platform. A former beach, occurring as a platform above the present high-water mark, and cut by waves at a time when sea level was higher relative to the land than at present. It may be due both to EUSTATIC changes in sea level and the raising of the land due to isostatic (*see* ISOSTASY) readjustments.

raised bog (raised mire). An area of OMBROGENOUS acid peatland with a convex profile found on level FLOOD PLAINS of mature rivers. They have their origin in TOPOGENOUS MIRE that has grown in height, so becoming isolated from ground water. *See also* BLANKET BOG.

raised mire. *See* RAISED BOG.

ramet. An individual member of a CLONE. *See also* ORTET.

rammed earth (pisé de terre). A building technique in which subsoil, preferably rather sandy and sometimes with the addition of a stabilizer of cement or other substance, is rammed into wooden shutters or made into blocks in a press. *See also* ADOBE.

Ramsar Convention on Wetlands of International Importance. A conference held in Ramsar, Iran, in 1971 led to the drafting of the Ramsar Convention, which is concerned with protecting the international chain of habitats used by migratory water birds. Over 200 sites, mainly in Europe and North Africa, are designated as areas to be protected. By 1982 the UK had 19 Ramsar sites.

random error. An error that distorts particular statistical results, but balances overall. *See also* BIAS.

random noise. A fluctuating quantity (of sound, electromagnetic radiation, etc.) whose amplitude distribution is GAUSSIAN. *See also* NOISE.

range. (1) Of an animal or group of animals, the area within which they seek food. (2) In agriculture, an area of unenclosed pasture within which livestock are allowed to graze at will. *See also* RANGE MANAGEMENT.

range management. The planning and management of the use of grazing land (*see* RANGE) in order to sustain maximum livestock production consistent with the conservation of the range resource.

rank. The stage reached by coal in the course of its carbonification. The chief ranks of coal, in order of increasing carbon content, are: lignite; subbituminous coal; BITUMINOUS COAL; ANTHRACITE.

Ranunculaceae. A family of DICOTYLEDONEAE, most of which are herbaceous perennials of northern temperate and arctic regions. Their floral structure shows primitive features (e.g., in the spiral arrangement of many free stamens and carpels in the buttercup), but many species (e.g., water crowfoot, *Ranunculus aquatilis*) are highly adapted to specialized habitats. Many (e.g., *Clematis, Delphinium, Anemone*) have showy flowers and are cultivated in gardens. Most Ranunculaceae are acrid, and they are often very poisonous because of the presence of ALKALOIDS. Some have been used in medicine (e.g., *Aconitum* as an analgesic). Buttercups (*Ranunculus repens, R. acris*) are poisonous weeds, but are normally avoided by grazing animals.

rape. *See* BRASSICA.

rapid-hardening cement. *See* CEMENT.

Raptores. *See* FALCONIFORMES.

rarity. The INTERNATIONAL UNION FOR CONSERVATION OF NATURE AND NATURAL RESOURCES has drawn up the following definitions of degrees of rarity, and all plants and animals in these categories are in need of special protection. (a) Endangered taxa, those whose numbers have been reduced to a critical level or whose habitats have been so drastically reduced that they are deemed to be in immediate danger of extinction if the causal factors continue operating. (b) Vulnerable taxa, those which are believed likely to move into the endangered category in the near future if the causal factors continue operating, because most or all of their populations are decreasing because of over-exploitation, extensive destruction of habitat or other environmental disturbance, those whose populations have been seriously depleted and whose ultimate security is not yet assured, or those whose populations are still abundant but under threat from serious adverse factors throughout their RANGE. (c) Rare taxa, those with small populations that are not at present endangered or vulnerable, but are at risk, usually because they are localized within restricted geographical areas or habitats, or are thinly scattered over a more extensive range.

Ratitae (ratite birds). A diverse, non-taxonomic group of flightless birds that have reduced wings and sternum, long legs and curly feathers. Many are large. They include ostriches (Africa and south-western Asia), rheas (South America), emus and cassowaries (Australasia), moas (extinct birds of New Zealand) and kiwis (New Zealand).

rattan. *See* PALMAE.

rattomorphia. The tendency to extrapolate into human situations information derived from observations of animal behaviour.

Raunkiaer's life forms. A classification of plants based on the type of organs (often buds) that they possess to survive unfavourable periods, and the position of these organs in relation to soil level. (a) Phanerophytes are woody plants (e.g., trees, shrubs) in which the perennating parts are more than 25 centimetres above ground level. (b) Chamaeophytes are woody plants in which the perennating parts are above the ground, but below the 25 centimetres level. (c) Hemicryptophytes are herbaceous plants in which the perennating parts are at soil level, often protected by dead portions of the plant. (d) Geophytes are herbaceous plants in which the perennating parts are below ground level. (e) Hydrophytes are herbaceous plants in which the perennating parts lie in water. (f) Helophytes are herbaceous plants of marshes, in which the perennating parts lie in mud. (g) Therophytes are herbaceous plants that survive unfavourable periods as seeds. Cryptophytes comprise the classes (geophytes, helophytes, hydrophytes) in which the perennating parts are covered by soil or water.

raw humus. *See* MOR.

ray. *See* CHARACTERISTIC IMPEDENCE.

Rayleigh number (Ra). The non-dimensional number that represents the interplay of forces tending to produce and subdue buoyant convection. Thus $Ra = (g \beta h^4)/(kV)$, where g is gravity, $\beta = (1/\rho) \times (\delta\rho/\delta z)$ where ρ is density and z is height, h is the depth of the layer of fluid, k is the thermal conductivity and V is the kinematic viscosity. ρ may be replaced by $(\alpha/T) \times (\delta T/z)$ where T is the absolute temperature and α the coefficient of thermal expansion.

Rayleigh scattering. The scattering of visible light by air molecules in a predictable and ISOTROPIC way. Particles of more than about 0.1 micrometre radius scatter light in a more complicated way, much of it in a forward direction (*see* MIE SCATTERING) relatively independently of the wavelength. They also absorb sunlight to some extent.

razorbills. *See* ALCIDAE.

realgar. *See* ARSENIC.

realists. In the West German GREEN PARTY, the faction that favours exerting influence by forming alliances with other, larger political parties, especially the SPD (socialists). *Compare* FUNDAMENTALISTS.

Recent. *See* GEOLOGICAL TIME, HOLOCENE.

receptacle. *See* FLOWER.

receptor. A sense organ (*see* CHEMORECEPTOR, EAR, EXTEROCEPTOR, EYE, INTEROCEPTOR, PROPRIOCEPTOR).

recessive. Applied to one of a pair of contrasted characters that is not developed in a heterozygous (*see* HETEROZYGOTE) individual (i.e. in the presence of the dominant gene). *Compare* DOMINANT.

recharge. The process of renewing underground water by infiltration during wet seasons. *See also* ARTIFICIAL RECHARGE.

recharge well. *See* ARTIFICIAL RECHARGE.

recumbent fold. A FOLD with a horizontal, or nearly horizontal, AXIAL PLANE.

recycling. The recovery and reuse of materials from wastes.

red algae. *See* RHODOPHYTA.

red beds. An assemblage of SEDIMENTARY ROCKS formed in a highly oxidizing environment so that the IRON present is in the ferric state. Red beds are probably indicative of an arid continental environment, and the term is often used not only to describe the red MARLS, SHALES and SANDSTONES of the New Red Sandstone, but associated BRECCIAS and EVAPORITES as well.

red blood cells (erythrocytes). *See* BLOOD CORPUSCLES.

red clay. A soft, deep-sea deposit rich in iron oxides and found at depths of over 200 fathoms (about 4 kilometres). It is made up of wind-blown dust (including volcanic particles), MANGANESE NODULES, dust from meteors and METEORITES, and insoluble organic remains such as sharks' teeth, whales' ear bones, etc.

Red Data Book. A collection of all the available data on species threatened with extinction, maintained by the International Union for Conservation of Nature and Natural Resources. *See also* RARITY.

redia. *See* TREMATODA.

red pepper. *See* SOLANACEAE.

red spider. *See* ACARINA.

red tide. *See* DINOFLAGELLATES.

reducer (decomposer). An organism (e.g., BACTERIA, FUNGI) that breaks down dead organic matter into simpler compounds.

reduction. *See* OXIDATION.

reduction division. *See* MEIOSIS.

reductionism. The belief that a system may be understood by a detailed examination of its components, based on dissection and analysis. *Compare* HOLISM, VITALISM.

redwood. *See* TAXODIACEAE.

reef. A ridge of sand, shingle, rock or coral

(*see* CORAL REEF) in the sea, whose upper level is at, or only slightly above or below, mean sea level.

reef ecosystem. The community of organisms living on and close to a CORAL REEF, and the reef itself. The water surrounding the reef is clear and contains very little nutrient. At the base of the reef ecosystem are the photosynthesizing ZOOXANTHELLAE and other algae living symbiotically (*see* SYMBIOSIS) with the coral POLYPS. They support large numbers of consumer organisms, including a wide variety of fishes. Boring sponges (*see* PORIFERA) and bivalves attack the surface and base of the reef. The polyps themselves are filter-feeders, subsisting mainly on material dislodged from the reef by other organisms. The reef itself traps sediments, providing additional nutrients. Although reefs occur in marine conditions that are the equivalent of deserts, they usually support rich and complex communities. The GREAT BARRIER REEF is said to support the richest ecosystem on Earth.

refection (autocoprophagy). An animal's habit of eating its own faeces, practised by some herbivorous animals (e.g., rabbits).

reflex. A simple form of animal behaviour in which a stimulus evokes a specific automatic and often unconscious response (e.g., touching a hot object causes immediate withdrawal of the affected part of the body). The response depends on the existence of an inborn nervous pathway – the reflex arc – in which nerve impulses pass from a sense organ along nerve fibres to an effector organ (e.g., muscle, gland) which brings about the response. *See also* CONDITIONED REFLEX.

refrigerant. A substance that is suitable as the working medium of a cycle of operations producing refrigeration. Such liquids include AMMONIA and freon (*see* CHLOROFLUOROCARBONS).

refugium. An area that has escaped great changes undergone by the region as a whole and so often provides conditions in which RELICT colonies can survive. An example is a

driftless area (*see* NUNATAK) which escaped the effects of glaciation because it projected above the ice.

reg. A desert region with a surface of pebbles. Reg is chiefly used in Algeria; serir is the term used in Libya and Egypt. *Compare* ERG, HAMADA.

Regional Convention for the Conservation of the Red Sea and Gulf of Aden. The international agreement, drawn up as part of the REGIONAL SEAS PROGRAMME, establishing an organization to monitor, prevent and control pollution from land-based sources in Saudi Arabia, Yemen Arab Republic, Yemen (Aden), Jordan, Sudan and Somalia entering the Red Sea, Gulf of Aqaba, Gulf of Suez, Suez Canal and Gulf of Aden.

regional metamorphism. *See* METAMORPHISM.

regional park. In British planning, an area that is intended primarily for recreational use and offers a wide range of recreation facilities. It is generally larger and more varied than a COUNTRY PARK, and is intended to serve a wider catchment area, but is smaller than a NATIONAL PARK and less emphasis is placed on its landscape value.

regional sea. A term used by the UN to designate seas that are land-locked or whose waters mix only slowly with those of the oceans, so they are especially susceptible to pollution from coastal states (e.g. Mediterranean Sea, Baltic Sea, Caribbean Sea, Persian Gulf, Malacca Straits). *See also* REGIONAL SEAS PROGRAMME.

Regional Seas Programme. A series of UNITED NATIONS ENVIRONMENTAL PROGRAMME plans to reduce pollution in REGIONAL SEAS, which are particularly vulnerable. Governments of countries bordering each sea are approached and international conferences held. When agreements have been reached and signed their implementation is passed to the governments concerned. The Regional Seas Programme covers the Mediterranean Sea, Caribbean Sea, Red Sea and

Gulf of Aden, south-west Pacific, south east Pacific and the seas off Kuwait, West and Central Africa, East Africa and East Asia.

regolith. Loose, unconsolidated and broken rock material covering BEDROCK

regosol. Weakly developed soil. *See also* REGOLITH, SOIL CLASSIFICATION.

regression. (1) The dependence of one variable upon another independent variable. Regression analysis seeks to find both the degree and mathematical form of the dependence which can be expressed in a regression equation. (2) The destruction of vegetation (e.g., by fire, grazing, etc.) and subsequent colonization at a lower level (e.g., the replacement of forest by grasses following the destruction of the trees). (3) *See* MARINE REGRESSION.

regressive. Applied to a body of water or to sediments associated with a lowering of sea level. *Compare* TRANSGRESSIVE. *See also* MARINE REGRESSION.

regulatory genes. *See* GENES.

rejuvenation. An interruption in the evolution of a landscape by TECTONIC movements, which alters its form and presents a new landscape for the agents of wind and water to sculpt.

relative abundance. The measure of the abundance of a species as an indication of the extent to which it is able to maintain itself under the prevailing conditions of environment and competition.

relative humidity. *See* HUMIDITY.

relative transpiration. The rate of TRANSPI-RATION per unit area from a plant surface, divided by the rate of evaporation from an equivalent area of open water surface under similar climatic conditions.

releaser stimulus. A STIMULUS that initiates an instinctive behaviour pattern in an animal. If the stimulus comes from a member of

the same species it is called a social releaser.

relict (relic). (1) Applied to an organism or group of organisms representing the surviving remnants of a population that was formerly more widespread or characteristic of the area. *See also* REFUGIUM. (2) Applied to sediments formed on the sea bed under different environmental conditions to those prevailing at present.

rem (roentgen equivalent man). The unit dose of IONIZING RADIATION that gives the same biological effect as that due to 1 roentgen of X-rays. The rem has been replaced by the SIEVERT (1 sievert = 100 rem). *See also* CURIE, RAD, ROENTGEN.

remote sensing. The obtaining of information at a distance from the subject being examined, usually from above. Instruments carried in ships (e.g., SONAR) scan the waters and seabed, those in aircraft or satellites scan the atmosphere and ground surface. Photographic techniques are not confined to the visible light spectrum (an alternative is IN-FRARED PHOTOGRAPH) nor to the representation of natural colours; by using false colours they can highlight contrasts between features of interest. *See also* LANDSAT, PHOTOGE-OLOGY, SPOT.

rendzina soils. Dark grey or black organic surface layers developed over soft, light CAL-CAREOUS material that is derived from CHALK, LIMESTONE or MARL. *See also* SOIL CLASSIFICA-TION, SOIL HORIZONS.

renewable resource. A RESOURCE that can be exploited without depletion because it is constantly replenished. This includes agricultural crops and fish, provided stocks are not over fished, and is extended to cover the energy of solar radiation, wind, waves and tides. *Compare* NON-RENEWABLE RESOURCE.

rentalism. That economic system, or part of an economic system, which is based on the short-term renting of goods rather than on private ownership. Rentalism tends to be characteristic of highly industrialized, affluent and mobile societies.

replica. In modelling (*see* MODEL), a complete reconstruction of the original in all its structural and functional details, which therefore cannot be called a model. Since it possesses all the features of the original, not just the essential ones, it offers no advantage over the original as a subject for study.

reproduction curve. The relationship between the numbers of a given stage in a particular generation ($n + 1$) plotted against the numbers of that stage in another generation (n).

Reptilia (reptiles). A class of essentially terrestrial vertebrates, whose modern representatives are poikilothermic (*see* POIKILOTHERMY) and scaly-skinned, with EMBRYOS protected by an AMNION and an ALLANTOIS. Most lay large eggs with leathery shells. Reptiles were the dominant land vertebrates in the MESOZOIC Era. *See also* CHELONIA, CROCODILIA, RHYNCHOCEPHALIA, SQUAMATA, THERAPSIDA.

reserves. *See* OIL FIELD, RESOURCE.

reservoir. (1) A natural or artificial lake used for the storage of water for industrial and domestic purposes and for the regulation of inland waterway levels. Service reservoirs store water for domestic supply purposes under cover and regulate diurnal fluctuations in demand. Impounding reservoirs provide storage to cover seasonal or year-to-year variations in inflow. Such reservoirs (feeder reservoirs) may supply water for domestic or industrial use or for regulating water levels in rivers and canals. Some impounding reservoirs, storing water for domestic use, contain relatively pure water which needs little further treatment before being piped directly to a supply network. (2) A natural underground container for fluids (e.g., water, crude oil, natural gas).

reservoir rocks. Rocks that form an underground RESERVOIR for fluids (e.g., water, NATURAL GAS, CRUDE OIL). Reservoir rocks are characterized by high POROSITY and PERMEABILITY, and are often SANDSTONES, LIMESTONES OR DOLOMITES.

residual oil. Residue after CRUDE OIL has been refined to produce LIQUEFIED PETROLEUM GAS, petrol, diesel oil, lubricants, KEROSENE, ASPHALT, etc. Generally it is sold for burning.

resinous. *See* LUSTRE.

resonance. The condition that occurs when a system is vibrating as a result of forced EXCITATION at a certain FREQUENCY. A diminution of the AMPLITUDE of vibration caused by altering the frequency of the exciting force indicates that the system is in resonance.

resonant. Capable of being excited into RESONANCE.

resonant frequency. The FREQUENCY at which RESONANCE occurs.

resource. A means that is available for supplying an economic want (e.g., land, labour). Minerals and fossil fuels are described as stock, resource or reserves. The stock of a substance is the total amount of that substance contained in the environment, much of which will be inaccessible or unprocessable by present-day technology. That part of the stock which could be used under specified social, economic and technological conditions is called the resource, and estimates of resources change with economic changes (e.g., alterations in prices), social changes and technological changes (e.g., a new technology might increase the quantity of a resource). The reserves are that part of the resource that can be exploited with current technology under current economic and social conditions. Reserves are further subdivided into proven, probable and possible recoverable reserves, all of which have been identified, but some of which will be paramarginal or submarginal, and undiscovered reserves which are either hypothetical resources or specified resources, their location being unknown or known, respectively.

respiration. (1) (external respiration) The process whereby oxygen is taken into an organism and carbon dioxide is given out. This often involves breathing, or respiratory movements (e.g., pumping air into and out of lungs or passing water across gills. (2) Internal respiration (tissue or cell respiration) comprises the chemical reactions from which an organism derives energy; glucose is oxidized to carbon dioxide and water, liberating 3.75 Calories per gram. This oxidation is carried out through a complicated series of oxidation-reduction reactions, (e.g., the CITRIC ACID CYCLE) under the control of ENZYME systems (e.g., OXIDASES, DEHYDRO-GENASES).

respiratory pigment. A substance that is contained in the blood and is capable of combining reversibly with oxygen, so carrying the gas from the respiratory organs to the tissues. *See also* HAEMOGLOBIN, HAEMO-CYANIN, RESPIRATION.

respiratory quotient (RQ). The ratio of the volume of carbon dioxide given off to the volume of oxygen used up by an organism during RESPIRATION. RQ varies according to the kind(s) of food being oxidized and indicates whether respiration is AEROBIC or AN-AEROBIC.

rest mass. The mass of a body when it is at rest relative to the observer. Mass varies with velocity, which becomes important as speeds (e.g., of atomic particles) approach the speed of light.

retarded timing (TC). A technique for improving fuel combustion in INTERNAL COM-BUSTION ENGINES, thereby reducing the formation and emission of pollutants.

retrovirus. One of a family of enveloped viruses (Retroviridae) which contain a single strand of RNA together with a few molecules of the ENZYME reverse transcriptase. Once inside a cell, the RNA can be transcribed into DNA which becomes incorporated in the DNA of the host. The host DNA then manufactures more viruses.

retting. A process in the extraction of certain vegetable fibres, (e.g., flax, jute) in which the plant stem is immersed in water (in a pond or on grass to be covered with dew

– dew retting) and the outer sheath is partially decomposed by BACTERIA. When washed and dried, the outer parts become brittle. Scutching breaks them into short pieces without damaging the fibre, and they are hackled (*see* HACKLE) by being drawn through a series of metal combs of increasing fineness until all the straw is removed and only the fibres remain. In flax preparation, the short fibres are called tow and the longer fibres line.

reverberation. Sound at a point that is increased by multiple reflection from surrounding surfaces and persists after the source ceases to emit sound.

reverberation time. The time required for reverberant sound (*see* REVERBERATION) of a given frequency to decay by 60 dB after the source ceases to emit sound waves. It is commonly calculated by measuring the first 30 dB decay and then extrapolating.

reverbertory furnace. A FURNACE in which the material that is heated is not mixed with the fuel. Its roof is heated by flames and the heat radiated down to the material. *See also* OPEN-HEARTH FURNACE.

reverse fault. A FAULT in which the FAULT PLANE DIPS to the upthrown side.

reverse osmosis. An industrial process for removing salts and other substances from water by forcing water through a SEMIPERME-ABLE MEMBRANE under a pressure that exceeds the OSMOTIC PRESSURE so that the flow is in the reverse direction to normal osmotic flow. Reverse osmosis is used to desalinate BRACKISH water and has been tested experimentally for the purification of water polluted by sewage effluent.

Rhaetian. A STAGE of the TRIASSIC System.

Rhaetic. *See* TRIASSIC.

rheophyte. A plant that grows in running water.

rheotaxis. *See* TAXIS.

Rhizobium. *See* ROOT NODULES.

rhizoid. A hair-like organ of attachment found in liverworts (HEPATICAE), mosses (MUSCI), fern (FILICALES) prothalli (*see* PROTHALLUS), some ALGAE and some FUNGI. Rhizoids may be formed from one or several cells.

rhizome (rootstock). An underground STEM, bearing buds and scale leaves, that lasts for more than one season and usually serves for both VEGETATIVE PROPAGATION and perennation. Many rhizomes (e.g., in iris) are stout, containing large amounts of stored food. The term rhizome is often reserved for horizontal underground stems (e.g., in mint), whereas rootstock is often confined to more or less erect structures (e.g., in primroses). *Compare* CORM, TUBER.

rhizomorph. A root-like strand of HYPHAE occurring in certain FUNGI (e.g., honey or bootlace fungus, *Armillaria mellia*) serving to spread the fungus and transport food materials.

Rhizophoraceae. A family of tropical trees, many of which (e.g., *Rhizophora bruguiera*) are MANGROVES. *Rhizophora* species develop many roots from the STEMS and branches to support the plants in the unstable mud. The SEEDS germinate while on the trees, and thus avoid being buried during their early development.

rhizoplane. The microenvironment of a root surface. *See also* RHIZOSPHERE.

Rhizopoda. A class of mainly free-living, freshwater and marine PROTOZOA, whose members usually feed and move by means of pseudopodia (*see* PSEUDOPODIUM). Rhizopoda may be entirely soft (e.g., *AMOEBA*) or have shells or skeletons. The CALCAREOUS shells of marine planktonic (*see* PLANKTON) groups of Foraminifera (e.g., *Globigerina*) on falling to the bottom form an important constituent of deep-sea oozes, and have contributed towards the formation of chalk. The siliceous skeletons of Radiolaria simi-

larly contribute to the formation of marine oozes and CHERT.

rhizosphere. The part of the soil immediately surrounding roots. Roots alter the nutrient status of the soil close to them by absorbing minerals and releasing other substances. This leads to an increase in the numbers of microorganisms and often alters the relative proportions of the different kinds of microorganisms present.

Rhodesia man. *See* HOMO.

rhodocrosite. *See* CARBONATE MINERALS.

Rhodophyceae. *See* RHODOPHYTA.

Rhodophyta (Rhodophyceae; red algae). A large, diverse, mostly marine group of ALGAE. Most are red, owing their colour to the presence of large amounts of phycoerythrin, which masks the CHLOROPHYLL and other pigments that they contain. Some species are microscopic, others membranous or filamentous, and often much branched. Red seaweeds are relatively small compared with the brown ones (*see* PHAEOPHYTA) and grow at greater depths or in rock pools too shady for the brown to thrive. Calcareous species are common, especially in tropical seas, where they often play a part in the formation of CORAL REEFS. Some yield AGAR-AGAR, other (e.g., laverbread, *Porphyra*; carragheen or Irish moss, *Chondrus crispus* and *Gigartina stellata*) are edible.

Rhus. A genus of trees and small shrubs (family: Anacardiaceae) whose popular name is sumac. *R. toxicodendron* is North American poison ivy; *R. vernicefera* yields lacquer. Other species are also grown for ornament.

Rhychocephalia. A small order of reptiles with many primitive features. The only living representative is the burrowing lizard-like *Sphenodon* (the tuatara of New Zealand), which has a pineal 'eye' in the roof of the skull.

Rhynchota. *See* HEMIPTERA.

rhyolite. A fine-grained to glassy ACIDIC IGNEOUS rock with a similar mineralogical and chemical composition to GRANITE. Many rhyolites are PORPHYRITIC and some show flow structure. Glassy rhyolites are termed OBSIDIAN or PITCHSTONE, and many have spherulites (spherical masses of radiating crystals) indicating partial devitrification. Rhyolites form near-surface INTRUSIONS and lava flows, although rhyolitic flows are not extensive at the present time; many ancient rhyolites have proved to be IGNIMBRITES.

ria. A drowned river valley produced by changes in sea level relative to the land after the river has worn its channel down to base level. *See also* EUSTATIC.

ribbon bomb. *See* BOMB.

ribbon worms. *See* NEMERTEA.

riboflavin (lactoflavin). A VITAMIN of the B group needed by animals (including humans) and some BACTERIA for the formation of certain respiratory COENZYMES. Liver, yeast, eggs and green vegetables are important sources for humans. Some bacteria living in the human intestine are able to synthesize riboflavin.

ribonucleic acid. *See* RNA.

ribose. *See* CARBOHYDRATES.

ribosomes (palade granules). Minute bodies, too small to be visible using a light microscope, present in the CYTOPLASM of all organisms and often attached to the ENDOPLASIC RETICULUM. Ribosomes consists of RNA, which is synthesized in the nucleoli (*see* NUCLEOLUS) of EUKARYOTIC cells, and PROTEIN. They are the site of protein synthesis, groups of ribosomes (polyribosomes, polysomes, ergosomes) probably being associated with single molecules of messenger RNA.

rice. *See* GRAMINAE, WILD RICE.

Richardson number. In a fluid (e.g., air), the ratio of the stabilizing force due to stratification to the destablizing effect of shearing motion, equal to $g\beta/(\delta u/\delta z)^2$ where $\beta = (1\delta\rho/\rho\delta z)$, ρ being the density, z the height and u the horizontal fluid velocity. It must be less than 1/4 if a stream is to be dynamically stable. It is named after Lewis Fry Richardson (1881–1953). the first person to attempt to forecast weather mathematically.

Richter scale. A scale for reporting the intensity of EARTHQUAKES, devised in 1935 by C.F. Richter, based on accurate measurements of the waves generated by the earthquake. The scale is logarithmic, from 1 to 10, such that an earthquake with a magnitude of 2 is a minor tremor, structural damage to buildings occurs at magnitude 6, and a major earthquake has a magnitude of 7 or more.

rickets. *See* VITAMIN D.

Rickettsia. A genus of very small BACTERIA (family: Rickettsiaceae) that cause severe diseases, including typhus fever in humans. Rickettsias are found inside the cells (they can live only inside cells) of ARTHROPODA such as ticks (ACARINA), lice (ANOPLURA) and fleas (APHANIPTERA), and are transmitted to mammals by the bites of these parasites. The vector of typhus is the human louse (*Pediculus humanus*). *See also* PARASITISM.

Ricinus. *See* EUPHORBIACEAE.

ridge (oceanic ridge). A submarine, broadly bilaterally symmetrical ridge with sloping sides sometimes, but not necessarily, in the middle of an ocean. Ridges are the site of EARTHQUAKE and volcanic activity and of heat flows higher than those of the average CRUST. According to the theory of PLATE TECTONICS they form the boundary between lithospheric plates where material is added to the plate edge by the process of SEA-FLOOR SPREADING. Some ridges are known as rises. There are also currently inactive ridges.

riffle. An area of shallow broken water in a river or stream.

rift valley. A linear valley bounded by NORMAL FAULTS. The longest rift valleys are developed along the axes of oceanic RIDGES, but the East African Rift Valley is the best known example.

rights of common. *See* COMMON LAND.

right whale. *See* BALAENOIDEA.

rigid pavement. *See* PAVEMENT.

rill. *See* EROSION.

rime. Ice deposited on solid surfaces caused by supercooled (*see* SUPERCOOLING) droplets (as in freezing fog) that freeze immediately on contact, so building up layers of ice sometimes to considerable thickness.

riparian. Applied to land bordering seashore, lake or river.

rip current. A localized, strong, outgoing surface or near-surface current which returns water seaward against the incoming surf.

rise. (1) *See* RAISE. (2) *See* RIDGE.

risk estimation. An actuarial technique, used in assessing the relative costs and benefits of a particular technology, that compares the actual recorded incidence of death or injury to humans to the number of people using the technology, extended over many years. The technique is statistically reliable, but does not account for PERCEIVED RISK.

river capture. A process in which a river system erodes the slopes that provide its drainage and so pushes back the WATERSHED (divide) between its own drainage basin and an adjacent one until a gap is breached and the river with the lower channel captures the headwaters of the other river, thus enlarging its own drainage basin. The river thus captured is said to be beheaded, and the point of diversion is known as the elbow of capture.

river terrace. The remains of alluvial FLOOD PLAINS of a river formed when the river channel ran at a higher level than at present. Raised sea levels of former times led to a higher base level down to which a river would cut its channel, but as sea levels descended relative to the land the river would cut a deeper channel, making a new, lower flood plain and leaving behind stepped alluvial terraces at the valley sides.

river zones. For biological purposes, rivers are divided into four zones. (a) The headstream, or highland brook, is often small and may be torrential (i.e. with water flowing at 90 centimetres per second or more). Temperature conditions may vary widely, and fish may be absent. (b) The trout beck is larger and more constant, although it may still be torrential. Trout may be the only permanent fish. (c) The minnow reach is still fairly swift, but patches of silt and mud collect in sheltered spots. Higher plants can gain a foothold, and the fish population is more varied. In Europe the minnow reach is sometimes known as the grayling zone. (d) The lowland reach is slow and meandering (*see* MEANDER), and coarse fish commonly appear. In Europe this is known as the bream zone. Some workers divide the lowland reach into two.

RMBK reactor. A Soviet NUCLEAR REACTOR, developed by scaling up a reactor designed originally to produce weapons-grade material, that uses unenriched uranium oxide fuel, a graphite MODERATOR and water for cooling, the coolant being carried in a complex series of pipes through the core. To reduce the risk of a water–gas reaction between graphite and steam the core is bathed in nitrogen, but the close proximity of graphite and water led to the rejection of the design on safety grounds in all countries except the USSR. In 1986 the USSR was believed to have 26 RMBK reactors in operation and several more under construction. It was an RMBK that failed at CHERNOBYL, after which the RMBK programme was halted, future plans being based on a Soviet PRESSURIZED WATER REACTOR design.

RMS value. *See* ROOT MEAN SQUARE VALUE.

Rn. *See* RADON.

RNA (ribonucleic acid). A single, long, unbranched chain of nucleotides. Each nucleotide is a combination of phosphoric acid, the monosaccharide ribose and one of the four nitrogenous bases (uracil, cytosine, adenine, guanine). The bulk of the cell's RNA occurs in the RIBOSOMES. The RNA is responsible for translating the structure of DNA molecules of the CHROMOSOMES into the structure of PROTEINS in a sequence of steps. Messenger RNA (mRNA) is built up on a length of DNA (a cistron) by specific base-pairing (adenine will link only with thymine or uracil and guanine only with cytosine). The messenger RNA molecule then moves from the chromosomes to the ribosomes. Here a specific POLYPEPTIDE molecule is built up. Its AMINO ACIDS are put in place by smaller molecules of RNA (transfer RNA, tRNA), of which there is a different kind for each amino acid. Each transfer RNA molecule becomes temporarily attached at one end to an amino acid molecule and at the other, by a distinctive set of three nucleotides, to a site on a messenger RNA molecule determined by specific base-pairing. In this way the polypeptide chain is assembled, its sequence of amino acids being determined by the sequence of nucleotides in the mRNA and ultimately in the DNA of the chromosomes. RNA forms the genetic material of some VIRUSES.

road base. *See* PAVEMENT.

robber flies. *See* BRACHYCERA.

robin's pincushion. *See* GALL.

roche moutonnée. A hillock or jointed (*see* JOINT) rock with one end smoothed by abrasion due to ice-borne debris and the other end with a stepped surface produced by the plucking out of blocks by the GLACIER. The smoothed (upstream) surface commonly shows GLACIAL STRIATIONS.

rock. Any naturally formed aggregate or mass of mineral matter.

rock crystal. *See* QUARTZ.

rock flour. Finely comminuted rock debris, especially that produced by the grinding effect of GLACIERS. Rock flour is a major constituent of BOULDER CLAY; the fine fraction of boulder clay is not rich in CLAY MINERALS when fresh, except when the rock flour was derived from ARGILLACEOUS rocks. *See also* ADOBE.

rock salt. *See* HALITE.

rod. (1) A specialized cell of the eye that consists of an inner and an outer segment. The outer segment abuts on the eye wall and consists of a stack of thin discs to which molecules of visual pigment are attached. It is probable that the first events in vision occur at these discs, from which effects are transmitted to the inner segment, connected to the outer by a cilium (*see* CILIA) with MITOCHONDRIA lying close to the cilium. The inner segment narrows behind the region containing the mitochondria to become a fibre with a swelling containing the nucleus, with a synapse (*see* NEURON) linking it to the bipolar cell and thence to the optic nerve. *Compare* CONE. (2) Straight stems of willow (osier) used in basket making. (3) An obsolete measure of length (also called pole or perch) equal to 5.5 yards (5 metres).

Rodentia (rodents). An order of placental mammals (EUTHERIA) that includes rats, mice, voles, squirrels, hamsters, porcupines, beavers, coypu, etc. Rodents have teeth adapted for gnawing, with one pair of large, chisel-shaped, persistently growing incisors in each jaw and no canines (*see* DENTAL FORMULA). The smaller species have a high reproductive potential, and because of this often become pests. They constitute a staple food for many carnivorous mammals, birds and reptiles. *See also* LAGOMORPHA.

roentgen. The unit formerly used to measure X- or gamma-radiation (*see* IONIZING RADIATION), being the amount of X- or gamma-radiation which will produce ions carrying one electrostatic unit of electricity of either sign (i.e. positive or negative) in 1 cubic centimetre of dry air. The unit is named after Wilhelm Konrad Roentgen (1845-1923), the discoverer of X-rays. It has been replaced in the SI system by coulombs per kilogram (of pure, dry air), 1 roentgen being equal to 2.58 x 10^{-4} C/kg. *See also* CURIE, RAD, REM.

root. The part of a VASCULAR PLANT that is usually underground, serving for anchorage and the absorption of water and minerals. Certain tropical epiphytic (*see* EPIPHYTE) orchids have green aerial roots, resembling stems or even leaves and having a photosynthetic function (*see* PHOTO-SYNTHESIS). Roots differ from stems principally in the absence of leaves and buds, and in the arrangement of the vas-cular tissue in a central core. The tip of an underground root is covered by a protective layer of cells (the root cap). A short distance behind this is a zone of root hairs, tubular outgrowths of the epidermal cells (*see* EPIDERMIS), which provide an extensive absorbing surface.

root cap. *See* ROOT.

root hairs. *See* ROOT.

root mean square value (RMS value). The effective value of a fluctuating quantity. It is calculated by squaring the values, averaging them, then extracting the square root.

root nodules. GALL-like swellings containing nitrogen-fixing (*see* NITROGEN FIXATION) BACTERIA (i.e. *Rhizobium*) on the roots of LEGUMINOSAE and a few other plants (e.g., alder, *Alnus*; bog myrtle, *Myrica*). The bacteria enter the roots from the soil (different strains infecting different species of plant) and cause the formation of the nodules. Here they multiply rapidly, using CARBOHYDRATES from the host plant and supplying it with organic nitrogen compounds formed during nitrogen fixation. Ultimately the nodule disintegrates, so bacteria are returned to the soil. Cultivation of leguminous crops, such as clover, increases the nitrogen content of the soil,

especially if the crop is ploughed into the ground. *See also* NITROGEN CYCLE.

rootstock. *See* RHIZOME.

rorquals. *See* BALAENOIDEA.

Rosaceae. A cosmopolitan family of DICOTYLEDONEAE, comprising herbs, shrubs and trees, and including many species with edible fruits. The fruits of *Prunus* species (i.e. plum, peach, cherry, almond, etc.) are DRUPES, and that of the bramble (*Rubus fruticosus*), is an aggregation of small drupes. The fruits of the strawber-ry (*Fragaria*) the apple (*Malus*) and the pear (*Pyrus*) are pomes. *See also* CRATAEGUS.

rose quartz. *See* QUARTZ.

rotational slump. *See* SLUMP.

rotenone. *See* DERRIS.

Rotifera (wheel animalcules, rotifers). A phylum of minute animals that move and feed by means of a crown of CILIA. They have no true COELOM. Most live in fresh water, a few occur in the sea. The great majority are free-swimming, a few are sedentary, colo-nial or parasitic. The fertilized eggs are very resistant to desiccation and can be dispersed by the wind.

Rotliegendes. The lower division of the PERMIAN System in north-western Europe.

rotor flow. *See* MOUNTAIN WAVES.

rottenstone. A much-weathered (*see* WEATHERING), but still coherent rock result-ing from the LEACHING out of one or more components. Most rottenstones are CALCAR-EOUS: fossiliferous SANDSTONES often weather to rottenstones.

roughness, aerodynamic. *See* AERODY-NAMIC ROUGHNESS.

roughness height. *See* AERODYNAMIC ROUGHNESS.

roundworms. *See* NEMATODA.

Royal Commission on Environmental Pollution. A UK statutory body appointed to advise the government on matters, both national and international, concerning the pollution of the environment. The Commission considers the adequacy of research in each area it studies and attempts to predict future environmental changes. Its first report was published in 1971, since when it has reported on industrial pollution, pollution of estuaries and coastal waters, air pollution, pollution by agriculture, the environmental effects of nuclear power, etc.

Royal Society for Nature Conservation (RSNC). A UK organization, founded in 1912 as the Society for the Promotion of Nature Reserves and later known as the Society for Nature Conservation, that acts as an umbrella body for the trusts for nature conservation (*see* NATURALISTS' TRUSTS) in every county of Britain and with a total membership approaching 170 000. In 1987 the trusts owned or managed a total of more than 1600 nature reserves, many of great scientific importance, with a total area of about 60 000 hectares.

Royal Society for the Protection of Birds (RSPB). A UK organization, founded in 1889, concerned with the conservation of birds and with popularizing bird-watching. It maintains nature reserves as bird sanctuaries throughout the UK and has prevented the loss of various breeding species (e.g., osprey).

RQ. *See* RESPIRATORY QUOTIENT.

RSNC. *See* ROYAL SOCIETY FOR NATURE CONSERVATION.

RSPB. *See* ROYAL SOCIETY FOR THE PROTECTION OF BIRDS.

rubber. *See* EUPHORBIACEAE, MORACEAE, *FICUS, HEVEA, MANIHOT.*

Rubiaceae. A very large, mainly tropical family of DICOTYLEDONEAE, including trees, shrubs and herbs, most of which have conspicuous, insect-pollinated flowers. The family includes the bedstraws (*Galium*) and madder (*Rubia*). Genera of economic importance are *Coffea*, whose seeds yield coffee, and *Cinchona*, trees native to the Andes, from the bark of which ALKALOID drugs such as quinine are extracted.

ruby. *See* CORUNDUM.

rudaceous. *See* RUDITE.

ruderal. Applied to plants that inhabit old fields, waysides or waste land.

rudite. A SEDIMENTARY ROCK with grains larger than 2 millimetres in diameter. Such sedimentary rocks are termed rudaceous. *See also* ARENITE, ARGILLITE.

Ruminantia (ruminants). The suborder comprising the even-toed ungulates (ARTIODACTYLA) that chew the cud (ruminate). The stomach is typically four-chambered, and there are no upper incisor teeth. Cattle, sheep, goats, deer, antelope and giraffes are members of this group.

runner. *See* STOLON.

run-off. Water from rain or snow that runs off the surface of the land and into streams and rivers.

ruralization. The migration of people in substantial numbers from urban to rural areas. Ruralization has been occurring in the USA and Western Europe since the 1970s. *Compare* URBANIZATION.

rushes. *See* JUNCACEAE.

rust fungi (Uredinales). An order of parasitic (*see* PARASITISM) BASIDIOMYCETES with complicated life histories, many of which are important plant PATHOGENS. *Puccinia graminis* causes black stem rust in cereals and uses barberry (*Berberis vulgaris*) as an intermediate host. *Uromyces fabae* causes broad bean rust. *See also* PHYSIOLOGICAL SPECIALIZATION.

rutile. A major ORE MINERAL of titanium (titanium dioxide, TiO_2) that is widespread in small amounts in many rocks and is produced, often with ILMENITE from PLACER deposits.

R wave. *See* EARTHQUAKE.

Ryazanian. *See* NEOCOMIAN.

rye (*Secale cereale*). An important cereal crop that was probably first domesticated in areas away from the main wheat-growing centres where it thrived as a weed among wheat crops, so leading early farmers to develop it as an alternative to wheat. *See also* GRAMINEAE.

S

S. *See* SULPHUR.

S₁ Second filial generation (*see* FIRST FILIAL GENERATION).

s. *See* SECOND.

saccharoidal. Having the appearance of sugar; a term used to describe an equigranular texture of white, or nearly white, rocks such as some MARBLES and some DOLOMITE rocks. In North America usage sucrosic is used.

Saccharomyces. *See* YEASTS.

saddle reef. An ORE BODY occupying the space between relatively COMPETENT BEDS in the hinge zone of a FOLD. Some gold–quartz veins form saddle reefs.

Safe Drinking Water Act, 1974. The US law that brings the purity of drinking water throughout the USA under federal supervision.

sago. *See* PALMAE.

Saharan green belt. Two belts of land along the southern edge of the Sahara Desert, extending from the Atlantic Ocean to the Red Sea, in which it is proposed that agriculture should be managed in order to halt the southward spread of the desert. Plants would be grown in both belts and protected against overgrazing. Cattle would be grazed in the northern belt, then moved south to lusher pastures for fattening. *See also* DESERTIFICATION, SAHEL.

Saharo-Arabian Floral Region. The part of the PALAEOTROPIC REALM that comprises Arabia and the area on the eastern side of the Persian Gulf.

Sahel (Sahelian region).   An area of semiarid lands bordering the southern Sahara and covering all or part of: Mauritania, Mali, Burkina Faso, Niger, Chad, Senegal, Ghana, Cameroon, Nigeria and the Central African Republic. The name is derived from the Arabic word for shore. Food production in the Sahel is critically dependent on seasonal rains, which in recent years have become increasingly unreliable, due mainly to climate change, so that food shortages are often widespread, sometimes amounting to famine.

St Anthony's fire. *See* ERGOT.

St David's Series. The second oldest SERIES of the CAMBRIAN System.

Salicaceae. A family of DICOTYLEDONEAE comprising trees and shrubs, mainly of northern temperate regions, including *Salix* (willow) and *Populus* (poplars). The flowers are grouped in male and female CATKINS, borne on different plants. Poplars are rapidly growing trees cultivated for timber that is used in matches, packing cases and paper pulp. White willow yields wood for cricket bats, and osiers are used in basket making.

Salientia. *See* ANURA.

salinity. The degree of concentration of salt solutions, determined by measuring the density of the solution using a salinometer (a type of hydrometer), by titration, by measuring the electrical conductivity of the solution, etc. *See also* EUHALINE, FRESH WATER, MESOHALINE, OLIGOHALINE, POLYHALINE.

salinization (salination). The accumulation

of soluble mineral salts near the surface of soil, usually caused by the CAPILLARY FLOW of water from saline GROUND WATER. Where the rate of surface evaporation is high, irrigation can exacerbate the problem by moistening the soil and causing water to be drawn from deeper levels as water evaporates from the surface. The evaporation of pure water leaves the salts behind, allowing them to accumulate, and they can reach concentrations that are toxic to plants, thus sterilizing the land.

salinometer. *See* SALINITY.

salivary gland chromosome. A MEGACHROMOSOME that occurs in the salivary glands of certain flies (*see* DIPTERA).

salpingectomy (tubal ligation). The operation for the sterilization of a female mammal in which the abdomen is opened and the fallopian tubes are tied, so preventing the ova from reaching the uterus.

salt. A chemical compound formed when the hydrogen ion of an acid is replaced by a metal or, together with water, when an acid reacts with a base. Salts are named according to the metal and the acid from which they are derived. Common salt is sodium chloride (NaCl, *See* HALITE).

saltation. (1) A leaping movement of sedimentary grains that are too heavy to be carried entirely by wind or water, but are bounced and rolled along the ground or stream bed by TURBULENCE, thereby inducing the movement of other grains with which they impact. (2) A sudden change. *Compare* STASIS.

Saltatoria. *See* ORTHOPTERA.

salt-dome. A structure that results from the upward movement of EVAPORITES and simultaneous downward movement of denser material lying over them. OIL FIELDS and gas fields are frequently associated with salt-domes, which are usually roughly circular to lozenge-shaped in plan and often several thousand feet in depth. *See also* DIAPIR.

salt field. An area underlain by workable deposits of salt.

samara. A winged ACHENE (e.g., a sycamore fruit or an ash key).

samphires. *See* CHENOPODIACEAE.

sand. *See* ARENITE.

sand dollar. *See* ECHINOIDEA.

sand dune. A mound or ridge of sand, blown by prevailing winds, and exhibiting a long windward slope and a steeper leeward slope. Unless stabilized (e.g., by vegetation) dunes will migrate in the direction of the prevailing wind as sand is blown up the windward slope and down the slip face of the leeward slope. In coastal regions, as a dune migrates landwards it may be followed by a second dune as new material is blown in from the beach.

sandpipers. *See* CHARADRIIDAE.

sandstone. A SEDIMENTARY ROCK consisting of cemented (*see* CEMENT) CLASTIC grains predominantly 0.06–2 millimetres in diameter.

Sangana. The nearest town, in Nigeria, to an oil well that blew out on 17 January 1980, releasing 30 000 tonnes of crude oil into the Niger River, contaminating drinking water and fish in villages with a total population of 30 000 – 50 000 people.

sanitary land-fill. A US term for the dumping of domestic refuse, compacted on site and covered regularly by a layer of earth. Microorganisms decompose the organic part of the refuse. This is engineered burial of refuse, but in many places the term is synonomous with a rubbish dump or waste tip. *See also* LAND-FILL.

Santonian. *See* SENONIAN.

Sapium. *See* EUPHORBICEAE.

saponin glycosides. *See* GLYCOSIDES.

sapphire. *See* CORUNDUM.

saprobe (saprobiont). An organism that feeds on dead or decaying organic matter.

saprobel. An amorphous material formed by the slow anaerobic decomposition of planktonic (*see* PLANKTON) remains, which may subsequently be converted to petroleum compounds after compression under accumulated sediment by processes that are not well understood.

saprobic classification (saprobien classification). A classification of river organisms according to their tolerance of organic pollution. (a) The polysaprobic group, including sewage fungus, BLOODWORMS and the rat-tailed maggot (*Eristalis tenax*), can live in grossly polluted water in which decomposition is primarily anaerobic. (b) The alpha-mesosaprobic group, including the water-louse (*Asellus*), can tolerate polluted water where decomposition is partly aerobic and partly anaerobic. (c) The beta-mesosaprobic group, including Canadian pondweed (*Elodea canadensis*), some caddis-fly larvae (TRICHOPTERA), the eel (*Anguilla anguilla*) and the three-spined stickleback (*Gasterosteus aculeatus*), can tolerate mildly polluted water. (d) The oligosaprobic group, including stone-fly nymphs (PLECOPTERA) and the river trout (*Salmo trutta fario*), are restricted to non-polluted water, which may contain the mineralized products of self-purification from organic pollution. *See also* BIOTIC INDEX.

saprobiont. *See* SAPROBE.

saprolite. An ancient soil, with rock deposits showing a gradual transition above hard bedrock, through deeply weathered (*see* WEATHERING) and leached (*see* LEACHING) zones which show the same structure as the BEDROCK, to a massive weathered zone and clay, beneath the modern soil.

sapropelic. *See* BOGHEAD COAL, COAL.

saprophagous (saprozoic). Applied to animals (e.g., house-fly larva) that feed on dead or decaying plant or animal material. *See also* SAPROPHYTE.

saprophyte. An organism (plant or PROTISTA) that obtains food in solution from the dead or decaying bodies of other organisms. Many FUNGI and BACTERIA are saprophytes, so are a few PROTOZOA (e.g., *Euglenia, see* EUGLENOPHYTA) and flowering plants (e.g., bird's nest orchid, *Neottia*). Saprophytes carry out the essential process of breaking down organic matter into simple substances such as carbon dioxide and nitrates, which are then available for synthesis by AUTOTROPHIC organisms into new organic matter. *See also* CARBON CYCLE, NITROGEN CYCLE.

saprozoic. *See* SAPROPHAGOUS.

sapwood. (1) In a woody plant, the outer layers of wood, which contain living cells. As well as providing mechanical support, sapwood conducts water and stores food. It is generally lighter in colour and less resistant to decay than HEARTWOOD. (2) In trees, a term applied to all wood with no clear distinction made between sapwood and heartwood.

Sargasso Sea. A roughly elliptical, fairly still area in the central North Atlantic Ocean (between about 20°N and 35°N and 30°W and 70°W) that lies inside a clockwise current of the Gulf Stream on the west side and other currents on the east. Floating matter, especially from the south-west, converges on the area, that contains large quantities of gulfweed (*Sargassum*) and supports many animals typical of the LITTORAL ZONE, some of which are not found anywhere else. The sea is particularly susceptible to pollution, especially from oil and floating debris. It was reported in 1969 by workers from the Woods Hole Oceanographic Institution, that oil globules were more plentiful than the gulfweed in the sea.

Sargassum. *See* SARGASSO SEA.

sarsen. A BOULDER of hard SANDSTONE occurring on the surface on the chalk downs of southern England. Sarsens are probably remnants of a widespread EOCENE deposit. Some have been erected to form stone circles (e.g., Stonehenge), although the inner circles at Stonehenge are formed from DOLERITE imported from the Preselly Hills in South Wales.

satellite. An object that orbits around a larger one. Artificial satellites orbiting the Earth are used for communications, the gathering of military intelligence, the monitoring of weather and other environmental phenomena, etc. *See also* EARTH RESOURCES TECHNOLOGY SATELLITE, LANDSAT, SPOT.

satin spar. A fibrous CALCITE or fibrous GYPSUM.

saturated adiabatic lapse rate. *See* WET ADIABATIC LAPSE RATE.

saturated fatty acid. *See* FATTY ACID.

saturation. A relative humidity of 100 percent measured by comparing the difference in readings between a dry bulb and WET BULB THERMOMETER. In unsaturated air, water evaporating from the wet bulb will lower its reading. In saturated air there will be no evaporation from the wet bulb, the water on the bulb will be at the same temperature as the surrounding air, thus the wet and dry bulb readings will be identical. The amount of water that air can contain as vapour varies with temperature (e.g., from 2 g/m^3 at -10°C to 51 g/m^3 at 40°C).

Sauria. *See* LACERTILIA, SQUAMATA.

Sauropsida. A term originally used to cover all living and extinct reptiles and birds. It is now applied to birds and living reptiles (e.g., dinosaurs, pterodactyls) together with fossil reptiles that have a similar skull structure, but it excludes the extinct mammal-like reptiles. *Compare* THEROPSIDA.

savannah. The extensive grassland BIOME of Africa south of the Sahara that supports a wide range of plants (mainly herbs and shrubs) and whose animal population in-

cludes many large herbivores and associated predators. The savannah may be in part a FIRE CLIMAX.

Savonius rotor. A vertical-axis windmill with high inertia and low efficiency, but which produces a more continuous output than conventional AEROGENERATORS. *See also* DARRIEUS GENERATOR, PANEMONE.

saw-flies. *See* HYMENOPTERA.

saxicolous (saxatile, saxicoline). Applied to organisms that live on or among rocks (e.g., saxifrages).

Sb. *See* ANTIMONY.

scale insects (plant bugs). HEMIPTERA (family: Coccidae), some of which are harmful parasites (*see* PARASITISM) of trees and shrubs (e.g., *Mytilapsis pomorum,* which causes apple scale), whereas others yield cochineal and shellac. Female scale insects are wingless and remain fixed to their food plant, being covered with a scale formed from cast-off EXOSKELETONS.

Scaphopoda (tooth-shells, tusk-shells). A small class of marine, burrowing MOLLUSCA with curved, tapering tubular shells, open at both ends, and prehensile tentacles round the mouth that are used to obtain food.

scheelite. A mineral that exhibits FLUORESCENCE. It is a tungstate of calcium ($CaWo_4$) and a major source of TUNGSTEN, commonly occurring with WOLFRAMITE and CASSITERITE in contact metamorphic (*see* METAMORPHISM) or hydrothermal deposits adjacent to ACID IGNEOUS bodies, and in PLACER deposits.

schist. A medium- to coarse-grained metamorphic (*see* METAMORPHISM) rock with a subparallel arrangement of flaky or ACICULAR minerals. Many schists are micaceous (*see* MICA) and have an undulose CLEAVAGE. Schists are named according to their most prominent minerals (e.g., hornblende schist, garnet-mica schist). Greenschists are chlorite schists, and blueschists have the blue AMPHIBOLE glaucophane.

Schistosoma. A genus of blood flukes (class: TREMATODA) that infest mammals, including humans, causing much disease (especially schistosomiasis or bilharzia) in Africa, Asia and South America. The larvae develop in various snails and enter the human body with drinking water or by burrowing through the skin.

schistosity. A texture of fairly coarsely crystalline metamorphic (*see* METAMORPHISM) rocks caused by the subparallel arrangement of either platy or elongated minerals. Schistosity is shown by SCHISTS and some GNEISSES.

schizocarpic. Applied to a dry fruit formed from two or more united CARPELS, which splits when mature into parts (mericarps), each of which represents the product of a single carpel. An example is the hollyhock fruit.

schizogenesis. Reproduction by cell division. *See also* SCHIZOGONY.

schizogony. A sexual reproduction by multiple fission, during which a single cell (schizont) produces many smaller cells (schizozoites). This occurs in many SPOROZOA.

Schizomycophyta. *See* BACTERIA.

schizont. *See* SCHIZOGONY.

schizozoites. *See* SCHIZOGONY.

schwingmoor. A mat of floating vegetation growing outwards over water from the shore.

sciophyll. *See* SKIOPHYLL.

sciophyte. *See* HELIOPHOBE.

sclereids. *See* SCLERENCHYMA.

sclerenchyma. A type of supporting tissue found in plants. It consists of cells whose walls are much thickened with cellulose or lignin. These cells occur singly or in groups,

and when mature usually contain no living PROTOPLASM. They are of two kinds: fibres, which are much elongated; sclereids (stone cells), which are not. Sclereids occur often in seed coats and fruits (e.g., forming the gritty particles in pear flesh). *See also* COLLENCHYMA.

sclerophyll. A plant with tough or leathery evergreen leaves (e.g., holly, pine).

scleroproteins. Stable fibrous PROTEINS (e.g., COLLAGEN, KERATIN) present in skeletal, protective and CONNECTIVE TISSUES of animals.

sclerotium. A hard, compact mass of fungal HYPHAE, capable of remaining dormant for long periods when conditions are unfavourable for growth. *See also* ERGOT.

scolex. *See* BLADDERWORM.

Scolytus scolytus. *See* DUTCH ELM DISEASE.

scoria. PYROCLASTIC material, ejected by a VOLCANO, that is dark in colour, basic in chemical composition and full of VESICLES. Scoria is at least partly glassy. Scoria between 4 and 32 millimetres in diameter are called cinders. Scoria is contrasted with PUMICE.

scorpion-flies. *See* MECOPTERA.

Scorpionidea (scorpions). VIVIPAROUS ARACHNIDA in which the posterior segments form a flexible tail bearing a terminal sting, which is used to paralyze their prey (mainly insects, spiders and other scorpions), which are frequently larger than themselves. Scorpions date from the SILURIAN, the earliest being marine.

scorpions. *See* SCORPIONIDEA.

Scottish Wildlife Trust. *See* NATURALISTS' TRUSTS.

SCP. *See* SINGLE-CELL PROTEIN.

scramble competition. The situation in

which a resource is shared equally among competitors.

scree. Slopes of frost-shattered rock material.

screwpines. *See* PANDANACEAE.

screw worm fly. *See* BIOLOGICAL CONTROL.

Scrophulariaceae. A cosmopolitan family of DICOTYLEDONEAE, most of which are herbaceous plants, including *Veronica, Verbascum* and *Antirrhinum.* Some members of the family (e.g., eyebright, *Euphrasia*; yellow rattle, *Rhinanthus*) are hemiparasites (*see* PARASITISM), which absorb food from the roots of grasses. Most Scrophylariaceae are poisonous. Digitalis (a GLYCOSIDE), obtained from the foxglove (*Digitalis*), is used to treat heart complaints.

scrub. A type of vegetation dominated by shrubs and containing few or no tall trees. Scrub may be the natural CLIMAX, but in the UK it is often a stage in the SUCCESSION, giving way to woodland. *See also* SCRUB FOREST.

scrubber. An apparatus used in sampling and in gas cleaning, in which gas is passed through a space containing wetted packing or spray. The liquid is used to remove solid or liquid particles or some gases from the carrier gas stream.

scrub forest. An inferior growth of shrubs and small or stunted trees. *See also* SCRUB.

scurvy. A deficiency disease caused by lack of VITAMIN C.

scutching. *See* RETTING.

Scyphozoa (Schyphomedusae). A class of CNIDARIA in which the POLYP stage is absent or inconspicuous and the MEDUSA is large and complex. The group consists mainly of the free-swimming jellyfishes, saucer- or bell-shaped animals with many tentacles around the margin, and a mouth at the end of a stalk on the underside. *See also* ACTINOZOA, HYDROZOA.

Scythian. *See* TRIASSIC.

Se. *See* SELENIUM.

sea-anemones. *See* CNIDARIA.

sea-beet. *See* BETA VULGARIS.

sea-breeze. *See* BREEZE.

sea-cows. *See* SIRENIA.

sea-cumbers. *See* HOLOTHUROIDEA.

sea-floor spreading. The concept that new material from the Earth's MANTLE is being added to the Earth's CRUST at the site of oceanic RIDGES, thus spreading some ocean floors. According to the theory of PLATE TECTONICS these are sites of constructive plate margins. Supporting evidence for sea-floor spreading comes from PALAEOMAGNETISM, seismic evidence, heat flow studies along oceanic ridges and RADIOMETRIC DATING. (Oceanic rocks and sediments have been shown to be geologically relatively young and progressively younger the closer they are to oceanic ridges.) It is estimated, for example, that the Atlantic is growing at the rate of about 4 centimetres per year, 2 centimetres per year being added to each ocean ridge flank. *Compare* CONTINENTAL DRIFT.

sea fog. Literally, FOG at sea, formed typically in moist tropical air in warm sectors of CYCLONES in early summer (June and July in the northern hemisphere) over colder areas of sea. In many coastal areas it is formed over land, drifts out to sea, where it remains by day, often crossing the coast in a sea breeze in the afternoon.

sea-gooseberries. *See* CTENOPHORA.

sealilies. *See* CRINOIDEA.

sea-mats. *See* POLYZOA.

sea-mount. An isolated, usually conical, submarine mountain rising at least 1000 metres above the sea floor. Sea-mounts occur on the CONTINENTAL RISES, ABYSSAL PLAINS and in the OCEANIC TRENCHES. Most rise abruptly from the bottom and are associated with large MAGNETIC ANOMALIES and submarine eruptions. They are thought to be of volcanic origin.

seasons. Divisions of the year according to regular changes in the character of the weather and, in the European languages, named by association with the annual life cycle of plants, winter being the period of dormancy, spring of sowing, summer of ripening and autumn of harvesting. In the higher latitudes, day length changes, related to progress toward a single maximum and a single minimum average temperature, regulate plant growth. Conventionally, in the northern hemisphere, winter is taken to cover the months December to February, spring March to May, summer June to August and autumn September to November, the seasons being reversed in the southern hemisphere. Actual weather experienced is often unseasonal, however, so that the divisions are purely arbitrary.

sea-slugs. *See* GASTROPODA.

sea-spiders. *See* PANTOPODA.

sea-squirts. *See* UROCHORDATA.

sea-urchins. *See* ECHINOIDEA.

seawater. *See* EUHALINE.

second. (1) As a angular measure, one-sixtieth of 1 minute, which is one-sixtieth of 1 degree of arc. (2) The SI unit of time, equal to the duration of 9 192 631 770 periods of the radiation corresponding to the transition between two hyperfine levels of the ground state of the caesium-133 atom; one-sixtieth of a minute.

secondary consumer. *See* FOOD CHAIN.

secondary economy. An economic system in which the principal activities, and so sources of wealth and employment, are connected with the processing of primary commodities (*see* PRIMARY ECONOMY) to produce industrially manufactured goods.

secondary enrichment. The natural enrichment of an ORE by material of later origin deposited from descending aqueous solutions. Commonly, the added material has been derived from an overlying oxidized deposit. The enrichment may involve: (a) the infilling of voids by further deposition of a pre-existing mineral or by deposition of a new mineral; or (b) the replacement of an ore mineral by a mineral richer in the valuable constituent.

secondary environmental quality standards. *See* ENVIRONMENTAL QUALITY STANDARDS.

secondary minerals. Those minerals (a) formed after consolidation of cooling MAGMA by reactions within still cooling rock or with circulating GROUND WATER, (b) formed as a result of metamorphic (*see* METAMORPHISM) activity, (c) formed by WEATHERING of IGNEOUS rock and (d) formed during DIAGENESIS of SEDIMENTARY ROCK (*see* AUTHIGENIC). *Compare* PRIMARY MINERALS.

secondary pollutant. A pollutant formed in the environment by the combination or reaction of other (primary) pollutants. *See also* SYNERGISM.

secondary rock. Rock formed by the alteration of pre-existing rock. SEDIMENTARY ROCK and metamorphic rocks (*see* METAMORPHISM) fall into this category. *Compare* PRIMARY ROCK.

secondary sexual characters. Characteristics, other than those of the actual reproductive organs, that differ according to the sex of an animal. They are produced as a result of HORMONES secreted by endocrine organs, including the testis or ovary. Examples are the antlers of a stag, the facial hair of a male human and differences in colour and size in the two sexes in birds.

secondary succession. *See* SUCCESSION.

secondary thickening. In a woody perennial plant, the formation of secondary VASCULAR tissue (*see* XYLEM, PHLOEM) as a result of the activity of cambium in the stems and roots. Most of the woody part of a tree is the product of secondary thickening. *See also* ANNUAL RINGS.

secondary treatment. A process to remove the amount of dissolved organic matter in waste water and to reduce further the suspended solids. *Compare* PRIMARY TREATMENT.

second filial generation. *See* FIRST FILIAL GENERATION

second trophic level. *See* FOOD CHAIN.

Second World. *See* THIRD WORLD.

secretion. (1) The passage of a substance elaborated by a cell, particularly that of a gland, to the outside of the cell. Usually the substance is passed through the PLASMA MEMBRANE, often against the concentration gradient, but sometimes the cell breaks down to liberate its contents. (2) A substance made and passed out by a cell (e.g., nectar in flowers, hydrochloric acid in the stomach of vertebrates).

sedentary. Untransported; as applied to soils, those weathered (*see* WEATHERING) from underlying rock or formed by the accumulation of organic matter *in situ.*

sedimentary rock. Rock formed by the accumulation of material in water, from the air or from glaciers. Sedimentary rock can be formed from fragments of pre-existing rocks (i.e. CLASTIC rocks such as SANDSTONE, SHALE, ARKOSE and BIOCLASTIC LIMESTONE), from organic remains (e.g., PEAT, COAL and some limestones), from chemical precipitates (i.e. the EVAPORITES such as GYPSUM and HALITE) or from fragments blown out of volcanoes (i.e. the PYROCLASTIC rocks such as TUFFS). Characteristically, sedimentary rocks have BEDDING PLANES, and some contain FOSSILS.

sedimentary structures. Various structural features within beds of SEDIMENTARY ROCK and on BEDDING PLANES, which are used to work (a) out the environment of deposition, (b) the upper side of the bed as originally deposited and (c) to determine palaeocurrent directions. Sedimentary structures within beds include CROSS-BEDDING, GRADED BEDDING and imbricated pebble beds (*see* IMBRICATE); and on bedding planes rain prints, ripple marks, desiccation cracks and various structures associated with TURBIDITES such as flute casts eroded by turbulent flow and tool marks made by moving objects.

sedimentation. (1) The separation of an insoluble solid from a liquid in which it is suspended by settling under the influence of gravity or centrifugation. (2) The settling of particles by gravity.

seed. A structure derived from a fertilized OVULE and consisting of an EMBRYO plant and its food store surrounded by a protective coat (testa). The food is either stored in the seed leaves (cotyledons) or outside the embryo in the ENDOSPERM.

seed plants. *See* SPERMATOPHYTA.

seepage tank. *See* SEPTIC TANK.

segmentation. (1) (cleavage) Repeated cell divisions following fertilization and resulting in the ZYGOTE of a plant or animal becoming an early EMBRYO. (2) (metameric segmentation, metamerism) The repetition of a series of fundamentally similar segments (metameres) along the length of the body in many metazoan animals (*see* METAZOA).This is particularly obvious in the ANNELIDA and ARTHROPODA. In the earthworm, the segments are visible externally, and each possesses a similar pattern of blood vessels, nerves, muscles, excretory organs, etc. Metameric segmentation is usually much modified, for instance by the development of the head. Vertebrate embryos clearly show metameric segmentation in muscular, skeletal and nervous systems.

segregation, law of (Mendel's first law).

Only one pair of contrasted characteristics (*see* ALLELOMORPHS) may be represented in a single GAMETE. Segregation is the result of the behaviour of CHROMOSOMES during MEIOSIS. *See also* INDEPENDENT ASSORTMENT.

seiche. An oscillation in the level of the water of a lake or inland sea. A strong wind can pile up the warm surface layer of water (EPILIMNION), which slides back over the lower layer (HYPOLIMNION) when the wind drops, then continues to oscillate for some time. Seiches can also occur as a result of temporary local fluctuations in the WATER TABLE.

seif. A longitudinal SAND DUNE that lies parallel to the wind direction.

seismograph. An instrument for recording shock waves generated by any transient disturbance of the Earth, whether natural (e.g., an EARTHQUAKE) or manmade (e.g., an explosion). *See also* ISOSEISMAL LINE.

seismonasty. *See* NASTIC MOVEMENT.

selection pressure. A measure of the effects of natural selection on the genetic composition of a population.

selective species. A species found most usually in a particular COMMUNITY, but also, rarely, in other communities.

selenite. A colourless, transparent variety of GYPSUM.

selenium (Se). A non-metallic element that exists in several allotropic forms (*see* ALLOTROPY). The silver–grey crystalline solid ('metallic') form varies in electrical resistance on exposure to light and is used in photoelectric cells and SOLAR CELLS, as well as in photographic light meters (*see* PHOTOVOLTAIC EFFECT). It is also used in the manufacture of rubber and some glasses. It occurs in nature as metallic selenides together with the sulphides of the metals. $A_r = 78.96$; $Z = 34$; SG 4.81; mp 217°C.

selenodont. Applied to cheek teeth with crescent-shaped ridges (cusps) on the crown. This type of tooth is characteristic of RUMINANTIA (e.g., cow). *Compare* BUNODONT, LOPHODONT.

self-fertilization. The fusion of GAMETES from the same individual or from RAMETS of a CLONE. *Compare* CROSS-FERTILIZATION.

self-pollination. The transference of POLLEN from a stamen to the stigma of the same FLOWER, or of a flower in the same plant. *Compare* CROSS-POLLINATION.

sematic coloration. *See* APOSEMATIC COLORATION.

semipermeable membrane. A membrane through which a solvent, but not certain dissolved or colloidal substances, may pass. See also COLLOID, OSMOSIS.

Senonian. A subdivision (usually ranking as a STAGE) of the CRETACEOUS System, and comprising the Coniacian, Santonian and Campanian.

sepal. *See* FLOWER.

septic tank. A watertight sedimentation tank for sewage in which solids settle and are decomposed anaerobically. The liquid effluent may be passed from this tank into the ground or into a seepage tank in which it is filtered through sand or gravel before release. A well-designed septic tank rarely requires emptying. *Compare* CESSPOOL.

septum. A partition.

Sequoia. *See* TAXODIACEAE.

Sequoiadendron. *See* TAXODIACEAE.

seration. A series of COMMUNITIES within a FORMATION or ECOTONE (e.g., the series of communities that may stretch across a valley from one hill top to the next).

serclimax. A stable plant COMMUNITY that persists at a stage before the SUBCLIMAX in the seral sequence. *Compare* POSTCLIMAX, PRE-

CLIMAX, PROCLIMAX.

sere. (1) A series of plant COMMUNITIES resulting from the process of SUCCESSION. *See also* CLISERE, HYDROSERE, LITHOSERE, PLAGIOSERE, PRISERE, PSAMMOSERE, SUBSERE, XEROSERE. (2) Any stage in a plant succession.

series. In stratigraphy, the major subdivision of a system. *Compare* STAGE.

serine. An AMINO ACID with the formula $CH_2OHCH(NH_2)COOH$ and a molecular weight of 105.1.

serir. *See* REG.

Serpentes. *See* SQUAMATA.

serpentine. A group of SILICATE MINERALS with the approximate formula $Mg_3(Si_2O_5)(OH)_4$. Serpentine is the main alteration product of PYROXENE and OLIVINE. Chrysotile, a fibrous serpentine, is one of the minerals worked for the manufacture of ASBESTOS. Nickeliferous serpentine, an important mineral, is called GARNIERITE.

serpentinite. A rock consisting of SERPENTINE minerals formed by METASOMATISM of various ULTRABASIC rocks. Serpentinites are cut, turned and polished to make ornamental objects.

serpentinization. *See* PNEUMATOLYSIS.

serule. A miniature SUCCESSION composed of minute or microscopic organisms that usually is engaged in the breaking down of organic matter to its simpler constituents.

service reservoir. *See* RESERVOIR.

sesquioxide. An oxide containing three oxygen and two metallic atoms. In soil these are chiefly iron and alumina (Fe_2O_3 and Al_2O_3, respectively).

sessile. (1) Not stalked. (2) Attached to a SUBSTRATE.

seston. The organisms (bioseston) and non-living matter (abioseston) swimming or floating in a water body.

Seveso. A village near Milan, Italy, that suffered as a result of the accidental discharge of DIOXIN (TCDD) following an explosion in July 1976, at the Icmesa factory which was manufacturing the herbicide 2,4,5-T. The population of Seveso (700 people) was evacuated, more than 600 domestic animals had to be destroyed, and contaminated vegetation had to be burned within an 8-kilometre (5-mile) radius. People heavily exposed to the dioxin suffered from CHLORACNE, but there were no human deaths and no evidence of lasting damage to health. A similar, but much smaller incident involving TCDD occurred several years earlier in Derbyshire, England, and as a result of the Seveso accident the Derbyshire factory was closed temporarily.

Seveso Directive. An EEC DIRECTIVE, agreed by environment ministers of the member states in December 1981, that obliges industries to notify the authorities, their own workers and local residents of stocks they hold of dangerous chemicals and that restricts the quantities of specified substances that may be stored within 500 metres (1640 feet) of one another. The directive arose and takes its name from the SEVESO incident.

sevin. *See* CARBARYL.

sewage fungus. A pale, slimy bacterial deposit that often occurs in water subject to organic pollution.

sex chromosomes. CHROMOSOMES that are responsible for determining the sex of an animal. There is a homologous pair (two X chromosomes) in the nuclei of the HOMOGAMETIC SEX, and a dissimilar pair (X and Y) or an unpaired X chromosome in the nuclei of the HETEROGAMETIC SEX. The X chromosome is usually larger than the Y chromosome and, unlike it, contains numerous GENES that show SEX LINKAGE. All of the GAMETES produced by the homogametic sex, but only half of those produced by the heterogametic sex, contain an X chromosome.

sex linkage. The occurrence of a character more frequently in one sex than in the other because the GENE controlling that character is carried only on the X chromosome (*see* SEX CHROMOSOMES). For instance, in humans, the male has a single X chromosome, and the effects of a RECESSIVE gene carried on this chromosome cannot be masked by a dominant ALLELOMORPH. This results in many more men than women manifesting sex-linked recessive characters such as haemophilia and red–green colour blindness.

sexual reproduction. Reproduction that involves the fusion of two haploid nuclei to form a ZYGOTE. The nuclei are often those of male and female GAMETES. *Compare* ASEXUAL REPRODUCTION.

sexual selection. A type of selection that DARWIN, CHARLES ROBERT, suggested might bring about the evolution of SECONDARY SEX-UAL CHARACTERS, such as brilliant plumage and courtship behaviour in birds. Selective mating, caused because a female chooses only males with certain characteristics, would result in these characteristics being transmitted to the offspring, to the exclusion of features possessed by rejected males.

SG. *See* SPECIFIC GRAVITY.

SGHWR. *See* STEAM-GENERATING HEAVY-WATER REACTOR.

shade plant. *See* HELIOPHOBE.

shale. A SEDIMENTARY ROCK, made up largely of CLAY, which splits along BEDDING PLANES. *Compare* MUDSTONE. *See also* OIL SHALE.

shear stress. In fluid flow, the force per unit area tending to produce shear (e.g., wind shear in the atmosphere). This stress is generated at surfaces as a result of viscosity and TURBULENCE.

sheep. *See* BOVIDAE.

sheep month. The amount of feed or grazing needed to maintain a mature sheep or a ewe and its suckling lamb in good condition for 30 days. This is usually considered to be equivalent to one-fifth of a cow month. *See also* ANIMAL UNIT.

shelf break. *See* CONTINENTAL SHELF.

shelterbelt. A windbreak, usually a stand of trees, planted so as to provide nearly continuous protection against the wind.

shield. A major continental block, usually of PRECAMBRIAN IGNEOUS and metamorphic rocks (*see* METAMORPHISM), that has been stable over a relatively long period of Earth history and has undergone only faulting (*see* FAULT) or gentle warping.

shield volcano (Hawaiian-type volcano). A central-vent VOLCANO with slopes of very low angle built up of flows of very fluid BASALTIC LAVA.

shifting cultivation. A primitive agricultural system in which an area of land, usually small, is cleared of natural vegetation and crops are sown. Cropping continues for several seasons (in the tropics often with more than one crop each year) until the combination of recolonization by wild plants and declining soil fertility cause crop yields to fall, when the operation is moved to a new site and the cycle begins again. With a number of sites available, shifting cultivators usually return to a site after an absence of about 12–15 years, by which time the land has recovered its fertility. The system requires no input of fertilizer and little weed or pest control, but uses a great deal of land. In regions where agricultural land is less plentiful than it was, shifting cultivators have been forced to accelerate their cycles (returning to a site after less than 12 years), and this has become a serious cause of soil EROSION. *See also* SWIDDEN FARMING..

shipworms. *See* TEREDO.

SHM. Simple harmonic motion (*see* SINE WAVE).

shock wave. In the atmosphere, a travelling wave in which the thermodynamic properties of the air change suddenly to a different value.

shore slaters. *See* MALACOSTRACA.

shore zonation. The division of the sea shore into zones, each of which supports a characteristic fauna and flora. (a) The splash zone lies above the extreme high-water level of spring tides, but may sometimes be drenched with spray. Small periwinkles and channel wrack (*Pelvetia canaliculata*) extend up to this zone. (b) The upper shore lies between the average high-tide level and the extreme high-water level of spring tides, so is only covered occasionally, and then for short periods. The number of species inhabiting this zone is relatively small. (c) The middle shore lies between the average low-tide level and the average high-tide level. It is an extensive zone, mostly covered by the sea twice a day. A large number of species inhabit this zone. (d) The lower shore extends from the extreme low-water level of spring tides up to the average low-tide level. It is uncovered only occasionally, and for short periods. Some species typical of the area below tide levels extend into this zone. (e) The sublittoral fringe lies below the extreme low-water level of spring tides and is never uncovered. The shallow water here has a greater range of temperature than the open sea. The upper part of the large oar weeds (*Laminaria*) typical of this zone on rocky shores may project above the surface at low water.

short-day plants. *See* PHOTOPERIODISM.

short-horned flies. *See* BRACHYCERA.

short-horned grasshoppers. *See* ACRIDIDAE.

shredding. *See* PULVERIZATION.

shrub. A low-growing, woody perennial, which, unlike a small tree, branches from the base. The term is often restricted to plants less than about 6 metres in height. *Compare* COPPICE.

shrub layer. *See* LAYERS.

shuttle box. A device used in behavioural experiments with animals, consisting of a box in which the animal shuttles from one end to the other in response to a stimulus.

Si. *See* SILICON.

sial. The discontinuous upper part of the crust of the Earth. Sial is derived from *si*licon and *al*uminium. The word is tending to be replaced by the less-committed term upper crust.

siblings (sibs). The offspring of the same parents.

sibling species (species air). Species that are very similar or even indistinguishable in appearance, but are distinct genetically and do not interbreed (e.g., chiffchaff and willow warbler, which are chiefly distinguishable by their song). Sibling species are probably examples of recent SPECIATION.

sibs. *See* SIBLINGS.

siccicolous (xerophilous). Drought resistant. *See also* XEROPHYTE.

siderite. (1) A METEORITE composed entirely of metal. (2) A CARBONATE MINERAL with the formula $FeCO_3$, often found as a GANGUE mineral in sedimentary iron ores, as nodules in CLAY and as a CEMENT.

Siegenian. The second oldest STAGE of the DEVONIAN System in Europe.

Sierra Club. A US-based conservation organization founded in 1892 by Warren Olney, John Muir and William Keith, with its head office in San Francisco, whose aim is to work in the USA and other countries to restore the quality of the natural environment and to maintain the integrity of ECOSYSTEMS. The Club arranges and sponsors outings, issues publications on wildlife and landscape conservation and campaigns on environmental issues, especially those concerning land-use, mainly within the USA.

sievert (Sv). The SI unit of RADIATION DOSE EQUIVALENT, equal to 1 joule of energy per kilogram of absorbing tissue. The sievert replaces the rem (roentgen equivalent man): 1 Sv = 100 rem.

sieve tube. *See* PHLOEM.

signal. *See* NOISE.

signal-to-noise ratio. *See* NOISE.

silcrete. *See* HARDPAN.

silencer. A sound-absorbing duct for exhaust systems, as on a car.

Silesian. The Upper CARBONIFEROUS. The Silesian usually ranks as a SERIES, and comprises the Westphalian and Stephanian Stages.

silica (SiO_2). Either a mineral of that composition (e.g., QUARTZ, CHALCEDONY, etc.) or the SILICATE MINERAL content of a rock, which is usually expressed chemically as the weight percentage of silica.

silicaceous sandstone. *See* ORTHOQUARTZITE.

silicate minerals. The major group of compounds in the Earth's CRUST, based on SiO_4 tetrahedra joined by the oxygen (which also bonds to the metal CATIONS) and classified on the arrangement of the tetrahedral units. The silica tetrahedra can be independent (e.g., OLIVINE), in rings (e.g., BERYL, MICA, talc), in chains (e.g., PYROXENES) or bands (e.g., AMPHIBOLES) or in a three-dimensional framework (as in QUARTZ and FELDSPAR), or two SiO_4 can share one oxygen. Aluminium can replace the silicon in some of the tetrahedra. The properties of the mineral are closely related to the atomic structure.

silicon (Si). A non-metallic element that occurs as a brown powder and as dark grey crystals. In nature it occurs in various SILICATE MINERALS and as SILICA. It is used in glass manufacture and in alloys and in silicone compounds used in lacquers, resins, lubricants and water-repellant finishes. A_r = 28.086; Z = 14; SG 2.42; mp 1420°C.

Silicon Valley. A valley to the south of San Francisco, California that contains a high concentration of factories manufacturing electronic equipment. Solvents and other chemicals seeping from factory stores are alleged to have contaminated GROUND WATER throughout the valley basin.

silicosis. A type of PNEUMOCONIOSIS caused by the inhalation of SILICA dust.

silicula. *See* SILIQUA.

siliqua. A type of dry fruit found in the CRUCIFERAE (e.g., *Cheiranthus*, wallflower), consisting of an elongated capsule, formed from two united carpels, which opens from below, exposing the seeds attached to a central framework. A silicula (e.g., the fruit of *Capsella bursapastoris*) is similar, except that it is short and broad.

silk. Very fine, strong PROTEIN threads secreted by various insects (e.g., LEPIDOPTERA, EMBIOPTERA, COLEOPTERA, TRICHOPTERA) and spiders (ARANEAE). The silk glands of the silkworm (*BOMBYX MORI*) are modified salivary glands. Embioptera have silk glands in the front legs, and those of spiders are situated in the abdomen. Silk is produced as a fluid, which is extruded through tubular spinnerets, and hardens on contact with the air. It is used in the construction of egg and pupal cocoons, webs, nets, 'balloons', etc.

silkworm moth. *See BOMBYX MORI.*

silky. *See* LUSTRE.

sill. A tabular IGNEOUS INTRUSION that is concordant (i.e. it follows a BEDDING PLANE in SEDIMENTARY ROCKS or the FOLIATION in metamorphic (*see* METAMORPHISM) rocks). Sills that cut across from one horizon to another are described as transgressive.

silt. Sediment made up predominantly of grains 0.002–0.06 millimetre in diameter.

Silurian. The third oldest PERIOD of the PALAEOZOIC Era, usually taken as beginning some time between 435 and 460 Ma. Silurian also refers to the rocks formed during the Silurian Period, called the Silurian System, which is normally divided into four series – Llandoverian, Wenlockian, Ludlovian and Pridolian or Downtonian. The Downtonian is placed by some geologists in the succeeding DEVONIAN System. The Silurian probably lasted 30 million years. The Silurian is zoned (*see* INDEX FOSSIL) using GRAPTOLITES.

silva. The aggregate of all the forest trees within an area.

silver (Ag). A white, rather soft metallic element that is extremely malleable and ductile and an excellent conductor of electricity. It occurs naturally as the metal and as ARGENTITE, ACANTHITE, horn silver and other compounds. It is used in jewellery and coinage, etc., and its compounds are used in photography. A_r = 107.87; Z = 47; SG 10.5; mp 960.5°C.

silver fish. *See* APTERYGOTA.

silver iodide. *See* ARTIFICIAL RAIN.

sima. The lower part of the Earth's CRUST, the word being derived from *si*licon and *ma*gnesium. The term is tending to be replaced by the less committed term lower crust.

simazine. An organic, soil-acting herbicide used to control germinating weeds in many crops and for total weed control on waste land, paths, etc. It is a long-lasting chemical that may prevent regrowth for as long as a year.

similarity. In fluid motion, a property of certain flows in which the flow pattern remains constant when the linear scale is changed. Examples are the momentum jet and the buoyant thermal, in which the development is conical and all of the flow properties are proportional to the distance from the origin.

Simmondsia chinensis. *See* JOJOBA.

simoon. *See* SIROCCO.

simple harmonic motion. *See* SINE WAVE.

simple interest. *See* LINEAR GROWTH.

Sinanthropus pekinensis. *See* HOMO.

Sinemurian. A STAGE of the JURASSIC System.

sine wave (sinusoidal wave). Simple harmonic motion (SHM); the path traced by a point that oscillates along a line about a central point, so that its acceleration towards the centre is always proportional to its distance from the centre. On a graph, tracing position against time, a sine wave may be described by the formula

$$y = r \sin 2\pi [(t/T) - (x/\lambda)]$$

where y = the distance of the point from the centre, t = time, r = amplitude, T = the period of the wave, λ = the wavelength and x = the distance the point has travelled from the centre in time t.

single-cell protein (SCP). PROTEIN derived from unicellular organisms (e.g., BACTERIA, YEASTS) grown on a HYDROCARBON substrate (e.g., crude oil, methanol, cellulose or wastes from food processing). SCP may be used directly as a meat extender, but in general it is used in livestock feedstuffs as a protein additive. *See also* NOVEL PROTEIN FOODS.

sinistral fault. A strike–slip FAULT, or a fault with a considerable component of strike–slip motion, in which the distant block shows the relative displacement to the left when viewed across the FAULT PLANE.

sink. (1) A receptacle, or receiving area, for materials translocated through a system (e.g., the oceans are the sink into which many water-borne pollutants are drained; the seed of a plant is the sink in which nutrients are stored). (2) In air pollution, a place or mechanism associated with the removal of air pollutants from the atmosphere. Examples are RAINOUT of PARTICLES, soil as an absorber of CARBON MONOXIDE, the OXIDATION of SULPHUR DIOXIDE to form SULPHUR TRIOXIDE and thence SULPHATE particles, etc.

sink hole. (1) A depression in marshy flat land, where water collects. (2) *See* SWALLOW HOLE.

Sino-Japanese Floral Region. The part of the HOLARCTIC REALM that comprises Japan, northern and eastern China, and the northern Himalayas.

sinter plant. An industrial plant in which metal particles are compressed into a coherent body under heat, but at temperatures below the melting point of the metal. Glass, ceramics and certain other non-metals may also be sintered. Sinter plants can be a source of PARTICLE emission.

sinusoidal wave. *See* SINE WAVE.

Siphonaptera. *See* APHANIPTERA.

Siphonopoda. *See* CEPHALOPODA.

Siphunculata. *See* ANOPLURA.

Sipunculoidea. *See* ANNELIDA.

Sirenia (sea-cows). An order of large aquatic herbivorous placental mammals (EUTHERIA) with flipper-like front limbs, vestigial hind limbs, a horizontal tail fin and little hair. Modern forms are the manatee of the Atlantic and the dugong of the Pacific and Indian Oceans. Sea-cows live along the coasts and in rivers in tropical and subtropical regions. Slow-moving, fully aquatic (their young are born in the water) and defenceless, the sea-cows have suffered from hunting and are endangered, but are also being considered for domestication since

their flesh is edible and by consuming large quantities of aquatic plants they may help control the excessive growth of aquatic plants in navigable waterways, preventing weeds from fouling the propellors of boats.

sirocco. A warm, dry wind that occurs in the Mediterranean area in spring, when it brings air from the central Sahara across sea that is much cooler than the desert. Siroccos are known locally as khamsin (in Egypt and Malta), leste (in Madeira and North Africa), simoom (in north-eastern Africa and Arabia) and leveche (in south-eastern Spain). Siroccos bring oppressive weather and often cause damage to vegetation.

site class. The productivity of a forest, measured a cubic metres of wood per hectare per year, calculated for a rotation of 100 years.

site of special scientific interest (SSSI). In Great Britain, an area of land which, in the opinion of the NATURE CONSERVANCY COUNCIL, is of special interest by reason of any of its flora, fauna, or geological or physiographical features. The Nature Conservancy Council must notify the owners, the occupier, the relevant local planning authority and the Secretary of State for the Environment about any such area. Sites of special scientific interest enjoy statutory protection.

site type. A group of COMMUNITIES that corresponds to a particular set of site characteristics.

SI units (Système International d'Unités). An internationally agreed system of units. The seven basic units are the METRE (m), KILOGRAM (kg), SECOND (s), AMPERE (A), KELVIN (K), MOLE (mol) and CANDELA (cd), with the RADIAN (rad) and STERADIAN (sr) added as supplementary units. Derived from these units are the hertz (Hz, *see* FREQUENCY), NEWTON (N), JOULE (J), WATT (W), COULOMB (C), VOLT (V), FARAD (F), OHM (Ω), WEBER (Wb), TESLA (T), HENRY (H), LUMEN (1m) and LUX (lx).

skarn. An impure LIMESTONE or DOLOMITE

that has undergone thermal METAMORPHISM and METASOMATISM, and usually contains calcium silicates and borosilicates. Many skarns are worked for sulphide minerals and manganese silicates.

skeletal muscle. *See* MUSCLE.

skiophyll (sciophyll). A plant with dorsiventral leaves (i.e. whose leaves lie more or less horizontally and have upper and lower sides differing in structure). *Compare* HELIOPHYLL.

skiophyte. *See* HELIOPHOBE.

slag. The non-metallic residue from the SMELTING of metallic ores that generally forms as a molten mass floating on the molten metal.

slaked lime. *See* CALCIUM HYDROXIDE.

slash-and-burn farming. *See* SWIDDEN FARMING.

slate. An ARGILLACEOUS metamorphic rock (*see* METAMORPHISM) with well-developed CLEAVAGE, but little recrystallization. Slates are formed under low-grade regional metamorphism. Slates subjected to thermal metamorphism develop new minerals in small patches, so producing SPOTTED SLATES, or long rods of chiastolite (a variety of andalusite). With increased temperature the cleavage is destroyed and a HORNFELS results. Slates are used for roofing, walling and hedging, and as an inert filler.

sleeping sickness. A disease of humans caused by *Trypanosoma gambiensis* and *T. rhodesiensis*, and transmitted by the tsetse fly (*Glossina*). *See also* CYCLORRHAPHA, TRYPANOSOMIDAE.

sleet. Precipitation of water and ice simultaneously.

slickenside. *See* FAULT PLANE.

slides. Of rocks, the movement downhill of massive whole rocks that slide along sloping BEDDING PLANES or JOINTS. Rock slides may be very rapid and catastrophic. *See also* SLUMP.

slime fungi. *See* MYXOMYCOPHYTA.

slime moulds. *See* MYXOMYCOPHYTA.

slope of front. The gradient of an approaching weather FRONT. A WARM FRONT has a surface (the boundary between two AIR MASSES) with a relatively shallow slope, around 1:100, whereas a COLD FRONT is steeper, at 1:50 over most of its length, steepening considerably near the surface.

slop tank. *See* LOAD ON TOP.

sloths. *See* EDENTATA.

sludge. (1) Thick mud, often greasy. (2) The suspended solid matter in industrial effluent or sewage after partial drying.

slugs. *See* GASTROPODA

slump. The movement downhill of a unit of weathered (*see* WEATHERING) rock or REGOLITH over a definite surface of failure, often lubricated by water. A slump may also be caused by water undercutting the base of a slope or by the faulty design of the cut of an embankment. If the surface of the failure is spoon-shaped, the slump is called a rotational slump.

smelting. The extraction of a metal from its ore by a process involving heat, generally with the reduction of the oxide of the metal using carbon, in a furnace, or the roasting or calcination of sulphide ores. Smelting works emit grit and, when sulphide ores are processed, SULPHUR DIOXIDE, as well as emissions from the combustion of fuels.

Smith, Adam (1723–90). A Scottish economist who advocated free trade and in *The Wealth of Nations* introduced the concept of 'the invisible hand'. Smith was much influenced by the school of Physiocrats (*see* PHYSIOCRACY), but divorced their main concept from its exclusive concern with agriculture and applied it to manufacturing industry.

smithsonite. *See* CARBONATE MINERALS.

smog. Originally, a contraction of smoke and fog, that characterized air pollution episodes in London, Glasgow, Manchester and many other cities (*see* LONDON SMOG INCIDENTS). The word was coined by H.A. Des Voeux in 1905. The great London smog of 1952 caused 4000 excess deaths and led to the enactment of the Clean Air Acts of 1956 and 1968, which provided for the banning of smoky fuels within specified areas. The word smog has since been applied to other air pollution effects not necessarily connected with smoke, such as Los Angeles smog, which arises from NITROGEN OXIDES and hydrocarbons emitted by motor vehicles, and the photochemical action of sunlight.

smoke. An AEROSOL of minute solid or liquid particles (most less than 1 micrometre in diameter) formed by the incomplete combustion of a fuel. In air pollution it is mainly associated with the burning of coal.

smoke stack. A chimney designed to remove emissions from an industrial plant or the boiler of a ship.

smooth muscle. *See* MUSCLE.

smudge pot. A smoke generator used to create an artificial fog with the object of preventing a ground frost (*see* DEW) or frost in a shallow layer of air that might damage fruit, etc. It acts by deepening the layer of air cooled by radiation at night.

smut fungi (Ustiginales). An order of parasitic (*see* PARASITISM) BASIDIOMYCETES with black spores that are the cause of many plant diseases. *Tilletia caries* infects wheat, causing stinking smut disease or bunt.

Sn. *See* TIN.

snails. *See* GASTROPODA.

snipe. *See* CHARADRIIDAE.

snow line. The altitude above which snow lies throughout the year. It varies from place to place depending on summer temperature, wind, snow amount, steepness of slopes, etc.

snow storm. Intense precipitation of ice crystals. In synoptic meteorology, a snow storm is classed as heavy if it exceeds 4 centimetres per hour.

soaring. The art of gaining height to remain airborne in motorless or non-flapping flight by the use of updraughts and upslope flow. It is practised by birds of prey and glider pilots.

social releaser. *See* RELEASER STIMULUS.

society. A group of plants that forms a minor CLIMAX community within a CONSOCIATION, and in which the DOMINANT species is different from that of the consociation.

Society for the Promotion of Nature Conservation (SPNC). *See* SOCIETY FOR THE PROMOTION OF NATURE RESERVES.

Society for the Promotion of Nature Reserves (SPNR). A British society, founded in 1912 by N.C. Rothschild, that owned and managed nature reserves in Britain. It was renamed the Society for the Promotion of Nature Conservation (SPNC) in 1976 and became the ROYAL SOCIETY FOR NATURE CONSERVATION (RSNC) in 1981.

sodium (Na). A soft, silvery–white metallic element that is very reactive, tarnishes rapidly in air and reacts violently with water to produce sodium hydroxide and hydrogen gas. Compounds are abundant and distributed widely, the most common being common salt (sodium chloride). It is an essential MICRONUTRIENT and is used as a coolant in most BREEDER REACTORS. A_r=22.9898; Z=11; SG 0.971; mp 97.5°C.

sodium chlorate . A translocated ($NaClO_3$) (*see* TRANSLOCATION) and SOIL-ACTING HERBICIDE used for total weed control on land not intended for cropping. It can persist in the soil for months after application and makes plant residues inflammable when dry.

It is not highly poisonous to mammals.

soft-energy paths. A term coined by Amory B. Lovins (and the title of his book, published in 1977) for a broad set of political and economic strategies whereby societies shift from a dependence on nuclear power and fossil fuels to intensive energy conservation and reliance on power derived from wind, wave and tidal sources, augmented by the burning of plant material grown for the purpose.

softwoods. Trees belonging to the GYMNO-SPERMAE or their timber. Almost all commercial softwoods are conifers. These do not possess vessels (TRACHEA) in their wood. *Compare* HARDWOODS.

soil. (1) Weathered (*see* WEATHERING), unconsolidated surface material in which plants anchor their roots and from which they derive nutrients and moisture. (2) Loose, unconsolidated material that can be moved without blasting. (3) Any material weathered *in situ*, thus including older weathered deposits that might be parent material for the present soil. *See also* SOIL CLASSIFICATION, SOIL HORIZONS, SURFACE DEPOSITS.

soil-acting herbicides. HERBICIDES that are applied to the soil and are absorbed by weeds before the emergence of aerial parts from the soil (e.g., DIURON, LINURON, SIMAZINE, DIALLATE, TRIALLATE, SODIUM CHLORATE).

Soil Association. A UK-based voluntary organization with worldwide connections and affiliations, concerned with the promotion of ORGANIC FARMING.

soil association. (1) A group of related soil types that form a pattern typical for the geographical region in which they occur. (2) A soil mapping unit, used where it is impracticable to record a SOIL SERIES.

soil classification. The taxonomic arrangement of soils, based on two main systems of classification. (a) Based on the concept of zonality (i.e. that commonly found soils can be associated with particular regions or zones of climate). There are three main orders, divided into suborders, world groups, series and types. (i) Order I Zonal. Belts of soil that correspond roughly with climatic zones (e.g., TUNDRA SOILS, PODZOLS (podsols), GREY–BROWN PODZOLS, BROWN FOREST SOILS, black earth soils (*see* CHERNOZEMS), PRAIRIE SOIL (brunizem), CHESTNUT SOILS, DESERT SOILS, LATERITIC SOILS. (ii) Order II Intrazonal. Soils where local conditions are more important than climate in development (e.g., RENDZINAS, GLEY SOILS. (iii) Order III Azonal or Skeletal. Young soils (e.g., blown sand, river ALLUVIUM, materials in screes and shingles). (b) The USDA Soil Taxonomy system gives ten major soil orders based on the present state of development of the soils. The orders are divided into suborders, great groups, subgroups, families and soil series: (i) Entisols, Young soils, without horizon development, that occur in all climates (e.g., LITHOSOLS). (ii) Vertisols. Clay-rich soils that swell and crack in seasonally wet and dry environments, thus mixing or inverting horizons (e.g., GRUMOSOLS). (iii) Inceptisols. Young soils with horizons weakly developed, occurring in variable climates, including brown earths and tundra soils. (iv) Aridisols. Soils of desert and arid regions with generally mineral profiles. The accumulation of salt, gypsum and carbonate is common. (v) Mollisols. Grassland soils characterized by a thick, organic-rich surface layer, covering a wide variety of profiles with strong structural development, including the brown chestnut, chernozem, prairie (brunizem), red prairie, rendzina and brown forest soils of the earlier definitions. (vi) Spodosols. Podzolized soils with a diagnostic iron oxide and/or organic-enriched B horizon underlying an ashy grey, leached (*see* LEACHING) layer in the A horizon, associated with a cool and cool–humid climate and a forest or heath vegetation cover. (vii) Alfisols. Relatively young, acid soils characterized by a clay-enriched B horizon, commonly occurring beneath deciduous forests and associated with humid, subhumid, temperate and subtropical climates. (viii) Ultisols. Deeply weathered, red and yellow clay-enriched soils, associated with humid temperate to tropical climates. (ix) Oxisols. Tropical and subtropical soils, intensely weathered (*see* WEATHERING) and including most lateritic and bauxitic (*see* BAUXITE) soils of earlier definitions. (x) Histosols. Organic soils developed largely by the accumulation of organic matter in a waterlogged site. There is a major climatic distinction, and the term includes BOG SOILS, half-bog and peat soils of earlier definitions. *See also* SOIL HORIZONS.

soil conservation. The management of a soil in order to minimize EROSION while maintaining or enhancing its ability to sustain crops.

soil creep. The slow movement of soil under the force of gravity. Soil creep produces such phenomena as terminal curvature of planar structures in underlying rock and deformed fence-lines and bulging walls.

soil drainage. The removal from the soil of water that is surplus to the requirements of the use to which the soil is to be put. Naturally, water will drain to the GROUND WATER, but where such drainage is insufficient for agricultural or other purposes (e.g., the WATER TABLE lies at or very close to the surface), the rate of drainage can be increased. Ditches dug along the upper boundary of a sloping field will reduce the amount of water entering the field from land higher up the slope. Mole drains are narrow tunnels made in the subsoil by a cylindrical implement attached to a blade. When dragged through the soil, the slit made by the blade closes, but the tunnel remains open. Tile drains are made from unglazed clay pipes, each about 25 centimetres long, laid in trenches dug to receive them on a bed of, and packed around with, small stones. Tile drains are laid end to end in straight lines, either running parallel to the slope to form a series of channels, or running parallel to the slope and often with side branches laid in a herring-bone pattern. Similar drains are sometimes improvised using pieces of corrugated iron, gorse or other woody material. Where the land is waterlogged, deep trenches may be dug to collect water which

is then pumped to a higher level to cross a WATERSHED, from where it can be fed into a river.

soil erosion. *See* EROSION.

soil flow. *See* SOLIFLUCTION.

soil horizons. Distinctive successive layers of soil produced by internal redistribution processes (*not* by sequential sedimentary deposition). Conventionally the layers have been divided into A, B and C horizons. The A horizon is the upper layer, containing HUMUS and is leached (*see* LEACHING) and/or eluviated (*see* ELUVIATION) of many minerals. The B horizon forms a zone of deposition and is enriched with CLAY MINERALS and iron/aluminium oxides from the A layer. The C layer is the parent material for the present soil and may be partially weathered (*see* WEATHERING) rock, transported glacial or alluvial material (*see* ALLUVIUM) or an earlier soil. Sometimes a D layer is quoted, which is the massive rock underlying the soil layers, and an O layer to designate the fresh organic litter on the surface of the ground. These layers have been subdivided by the use of numbers to indicate minor differentiation with the horizons: A_0, fresh organic litter; A_1, organic-rich A layer; A_2, leached A layer; A_3, A layer grading into B; B_1, B layer grading into B_2; B_2, depositional layer; B_3, B horizon grading into C; C, weathered parent material. Latterly, and coming into favour, is a more detailed nomenclature for the sublayers, using lower-case letters to follow the capital letter that designates the kind of horizon: P, a ploughed or cultivated surface horizon (e.g., A_p); b, buried horizon (e.g., A_b); g, waterlogged gleyed (*see* GLEY) layer (e.g., A_g); e, leached acid horizon (e.g., A_e); h, accumulation of organic matter (e.g., B_h); ca, accumulation of calcium carbonate (e.g., B_{ca}, C_{ca}); s, enriched with translocated SESQUIOXIDES of iron and aluminium (e.g., B_s); ir, accumulation of iron (e.g., B_{ir}); sa, accumulation of soluble salts (e.g., B_{sa}, C_{sa}).

soil map. A map that shows the distribution of soil types in relation to other features of the land surface.

soil particle size.

	Old international (mm)	American (mm)
Coarse sand	0.2–2.00	0.2–2.00
Fine sand	0.02–0.2	0.05–0.2
Silt	0.002–0.02	0.002–0.05
Clay	<0.002	<0.002

See also TEXTURE.

soil phase. A local variation of SOIL TYPE or SOIL SERIES based on some unusual condition of soil (e.g., stoniness, slope, salinity, etc.).

soil profile. A vertical cross-section of SOIL HORIZONS, not including the D layer.

soil series. The basic soil mapping unit composed of soils similar in structure, colour and depth, developed on a uniform or similar soil parent material. The surface layer may show differences. *See also* SOIL ASSOCIATION, SOIL PHASE, SOIL TYPE.

soil type. A subdivision of a SOIL SERIES based on the texture (i.e. SOIL PARTICLE SIZE) of the surface SOIL HORIZON.

Solanaceae. A family of DICOTYLEDONEAE that are herbs, shrubs and small trees of tropical and temperate regions. Economically important genera include *Nicotiana* (tobacco), *Capsicum* (red pepper) and *Solanum*. *S. tuberosum*, native to South America, is cultivated for its stem tubers (potatoes) and *S. lycopersicum* is cultivated for its fruit (tomatoes). *Atropa belladonna* (deadly nightshade, dwale) is very poisonous, containing the ALKALOIDS atropine and hyoscyamine. *See also* GLYCOSIDES.

solanin. A naturally occurring ALKALOID poison found in some members of the SOLANACEAE. *See also* GLYCOSIDES.

Solanum. *See* SOLANACEAE.

solar cell. A device for converting sunlight into electrical power using a semiconductor sensitive to the PHOTOVOLTAIC EFFECT. Solar cells are used on space satellites to power electronic equipment, and as their price falls they may come to be used to provide energy on the Earth.

solar collector. A dark-coloured (ideally matt black) surface used to absorb solar heat. The heat is then transferred, usually to water that flows beneath the surface. Solar collectors can be used to heat water or to provide, less usefully, space heating.

solar constant. The energy flux per unit area due to SOLAR RADIATION, which would be measured outside the Earth's atmosphere at the mean distance of the Earth from the Sun. Its value is about 139.6 mW/cm^2.

solar farm. A suggested power utility, based in a desert and covering a considerable area, where large amounts of electrical energy would be generated from solar energy.

solar furnace. A device for generating large amounts of energy from SOLAR RADIATION, using mirrors to focus heat rays so that intense heat is collected in a small area.

solarimeter. An instrument for measuring total SOLAR RADIATION per unit area received on the ground.

solarization. The inhibitory effect of very high intensities of light upon PHOTOSYNTHESIS.

solar plexus. *See* AUTONOMIC NERVOUS SYSTEM.

solar power. The extraction of useful energy from SOLAR RADIATION. The most successful examples so far are SOLAR CELLS used in satellites and SOLAR COLLECTORS used to heat water. In domestic applications, solar heating is at a disadvantage because the annual cycle of solar radiation intensity is out of phase with the cycle of heating demand. Since it is solar heat that provides the energy to power weather systems, wind power is sometimes classed (e.g., in the USA, but not in the UK) as solar energy, and since the movement of ocean currents and waves is also caused partly by the weather, wave power may also be counted as derived partially from solar energy.

solar radiation. Electromagnetic waves emitted by the Sun. In space outside the Earth's atmosphere, this radiation covers a wide range of wavelengths, but absorption in the STRATOSPHERE restricts the spectrum received at the ground to certain limited bands which include the range of visible light wavelengths.

solar satellite. A device by which solar energy is converted into electrical energy in space and then transmitted to Earth. Designs for solar satellites envisage an array of SOLAR CELLS, several kilometres in span, and a central generating and transmission unit, all held in geostationary orbit. The power would be converted into microwaves and beamed to one or more receiving stations on the ground, where the microwaves would be converted back into electricity and fed into the supply network.

sol brun acide. An acid forest soil with a strongly leached (*see* LEACHING) upper brown layer, but without much accumulated iron, aluminium or clay compounds in the depositional zone. *See also* SOIL HORIZONS.

solenocyte. *See* NEPHRIDIUM.

solfatara. A volcanic vent emitting primarily sulphurous and aqueous vapours. Solfataras are a variety of FUMAROLE.

solifluction (solifluxion, soil flow). The gradual downslope movement of particles at the Earth's surface. Some authors restrict solifluction to processes controlled by freezing and thawing and the production of such deposits as COMBE ROCK.

soligenous (minerotrophic). Applied to wet peatlands such as FENS that are supplied with ground water. *Compare* OMBROGENOUS.

solum. The soil mantle, comprising the organic layer, the leached (*see* LEACHING) layer and the layer of deposition, but not including the parent material (i.e. the A and B horizons). *See also* SOIL HORIZON.

solution. In soil, water entering the ground encounters organic matter containing carbon dioxide and acid ions which become dissolved in the water. The acids thus produced dissolve mineral matter as they percolate downwards. Most commonly, a weak solution of carbonic acid (H_2CO_3) is formed which rapidly dissociates to hydrogen and bicarbonate ions (H^+ and HCO_3^-). This is a solvent of some rock, particularly LIMESTONE, CHALK and the cementing matrix of some SANDSTONES.

soma. All the cells of an organism other than the reproductive cells.

somatic. Pertaining to the SOMA or to the body wall of an animal.

somatic mutation. *See* MUTATION.

somatotrophic hormone. *See* GROWTH HORMONE.

sonar. An acronym for *so*und *na*vigation *r*anging, describing a technique, and the associated devices, for investigating an environment by emitting sound waves, detecting their reflections and measuring the time that elapses between emission and reception. Sonar is used for locating underwater objects. *See also* ECHO.

sonde. A device for obtaining direct measurements of the condition of the atmosphere at various altitudes. It comprises a lifting device, such as a balloon or rocket, and a set of transducers with either a recorder or, more frequently, a radio transmitter.

sone. A unit of loudness, designed to give a scale of numbers roughly proportional to the loudness. *Compare* PHON.

song birds. *See* PASSERIFORMES.

sonic boom. The transient noise heard by a stationary observer as an object (especially an aircraft) passes nearby travelling above the SPEED OF SOUND. The cause of the sound is the passage of a shock wave with its associated pressure pulses. For supersonic aircraft, the pressure pulse is of the order of 0.0705 kilograms per square centimetre (1 pound per square inch), but it may fluctuate above and below this level. Strong sonic booms can damage buildings and break windows, whereas their sudden onset can be disturbing to people even at quite modest pressure levels. *See* MACH NUMBER, SST.

Sorghum vulgare. *See* GRAMINEAE.

sorting. According to nature of transportation, wind or water can fractionate a homogeneous rock debris into grains of similar sizes. The sorting of a sediment refers to the range of grain sizes within a particular sample. The term well-sorted has a different meaning in geology and engineering. To a geologist a well-sorted sediment has a preponderance of just one grain size. To an engineer, well-sorted grains means an even mix of range of sizes, referred to as a graded aggregate.

sound, speed of. *See* SPEED OF SOUND.

sounding balloons. *See* BALLOONS.

sound level. The value of sound in DECIBELS. *See also* LOUDNESS.

sound power level. The total energy per second emitted by a sound source expressed in DECIBELS.

sound pressure level. The effective value (i.e. ROOT MEAN SQUARE VALUE) of pressure fluctuations above and below atmospheric pressure caused by the passage of a sound wave expressed in DECIBELS.

sound propagation. SOUND WAVES propagate in gases in the form of small variations in pressure and displacement in the direction of propagation. A sound source is a generator of pressure fluctuations, which are communicated from each small element of the gas to the next by the acceleration of gas due to transient higher pressure gradients leading to compression of the adjacent gas and consequently higher pressures there. *See also* SPEED OF SOUND.

sound shadow. The acoustical equivalent of a light shadow.

sound waves. Pressure and displacement waves in material media. From point sources in uniform media the waves are spherical and propagate radially. *See also* SOUND PROPAGATION, SPEED OF SOUND.

source rock. The geological formation in which petroleum or minerals originated.

sour gas. NATURAL GAS containing high levels of HYDROGEN SULPHIDE. The hydrogen sulphide is separated and used as a source of sulphur. *Compare* SWEET GAS.

South African Floral Region. The part of the AUSTRAL REALM that comprises South Africa, Botswana and Namibia.

South Asian Cooperative Environment Programme. An agreement, drawn up under the auspices of the UNITED NATIONS ENVIRONMENT PROGRAMME in 1981, under which Afghanistan, Bangladesh, India, Iran, the Maldives, Nepal, Pakistan and Sri Lanka proposed to study environmental problems in their region and to support environmental projects.

South-East Asian Floral Region. The part of the PALAEOTROPIC REALM that comprises southern Burma, northern and central Malaysia and Indochina.

southerly buster. The name given to a burst of air of polar origin entering Australia behind a COLD FRONT and with a strong wind.

Southern Arch. *See* ARCH CLOUD.

Southern Realm. *See* AUSTRAL REALM.

South Oceanic Floral Region. The part of the AUSTRAL REALM that comprises the islands of the South Atlantic and Indian Oceans south of latitude 50°.

soya (*Glycine max,* soya bean). An annual herb, belonging to the family LEGUMINOSAE, that grows as an erect bushy plant to a height of 45 centimetres to 2 metres and bears rough, brownish hairs on its leaves and pods. The pods (up to 5–7 centimetres long) are constricted between the two to four seeds that each contains. The seeds (beans) may be off-white, green, brown, yellowish or black, and they are grown for their high protein content as food for livestock and humans. The plant also yields an oil used for cooking, as a salad oil, in the manufacture of margarine and in the manufacture of soaps, plastics, paints and other products. The young shoots of beans are important in Chinese cuisine (bean sprouts), its seeds can be dried and ground to make a flour used to increase the protein content of other flours. It is used in ice cream manufacture, and a 'milk' extracted from the seeds is used in Chinese and Japanese cuisine and is recommended for invalids. Soy sauce, made from fermented beans, is used widely as a condiment.

spadix. A spike bearing flowers, sometimes sunken, and enclosed in a SPATHE.

spangle gall. *See* GALL.

spathe. A large sheathing BRACT.

special waste (dangerous waste, hazardous waste, intractable waste). In England and Wales, under the Control of Pollution Act, 1974, waste that cannot be disposed of under the terms of a disposal licence issued by the Secretary of State for the Environment, but is subject to additional regulations. Depending on the waste, these may include the requirement that the location of the waste be notified to the local authority, directions to the disposal authorities setting out the procedures to be followed, limits to the quantities kept at a particular place, instructions for storage and the keeping of records of the waste setting out its composition and amounts handled. In some cases special wastes may be disposed of only by or under the direct supervision of agencies appointed for the purpose, or by officials of the Department of the Environment. To qualify as special waste a substance must be suffi-

ciently toxic to present a hazard to human health or risk of serious environmental contamination.

speciation. The evolutionary process by which a new species is formed. *See also* CLASSIFICATION.

specient. An individual member of a species.

species. *See* CLASSIFICATION.

speciesism The attitude that the interests of one's own species should be given priority over those of all other species. *See also* ANTHROPOCENTRISM.

species pair. *See* SIBLING SPECIES.

specific gravity (SG). Relative density, being the ratio of the density of a substance at a particular temperature to the density of water at the temperature of its maximum density (i.e. 4°C). Numerically, the specific gravity is equal to the density in grams per cubic centimetre, but it is stated as a pure number.

specific humidity. *See* HUMIDITY.

spectrum of turbulence. The variation of TURBULENCE intensity with FREQUENCY. Atmospheric turbulence may be thought of as being composed of oscillations in local velocity with a wide range of time and length scales (and therefore of frequency). The spectrum of these oscillations is the ROOT MEAN SQUARE VALUE of the velocity at a given point and in a given direction per unit bandwidth, centred on the various frequencies contributing to the motion.

specularite. Specular iron ore; HAEMATITE in the form of metallic black crystals.

speed of sound. In air, the speed of propagation of sound waves is about 332 metres per second at 0°C. Generally, in a perfect gas, the speed of sound is given by $a = \sqrt{(\gamma RT)}$, where γ is the ratio of specific heats C_p/C_r, R is the gas constant and T is the absolute temperature (i.e. the temperature in KEL-VINS).

spell of weather. A convenient concept in the study of the persistence of weather types. It may be defined for the purpose of statistical studies as a period of consistent type of *n* days, *n* being 5 or 10, or a number sufficiently large to make the spell a notable event. Thus a spell of dry weather would be longer than a spell of fog.

spermagone. *See* SPERMOGONIUM.

spermagonium. *See* SPERMOGONIUM.

Spermaphyta. *See* SPERMATOPHYTA.

spermatia. *See* SPERMOGONIUM.

spermatocyte. (1) A cell that gives rise to a SPERMATOZOID without cell divisions. (2) A primary spermatocyte gives rise by MEIOSIS first to two secondary spermatocytes, then to four spermatids. These become spermatozoa (*see* SPERMATOZOON), usually developing flagella (*see* FLAGELLUM).

spermatogenesis. The.cell divisions and changes resulting in the formation of SPERMATOZOIDS or spermatozoa (*see* SPERMATOZOON).

spermatogonium. A cell in a testis that gives rise to SPERMATOCYTES.

spermatophores. Packets of spermatozoa (*see* SPERMATOZOON) produced by certain animals (e.g., newts, some CRUSTACEA and MOLLUSCA) with internal fertilization.

Spermatophyta (Spermaphyta, Phanerogamia, seed plants). Plants that produce seeds. The group comprises the GYMNOSPERMAE and the ANGIOSPERMAE.

spermatozoid (antherozoid). A male GAMETE produced by lower plants (ALGAE, BRYOPHYTA, PTERIDOPHYTA) that is able to move by means of a FLAGELLUM.

spermatozoon (pl. spermatozoa). A male

GAMETE, produced by an animal. Most spermatozoa can move by means of a FLAGELLUM.

spermogonium (spermogonium, spermagone, pycnium). A hollow organ inside which non-motile male GAMETES (spermatia) are formed in some FUNGI (e.g., RUST FUNGI). Spermatia are transferred by the wind or insects to other HYPHAE.

sphaeroidal weathering (onion weathering). The WEATHERING of a rock such that concentric layers separate off a less-weathered core. It is commonly seen in well-jointed (*see* JOINT) rocks like BASALTS and DOLERITES in which water penetrates along the joints attacking each block from all sides.

sphagnicolous. Inhabiting bog moss (*Sphagnum*).

sphagniherbosa. A plant community growing on peat and containing large amounts of bog moss (*Sphagnum*).

***Sphagnum* moss.** *See* BOG.

sphalerite (zinc blende, ZnS). The mineral zinc sulphide, in which CADMIUM commonly substitutes for ZINC; the most important ore of both zinc and CADMIUM. Sphalerite commonly occurs, with GALENA, in HYDROTHERMAL deposits, in sedimentary stratiform deposits and in metasomatized (*see* METASOMATISM) LIMESTONES.

Spheniscidae (penguins). Aquatic birds of the southern hemisphere that have lost the power of flight and swim mainly by means of the forelimbs which are modified as flippers. Many make no nest, and some incubate their eggs by carrying them on their webbed feet. *See also* SPHENISCIFORMES.

Sphenisciformes. An order that includes the family SPHENISCIDAE (penguins). The penguins have no close living relative, but may be allied distantly to the petrels (PROCELLARIFORMES).

Sphenodon. *See* RHYNCHOCEPHALIA.

Sphenophyllales. An order of fossil PTERI-DOPHYTA that flourished during the CARBON-IFEROUS and PERMIAN Periods. They were shrubby or herbaceous plants with grooved stems and whorls of leaves. SPORES were formed in terminal CONES.

Sphenopsida. A group of vascular plants (TRACHEOPHYTA) that comprises the horsetails (EQUISETALES) and SPHENO-PHYLLALES.

spherulite. *See* RHYOLITE.

spiders. *See* ARANEAE.

spike. An unbranched, elongated flower head, bearing flowers that have no stalks.

spikelet. A unit of the flower head of a grass (GRAMINEAE), usually bearing two GLUMES and one or more florets.

spilite. A sodium-rich BASALT.

spindle. *See* MEIOSIS, MITOSIS.

spindle bomb. *See* BOMB.

spindle tree. *See* APHIDIDAE.

spinel. A group of minerals with the general formula AB_2O_4, where A is magnesium, iron, zinc, manganese or nickel, and B is aluminium, iron or chromium. Spinels are formed in IGNEOUS and metamorphic rocks (*see* METAMORPHISM) and include CHROMITE ($FeCr_2O_4$), the main ore mineral of chromium and MAGNETITE (Fe_3O_4). Most of the gem spinels are spinel *sensuo stricto*, $MgAl_2O_4$. Spinel can become concentrated in PLACER deposits.

spiny anteaters. *See* MONOTREMATA.

spiny-headed worms. *See* ACANTHOCEPH-ALA.

spiracle. *See* STIGMA.

spirellus. Any roughly spiral bacterium (*see* BACTERIA).

spirochaete (spirochete). A spirally twisted unicellular bacterium that moves by undulating. Some species are free-living, others are parasitic (*see* PARASITISM), causing such diseases as syphilis.

splash erosion. *See* EROSION.

splash zone. See SHORE ZONATION.

splenic fever. *See* ANTHRAX.

spodic. Applied to a soil layer enriched in SESQUIOXIDES and/or organic matter, with or without CLAY. *See also* SOIL CLASSIFICA-TION.

Spodosols. *See* SOIL CLASSIFICATION.

sponges. *See* PORIFERA.

spongy mesophyll. *See* MESOPHYLL.

spoonbills. *See* CICONIIFORMES.

sporangium. A plant organ inside which asexual spores are formed. *See also* MYXOMYCOPHYTA, SPOROPHYLL.

spore. A reproductive body, consisting of one or several cells, produced by plants, bacteria and PROTOZOA. Spores are usually microscopic and are often produced in vast numbers. They are widely dispersed and can effect a rapid increase in the population of a species. Some are thick-walled resting spores which can survive unfavourable conditions. In flowering plants (SPERMATOPHYTA), the spores are the POLLEN grain (microspore) and EMBRYO SAC (megaspore). *See also* HETEROSPORY, HOMOSPORY.

sporogonium. The spore-producing structure (SPHOROPHYTE) of BRYOPHYTA.

sporophyll. A leaf, or structure derived from a leaf, on which sporangia (*see* SPORAN-GIUM) are borne. In some plants (e.g., bracken) sporophylls are like ordinary leaves, but in other plants they are much modified. The stamens and carpels of flowering plants are sporophylls.

sporophyte. A SPORE-producing generation in the life cycle of certain multicellular plants. The spores develop without fertilization into gametophytes, the alternate generation (*see* ALTERNATION OF GENERATIONS).

Sporozoa. A class of parasitic (*see* PARASITISM) PROTOZOA that form large numbers of SPORES. Most Sporozoa have complicated life cycles involving alternation of sexual and asexual reproduction (*see* ALTERNATION OF GENERATIONS). Sporozoa cause COCCIDIOSIS and malaria (*see* PLASMODIUM), red water fever in cattle (transmitted by a tick) and diseases in the silkworm and honey bee.

SPOT (Système Probatoire d'Observation de la Terre). A French observation satellite, launched in 1986, which provides images of the Earth for use by farmers, geologists and land-use planners. It has a resolution of 10 metres in black and white. *Compare* LANDSAT.

spotted slate. SLATE with darker inclusions of new minerals, found in the outer part of a METAMORPHIC AUREOLE. Inclusions are usually small, spherical or ovoidal in shape, but may be rectangular, and they are due to the recrystallization of minerals by thermal METAMORPHISM.

spray. Liquid droplets greater than 10 micrometres in size, created by mechanical disintegration processes. Spray is a source of salt particles in the atmosphere. The smallest drops are formed when tiny air bubbles break the surface. These remain airborne long enough for evaporation of their water to occur before they fall back to the sea. Drops in spray from wave crests are too large to make a significant contribution.

springtails. *See* APTERYGOTA.

spurges. *See* EUPHORBIACEAE.

Sputnik. A Soviet earth-orbiting satellite, launched in 1957; the first Earth-orbiting satellite, whose success stimulated the US space research programme.

squall. A strong wind that begins suddenly, lasts for several minutes, then dies away rather more slowly. The term is used also to describe a kind of storm characterized by a series of squalls (i.e. a succession of wind gusts) each lasting from a few seconds to a minute and with speeds that may be half as great again as the average wind speed. Windspeeds of 50–100 kilometres per hour are common in squalls, and gusts of up to 160 kilometres per hour have been known. *See also* LINE SQUALL.

Squamata (Lepidosauria). An order of reptiles comprising lizards (Sauria, sometimes called Lacertilia), snakes (Serpentes, sometimes called Ophidia) and some extinct groups. Snakes are limbless, elongated and have a jaw arrangement enabling the mouth to be opened extremely widely. Locomotion is by lateral undulations of the body, helped by ventral scales attached to long, movable ribs. Lizards have a normal-sized jaw gape, and only a few species (e.g., the slow-worm, *Anguis fragilis*) lack limbs.

squids. *See* CEPHALOPODA.

Sr. *See* STRONTIUM.

sr. *See* STERADIAN.

SSSI. *See* SITE OF SPECIAL SCIENTIFIC INTEREST.

SST (supersonic transport aircraft). Civil aircraft designed to fly at speeds in excess of the SPEED OF SOUND over part of their route. Development of SSTs has proceeded in the USA, the USSR and in the combined Anglo-French Concorde, although only the Concorde operates regular scheduled services. SST development has caused great controversy, particularly with regard to the cost of the development programme and the likelihood of the aircraft being able to compete commercially with much cheaper, slower aircraft, although experience with Concorde operations has shown that a demand exists. Environmental objections arise mainly from noise at take-off and sonic boom (limited, but not eliminated by restrictions on supersonic flight over populated land areas). Early fears of possible chemical reactions between exhaust gases and atmospheric gases at high operating altitudes were largely unfounded. *See also* LANDING AND TAKE-OFF CYCLE.

stability. *See* DYNAMIC STABILITY, STATIC STABILITY.

stabilization. The increase of dominance (*see* DOMINANT) that ends in a stable CLIMAX, produced by the invasion of species leading to the establishment of a population most completely fitted for the prevailing conditions. This degree of adaptation to conditions ensures that the climax is permanent and can be changed only by a change in the conditions to which the population is fitted.

stabilizer. A trace chemical added to a plastic to reduce its degradation by oxidation and the action of ULTRAVIOLET RADIATION.

stable population. A population in which births and deaths are in balance, so that, discounting migrations, the size of the population remains constant.

stage. In stratigraphy, a subdivision of a SERIES.

stagnation point. When a fluid flow divides to pass by an object, there exists on the upstream (or upwind) side a point (or in two-dimensional flow a line) on the surface of the object at which the velocity of flow is zero.

stamen. *See* FLOWER.

staminate. Applied to FLOWERS that have stamens, but not carpels, and so are male. *Compare* PISTILLATE.

stand. (1) An aggregation of plants that is more or less uniform in species composition, age and condition, and is distinguishable from adjacent vegetation. (2) In forestry, the amount (usually expressed as volume) of standing timber per unit area.

standing crop. The amount of living matter (usually expressed as BIOMASS) present in a population of one or more species within a given area.

standing wave. In a fluid, a wave or distortion in the flow whose shape is stationary in relation to the surface (e.g., a MOUNTAIN WAVE).

starch. *See* CARBOHYDRATES.

starfishes. *See* ASTEROIDEA, CROWN OF THORNS STARFISH.

stasis. A period of no change. *Compare* SALTATION.

static electricity. Electricity that is at rest, rather than flowing. If the electrons within a substance are at rest, there is no electrostatic force present in the substance to move them (i.e. there is no electrical potential between any two points within the substance). However, should this substance enter the electrostatic fields of another substance, a potential will be created between the two substances, and there will be a consequent flow of current.

static reserve index. The length of time for which the known reserves of a RESOURCE will last if the rate at which they are used remains constant, calculated by dividing the amount of the reserve by the amount used during each time period. If the rate of use is increasing exponentially (*see* EXPONENTIAL GROWTH), then the exponential reserve index applies, calculated by comparing the reserves with a rate of use that increases exponentially.

static stability. In the atmosphere, the condition in which small vertical displacements of a parcel of air cause gravitational restoring forces in the absence of horizontal wind. This is the case if the LAPSE RATE is greater than the ADIABATIC LAPSE RATE. *Compare* DYNAMIC STABILITY.

station. The geographical location in which an organism or community occurs, making no reference to environmental factors.

statocyst (otocyst, lithocyst). An organ of balance, present in many invertebrates, that consists of a fluid-filled sac containing one or more granules (statoliths, otoliths, lithites) of lime or a similar substance that shift about as the animal moves and stimulate the sensory cells lining the sac.

statolith. *See* STATOCYST.

steam fog. A FOG that forms when cold air overlies sufficiently warm water. The air just above the surface is saturated and at the water temperature, so that on mixing with air at higher levels condensation may occur. The density profile is unstable so that free convection occurs and the fog appears as vertical streaks.

steam-generating heavy-water reactor (SGHWR). A NUCLEAR REACTOR that uses enriched URANIUM as a fuel, boiling light water as a coolant and heavy water as a MODERATOR.

steel. An alloy of iron with small amounts of carbon. *See also* ALLOY STEEL, BESSEMER PROCESS.

steering of storms. The movement of storms that must be understood if their tracks are to be predicted. The most useful rule is that storms travel in the direction of the THERMAL WIND, or with the wind at about the mid-height of their circulation.

Stefan's law. The law which states that the total radiation emitted by a BLACK BODY is proportional to the fourth power of its absolute temperature (i.e. its temperature in KELVINS). The law is named after the Austrian physicist Josef Stefan (1835–93).

stele (vascular cylinder). The core or (in some ferns, FILICALES) network of vascular tissue in a root or stem. It is made up of XYLEM and PHLOEM with accompanying ground tissue, and is surrounded by the ENDODERMIS, when this is present.

stem. That part of a VASCULAR PLANT which bears leaves, buds and often reproductive structures (e.g., flowers). The VASCULAR BUNDLES in a stem are scattered or arranged in a ring. Most stems grow above the ground, but RHIZOMES are subterranean and their leaves are reduced to scales.

stenohaline. Applied to organisms that are able to tolerate only narrow range of salinity (i.e. a small variation of OSMOTIC PRESSURE) in the environment. *Compare* EURYHALINE.

stenothermous. Applied to organisms that are able to tolerate only a small variation of temperature in the environment. *Compare* EURYTHERMOUS.

stenotopic. Applied to organisms with a restricted distribution. *Compare* EURYTOPIC.

Stephanian. The youngest STAGE of the CARBONIFEROUS System in Europe.

steppe. An extensive treeless plain, the grassland BIOME of Asia, and equivalent to the PRAIRIE and pampas of the Americas.

steradian (sr). The supplementary SI unit of solid angle, being that angle which encloses a surface on a sphere equal to the square of the radius of the sphere.

Sterculiaceae. A family of DICOTYLEDONEAE, most of which are tropical herbs, shrubs and trees. The seeds of the tropical American genus *Theobroma* yield cocoa and cocoa butter. The tree *Cola* is the source of kola nuts, which contain caffeine and are an important article of trade in West Africa.

stereo pictures. Two photographs of the same object taken from different positions (e.g., by an overflying aircraft which takes two pictures a few seconds apart or, more usually, by a camera with two lenses a fixed distance apart and two films, whose shutters operate simultaneously). Stereo pictures are used to obtain a three-dimensional appreciation of a scene (e.g., to measure the height of buildings and hills).

stereoscopic vision. *See* BINOCULAR VISION.

stereotaxis. *See* TAXIS.

stereotropism. *See* HAPTOTROPISM.

Stevenson screen. A standardized wooden container in which surface weather measurements are made. It is of louvred construction so that the air mass may pass freely through it, while thermometers are shielded from direct sunlight. It is placed so that the thermometers are 1.25 metres above ground and is painted white all over. It was designed by the engineer Thomas Stevenson.

STH. *See* GROWTH HORMONE.

stick-insects. *See* PHASMIDA.

stigma. (1) The terminal part of a carpel (*see* FLOWER), which receives the POLLEN. (2) (eye spot) A light-sensitive pigmented ORGANELLE, found in many flagellate Protozoa (FLAGELLATA) and microscopic, motile green algae (CHLOROPHYTA). (3) (spiracle) The external opening of a trachea (breathing tube) in an insect. (4) A small coloured area on the clear wings of some insects (e.g., dragonflies).

still. An apparatus for DISTILLING, used to prepare alcoholic drinks, distilled water and other purified liquids.

stimulus. A change in the internal or external environment of an organism that evokes a response in the organism, but does not provide energy for this response.

stingy bark. *See* EUCALYPTUS.

stinking smut disease. *See* SMUT FUNGI.

stipule. An outgrowth that may be leaf-like or scaly at the base of a petiole (leaf stalk).

stochastic. In statistics, applied to a random element (e.g., a stochastic process is one in which some element of chance occurs).

stock. (1) A discordant IGNEOUS INTRUSION

with an area of OUTCROP of a few square kilometres to tens of square kilometres. Some stocks are probably CUPOLAS of BATHOLITHS. *See also* BOSS. (2) *See* RESOURCE.

Stockholm Conference. *See* UNITED NATIONS CONFERENCE ON THE HUMAN ENVIRONMENT.

stockwork. An ore deposit consisting of a large-scale mass of closely spaced, narrow mineralized veins. Stockworks are commonly worked by OPEN-CAST MINING.

stoichiometric. Pure. A chemical compound is stoichiometric when its component elements are present in the precise proportions represented by its chemical formula. A stochiometric mixture is one that will yield a stoichiometric compound on reaction (i.e. it consists of elements in exactly the proportion required to yield a stoichiometric compound utilizing all the components of the mixture).

stoker. Originally a man, now more commonly a machine, for feeding coal or other combustible material into a furnace and supporting it there during combustion.

stolon. (1) (runner) A short-lived horizontal stem (e.g., a strawberry runner) that produces roots and shoots, so vegetatively propagating (*see* VEGETATIVE PROPAGATION) the plant. The term is usually restricted to stems that creep above the ground. (2) A root-like part of a colony of animals (e.g., some HYDROZOA) that serves for anchorage. (3) An outgrowth of the body of some sea-squirts that produces the new individuals by budding.

stomata (sing. stoma). Pores in the EPIDERMIS of a plant through which exchange of gases (water vapour, oxygen, carbon dioxide) takes place. Stomata are particularly abundant in leaves. Each pore is surrounded by two crescent-shaped CHLOROPHYLL-containing guard cells. Absorption and loss of water causes alterations in the shape of these cells, leading to variations in the size of the stomatal aperture.

stone. A METEORITE composed mainly of SILICATE MINERALS.

Stone Age. That period in the development of human societies in which tools and implements are made from stone, bone or wood, and no metals are used. The date of the period varies widely from place to place, and the term is often used loosely to describe the cultures of present-day preagricultural peoples, although these may (and often do) use imported metal artifacts. *Compare* PALAEOLITHIC, MESOLITHIC, NEOLITHIC.

stone cells. *See* SCLERENCHYMA.

stoneflies. *See* PLECOPTERA.

stoneworts. *See* CHAROPHYTA.

stony iron. A METEORITE rich in both nickel–iron alloys and SILICATE MINERALS.

stope. *See* STOPING.

stoping. (1) A process postulated to occur during IGNEOUS INTRUSION, particularly of GRANITES, whereby MAGMA forces its way along fissures in the COUNTRY ROCK, so forcing blocks to fall into the magma chamber and be assimilated. (2) In mining, the method of winning ore from a steeply inclined LODE by driving horizontal tunnels (called drives) along the lode, then extracting the ore above or below the drive (called overhand or underhand stoping, respectively), thus forming caverns called stopes.

storks. *See* CICONIIFORMES.

storm surge. An unusual variation in the AMPLITUDE of the tide, caused by atmospheric factors (e.g., wind and pressure gradients).

stoss. The direction from which ice has come. Stoss-and-lee topography is a landform showing rocks with smoothly abraded slopes on one side and broken, steep slopes on the other. *Compare* LEE.

stoss-and-lee topography. *See* STOSS.

strain. (1) A subspecific group in which the organisms are not sufficiently different genetically from the rest of the species to form a variety. (2) The distortion of material, usually producing stress (i.e. internal forces). In an elastic solid the stress is proportional to the amount of the strain, in a viscous fluid to the rate of strain. Plastic and viscoelastic materials have more complex relationships.

strata (sing. stratum). (1) *See* LAYERS. (2) *See* STRATIFICATION.

stratification. The arrangement of material in discrete layers (strata, one above the other (e.g., in the formation of SEDIMENTARY ROCKS). In the atmosphere, the forming of stable horizontal layers that do not intermingle, because the LAPSE RATE is less than the ADIABATIC LAPSE RATE.

stratocumulus. CUMULUS clouds in a layer or a layer, often complete, of clouds generated by CONVECTION, which occurs when convection is confined by an INVERSION at the top.

stratopause. The boundary between the STRATOSPHERE and the MESOSPHERE, which is at a height of about 50 kilometres.

stratosphere. The region of the atmosphere above the TROPOSPHERE in which temperature increases with height. There is therefore little mixing. The stratosphere contains relatively large amounts of OZONE formed by absorption of ULTRAVIOLET RADIATION.

stratovolcano (composite volcano). A VOLCANO that emits both TEPHRA and LAVA, and builds up a steep-sided cone with irregularly alternating layers.

streak. The colour of a powdered mineral. Streak is normally determined by drawing the mineral across an unglazed porcelain tile. The colour of the streak is far less variable than the colour of the mineral in coarser form.

streak clouds. Fibrous patches of cloud

elongated in the direction of the WIND SHEAR.

streamline. A line drawn through a fluid in motion that is always parallel to the local direction of flow.

stream tin. CASSITERITE that occurs as detrital grains in alluvial deposits (*see* ALLUVIUM).

Strepsiptera (stylopids). An order of minute insects (ENDOPTERYGOTA) whose larvae are mostly parasitic (*see* PARASITISM) on other insects such as plant-bugs and bees, often rendering the host sterile. The adult females are grub-like and usually remain inside their hosts, but the males are free-living insects with vestigial forewings.

Streptomyces griseus. See STREPTOMYCIN.

streptomycin. An ANTIBIOTIC produced by the bacterium *Streptomyces griseus*. It is used in the treatment of diseases such as tuberculosis and as a fungicide for the control of downy mildew on hops. *See also* ACTINOMYCETALES.

Streptoneura. *See* GASTROPODA.

striated muscle. *See* MUSCLE.

Strigiformes (owls). Birds of prey that are specialized for hunting at night. They have large heads, short necks, hooked beaks and powerful claws. The eyes cannot be moved in their sockets and the whole head swivels to compensate. There are large external ear flaps. Owls fly silently, probably detecting prey (small mammals, birds and insects) mainly by sound. Most nest in cavities.

strike. (1) The direction of the line of intersection of an inclined rock surface (e.g., BEDDING PLANE, VEIN, JOINT, CLEAVAGE PLANE, FAULT, SCHISTOSITY) with a horizontal plane. Strike is at right angles to the direction of DIP. (2) Less precisely, an indication of the general trend of strata (*see* STRATIFICATION).

strike–slip fault. *See* FAULT.

striped muscle. *See* MUSCLE.

stripes. A linear arrangement of stones down a slope, found in regions of present or former PERMAFROST. Stripes grade into POLYGONS with a decrease of slope.

strip mining (open-cut mining). The working of coal or an ore by removing the OVERBURDEN to expose the ore body, which is removed and may be partly processed on site. The technique is used especially where the rock contains so little of the required substance that conventional techniques of tunnelling along veins cannot be used. The effect on the landscape is considerable, with some minerals the pollution may be high, and restoration of the land when work has finished may be difficult.

strobiliation. The repeated formation of similar structures (e.g., proglottides of a tapeworm, *see* CESTODA; larvae of some jellyfish) that remain connected for a time, but after a period of development drop off the chain one by one. *See also* SEGMENTATION.

strobilus. *See* CONE.

stroma. (1) CONNECTIVE TISSUE binding together the components of an organ in an animal. (2) A mass of fungal HYPHAE from which, or inside which, fruiting bodies are formed. (3) The colourless part of a CHLOROPLAST.

strombolian volcano. *See* VOLCANO.

Strong, Maurice. *See* UNITED NATIONS ENVIRONMENT PROGRAMME.

strontium (Sr). A reactive metallic element that resembles CALCIUM and can replace it in biological processes. It occurs as CELESTITE and strontianite. The radioisotope strontium-90 is present in the fall-out from nuclear explosions and is a health hazard because it can be deposited in bone, replacing calcium. Strontium-90 has a HALF-LIFE of 28 years. $A_r = 87.62$; $Z = 38$; SG 2.6; mp 757°C.

structural genes. *See* GENES.

structural terrace. A local flattening in an otherwise uniformly dipping (*see* DIP) series of BEDS. *Compare* MONOCLINE.

structure. In ECOLOGY, the spatial and other arrangements of species within an ECOSYSTEM. The structure takes account of the composition of the biological community, including species, numbers, BIOMASS, life cycle and spatial distribution; the quantity and distribution of the non-living materials (nutrients, water, etc.); the range, or gradient, of conditions such as temperature, light, etc.

strychnine. *See* STRYCHNOS.

Strychnos. A genus of tropical DICOTYLEDONEAE (family: Strychnaceae) comprising trees and climbing shrubs whose seeds yield the poison strychnine. The bark of *Strychnos toxifera* yields curare, used on poison arrows by South American Indians. The seeds of *S. potatorum* are used to purify water by precipitation.

style. *See* FLOWER.

stylopids. *See* STREPSIPTERA.

subbase. *See* PAVEMENT.

subbituminous coal. *See* RANK.

subclimax. (1) The stage that precedes the CLIMAX in a complete SERE. (2) A plant COMMUNITY that is stabilized at the subfinal stage in the succession by EDAPHIC or other factors that arrest development into the CLIMATIC CLIMAX. *Compare* POSTCLIMAX, PRECLIMAX, PROCLIMAX SERCLIMAX.

subcloud layer. The layer of air below CLOUD BASE, used sometimes to mean the shallow, stable layers of air immediately beneath convection clouds, at other times to mean all the air between cloud base and the Earth's surface.

subdominant. (1) Species that may appear more abundant than the true DOMINANT in a CLIMAX at particular times of the year (e.g.,

trees and shrubs are more conspicuous in a savannah than are the grasses that are the true dominants). (2) Species occurring abundantly, but at a lower frequency than the dominant species.

subduction zone. According to the theory of PLATE TECTONICS, those belts on the Earth where lithospheric plates are descending into the MANTLE.

suberin. *See* SUBERIZATION.

suberization. The laying down of suberin, the waterproof substance in the walls of CORK cells. *See also* BARK.

subharmonic. A HARMONIC of a FREQUENCY that is an integral number of times lower than the FUNDAMENTAL FREQUENCY in a periodic wave.

subimago. *See* EPHEMEROPTERA.

sublimation. The change in a substance between the solid and the gaseous phase without passing through the liquid phase.

sublittoral fringe. *See* SHORE ZONATION.

sublittoral zone. (1) That part of a lake which is too deep for rooted plants to grow. *Compare* LITTORAL. (2) The zone of a sea lying below the intertidal zone and extending to the limit of the CONTINENTAL SHELF. The part of the NERITIC zone lying beneath the littoral zone. *See also* SHORE ZONATION.

submetallic. *See* LUSTRE.

subsequent. Applied to rivers whose courses follow channels cut by themselves into easily erodable rock formations. *Compare* CONSEQUENT.

subsere. The series of plant communities making up the stages in a secondary SUCCESSION, or any one of these stages.

subsidence. In the atmosphere, the gradual descent of an AIR MASS. Near the ground, the rate of descent is lower, and the flow spreads horizontally. Adiabatic compression occurs and the descending air is warmed, producing a stable LAPSE RATE. This is called the subsidence inversion. Subsidence is usually associated with anticyclonic conditions (*see* ANTICYCLONE).

subsidence inversion. A high-level INVERSION. *See also* SUBSIDENCE.

sub-song. *See* BIRD SONG.

subsonic. Applied to a flow if the relative speed between all parts of the flow and any solid boundary is less than the local SPEED OF SOUND in the fluid.

subspecies. A group of organisms genetically distinct from the rest of their species, which has arisen through partial or recent reproductive isolation. Subspecies are often regarded as incipient species still capable of interbreeding within their species.

substituent. *See* SUBSTITUTION PRODUCT.

substitution product. (1) A compound obtained by replacing an atom or group of atoms by another atom or group of atoms within a molecule, the new atom or group being known as the substituent. (2) A product that enters a market in place of another, but satisfies the same consumer want.

substrate. (1) (substratum) The surface to which an organism is attached or upon which it moves. (2) The material on which a microorganism grows (e.g., culture medium, host organism). (3) The particular substance or group of substances that an ENZYME activates.

substratum. (1) The soil layer beneath the SOLUM. The substratum either conforms (horizon C) or is unconforming (horizon D). *See also* SOIL HORIZONS. (2) *See* SUBSTRATE.

subsun. A bright spot of light seen where the Sun is reflected in the horizontal upper surfaces of crystals in ice clouds below the observer.

subtropical high. A feature of the GENERAL CIRCULATION: a region of high pressure found in latitudes intermediate between the tropical and temperate zones. The Azores high is in this region.

subtropical jet. A JETSTREAM or wind maximum at the TROPOPAUSE blowing from the west at around 30° latitude above places where there is predominantly a descent of air and a dry climate.

succession. The progressive natural development of vegetation towards a CLIMAX, during which one COMMUNITY is gradually replaced by others. A primary succession starts at sites (e.g., sand dunes, lava flows) that have not previously borne vegetation. A secondary succession is one that follows the destruction of part or all of the original vegetation of an area. A natural succession has two components: the physiographic in which living organisms respond to topographical features; the biotic, in which organisms react with one another. *See also* SERE.

successive percentage mortality (apparent mortality). The mortality in a population at each developmental stage (e.g., age group) expressed as a percentage of the number alive at the beginning of the stage.

succulent. A plant, or applied to the leaves of a plant, that has enlarged tissues that hold water or sugar.

sucking-lice. *See* ANOPLURA.

sucrose. *See* CARBOHYDRATES.

sucrosic. See SACCHAROIDAL.

Sudanian–Sindian Floral Region. The part of the PALAEOTROPIC REALM that comprises the SAHEL region of Africa, Sudan and north-west India.

sudd. A floating mass of vegetation on the River Nile.

suffruticose (suffrescent, suffrutescent). Applied to perennial plants that are woody at the base but herbaceous above and that do not die down to ground level in the winter. A distinction may be drawn between suffrescent and suffruticose, the former referring to less woody plants, the latter to more shrublike (fruticose, *see* FRUTESCENT) ones.

sugar-cane. *See* GRAMINEAE.

sugar-beet. *See* BETA VULGARIS.

sullage water. *See* COMBINED SEWER.

sulphate (SO_4^{2-}). The ion formed when SULPHUR TRIOXIDE (SO_3) reacts with water to form SULPHURIC ACID (H_2SO_4). This ion is often found in the atmosphere in other chemical compounds (e.g., ammonium sulphate, $(NH_4)_2SO_4$), a major constituent of Teesside mist, caused by the combination in the air of sulphur trioxide and ammonia and water. Sulphate ions are also released in volcanic eruptions, together with SULPHUR DIOXIDE, especially the eruptions of MOUNT TAMBORA (1815), KRAKATOA (1883) and MOUNT AGUNG (1963).

sulphate-resistant cement. *See* CEMENT.

sulphite (SO_3^{2-}). The ion formed notably in the reaction of SULPHUR DIOXIDE (SO_2) with water to form sulphurous acid (H_2SO_3).

sulphur (brimstone, S). An element that is deposited from volcanic VENTS and FUMAROLES and also is found in SEDIMENTARY ROCKS, particularly with GYPSUM and LIMESTONE, and associated with SALT-DOMES. Native sulphur is the main source of sulphur for the sulphuric acid industry, followed by sour gas (natural gas containing hydrogen sulphide) and PYRITES. Sulphur is an essential plant MACRONUTRIENT. $A_r = 32.064$; $Z = 16$; SG 2.07; mp 112.8°C; bp 444.6°C.

sulphur dioxide (SO_2). A constituent of products of combustion of a wide range of fuels, but particularly heavy fuel oil and coal. It is widely used as an index of the level of air pollution, but epidemiological research has so far failed to prove that it has harmful effects on humans in even the high-est concentrations found in the atmosphere. Its importance lies mainly in its ease of measurement (by acidity) and its dependence on fuel consumption and atmospheric dilution, on which the levels of many other pollutants also depend. It has been implicated as a causal agent in ACID RAIN. It is released, with SULPHATES, in volcanic eruptions. *See also* SULPHUR TRIOXIDE.

sulphuric acid (H_2SO_4). In the atmosphere, an acid formed by the combination of SULPHUR TRIOXIDE with water to form a relatively stable mist of acid droplets. SULPHUR is an essential MACRONUTRIENT for plants, and so atmospheric sulphuric acid contributes nutrients to soils. It also tends to increase soil acidity, and in soils that are naturally acid this may require the use of LIME to raise the pH. Some Scandinavian soils have been damaged in this way partly as a result of industrial air pollution originating in the UK, Germany (East and West) and Poland. *See also* ACID RAIN, TRANSFRONTIER POLLUTION.

sulphur trioxide (SO_3). In the atmosphere, sulphur trioxide is formed by the oxidation of SULPHUR DIOXIDE in the reaction

$$2SO_2 + O_2 \rightarrow 2SO_3$$

It is believed that this reaction may occur more readily in the surface of water droplets, leading immediately to the formation of SULPHURIC ACID.

sumac. *See* RHUS.

summer. *See* SEASONS.

sundew. *See* DROSERACEAE.

sun dog. *See* MOCK SUN.

sun-dried brick. *See* ADOBE.

sunn hemp (*Crotalaria juncea*). A tropical crop grown for its fibre. It is unrelated to true hemp (Cannabis sativa).

sun pillar. A vertical streak of light above the Sun, usually seen at sunrise or sunset. It is caused by reflection from ice crystals.

sun plant. *See* HELIOPHYTE.

sunrise. Defined in meteorology as the moment when the upper edge of the Sun appears to rise above the apparent horizon on a clear day. *Compare* SUNSET.

sunset. In meteorological convention, the moment when the upper edge of the Sun appears to fall below the apparent horizon on a clear day. Effects of refraction cause the apparent position to be about 34' above the true position. *Compare* SUNRISE.

sunspot. A local disturbance on the visible surface of the Sun that causes an increase in SOLAR RADIATION and consequently alters the Earth's MAGNETOSPHERE and IONOSPHERE. The intensity of sunspot activity varies in an 11-year cycle, and it appears possible that this may cause a similar cycle in global meteorology, but this yet has to be proved.

supercooled fog. *See* FROZEN FOG.

supercooling. A metastable state of a liquid in which it remains in the liquid phase although at a lower temperature than the freezing point. It often occurs in cloud droplets in the atmosphere.

superfluent. An animal species of similar importance to that of a SUBDOMINANT plant species.

Superfund. In the USA, a federal fund set up to finance the cleaning up of sites contaminated with hazardous wastes.

supergene. Applied to mineral deposits or processes involving downward percolating water or aqueous solutions. SECONDARY ENRICHMENT is a special case of a supergene process. *Compare* HYPOGENE.

superior image. An optical reflection at an INVERSION giving the appearance of 'castles in the air'. They are most common in winter over land when the ground is snow-covered

and the weather anticyclonic (*see* ANTICYCLONE) or over closed seas (e.g., the Mediterranean Sea) in summer.

supernumerary bow. The additional RAINBOWS seen inside the primary (42°) rainbow when the range of raindrop sizes is small, the order of the colours being the same.

superorganism. *See* EPIORGANISM.

superparasite. *See* PARASITISM.

Superphénix. A French-based, but internationally funded commercial BREEDER REACTOR located at Creys-Malville, near Lyons. *See also* PHÉNIX.

superphosphate. *See* PHOSPHATE.

superposition. The arithmetical combination of two or more waves at successive instants or points.

supersaturation. A metastable state of moist air in which there is more vapour present than is required to saturate the air. This can occur in cooling air if there are too few condensation nuclei (*see* CONDENSATION NUCLEUS) to act as initiating sites for condensation of droplets. *See also* SATURATION.

supersonic transport aircraft. *See* SST.

supraneustronic. *See* NEUSTRON.

supraorganism. *See* EPIORGANISM.

surface-active agent (surfactant). A substance that causes a liquid to spread more readily on a solid surface, mainly by reducing surface tension. Many detergents are surface-active agents.

surface analysis. The analysis of a surface chart (e.g., the identification of pressure systems, AIR MASSES and FRONTS from surface observations).

surface deposits (surficial deposits). Unconsolidated material that covers the BEDROCK, either weathered (*see* WEATHERING) from bedrock *in situ* (residual) or weathered from one area and transported to another by wind (AEOLIAN DEPOSITS, e.g., sand dunes and loess), by water (e.g., ALLUVIUM), by ice (e.g., till (*see* BOULDER CLAY) or GLACIAL DRIFT), or by gravity (*see* COLLUVIUM).

surface flow. *See* EROSION.

surface pressure. Normally, the atmospheric pressure inside a STEVENSON SCREEN (i.e. effectively at ground level).

surface pressure chart. A chart of surface atmospheric pressure plotted as ISOBARS over a geographical area. It may include other surface parameters.

surface water. *See* GROUND WATER.

surface wind. Conventionally, the wind at a height of 10 metres above a flat, smooth piece of ground unaffected by obstructions.

surfactant. *See* SURFACE-ACTIVE AGENT.

surficial deposits. *See* SURFACE DEPOSITS.

suspended sediment load. Particles small enough to be carried in suspension by moving water.

suspended solids. (1) Fine dust particles distributed as an AEROSOL in the air. (2) Small particles of solids distributed through water.

sustainable capacity. *See* SUSTAINABLE YIELD.

sustainable development. Economic development that can continue indefinitely because it is based on the exploitation of RENEWABLE RESOURCES and causes insufficient environmental damage for this to pose an eventual limit.

sustainable yield (sustainable capacity). The maximum extent to which a RENEWABLE RESOURCE may be exploited without depletion. For example, water may be abstracted from an AQUIFER sustainably provided the

rate of abstraction does not exceed the rate of replenishment; fish can be caught sustainably provided the number caught does not exceed the number of young that survive to join the stock.

Sv. *See* SIEVERT.

swallow hole (sink hole). A funnel-shaped hole in LIMESTONE caused by SOLUTION of the rock when rain water, with dissolved carbon dioxide from the soil, drains through fissures and enlarges them. By continued solution the holes may be connected with the formation of vast underground caverns.

swamp. An area that is saturated with water for much of the time, but in which the soil surface is not deeply submerged. Used in a restricted sense, the term implies an area characterized by woody vegetation, but it is often used more widely to include MARSH, BOG, etc.

swans. *See* ANATIDAE.

S wave. The secondary wave that reaches a SEISMOGRAPH from an EARTHQUAKE. S waves are shear waves and so do not travel through liquids. The absence of S waves beyond 105° of the arc from the EPICENTRE of an earthquake indicate the fluid nature of the Earth's CORE. *Compare* P WAVE.

sweet gas. NATURAL GAS containing little HYDROGEN SULPHIDE. *Compare* SOUR GAS.

swidden farming (slash-and-burn farming). A system of primitive agriculture in which areas of natural vegetation are cleared by felling trees, cutting back shrubs and then burning off the stumps and herbaceous plants. The area is sown to crops for several years in succession, and when yields begin to fall due to the exhaustion of soil nutrients, or invading wild plants depress crops, the farmers move to a new area where the process is repeated. Thus each community moves every few years to a new site, eventually returning to the point from which they began. This type of agriculture was practised extensively in Europe in NEOLITHIC times and more recently in many areas of the tropics. *See also* LANDNAM, SHIFTING CULTIVATION.

swifts. *See* APODIDAE.

syenite. A coarse-grained, ALKALINE, INTERMEDIATE IGNEOUS rock characterized by the presence of alkali FELDSPAR, with or without FELDSPATHOIDS and such FERROMAGNESIAN MINERALS as BIOTITE, HORNBLENDE, AUGITE and more alkaline types. Syenites are the plutonic (*see* PLUTON) equivalent of TRACHYTES.

sylvite. An EVAPORITE mineral (KC1), which is one of the major sources of potassium, an essential fertilizer. *See also* CARNALLITE, MACRONUTRIENTS.

symba process. A process for treating starch wastes from the food industry using the fungus *Endomycopsis fibuliger* to hydrolyze (*see* HYDROLYSIS) the starch, then the yeast *Torula* to produce proteins. *See also* NOVEL PROTEIN FOODS.

symbiont. A symbiotic organism (*see* SYMBIOSIS).

symbiosis. A close and mutually beneficial association of organisms of different species. The occurrence of cellulose-digesting protozoans (*see* PROTOZOA) in the guts of wood-eating cockroaches and termites (*see* ISOPTERA) is a symbiotic relationship, as the insects cannot digest cellulose unaided and the Protozoa cannot live independently. *See also BACILLUS SUBTILIS*, COMMENSALISM, MYCORRHIZA, PARASITISM, ROOT NODULES.

symmetrical fold. A FOLD in which both limbs dip away from the AXIAL PLANE.

sympathetic nervous system. *See* AUTONOMIC NERVOUS SYSTEM.

sympatric. Applied to different species or subspecies whose areas of distribution overlap or coincide. *Compare* ALLOPATRIC.

Symphyta. *See* HYMENOPTERA.

syn-. Prefix meaning with, together, along with, at the same time.

synanthrope. A plant or animal often found associated with humans, human dwellings or other human artifacts (e.g., house martin, stinging nettle).

synapse. *See* CENTRAL NERVOUS SYSTEM, NEURON.

synapsis. The pairing of CHROMOSOMES during MEIOSIS.

syncarp. A multiple fruit, made up of many small fruits united together.

synchorology. The study of the distribution ranges of plant communities, phytosociological regions (*see* PHYTOSOCIOLOGY), vegetation complexes, geographical complexes and contemporary plant migration patterns.

synchronous satellite. A SATELLITE that occupies such an orbit that its position remains fixed relative to a particular point on the Earth's surface. Such satellites are often used for communications.

syncline. A FOLD that is concave downwards, with the youngest rocks occupying the inner core. *Compare* ANTICLINE.

synclinorium. A compound SYNCLINE in which the limbs are folded. The term is used for large-scale structures.

syncytium. A mass of PROTOPLASM containing many nuclei and enclosed by a single PLASMA MEMBRANE. The term is applied only to animal tissues (e.g., muscle fibres). *See also* COENOCYTE, PLASMODIUM.

syndynamics. The study of successional changes (*see* SUCCESSION) in plant COMMUNITIES, their causes and trends.

synecology (biocoenology). The study of the relationships between COMMUNITIES of species and their environment. *Compare* AUTECOLOGY.

synergism. (1) The combined environmental effect of two or more pollutants that react together in such a way as to affect living organisms differently from the way either or any of them would alone, or all of them would if their effects were added together. (2) The combined effects of agents such as HORMONES or drugs when they act in the same direction on living systems. Synergism may result in an effect that is greater than the sum of the effects of the agents when they act individually. (3) The coordinated action of pairs or sets of muscles which contract together to produce a particular movement. *See also* ANTAGONISM.

syngameon. A group of species among which hybridization (*see* HYBRID) occurs.

syngamy. The union of two GAMETES during FERTILIZATION.

syngeneic. *See* ISOGENEIC.

synmorphology. The ECOLOGY of plant COMMUNITIES.

synnecrosis. Mutual death.

synphylogeny. The study of the historical and evolutionary trends and changes within plant COMMUNITIES.

synrock. A synthetic, rock-like material invented in Australia into which intermediate-level radioactive waste (*see* INTERMEDIATE-LEVEL WASTES) is incorporated for final disposal. It is claimed to be more resistant to leaching of RADIONUCLIDES than borosilicate glass, which is the alternative on which most waste disposal plans are based.

syntype (cotype). Any specimen used to designate a species when neither TYPE SPECIMEN nor PARATYPE has been selected.

synusia. A group of plants, all of the same general form, occurring together in the same habitat (e.g., an aggregation of floating herbs, the tree layer in a wood).

Syringia vulgaris. *See* OLEACEAE.

system. (1) An assemblage or combination of things or parts forming a complex or unitary interacting whole. (2) A sequence of strata deposited during a geological period. The systems were originally defined by geologists working in different regions in the 18th and 19th centuries. Some systems were established on distinctive rock types and others on distinctive faunal content, with boundaries usually chosen at distinct breaks in the geological record. Nowadays, the systems and the subdivisions (i.e. SERIES, STAGE and substage) within them are recognized on the basis of distinctive fossils.

Système International d'Unités. An internationally agreed system of units (*see* SI UNITS).

systemic insecticides. Insecticides that are taken in and then translocated (*see* TRANSLOCATION) throughout a plant, rendering the whole surface lethal to insect pests, even if only a small area has been treated. These insecticides may be less harmful to wildlife than contact poisons because insects not actually feeding on the treated plants may escape their effects.

T

2,4,5-T (2,4,5-trichlorophenoxyacetic acid). A translocated HORMONE weed killer used to control woody weeds, and of great value in preventing regrowth of scrub after clearance. *See* TRANSLOCATED HERBICIDES.

T. (1) Trillion (in the American sense of one million million, i.e. 10^{12}); used in conjunction with 'cf' in figures for natural gas reserves to signify trillions of cubic feet. (2) *See* TERA-. (3) *See* TESLA. (4) *See* TRITIUM.

T_1. First trophic level (*see* FOOD CHAIN).

T_2. Second trophic level (*see* FOOD CHAIN).

Tabadinae. *See* BRACHYCERA.

Tachinidae. *See* CYCLORRAPHA.

taconite. *See* BANDED IRONSTONE.

taiga. Forest (usually coniferous) adjacent to arctic TUNDRA.

tailings. Those portions of washed ore that are considered too poor to be treated further.

talus. The slope formed by fallen rock debris or slide rock at the foot of a cliff.

Tambora, Mount. *See* MOUNT TAMBORA.

tangent arc. A coloured arc commonly seen at the highest point of a HALO round the Sun, which has opposite curvature to the halo when the Sun is low. Tangent arcs may be seen at the bottom of a halo. They occur when there is a predominance of crystals with their axes horizontal. The circumferential arc is sometimes thought to be the very rare upper tangent arc of a 46° halo because it is in about the same position.

tangential velocity (circumferential velocity). The component of velocity along the tangent in a curved flow.

Tanio. An oil tanker that sank in a storm in the English Channel on 7 March 1980, with a cargo of 24 000 tonnes of crude oil. The ship broke in half releasing 3000 tonnes of oil, and for several days the stern section leaked about 7 tonnes a day. The incident caused severe pollution to beaches in Britanny.

tapeworms. *See* CESTODA.

tapioca. *See* MANIHOT.

tap root system. A root system with a prominent vertical main root (tap root) bearing numerous small lateral roots. Some tap roots (e.g., carrot) become swollen with stored food. *Compare* FIBROUS ROOT SYSTEM.

Tardigrada (bear animalcules, water bears). A phylum of minute coelomate (*see* COELOM) animals found in damp moss, ponds, etc. They resemble ARACHNIDA because of their four pairs of stumpy legs, but are much more primitive evolutionarily, although they are probably related to the ARTHROPODA.

tarn. A small lake at high altitude (e.g., in a CIRQUE).

tarsals. *See* PENTADACTYL LIMB.

tar sand. A surface or near-surface sand or SANDSTONE containing a high percentage of very viscous natural HYDROCARBONS. Some such deposits probably represent the remains of OIL FIELDS exposed by erosion and from which the lighter hydrocarbons have evaporated.

Taxaceae. A small family of much-branched trees and shrubs (GYMNOSPERMAE), including the yew (*Taxus*). In the yew the solitary seed is surrounded by a fleshy red cup (aril) attractive to birds, which disperse the seeds. The leaves and seeds are very poisonous to mammals, and the wood is valuable. Some yews in English churchyards have girths of 9–10 metres, and are about 1000 years old.

Taxales. An order of GYMNOSPERMAE containing a single family, TAXACEAE.

taxis. A locomotory response of a cell or organism in which the direction of movement is oriented with relation to the stimulus (*compare* KINESIS). In chemotaxis the stimulus is chemical (e.g., a PHEROMONE emitted by a female moth to attract a male or sugar exuding from the female sex organ (ARCHEGONIUM) of a moss plant to attract SPERMATOZOIDS). In phototaxis (heliotaxis) the stimulus is light, in geotaxis it is gravity, in thermotaxis it is heat, and in rheotaxis it is flowing movements in the immediate environment (e.g., water currents). In thigmotaxis (stereotaxis) the stimulus is touch, which sometimes causes inhibition of movement leading to the close contact of an organism with a solid surface.

Taxodiaceae. A small family of coniferous trees, some of which are extremely large and yield valuable timber. The redwood (*Sequoia sempervirens*) grows up to 102 metres high and 8.5 metres thick. The mammoth tree (*Sequoiadendron giganteum*) of California reaches 10.5 metres in thickness, the tallest is 96 metres, and the age of the largest is about 1500 years. The oldest redwood has been dated as 3212 years old. *See also* BRISTLE-CONE PINE.

taxon. Any named taxonomic group, whatever its rank (e.g., phylum, genus, variety, etc.). *See also* CLASSIFICATION.

taxonomy. The science of classification,

usually restricted to mean the classification of plants and animals. The word is derived from the Greek *taxis* meaning arrangement and *nomos* meaning law. Taxonomy arranges plants and animals into hierarchical groups.

TC. *See* RETARDED TIMING.

TCDD. *See* DIOXIN.

TDI. *See* TOLYLENE DIISOCYANATE.

tea. *See* CAMELLIA.

teak. *See* TECTONA.

tear fault. *See* FAULT.

technological fix. A solution to a problem that is based on technology, the term often being used in a pejorative sense as an apparent or simplistic technological solution to a complex human problem whose benefits may be only cosmetic. *See also* CORNUCOPIAN PREMISES.

technosphere. That part of the physical environment built or modified by humans. *Compare* PSYCHOSPHERE.

Tectona grandis (teak). A deciduous tree (family: Verbenaceae) cultivated in India, Java, etc. for its very hard, durable timber, extensively used in shipbuilding. In India the wood is dried out by removing a ring of bark and SAPWOOD from the standing tree, which dies and is left standing for two years before felling.

tectonic. Applied to deformation of the Earth's CRUST through warping, folding (*see* FOLD) and faulting (*see* FAULT).

tectonic creep. Slight, and apparently continuous, movement along a FAULT.

Teesside mist. *See* SULPHATE.

tektite. A piece of natural glass, probably of extraterrestrial origin or produced from surface materials by the energy of high-velocity meteorite impacts, usually weighing a few grams, and resembling OBSIDIAN, but with a distinctive shape, such as that of a tear drop, dumb-bell or button. Tektites are found in groups, often far from a volcanic source, and unlike meteorites they are geologically young, with ages of up to 30 million years.

TEL. *See* TETRAETHYLLEAD.

teleoptiles. The feathers of an adult bird, including down feathers, contour feathers, quills and hair-like filoplumes. *Compare* NEOSSOPTILES.

Teleosti. A large group of bony fishes containing all the present-day ACTINOPTERYGII except for a few primitive species (e.g., sturgeon and gar-pike). *See also* CLUPEIDAE, CYPRINIDAE, GADIDAE.

telolecithal (megalecithal). Applied to eggs (e.g., those of birds) that contain much yolk. *Compare* MICROLECITHAL.

telophase. *See* MITOSIS.

temperature coefficient. *See* Q_{10}.

temperature inversion. *See* INVERSION.

tendon. CONNECTIVE TISSUE, associated closely with MUSCLES, that transmits the mechanical force of muscle contraction to the bones, which function as levers.

tendril. A modified stem, leaf or part of a leaf used by climbing plants for attachment. Some tendrils (e.g., the modified leaflets of the sweet pea, *Lathyrus odoratus*) twine; others (e.g., the branch tendrils of virginia creeper, *Ampelopsis veitchii*) end in adhesive pads.

tensiometer. An instrument used to measure the amount of water in the plant root area of a soil.

teosinte (*Zea mexicana,* wild maize). A member of the grass family (*see* GRAMINEAE) that still occurs in Central America and is the ancestor of domesticated maize, with which it interbreeds locally at random, producing crops that display HYBRID VIGOUR.

tepal. *See* FLOWER.

tephigram. A diagram that shows the temperature and ENTROPY at different levels in the atmosphere.

tephra. A collective name for all pyroclastic material ejected from a volcanic vent into the air. Tephra includes volcanic dust, ASH, CINDERS, LAPILLI, SCORIA, PUMICE, BOMBS and blocks.

tera- (T). The prefix used in conjunction with SI units to denote the unit x 10^{12}.

teratogen. A substance that produces deformation of the foetus in the womb.

terbutryne. A SOIL-ACTING HERBICIDE used for the control of grasses and broadleaved weeds in crops. It is also used to kill floating and submerged plants, including algae, in still and slow-flowing water.

Teredo (shipworms). A genus of lamellibranch (*see* LAMELLIBRANCHIATA) Mollusca with long, slender bodies and much reduced shells, by means of which they bore into wood. They cause great damage to ships, piers, wharves, etc.

terminal curvature. The bending-over of the upper end of rock layers caused by CREEP of the overlying soil.

terminal moraine. *See* MORAINE.

termites. *See* ISOPTERA.

termiticole (termitiphile, termitophil). An organism that lives with termites (*see* ISOPTERA) inside their nests. Examples include certain FUNGI and insects.

terns. *See* LARIDAE.

terrace. (1) *See* STRUCTURAL TERRACE. (2) *See* RIVER TERRACE.

terrapins. *See* CHELONIA.

terra rossa. Red earths developed on LIME-STONE due to accumulations of insoluble iron oxides.

terrigenous. Applied to sediments derived from the land.

terriherbosa. Herbaceous vegetation growing on dry land.

territory. An area that an animal or group of animals defends, mainly against members of the same species. *Compare* RANGE. *See also* BIRD SONG.

Tertiary. That part of GEOLOGICAL TIME from the end of the CRETACEOUS to the beginning of the PLEISTOCENE, comprising the PALAEO-CENE, EOCENE, MIOCENE, OLIGOCENE and PLIO-CENE. Tertiary also refers to rocks formed during this time. In some usages Tertiary is synonymous with CENOZOIC, whereas in others Tertiary and QUATERNARY together make up the Cenozoic.

tertiary economy. An economic system in which the principal activities, and so sources of wealth and employment, are connected with the provision of services to the primary and secondary economic sectors. *Compare* PRIMARY ECONOMY, QUATERNARY ECONOMY, SECONDARY ECONOMY.

tertiary treatment. A third stage in the treatment of waste water, usually involving the removal of soluble plant nutrients which might cause EUTROPHICATION were they released into still or slow-moving fresh water. *See also* ADVANCED WASTE TREATMENT, PRIMARY TREATMENT, SECONDARY TREATMENT.

tesla (T). The derived SI unit of magnetic FLUX density, being the density of 1 WEBER of magnetic flux per square meter. It is named after Nikola Tesla (1870–1943).

testa. *See* SEED.

testosterone. *See* ANDROGENS.

Testudinae. *See* CHELONIA.

2,3,7,8-tetrachloro dibenzo-*p*-dioxin. *See* DIOXIN.

tetrachloroethylene. A solvent, used in the electronics industry and as a dry-cleaning fluid, and suspected of causing cancer in humans.

tetraethyllead (TEL). A compound of lead added to petrol (*see* ANTIKNOCK ADDITIVE) to increase its octane number and so reduce knocking. It is the major source of airborne lead.

tetrahedrite ($Cu_{12}Sb_4S_{13}$). A copper-containing antimony sulphide mineral that is an important ore of both COPPER and ANTIMONY. It is found in HYDROTHERMAL VEINS.

tetraploid. Applied to a POLYPLOID organism or cell that has four times the haploid number of CHROMOSOMES. *Compare* ALLOTETRAPLOID.

Tetrapoda. The essentially terrestrial vertebrates (Amphibia, reptiles, mammals) that characteristically possess two pairs of limbs.

textural maturity. *See* TEXTURE.

texture. The size, shape and arrangement of the particles that make up a surface deposit, rock or soil (in soil the term generally applies only to size). The degree of SORTING and rounding of the component particles of sediments is referred to as their textural maturity.

textured vegetable protein. *See* NOVEL PROTEIN FOODS.

Th. *See* THORIUM.

thallium (T1). A white, malleable, metallic element used in ALLOYS. Its compounds are all extremely poisonous and are used in pesticides. $A_r = 204.37$; $Z = 81$; SG 11.85; mp 303.5°C.

Thallophyta. A former large division of plants, comprising the primitive forms (AL-GAE, FUNGI, BACTERIA, slime fungi (*see*

Myxomycophyta) and lichens) in which the plant body (thallus) is organized in a simple way, without differentiation into root, stem and leaf. The name Thallophyta is no longer used in taxonomy.

thallus. A simple plant body that does not show differentiation into root, stem and leaves. Some thalli are unicellular. Multicellular ones are filamentous or ribbon-like, branched or unbranched.

thanatocoenosis. An assemblage of fossils consisting of the remains of organisms brought together after death. *Compare* biocoenosis.

Thea. *See* Camellia.

thecodont. Having teeth set in sockets in the jaw bones as, for instance in mammals. *Compare* acrodont, pleurodont.

theodolite. An instrument used particularly in surveying for measuring horizontal and vertical angles.

Therapsida. A group of extinct reptiles that originated in the Permian Period and were ancestral to mammals. Mammal-like tendencies which evolved in the Therapsida include the differentiation of teeth for various functions, the turning of the limbs under the body and enlargement of the brain case.

therm. A unit of heat, being the heat energy required to raise 1000 pounds of water through 100 degrees Fahrenheit. It is equal to 10^5 British thermal units or 1.055×10^8 joules.

thermal oxide reprocessing plant (THORP). A factory in which spent uranium oxide fuel from thermal nuclear reactors is removed from its containers and processed in order to extract its remaining uranium-235 and plutonium, which can then be used to fuel other reactors.

thermal pollution. The raising of the temperature of part of the environment by the discharge of substances whose temperature is higher than the ambient (e.g., thermal pollution of rivers from the discharge of cooling waters). In fresh water, the amount of free dissolved oxygen tends to decrease with increasing temperature, so affecting living organisms.

thermal soaring. The art of circling in thermals to gain height. It is used by some birds such as gulls, vultures, some birds of prey, and by swallows and swifts, which feed on insects carried up in thermals.

thermal stratification. The existence within a water body of a succession of well-differentiated layers, each with a different temperature, and lying at various depths, the coldest at the bottom. Inverse stratification occurs beneath ice. *See also* epilimnion, hypolimnion, meromixis, thermocline.

thermal wind. The increase in the geostrophic wind with height due to horizontal temperature gradients.

thermionic valve (tube). A system of electrodes arranged in an evacuated glass or metal envelope.

thermocline. The layer of water in a lake that lies between the epilimnion and the hypolimnion. Within the thermocline the temperature decreases rapidly with increasing depth (usually by more than 1°C for each metre). *See also* thermal stratification.

thermodynamics, laws of.
zeroth law. Two objects are in thermal equilibrium when no heat passes between them when they are placed in contact with one another.
first law of thermodynamics. Energy can be neither created nor destroyed, but it can be changed from one form to another (the conservation of energy).
second law of thermodynamics. When two bodies at different temperatures are placed in contact, heat will flow from the warmer to the cooler, and there is no continuous, self-sustaining process by which heat can be transferred from a cooler to a warmer body. Thus energy tends to become distributed

evenly throughout a closed system (i.e. the ENTROPY of a close system increases with time.
third law of thermodynamics. The absolute zero (0K) can never be attained.

thermograph. An instrument for recording changes in temperature as a line on a rotating drum.

thermonasty. *See* NASTIC MOVEMENT.

thermophilic microorganisms. Microorganisms that grow well at temperatures over 45°C. Examples include BACTERIA in hot springs, manure heaps and fermenting hay ricks. *Compare* MESOPHILIC MICROORGANISMS, PSYCHROPHILIC MICROORGANISMS.

thermosphere. That part of the upper atmosphere in which temperature increases with height.

thermotaxis. *See* TAXIS.

therophyte. *See* RAUNKIAER'S LIFE FORMS.

Theropsida. A wide term that covers all the mammals and mammal-like extinct reptiles (e.g., THERAPSIDA). These groups all have a similar skull structure. *Compare* SAUROPSIDA.

thiamine (aneurine, vitamin B$_1$). A VITAMIN of the B group that is needed by many animals (e.g., vertebrates, insects) for the formation of a COENZYME concerned in CARBOHYDRATE metabolism. Some BACTERIA, FUNGI and PROTOZOA are able to synthesize thiamine. Rich sources are germs of cereals, leguminous seeds and yeast. Deficiency in humans causes beri-beri, a disease affecting the nerves, which is prevalent among populations that subsist on polished rice.

thigmotaxis. *See* TAXIS.

thigmotropism. *See* HAPTOTROPISM.

thin section. A fragment of rock or mineral that has been ground to a thickness usually of 0.03 millimetres and mounted in glass as a microscopical slide. Nearly all minerals, except for most oxides and sulphides, are transparent at this thickness. Thin sections are usually viewed through a petrographic (petrological) microscope which has two polarizing filters, one on each side of the specimen. The optical properties, as viewed through the microscope, are diagnostic of the rock from which the slice was taken.

thiocarbamates. See CARBAMATES.

thiodan. *See* ENDOSULFAN.

thiophos. *See* PARATHION.

Third World. A term introduced in the late 1940s to describe the developing countries (*see* ECONOMIC DEVELOPMENT). The world was divided into three: the First World comprised the countries whose economies were governed to a large extent by the free market; the Second World of those (socialist) countries whose economies are planned centrally. Only the term Third World remained in use, describing those countries whose economies were at a primary level (*see* PRIMARY ECONOMY) and where annual per capita incomes were below a certain threshold. More recently a Fourth World has been added by dividing the old Third World into those countries (e.g., the OPEC countries) that have natural resources of high value (Third World) and those developing countries (Fourth World) that have not. As countries of the Third World industrialize their economies they enter either the old First (e.g., South Korea, Taiwan) or Second Worlds. Although the term Third World remains in popular and informal use, it is no longer used officially by governments or intergovernmental agencies because of its vagueness and because it is considered patronizing.

thixotrophy. An infinitely reversible property of certain CLAY MINERALS (e.g. MONTMORILLONITE). When mixed with water the mineral forms a GEL and on agitation becomes a highly viscous fluid. This property has applications in civil engineering.

thorium (Th).   A dark grey, radioactive metallic element used in ALLOYS, lamp filaments and in NUCLEAR REACTORS. It occurs in MONAZITE. A_r = 232.038; Z = 90; SG11.2; mp 1845°C.

THORP. *See* THERMAL OXIDE REPROCESSING PLANT.

thread cell. *See* NEMATOBLAST.

threadworms. *See* NEMATOMORPHA.

Three Mile Island.   An island in the Susquehanna River, near Harrisburg, Pennsylvania, that is the site of a NUCLEAR REACTOR which failed on 28 March 1979 in the most serious incident experienced by the nuclear industry up to that time. By a combination of human and mechanical error, water flooded the reactor building, the core overheated, possibly with a partial meltdown of fuel elements, and it was 16 hours before the situation was brought under control. There was a small release of radioactive gases, but the public was not harmed.

threonine. An AMINO ACID with the formula $CH_3CHOHCH(NH_2)COOH$ and a molecular weight of 119.1.

threshold limit value (TLV). The concentration of an air-borne contaminant to which workers may be exposed legally day after day without any adverse effect. Several defined TLV levels have been found later to cause adverse health effects.

threshold of audibility.   The minimum SOUND PRESSURE LEVEL at which a person can hear a sound of a given FREQUENCY.

threshold of pain (threshold of feeling).   The minimum SOUND PRESSURE LEVEL at which a person begins to experience pain in the ear from a sound of a given FREQUENCY.

threshold shift. An alteration, temporary or permanent, in a person's THRESHOLD OF AUDIBILITY.

thrips. *See* THYSANOPTERA.

thrombocyte. *See* BLOOD PLATELETS.

throw. The vertical displacement between the upthrown and downthrown sides of a FAULT.

thrust. A low-angle REVERSE FAULT.

thunderflies. *See* THYSANOPTERA.

thymine. *See* DNA.

thymonucleic acid. *See* DNA.

thyroid gland. *See* THYROXIN.

thyroxin (thyroxine). An iodine-containing AMINO ACID-based HORMONE, produced along with a similar hormone (triiodothyronine) by the thyroid gland, which is situated in the neck region of vertebrates. Thyroid secretion increases the rate of oxidative processes and stimulates growth in young animals. Deficiency (due sometimes to lack of iodine in the diet) results in physical and mental stunting (cretonism in humans). In AMPHIBIA, thyroid secretion initiates METAMORPHOSIS in the tadpole. *See also* PAEDOGENESIS.

Thysanoptera (Physopoda; thrips, thunderflies).   An order of minute slender insects (division: EXOPTERYGOTA) commonly found in flowers. They usually have two pairs of narrow fringed wings. Most feed on plant juices and may be present in such large numbers that they become serious agricultural pests, damaging peas, wheat and other crops. Some transmit plant diseases.

Thysanura. *See* APTERYGOTA.

ticks. *See* ACARINA.

tidal power. Mechanical power, which may be converted to electrical power, generated by the rise and fall of ocean tides. The possibilities of utilizing tidal power have been studied for many generations, but the only feasible schemes devised so far are based on the use of one or more tidal basins, separated from the sea by dams (known as barrages), and of hydraulic turbines through which water passes on its way between the basins and the sea. The world's largest tidal power plant is located on the estuary of the River Rance, in Brittany, France, completed in the 1960s, which generates 544 kilowatt-hours per year. The disadvantages of tidal power generation are the very high capital cost of the barrages and possible ecological disturbance, but the major limitation imposed by the tidal cycle can be overcome by generating energy from both the filling and emptying of the basins to provide base-load power, and by using base-load power at times of low demand to pump water to higher reservoirs, from which it can be released to meet peak load demands (*see* PUMPED STORAGE).

tidal wave. *See* TSUNAMI.

tight fold. *See* FOLD.

tile drain. *See* SOIL DRAINAGE.

Tiliaceae. A family of DICOTYLEDONEAE, mostly trees and shrubs of tropical and temperate regions. *Tilia*, a genus including the European limes and the American basswood, is visited by bees for the nectar secreted by the sepals and is a valuable source of honey as well as of timber.

till. *See* BOULDER CLAY.

tillite. Lithified BOULDER CLAY.

tilth. (1) Cultivation of soil. (2) That part of the soil which is affected by cultivation. (3) The physical condition of the soil with reference to its suitability for cultivation. Soil with a good tilth is friable and porous, with a stable, granular structure.

tiltmeter. An instrument that is used to measure changes in slope of the ground surface. Tiltmeters measure vertical displacement and are used to indicate impending EARTHQUAKES or volcanic activity.

Times Beach. A town near St Louis, Missouri that in February 1983 was bought in its entirety by the US Federal Government and declared unfit for human habitation. In the 1970s chemical wastes mixed with used oil had been used to control dust on dirt and gravel roads. Severe flooding in December 1982 exposed inhabitants to high levels of DIOXIN, revealing extensive contamination of the soil, and the temporary evacuation because of the floods was made permanent.

tin (Sn). A silvery–white metallic element that is soft, malleable and ductile (*see* DUCTILITY). At ordinary temperatures it is unaffected by air or water. It exists in three allotropic forms (*see* ALLOTROPY) and occurs as CASSITERITE. It is used for tin-plating and in many ALLOYS. $A_r = 118.69$; $Z = 50$; SG 7.31; mp 231.85°C.

tincture of iodine. *See* IODINE.

tissue. A group of cells with similar structure and function (e.g. muscle tissue in animals, CORK tissue in plants).

tissue culture (explantation). A method of keeping CELLS alive after their removal from the organism, using a suitable medium containing the correct balance of salts, pH, oxygen and food kept at the right temperature.

tissue respiration. *See* RESPIRATION.

titration. An operation that forms the basis of volumetric analysis, in which a measured amount of a solution of one reagent is added, a little at a time, to a known amount of a solution of another reagent until they cease to react with one another, the second reagent having been entirely consumed.

Tl. *See* THALLIUM.

TLV. *See* THRESHOLD LIMIT VALUE.

toads. *See* ANURA.

toadstools. *See* AGARICACEAE.

Toarcian. A STAGE of the JURASSIC System.

tobacco. *See* SOLANACEAE.

tocopherol. *See* VITAMIN E.

toddy. *See* COCOS NUCIFERA.

Tokamak. A magnetic container used in experimental FUSION REACTORS that was invented in the USSR and proved superior to US or European designs so that it is now incorporated in several western experimental reactors, most notably the Princeton TOKAMAK FUSION TEST REACTOR.

Tokamak Fusion Test Reactor. A US research project into nuclear fusion whose largest working fusion machine is the Princeton Large Torus.

Tolba, Mostafa K. *See* UNITED NATIONS ENVIRONMENT PROGRAMME.

tolylene diisocyanate (TDI). A chemical that causes chest irritation, leading to sensitization, after which the victim cannot tolerate even the smallest atmospheric concentrations, suffering symptoms akin to acute asthma. In the UK, the THRESHOLD LIMIT VALUE is 0.02 ppm for not more than 20 minutes, or 0.005 ppm for eight hours. Diisocyanates are produced by the chemical industry (e.g., in the manufacture of polyurethane foams) and are used in the rubber industry.

tomato. *See* SOLANACEAE.

tombolo. A spit or bar that joins an island to the mainland or to another island.

tone. A sound of definite PITCH.

tongue worms. *See* PENTASTOMIDA.

tonoplast. The plasma membrane that surrounds the VACUOLE in plant cells. *See also* PLASMALEMMA.

toothed whales. *See* ODONTOCETI.

tooth-shells. *See* SCAPHOPODA.

topogenous mire. A MIRE that develops in such places as depressions and coastal plains, where local relief results in a permanently high WATER TABLE.

topography. A detailed description of the natural and artificial surface features of an area.

topotype. A POPULATION in one geographical region whose characteristics differ from those of another region. A topotype may be extraclinal if it does not fall within a geographical gradient or intraclinal if it is characteristic of a particular geographical gradient.

topset. A flat-lying sedimentary layer that overlies the FORESET BEDS of a DELTA.

topsoil. (1) The surface layer of soil, usually to a depth of about 30 centimetres, which is the depth disturbed by ploughing or other cultivation (i.e. the tilth). This layer is enriched with HUMUS. (2) The A SOIL HORIZON in any soil.

tornado. *See* BATH PLUG VORTEX.

Torrey Canyon. *See* PANHONLIB GROUP.

tortoises. *See* CHELONIA.

Torula thermophila. A YEAST that is grown on sugars to produce NOVEL PROTEIN FOODS.

total population curve. The total POPULATION DENSITY of individuals of all stages (i.e. age groups) plotted against time.

tourmalinization. *See* DEUTERIC.

Tournaisian. The oldest STAGE of the CARBONIFEROUS System in Europe.

tow. *See* RETTING.

toxicology. The study of poisons.

Toxic Substances Control Act. The US law that seeks to protect the public from avoidable exposure to hazardous substances. It requires industries to notify the ENVIRON-

MENTAL PROTECTION AGENCY of stocks of hazardous substances held in storage and with details of the movement and disposal of such substances.

toxic waste. See INDUSTRIAL WASTE.

trace elements (essential elements). Elements that are necessary in extremely small amounts for the proper functioning of METABOLISM in plants and animals. Most are probably constituents of ENZYMES. Higher plants need traces of copper, zinc, boron, molybdenum and manganese. Heart rot of sugar-beet is produced by lack of boron. Deficiency of cobalt causes disease in cattle and sheep. See also MICRONUTRIENTS.

tracer. A substance mixed with or attached to a given substance to enable the distribution or location of the latter to be determined subsequently. Tracers can be physical, chemical, radioactive or isotopic.

trachea. (1) In plants, a non-living tube-like element of XYLEM, derived from a row of cylindrical cells (vessel elements) by the more or less complete breakdown of adjacent end walls. The longitudinal walls of the vessels are thickened with deposits of LIGNIN in spiral, annular or reticulate patterns, and the vessels often communicate by means of minute pits in these thickened walls. Vessels conduct water and dissolved mineral salts and provide mechanical support. They occur mainly in flowering plants. See also TRACHEID. (2) The wind pipe of vertebrates. (3) A branching cuticle-lined tube in an insect that conducts air from an external opening (spiracle) directly to the tissues.

tracheid. In plants, a non-living element of XYLEM formed from a single elongated cell, with tapering ends. The walls are thickened with LIGNIN. Tracheids usually communicate by means of pits in the walls. They conduct water and dissolved mineral salts and provide mechanical support. The conducting tissue in the wood of conifers (see CONIFERALES) is entirely made up of tracheids, which also occur along with vessels in some parts of flowering plants. See

also TRACHEA.

Tracheophyta. A modern term for the vascular plants (PTERIDOPHYTA and SPERMATOPHYTA of the older classifications). The group comprises the PSILOPSIDA, Lycopodiales, SPHENOPSIDA and PTEROPSIDA.

trachyte. A fine-grained ALKALINE INTERMEDIATE rock. Trachytes characteristically show trachytic texture (i.e. the parallel alignment of lath-like FELDSPAR crystals). Trachytes are the volcanic equivalent of SYENITES.

trade cumulus. CUMULUS clouds in the trade-winds that carry up moisture, converting the air from dry polar into moist tropical.

trade inversion. The INVERSION, or stably stratified layer, that lies at the top of the trade-winds, which blow from east-north-east, and above which the wind is usually from a westerly point. The height increases as the air moves over warmer water. It may be at around 1 kilometre off north-west Africa, rising to 3 or 4 kilometres as the air approaches the West Indies, where the convection breaks through to make showers, thunderstorms and seasonal hurricanes.

trade-winds. See GENERAL CIRCULATION.

Tragedy of the Commons. The title of an essay by Garrett Hardin, published in *Science*, based on *Two Lectures on the Checks to Population* by William Foster Lloyd (1794–1852) first published in England by the Oxford University Press in 1833, which Hardin republished in part in his *Population, Evolution, and Birth Control* (W.H. Freeman, 1964). Lloyd sought to rebut the concept of the INVISIBLE HAND by showing that any resource that is the property of all (a commons) may be over-exploited, since each user seeks to increase his profit from it. The profit accrues to the individual, but the cost is shared among all. Thus, for all users, the benefit of increasing use exceeds the costs until the shared resource is reduced, possibly suddenly, to a level that prevents any further use by any of the commoners. Hardin's essay attracted wide attention during the late 1960s and was reprinted in many anthologies.

tramontana. *See* MISTRAL.

transad. Closely related organisms that have become separated by an environmental barrier (e.g., the caribou of North America and the reindeer of Europe).

transduction. The transfer by BACTERIOPH-AGES of genetic material from one bacterium to another.

transect. A cross-section of an area, used as a sampling line for recording the vegetation.

transfer RNA. *See* RNA.

transformation. (1) The acquisition by a bacterium of genetic material (*see* DNA) from a related strain. Characteristics such as resistance to ANTIBIOTICS may be modified by growing BACTERIA in the presence of killed cells or extracts of other bacteria. (2) An inherited change in cultured animal cells produced by VIRUSES or other agents.

transform fault. A strike slip FAULT offsetting a spreading ridge axis. In the offset zone the motion along the FAULT PLANE is in the opposite direction to the displacement. Transform faults probably originate as irregularities in the fracture of the crust at the initial separation of two plates. *See also* PLATE TECTONICS.

transfrontier pollution. The crossing of international frontiers by pollutants, in water or air, and whose effects can be mitigated only as a result of international agreement, since damage is not caused in the country of origin. Examples include the pollution of certain European rivers (e.g., Danube, Rhine), REGIONAL SEAS and air pollutants (e.g., sulphur dioxide) carried from industrial areas in Europe.

transgressive. (1) A species that overlaps from one COMMUNITY to another. (2) Applied

to a body of water or sediments associated with a spreading of water over the land.

transgressive sill. *See* SILL.

translocated herbicides. HERBICIDES that spread throughout a plant after absorption through the roots or leaves (e.g., 2,4,5-T, 2,4,-D, DICHLORPROP, DALAPON, MCPA, MECOPROP, BARBAN, ASULAM, LINURON).

translocation. (1) The breaking away of any part of a chromosome and its attachment to another chromosome. *See also* MUTATION. (2) The transport of materials (e.g., mineral salts, organic materials) from one part of a plant to another, mainly via the XYLEM and PHLOEM in higher plants. *See also* HORMONE WEEDKILLERS, SYSTEMIC INSECTICIDES, TRANSLOCATED HERBICIDES.

transmission loss. A measure of the sound insulation of a wall, partition or panel, equal to the difference in sound level on either side of the wall, etc. or, if the intensity of the waves on either side is expressed as a ratio of incident to transmitted sound, the transmission loss is equal to ten times the logarithm of that ratio.

transpiration. The loss of water vapour from a plant, mainly through the STOMATA and to a small extent through the CUTICLE and LENTICELS. Transpiration results in a stream of water, carrying dissolved minerals salts, flowing upwards through the XYLEM.

Transport 2000. A UK voluntary environmental organization that campaigns against road building, urging instead that more passengers and freight be carried by an expanded rail network. The organization was formed in 1973 by the Conservation Society (disbanded in 1987), trades unions representing rail workers and British Rail management.

transuranic elements. Elements beyond uranium in the PERIODIC TABLE (i.e. elements with atomic numbers greater than 92). They do not occur in nature, but are produced by nuclear reactions and all are radioactive.

traveller's joy. *See* CLEMATIS VITALBA.

travelling wave. A motion in which a fixed pattern of displacements or wave travels longitudinally, while any one particle involved in the motion does not travel in the mean, but executes an oscillatory motion. In the atmosphere, examples are gravity waves, sound waves and ATMOSPHERIC TIDES.

travertine. A form of TUFA deposited by hot springs. It is used as an ornamental stone.

tree. A large woody, PERENNIAL plant with a single stem (bole). The term is often restricted to plants at least 6 metres in height. *Compare* SHRUB. *See also* BROADLEAVED TREES, CONIFERS, HARDWOODS, SOFTWOODS.

tree ferns. *See* FILICALES.

tree layer. *See* LAYERS.

tree ring dating. *See* DENDROCHRONOLOGY.

trellis drainage. A river system resembling a trellis pattern and characteristic of areas of folded (*see* FOLD) SEDIMENTARY ROCKS where tributaries cut channels through less resistant beds.

Tremadoc Series. In the UK, the youngest SERIES of the CAMBRIAN System, but in most countries placed at the bottom of the ORDOVICIAN System.

Trematoda (flukes). A class of parasitic flatworms (PLATYHELMINTHES) which have leaf-like bodies, hooks and/or suckers and thick CUTICLES. Some are ectoparasites (*see* PARASITISM) of aquatic animals, living for instance on the gills of fish. Others are endoparasites of vertebrates (e.g., the blood fluke, *Schistosoma*; the liver fluke of sheep, *Fasciola hepatica*). The various larval stages (miracidium, redia, cercaria) of endoparasitic flukes require at least one intermediate host, including a mollusc (e.g., the water snail, *Limnaea truncatula*, for *F. hepatica*).

trench. *See* OCEANIC TRENCH.

tri. *See* DEGREASING.

triage. An approach to the medical treatment of casualties following a catastrophe (e.g., thermonuclear attack) that exceeds the capacity of the services available. Casualties are divided into three categories: (a) those who will die, even with treatment; (b) those who will die without treatment, but recover with treatment; (c) those who will recover without treatment. Priority is given in the order (b), (c), (a). Some writers have suggested adapting the approach to the allocation of aid (e.g., food aid) to developing countries.

triallate. SOIL-ACTING HERBICIDE of the thiocarbamate (*see* CARBAMATES) group used to control wild oat and black-grass in cereal, pea, bean and carrot crops. It can be irritating to the skin and is harmful to fish.

Triassic. The oldest PERIOD of the MESOZOIC Era, usually dated as beginning some time between 225 and 245 Ma, and lasting about 30 my. Triassic also refers to the rocks formed during the Triassic Period. These are called the Triassic System, which in Europe is divided into three SERIES – the Lower, Middle and Upper – which correspond approximately to the original threefold division into Bunter, Muschelkalk and Keuper. The three series are divided into STAGES: the Lower has one stage, the Scythian; the Middle two, the Anisian and Ladinian; and the Upper three, Karnian, Norian and Rhaetian. In the UK the Triassic is divided into Bunter, Keuper and Rhaetic.

tribe. (1) A group of people who recognize a common leader or leaders, who usually claim a common ancestry, and many of whom are related. (2) A group of closely related genera within a plant family. For instance, within the grass family (GRAMINEAE) the tribe Festuceae contains *Festuca, Dactylis, Lolium, Poa* and other genera.

tricarboxylic acid cycle. *See* CITRIC ACID CYCLE.

trichloroethane (trichloroethylene, tri,

trike, $CHC1CC1_2$). A solvent that is used in the electronics industry and as a dry-cleaning fluid. It is suspected of causing cancer in humans. *See also* DEGREASING.

trichloromethane. *See* CHLOROFORM.

2,4,5-trichlorophenoxyacetic acid. *See* 2,4,5-T.

Trichoptera (caddis flies). An order of insects (division: ENDOPTERYGOTA), almost all of which have aquatic larvae. The adults have two pairs of wings covered with minute hairs, long, slender antennae and reduced mouthparts, so that they rarely feed. Many caddis larvae build tubular cases of sand, snail shells or plant material, others build silk nets in which food is trapped. Trichoptera form an important item of food for freshwater fish.

trickling filter. *See* PERCOLATING FILTER.

triiodothyronine. *See* THYROXIN.

trike. *See* DEGREASING.

Trilobita. A class of marine PALAEOZOIC ARTHROPODA, with flattened oval bodies divided longitudinally into three lobes, a single pair of antennae and numerous similar forked (biramous) appendages. *See also* INDEX FOSSIL.

triple point. *See* KELVIN.

triple superphosphate. *See* PHOSPHATE.

triploblastic. Applied to animals whose bodies are made up of three layers of cells: ectoderm, mesoderm, endoderm (*see* GERM LAYERS). All METAZOA except coelenterates are triploblastic. *Compare* DIPLOBLASTIC.

triploid. Applied to a POLYPLOID cell or organism with three times the haploid number of CHROMOSOMES.

tripton. The non-living matter suspended in a body of water; the abioseston. *See also* SESTON.

trisomic. Applied to cells or organisms showing a kind of aneuploidy (*see* ANEUPLOID) in which one kind of CHROMOSOME is represented three times instead of twice as is normal. Mongolism in humans is caused by the presence of an extra chromosome (number 21).

Triticum durum. *See* DURUM.

tritium (T). A radioactive ISOTOPE of HYDROGEN, with mass number 3 and atomic mass 3.016. It occurs in natural hydrogen as one part in 10^{17}, but it can be made artificially in nuclear reactors. Its HALF-LIFE is 12.5 years. The nucleus of a tritium atom is called a triton. *See also* FUSION REACTOR.

triton. *See* TRITIUM.

trochophore (trochosphere). A type of planktonic (*see* PLANKTON) larva characteristic of many aquatic invertebrates (e.g., bristle-worms, some MOLLUSCA and POLYZOA). The larva swims by means of CILIA, a ring of which encircles the mouth.

trophallaxis. The exchange of food or secretions (e.g., saliva) between animals, especially in social insects. In ants and wasps the worker, for example, taking food to the grub receives in return a drop of saliva.

trophic level. (1) The nutrient status of a water body, especially as regards the levels of NITRATE and PHOSPHATE. *See also* DYSTROPHIC, EUTROPHIC, MESOTROPHIC, OLIGOTROPHIC. (2) *See* ENERGY FLOW, FOOD CHAIN.

trophobiont. *See* TROPHOBIOSIS.

trophobiosis. A type of SYMBIOSIS in which two organisms (trophobionts) of different species feed one another. Some species of ants, for example, rear aphids on plant roots inside their nests and feed on the HONEYDEW secreted by the aphids.

trophogenic region. The region of a water body where organic material is produced by PHOTOSYNTHESIS. *See also* COMPENSATION POINT, PHOTIC ZONE.

tropical cyclone. A violent storm, with a very small area of low pressure at the centre, around which the ISOBARS are almost circular, very close together and the winds extremely violent. They may be 500–600 kilometres in diameter, with the wind, cloud and general weather pattern similar all around them. At the very centre (i.e. the eye of the storm) winds are light, skies are clear or almost clear, and there is no precipitation. The term cyclone is used in the Indian Ocean and Bay of Bengal, hurricane in the Caribbean, typhoon in the China Sea and willy nilly in Western Australia. *See also* BATH PLUG VORTEX.

tropical rain forest. *See* FOREST.

tropina. SINGLE CELL PROTEIN used as a feed supplement for livestock and made in France by British Petroleum Ltd from YEASTS (*Candida* species) grown on paraffins (gas oil containing waxes). After extracting the yeast protein, the wax-free paraffins are returned for refining.

tropism. (1) A growth response by plants or sedentary animals, which produces curvatures whose direction is determined by the location of the stimulus. *See also* GEOTROPISM, HAPTOTROPISM, HYDROTROPISM, NASTIC MOVEMENT, PHOTOTROPISM. (2) A formerly used synonym for TAXIS.

tropoparasite. *See* PARASITISM.

tropopause. The boundary between the TROPOSPHERE and the STRATOSPHERE, at a height that varies from day to day, but averages about 17 kilometres over the equator, decreasing to about 6 kilometres over the poles. There is very little exchange of air across the tropopause because its boundary is defined as the level at which temperature ceases to decrease with height and therefore rising air, and substances carried in it, cease to rise when they reach it.

tropopause break. A phenomenon that occurs when a CYCLONE induces the folding-over of the TROPOPAUSE, with the tropopause higher on the warm side of a front by perhaps a few thousand metres.

tropophyte. A plant that lives under moist conditions for part of the year and under dry conditions for the rest of the time. An example is a deciduous tree, which sheds it leaves for the dry season or for the winter, when water may not be available because it is frozen.

troposphere. The lowest level of the atmosphere, in which temperature decreases with height and in which weather phenomena occur, bounded by the land or sea surface below and by the TROPOPAUSE above.

trout beck. *See* RIVER ZONES.

true fat. *See* FAT.

truewood. *See* HEARTWOOD.

truncated spur. The steep walls of the U-shaped trough formed by a GLACIER as spurs and projections are abraded away by the advancing ice.

Trypanosomidae. A family of parasitic (*see* PARASITISM) Protozoa (*see* FLAGELLATA), some of which cause serious disease in humans and domestic animals. Oriental sore and kala-azar are caused by *Leishmania,* which attacks the cells lining the blood vessels and is transferred by sandflies of the genus *Phlebotomus. Trypanosoma gambiensis* and *T. rhodesiensis* cause sleeping sickness and are transmitted to humans by the bite of the tsetse fly (*Glossina*). These species of *Trypanosoma* and *T. brucei,* which causes African cattle sickness, are non-pathogenic in antelopes.

tryptophan. An AMINO ACID with the formula $C_6H_4NHCH_2CH_2CH(NH_2)COOH$ and a molecular weight of 204.2.

tsetse fly. *See* CYCLORRAPHA, TRYPANOSOMIDAE.

tsunami. A sea-wave produced by a submarine EARTHQUAKE or volcanic explosion that travels across the ocean at a velocity of several hundred kilometres per hour, but with a low wave-height. On reaching a shelving shore the energy of the tsunami is concentrated into a hugely destructive series of waves several metres high. Tsunamis are misleadingly called tidal waves. In Japanese tsunami means harbour wave.

tuatara. *See* RHYNCHOCEPHALIA.

tubal ligation. *See* SALPINGECTOMY.

tubenoses. *See* PROCELLARIIFORMES.

tuber. A swollen food-storing portion of an underground STEM (e.g., potato) or ROOT (e.g., dahlia). Tubers are organs of perennation and VEGETATIVE PROPAGATION which last only one year, and do not give rise to those of the succeeding year. *Compare* CORM, RHIZOME.

Tubulidentata (Orycteropodidae). An order of placental mammals (EUTHERIA) containing only *Orycteropus*, the aardvark (earth pig or African anteater). *Orycteropus* lives on termites, has a long tongue and an elongated snout like other anteaters (EDENTATA and PHOLIDOTA), but possesses unique peg-like teeth with tubes in the dentine.

tufa. A SEDIMENTARY ROCK composed of CALCIUM CARBONATE or SILICA and deposited from aqueous solution by evaporation of circulating GROUND WATER or water from springs or in lakes.

tuff. Lithified PYROCLASTIC rock with a preponderance of fragments less than 2 centimetres in diameter.

tundra. Treeless arctic and alpine regions that may be bare of vegetation or may support mosses, lichens, herbaceous plants and dwarf shrubs. *See also* TUNDRA SOILS.

Tundra Soils. Shallow soils with permanently frozen subsoil which impedes drainage, so relatively little moisture is available. Where plants (e.g., lichens, mosses, herbs and shrubs) can colonize, organic matter accumulates and a dark peaty layer overlies an anaerobic greyish layer. Profiles generally are poorly developed, and the soils grade between the polar desert soil of arid northern regions to arctic BROWN FOREST SOILS in better drained, more humid upland regions. *See also* SOIL CLASSIFICATION.

tungsten (W). A grey, hard, ductile, malleable metallic element that resists corrosion. It is used in ALLOYS, as tungsten carbide for hard tools and in electric lamp filters. A_r = 183.85; Z = 74; SG 19.3; mp 3410°C. *See also* SCHEELITE, WOLFRAMITE.

Tunicata. *See* UROCHORDATA.

Turbellaria. A class of mainly free-living flatworms (PLATYHELMINTHES) that move by means of CILIA which cover the body. Some (e.g., *Planaria*) live in fresh water, others are marine and a few are terrestrial (e.g., *Bipalium kewense,* a tropical form reaching 30 centimetres in length, sometimes found in greenhouses).

turbidite. Deposits from a moving slurry of sediment and water (called a turbidity current or density current). Experimental turbidity currents produce SEDIMENTARY STRUCTURES within each BED, such as GRADED BEDDING, and, on the base of each bed and underlying bed, such as grooves, flute cases, load casts and tool marks. These features are shown by many natural deposits, from unlithified graded deposits in the ABYSSAL PLAINS to PRECAMBRIAN SANDSTONES.

turbidity. The cloudiness in a fluid caused by the presence of finely divided, suspended material.

turbidity current. *See* TURBIDITE.

turbine. A machine that converts the energy in a stream of fluid into mechanical energy by passing the stream through a system of fixed and/or moving fan-like blades, causing them to rotate. Turbines have wide uses in large-scale power generation (usually employing steam-driven turbines), small-scale power generation (e.g., Pelton wheel), jet aircraft propulsion (gas turbines deriving power from a stream of heated air), marine engines, etc.

turbulence. An irregular movement of a fluid in which, in general, no two particles of the fluid follow the same path. Most natural movement of fluids is turbulent rather than laminar, and in air turbulence is a major cause of mixing.

turgor pressure. The HYDROSTATIC PRESSURE within a plant cell exerted against the CELL WALL as a result of the intake of water by OSMOSIS. Turgor pressure plays a large part in the mechanical support of plant tissues. When water loss exceeds intake, turgor pressure falls and wilting follows. Some plant movements, such as seismonasty (*see* NASTIC MOVEMENT) and the opening and closing of STOMATA, are caused by variations in turgor pressure. The turgor pressure of vacuolated (*see* VACUOLE) cells in the NOTOCHORD of chordates (*see* CHORDATA), acting against the notochord sheath, creates a stiff, flexible supporting rod.

Turin Project. A scientific study conducted in the 1980s in Turin, Italy, to determine the effects on human health of exposure to sublethal doses of lead.

turions. Detachable winter buds formed by many water plants, enabling them to survive the winter.

turnip. *See* BRASSICA.

turn-over. (1) The continuous, balanced process of generation and loss of cells or molecules in living systems. The turn-over time is the time needed for the replacement by turn-over of the cells or molecules equivalent to the total BIOMASS of a population, or the time taken for an individual organism to mature, die and undergo decomposition. (2) *See* OVERTURN

Turonian. A STAGE of the CRETACEOUS System.

turtles. *See* CHELONIA.

Tuscacora rice. *See* WILD RICE.

tusk-shells. *See* SCAPHOPODA.

TV dinner. A meal whose PROTEIN content is derived from textured vegetable protein (TVP) and from the vegetables present in the meal in their traditional form. The meal contains no meat. *See also* NOVEL PROTEIN FOODS.

TVP. Textured vegetable protein (*see* NOVEL PROTEIN FOODS).

twister. In US usage, a tornado (*see* BATH PLUG VORTEX) or WATER SPOUT.

type specimen (holotype). The original specimen from which the description of a new species is made. *Compare* LECTOTYPE, NEOTYPE, PARATYPE.

typhoon. A TROPICAL CYCLONE in the China Sea.

typhus. *See* ANOPLURA.

tyrosinase. *See* ALBINISM.

tyrosine. An AMINO ACID with the formula $C_6H_4OHCH_2CH(NH_2)COOH$ and a molecular weight of 181.2.

U

U. *See* URANIUM.

UKAEA. *See* UNITED KINGDOM ATOMIC ENERGY AUTHORITY.

Ulex. Gorse, furze or whin (*see* LEGUMINOSAE).

Ulmus (elms). A genus of tall deciduous trees (family: Ulmacaea) with winged fruit that yield valuable timber. *Ulmus procera* is the English elm, common in hedgerows and woods in the south of England. DUTCH ELM DISEASE has drastically reduced the elm population in the UK.

ultimate strength. *See* FRACTURE.

Ultisols. *See* SOIL CLASSIFICATION.

ultrabasic. Applied to IGNEOUS rocks almost entirely composed of FERROMAGNESIAN MINERALS to the virtual exclusion of QUARTZ, FELDSPAR AND FELDSPATHOIDS (i.e. to rocks containing less than 45 percent SILICA). Ultrabasic rocks are grouped into: (a) PERIDOTITES (e.g., DUNITE and KIMBERLITE), consisting essentially of OLIVINE with or without accessory ferromagnesian minerals, and without feldspar; (b) perknites, consisting essentially of ferromagnesian minerals other than olivine; (c) picrites, consisting of over 90 percent ferromagnesian minerals and up to 10 percent feldspar. Picrite is thus gradational into basic rocks such as GABBRO. Purely ultrabasic INTRUSIONS are rare and occur late in an OROGENY. Ultrabasic rocks are normally found in association with basic rocks in layered igneous intrusions. Such intrusions often contain economic deposits of CHROMITE, ILMENITE, MAGNETITE and the platinum group of metals (platinum, iridium, osmium, palladium, rhodium and ruthenium). *Compare* ULTRAMAFIC.

ultra- low-volume sprayer (ULV sprayer). A sprayer in which a small electric motor pumps a highly concentrated solution of PESTICIDE on to a rapidly spinning toothed disc. The liquid moves centrifugally across the surface of the disc and leaves from the points of the teeth as fine filaments, which break almost instantly into droplets whose size is determined by the speed of rotation of the disc. The sprayer delivers pesticide as a fine mist that drifts onto a crop, coating plants on all exposed surfaces. It uses 1–10 percent of the amount of pesticide needed by a conventional sprayer, and very little falls to the ground.

ultramafic. Applied to IGNEOUS rocks rich in the MAFIC (dark-coloured FERROMAGNESIAN) minerals. *Compare* ULTRABASIC.

ultraviolet radiation (UV radiation). Electromagnetic radiation in the wavelength range of about 4×10^{-5} to 5×10^{-9} m, placing it between visible light and X-rays in the spectrum. The UV spectrum is divided further by wavelengths into A, B and C bands, and certain wavelengths that affect cells are known in the USA as DUV (damaging UV). Much of the UV radiation from the Sun is absorbed in the OZONE LAYER of the Earth's atmosphere. UV radiation affects photographic films and plates, and in the skin of some animals, including humans, it acts on ERGOSTEROL to produce VITAMIN D. UV is non-ionizing radiation with less penetrative power in a column of water than visible light, but within cells it can cause CHROMOSOME breaks. In nature, most organisms are well protected against harmful UV. In humans, the radiation is absorbed by MELANIN in the skin, causing the melanin to darken, but excessive exposure of fair-skinned people can cause non-melanoma skin cancers. UV radiation can be produced artificially by mercury vapour lamps.

ULV sprayer. *See* ULTRA-LOW-VOLUME SPRAYER.

Umbelliferae. A cosmopolitan family of DICOTYLEDONEAE, most of which are herbs with stout, hollow stems. The INFLORESCENCE is usually umbrella-shaped, with many small flowers massed together, making them attractive to the insects that pollinate them. Many umbellifers are cultivated as food crops (e.g. *Daucus,* carrot; *Pastinacea,* parsnip; *Apium,* celery). Many are poisonous (e.g., *Conium maculatum,* hemlock).

UNCLOS. *See* UNITED NATIONS CONFERENCE ON THE LAW OF THE SEA.

unconditioned reflex. *See* CONDITIONED REFLEX.

unconformity. (1) A gap in the geological record marked by a lack of continuity in the deposition of SEDIMENTARY ROCKS (i.e. deposition ceased for a time) between two BEDS that are in contact. (2) The surface of contact between two such strata.

UNCTAD. *See* UNITED NATIONS CONFERENCE ON TRADE AND DEVELOPMENT.

undergrowth. The shrubs, saplings and herbaceous plants in a forest.

understorey. The lower layer of trees in a two-layered woodland (e.g., COPPICE under standards) or the lower storey of a two-storeyed high forest.

UNDP. *See* UNITED NATIONS DEVELOPMENT PROGRAMME.

UNEP. *See* UNITED NATIONS ENVIRONMENT PROGRAMME.

UNESCO. *See* UNITED NATIONS EDUCA-

TIONAL, SCIENTIFIC AND CULTURAL ORGANIZATION.

Ungulata (ungulates). Hoofed mammals, including the ARTIODACTYLA and PERISSODACTYLA.

unguligrade. Applied to mammals that walk on the tips of the digits, which end in hooves. *Compare* DIGITIGRADE, PLANTIGRADE.

uniformitarianism. The principle that processes active today also acted in the past producing the same results. Uniformitarianism is often encapsulated in the statement 'the present is the key to the past'.

uniformity. In ECOLOGY, the general tendency of an ASSOCIATION to have a uniform distribution of component species within it, producing a uniformity of the association as a whole, so providing one of the most distinctive characteristics of that association.

union. A group of plants that are ecologically significant within the total flora because they usually appear together whenever environmental conditions are suitable, their requirements being similar or complementary.

uniovular twins. *See* MONOZYGOTIC TWINS.

uniparous. *See* MONOTOCOUS.

unisexual. (1) Applied to MONOECIOUS or DIOECIOUS plants that have their stamens and carpels in separate flowers. (2) Applied to an animal that produces either male or female GAMETES. *Compare* HERMAPHRODITE.

United Kingdom Atomic Energy Authority (UKAEA). The statutory body in the UK that is responsible for all aspects of atomic energy and for the commercial and scientific use of radioisotopes. The UKAEA is also involved in more general research into energy use and resources. It is established under the Atomic Energy Act, 1946. *See also* MAJOR HAZARD INCIDENT DATA SERVICE.

United Nations Centre for Human

Settlements. *See* UNITED NATIONS CONFERENCE ON HUMAN SETTLEMENTS.

United Nations Commission on Human Settlements. *See* UNITED NATIONS CONFERENCE ON HUMAN SETTLEMENTS.

United Nations Conference on Desertification. An international conference, held in Nairobi in 1977, to discuss the problem of DESERTIFICATION and to devise plans to combat it.

United Nations Conference on Human Settlements. An international conference, held in Victoria, British Columbia, Canada, in 1976 to debate urban problems and especially the rapid URBANIZATION occurring in the THIRD WORLD. It led to a United Nations Habitat and Human Settlements Foundation, established in 1978 with headquarters in Nairobi, Kenya, and a United Nations Centre for Human Settlements in New York, where a United Nations Commission on Human Settlements held its first meeting in April 1978.

United Nations Conference on the Human Environment. A conference held in Stockholm in the summer of 1972, attended by most United Nations member governments, except for the USSR and the East European states, but including the People's Republic of China. *See also* DECLARATION ON THE HUMAN ENVIRONMENT, EARTHWATCH, UNITED NATIONS ENVIRONMENT PROGRAMME.

United Nations Conference on the Law of the Sea (UNCLOS). A series of international meetings, held under the auspices of the United Nations between 1974 and 1982, with the aim of establishing new international law to regulate the exploitation of the natural resources of the seas while maintaining the right of ships to free passage. The Conferences ended by producing a LAW OF THE SEA CONVENTION. *See also* EXCLUSIVE ECONOMIC ZONE.

United Nations Conference on Trade and Development (UNCTAD). A principal United Nations forum in which developed and developing countries meet to discuss matters relating to trade and development.

United Nations Consultative Committee on the Ozone Layer. The United Nations body, established in 1977, that seeks to achieve international agreement on measures to prevent the depletion of the OZONE LAYER. In particular, it aims to reduce the production and use of CHLOROFLUOROCARBONS. Since 1985 its activities are specified by the CONVENTION TO PROTECT THE OZONE LAYER.

United Nations Development Programme (UNDP). The United Nations agency that acts as the main channel for multilateral technical and pre-investment funding for THIRD WORLD countries. The funding is then administered with the cooperation of the appropriate specialist UN agency.

United Nations Educational, Scientific and Cultural Organization (UNESCO). A United Nations agency, founded in 1945, to support and complement the efforts of member states to promote education, scientific research and information, and the arts, and to develop the cultural aspects of world relations. It holds conferences and seminars, issues publications, promotes research and the exchange of information and provides technical services. It is financed by its own budget and also draws on funds pledged by member states to the UNITED NATIONS DEVELOPMENT PROGRAMME. Its headquarters are in Paris. *See also* MAN AND THE BIOSPHERE PROGRAMME.

United Nations Environment Programme (UNEP). The United Nations agency charged with the coordination of intergovernmental measures for environmental monitoring and protection. It was formed after the 1972 UNITED NATIONS CONFERENCE on the HUMAN ENVIRONMENT. UNEP's first Executive Director was Mr Maurice Strong, who had been Secretary General of the 1972 conference. He was succeeded, in 1976, by Dr Mostafa K. Tolba. UNEP operates the EARTHWATCH programme, which includes the GLOBAL

ENVIRONMENTAL MONITORING SYSTEM and the GLOBAL RESOURCE INFORMATION DATABASE, and funds EARTHSCAN. Its headquarters are in Nairobi, Kenya.

United Nations Habitat and Human Settlements Foundation. *See* UNITED NATIONS CONFERENCE ON HUMAN SETTLEMENTS.

United Nations Research Institute for Social Development (UNRISD). A United Nations agency, based in Geneva, engaged in social research, especially into the effects of development policies in developing countries.

unit membranes. A general term used to describe all osmotically active (*see* OSMOSIS) cell membranes (*see* PLASMA MEMBRANE) because they have a universal molecular architecture, regulating the quantity of materials being exchanged across it by altering its thickness, rather than by making any fundamental change in its composition.

universal oil product (UOP). A US process for removing SULPHUR DIOXIDE from FLUE GASES, using a WET SCRUBBER, followed by extraction of the sulphur and regeneration of the reagent.

univoltine. Applied to an organism that produces only a single generation in a year. *Compare* BIVOLTINE, MULTIVOLTINE.

UNRISD. *See* UNITED NATIONS RESEARCH INSTITUTE FOR SOCIAL DEVELOPMENT.

unsaturated fatty acid. *See* FATTY ACID.

unsettled weather. Weather that exhibits considerable changes from sunshine to cloud and rain within the passage of hours. It is typical of the westerly type in temperate climates.

unsorted. Applied to rock debris composed of particles with a wide range of sizes (*see* SORTING).

UOP. *See* UNIVERSAL OIL PRODUCT.

updraught. Air rising in a CONVECTION current.

upper air contours. Upper air maps are drawn showing the contours of constant height of an isobaric surface rather than ISOBARS at constant height because of greater ease of construction and simpler mathematical use.

upper crust. *See* SIAL.

upper shore. *See* SHORE ZONATION.

uracil. *See* RNA.

uraninite. The mineral uranium oxide (UO_2) found in GRANITES and PEGMATITES, in HYDROTHERMAL VEINS, and also in some SEDIMENTARY ROCKS. Uraninite is the chief ore of URANIUM.

uranium (U). A naturally occurring, radioactive, hard, white metallic element, the natural element consisting of 99.28 percent $^{238}_{92}U$, which has a HALF-LIFE of 4.5×10^9 years, and 0.71 percent $^{235}_{92}U$, which has a half-life of 7.1×10^8 years (*see* URANIUM ENRICHMENT). Uranium-235 is used as a fuel in NUCLEAR REACTORS. $A_r = 238.03$; $Z = 92$; SG 19.05; mp 1150°C.

uranium enrichment. The improvement of the properties of natural URANIUM by removing the diluent uranium-238, which comprises 99.28 percent of the natural element, in order to increase the concentration of uranium-235, which is able to sustain a nuclear CHAIN REACTION, and so can be used as fuel in a NUCLEAR REACTOR.

urbanization. The migration of people in substantial numbers from rural to urban areas. Urbanization is characteristic of regions in the early stages of industrialization and is marked in THIRD WORLD countries. *Compare* RURALIZATION.

urban node. A planning concept in which specific areas are identified for development to city centre building and population density. This allows services (e.g., DISTRICT

HEATING) to be provided more conveniently and efficiently than is possible where communities are more dispersed. *Compare* URBAN VILLAGE.

urban village. A planning concept in which new towns are designed to combine workplaces, leisure facilities and homes within the same small area so that all parts of the town are within walking distance of one another. *Compare* URBAN NODE.

urea ($CO(NH_2)_2$). A soluble breakdown product of PROTEINS that is excreted by many vertebrates. It is also produced by some plants. *See also* DEAMINATION, UREOTELIC.

urea formaldehyde. An adhesive or thermosetting plastic, made from UREA and FORMALDEHYDE. As a plastic foam it is used as an insulating material in cavity walls. This use was banned in the USA in February 1982, when it was found that urea formaldehyde could emit formaldehyde.

urea herbicides. A group of herbicides (e.g., DIURON and LINURON) that control weed seedlings by inhibiting PHOTOSYNTHESIS. They persist for long periods in the soil, but are not very toxic to mammals.

Uredinales. *See* RUST FUNGI.

ureotelic. Applied to animals whose main nitrogenous excretory product is UREA; fish, Amphibia and mammals are ureotelic. During their embryonic development dissolved urea diffuses away into the surrounding water or is removed in the maternal blood stream. *Compare* URICOTELIC.

uric acid. A complex nitrogenous compound; an almost insoluble breakdown product of PROTEINS and nucleic acids (*see* DNA, RNA). It is excreted in large amounts by URICOTELIC animals and in small amounts by

some other animals.

uricotelic. Applied to animals whose main nitrogenous excretory product is URIC ACID. Birds, insects, terrestrial snails and most reptiles are uricotelic. These animals develop inside shells (*see* CLEIDOIC EGG) and must therefore store their nitrogenous excretory product in an insoluble form. *Compare* UREOTELIC.

Urochordata (Urochorda, Tunicata, sea-squirts). A group of marine CHORDATA whose larvae are tadpole-like, with a NOTOCHORD in the tail. Most adult urochordates are sedentary, with no notochord, a reduced nervous system and an enormous perforated pharynx through which water, bearing food particles, is passed by ciliary action (*see* CILIA). The body is surrounded by a horny or gelatinous coat (the tunic or test). Many sea-squirts are colonial.

Urodela (Caudata). An order of AMPHIBIA that includes the newts and salamanders. Urodeles have a well-developed tail and two pairs of limbs of more or less equal sizes. Gills may persist in the adults (e.g., in the axolotl, *see* NEOTONY).

US AID (US Agency for International Development). The major US agency concerned with AID to developing countries.

Usonia. The name of an idealized city planned by FRANK LLOYD WRIGHT.

Ustiginales. *See* SMUT FUNGI.

utopianism. A form of planning based on the construction of ideal futures, together with assessments of their feasibility and strategies by which they may be realized.

UV light. *See* ULTRAVIOLET RADIATION.

V

V. *See* VOLT.

vacuole. A fluid-filled space inside the CY-TOPLASM, more common in plant than in animal cells. The vacuole in many plant cells occupies most of the cell and contains cell sap. Many PROTOZOA contain contractile vacuoles which collapse periodically, expelling excess water from the cell. Food vacuoles of Protozoa and other animals are spaces where ingested food particles are digested. *See also* TURGOR PRESSURE.

vadose. Applied to the region between the ground surface and the WATER TABLE, where the soil is unsaturated and most biological activity is concentrated.

vadose water (gravitational water). Water moving in aerated ground above the WATER TABLE. Literally it means wandering water. *Compare* PHREATIC WATER.

Vaiont. An area in Italy that was the site of one of the most severe manmade disasters in modern times. On the night of 9 October 1963, a sudden and very large and rapid rockslide completely filled the Vaiont reservoir causing large waves. As a result some 3000 people died. The disaster was caused by two weeks of heavy rainfall, which raised the water level in the cavernous surrounding rocks, which were inherently weak and whose limestone beds and clay layers offered little frictional resistance to sliding. The high water level in the reservoir saturated part of the side, further increasing buoyancy and decreasing friction. No witness of the incident survived.

Valanginian. *See* NEOCOMIAN.

valine. An AMINO ACID with the formula $(CH_3)_2CHCH(NH_2)COOH$ and a molecular weight of 117.1.

valley fog. FOG formed in air that has cooled on hillsides by radiation at night. Frequently the air containing the fog stagnates in a valley by day because the fog reflects a significant fraction of sunshine and cools by radiation at infrared wavelengths.

valley glacier. *See* GLACIER.

valley train. A body of OUTWASH DEPOSITS that partly fills a valley.

valve. (1) A device for controlling the passage of a fluid through a pipe or tube. (2) A membranous structure that permits a liquid (e.g., blood) to flow in one direction only. (3) The shell of a mollusc (*see* MOLLUSCA) or each complete part of the shell of a bivalve mollusc. (4) A segment of a dehiscent fruit. (5) *See* THERMIONIC VALVE.

vanilla. *See* ORCHIDACEAE.

variable. (1) A quantitative characteristic of an individual (e.g., height or weight in humans). *See also* ATTRIBUTE, VARIATE. (2) Any quantity that can have more than one value.

variate. A quantity that can have one of a range of specific values, each with a specific probability.

variation. Differences that exist between individuals of the same species at corresponding stages in the life cycle. Variation is a reflection of the genetic constitution of the individuals and of the environmental influences brought to bear upon them.

continuous variations. (a) Relatively small variations, not due to MUTATION. (b) Variations that show gradation from one extreme to another, with a preponderance of intermediate types (e.g., stature in humans). The type of variation is caused by the combined action of a number of GENES, each having a small effect (*see* MULTIPLE FACTORS).

discontinuous variations. (a) Mutations. (b) Variations in which the expression of a particular character differs sharply among individuals, and there is no range of intermediate types (e.g., sex differences). This type of variation is caused by the action of a few (often two) genes which exert a large effect.

variety. (1) A group of plants or animals that differ distinctly from others within the same SUBSPECIES. (2) A term used loosely for a VARIATION of any kind within a plant or animal species.

Variscan. Usually understood today to refer to processes and products of the mountain building in Europe from the CARBONIFEROUS to TRIASSIC Periods, inclusive. Originally Variscan was restricted to the HERCYNIAN of central Europe, but Hercynian and Variscan are now usually considered interchangeable terms.

varves. Rhythmically laminated sediments where, due to special conditions, distinctive layers have been deposited in the course of a single year (e.g., the coarse and fine sediments deposited after the annual melting of an ice sheet, or different types of sediment deposited in a lake bottom during summer and winter). These laminae can be counted like the annual rings of a tree (*see* DENDRO-CHRONOLOGY) and used for estimating the time taken for the entire sediment to be deposited.

vascular. Applied to tissues that conduct liquids (e.g., water, blood) and to plants that possess vascular tissue. *See also* VASCULAR BUNDLE, VASCULAR PLANTS.

vascular bundle. A strand of conducting tissue in a higher plant, consisting mainly of XYLEM and PHLOEM.

vascular cryptogam. *See* CRYPTOGAMIA.

vascular cylinder. *See* STELE.

vascular plants. 'Higher' plants; characterized by the possession of true roots, stems and leaves containing specialized vascular tissue through which liquids are conducted. The group includes the PTERIDOPHYTA and SPERMATOPHYTA. *See also* TRACHEOPHYTA.

vasectomy. Male sterilization, in which the vas deferens is cut and the two ends sealed, so preventing the release (but not the manufacture) of sperm.

Vavilov, Nikolai Ivanovich (1887–1943). A Russian plant geneticist who made a comprehensive study of the origin of cultivated plants and proposed centres of diversity and centres of origin. He argued that if the rate of natural MUTATION among plant species is constant with time, the longer a species occupies a particular site the greater the genetic diversity that will exist among its individual members. Thus, the fewer the VARIATIONS, the shorter the time the species must have been present. From this he located a series of areas in all parts of the world where particular species of crop plants are found in the greatest diversity. From this he attempted to argue that these centres of diversity were also centres of origin from which knowledge of the cultivation of those particular species had spread by a process of cultural diffusion. This conclusion could not be supported on the evidence, but Vavilov made a major contribution to knowledge of plant genetics. He was removed from all his posts at LYSENKO, TROFIM DENISOVICH, rose in favour in the USSR and is believed to have died in prison in Siberia.

VCM. *See* VINYL CHLORIDE MONOMER.

vector. (1) Any physical quantity that cannot be described completely without reference to a direction (e.g., VELOCITY). (2) An organism that conveys a parasite (*see* PARA-

SITISM) from one host to another (e.g., *Anopheles* mosquito, which transmits malaria).

veering wind. *See* BACKING WIND.

vegetation. The plants of an area considered in general, or as COMMUNITIES, but not taxonomically; the total plant cover in a particular area, or on the Earth as a whole.

vegetation study. The study of VEGETATION. The study may be qualitative (of the species and characteristics of the plants that make up a COMMUNITY) or quantitative (of the distribution of particular plants).

vegetative propagation (vegetative reproduction). (1) Asexual reproduction in plants or animals.   (2) Asexual reproduction in plants by means of part of the plant other than a SPORE (e.g., BULB, CORM, TUBER, RHIZOME). *See also* GEMMATION.

veil of cloud. A layer of almost textureless cloud that spreads characteristically across the sky ahead of the WARM FRONT of an advancing CYCLONE. It is caused by the slow ascent of air over a wide area, and HALOS are often seen around the Sun in it.

vein. (1) A tabular or sheet-like body of a mineral or minerals found in a rock body. Most veins of ore minerals (lodes), lie along JOINTS or FAULT PLANES.   (2) A blood vessel which carries blood towards the heart.

veliger. A free-swimming larval stage (*see* LARVAE) of many aquatic MOLLUSCA that feeds by means of CILIA and has the rudiments of adult organs, including the shell.

velocity. Speed and direction. The wind velocity is the wind speed and the direction from which the wind blows (the reciprocal of the direction of flow). Velocity is sometimes used informally (and incorrectly) to mean speed alone.

vent. The opening through which a VOLCANO ejects material.

vent agglomerate. *See* AGGLOMERATE.

vent breccia. *See* BRECCIA.

ventifact. A pebble faceted by abrasion by wind-blown sand. A DREIKANTER and an EINKANTER are special forms.

ventral. (1) The part of an animal or organ at, or nearest to, the side normally directed downwards (i.e. the side furthest from the spine). In bipedal mammals the ventral surface is directed forwards. *Compare* DORSAL. (2) *See* ADAXIAL.

Venturi effect. The acceleration of a fluid stream as it passes through a constriction in a channel (e.g., a narrowing of river banks, a narrow nozzle on a pipe or an aerofoil section). The acceleration is associated with a reduction in the pressure in the fluid. In the flow of air over an aerofoil this produces lift. It is named after G.B. Venturi (1746–1822) who invented the Venturi tube to exploit this phenomenon in order to obtain a fine mixture of a liquid and a gas, so providing the basis of the modern carburettor.

verfluent. A minute or microscopic animal member of a BIOME.

Vermes. An outmoded term covering all worm-like animals.

vermiculite. Substance that occurs naturally as a magnesium iron hydroaluminosilicate, with water molecules held between the silicate sheets. The flaky natural material is passed momentarily through a furnace at 1000°C which brings the particle temperature to about 230°C. The steam produced causes the mineral to exfoliate (*see* EXFOLIATION) to a low-density material which is used in thermal and acoustic insulation, packaging and horticulture, and as an inert carrier for animal feedstuffs.

vernalization. The advancement of flowering by subjecting a plant to abnormal temperatures or by exposure to GIBBERELLIN. For example, spring-sown 'winter' varieties of cereals, which would not normally flower the same year, can be made to produce a crop in the summer of the same year by exposing the germinating seed to temperatures just above 0°C for a few weeks. When sown in the winter these plants are subjected naturally to similar temperatures.

Vertebrata (Craniata). Animals with a skull and (except perhaps in some extinct forms) a vertebral column made of BONE or CARTILAGE. The vertebral column is developed only imperfectly in lampreys (AGNATHA). Vertebrates form a subphylum of the CHORDATA and comprise Agnatha, true fishes, AMPHIBIA, REPTILIA, birds (AVES) and MAMMALIA.

vertical. Perpendicular to a horizontal surface (e.g., of static water). A perpendicular line does not point exactly to the geometrical centre of the Earth, because the Earth is oblate due to its rotation. Minor variations are due to the pressure of large mountains or significant variations in density of the Earth's CRUST.

Vertisols. See GRUMOSOLS, SOIL CLASSIFICATION.

vesicle. (1) A small, rounded or pipe-like cavity in a fine-grained or glassy volcanic rock, formed by the trapping of a gas bubble as the lava solidified. Where these cavities have been filled with SECONDARY MINERALS they are known as amygdales. (2) A blister. (3) A small bladder or other hollow structure.

vesicular. Possessing internal hollow structures (i.e. VESICLES).

vessel elements. See TRACHEA.

vestigial. Applied to a structure, function or behaviour pattern that has diminished during the course of evolution, leaving only a trace (e.g., the vermiform appendix of humans, the much reduced 'wing' bones of the kiwi).

Vesuvian volcano. See VOLCANO.

vibration. See FORCED OSCILLATION.

vicariad (vicarious species). A member of a

group of closely related species that are ecologically equivalent and whose distribution is ALLOPATRIC (e.g., the European reindeer and the caribou of North America).

villagers. A small group of Soviet writers, formed in the 1980s, that comments on environmental problems (e.g., the KARA-BOGAZ-GOL dam) and acts as a pressure group for environmental protection. In 1987 it formed a Soviet equivalent of GREENPEACE as a commission of the Soviet Peace Committee.

vinyl chloride monomer (VCM, chloroethene, CH_2:CHC1). A compound used in the manufacture of POLYVINYL CHLORIDE (PVC) and believed to cause cancer. The amount of VCM remaining in PVC has been strictly regulated in most countries since 1977. In the EEC, when PVC comes into contact with food it must not contain more than 1 milligram of VCM per kilogram.

virement. The diversion of funds to purposes for which they were not allocated, especially the spending of surplus moneys remaining after a funded project has been completed.

virga. Trails of ice crystals falling from a cloud.

virion. A mature virus.

viruses. Submicroscopic agents that infect plants, animals and bacteria, and are unable to reproduce outside the tissues of the host (*see* PARASITISM). A fully formed virus consists of nucleic acid (DNA or RNA) surrounded by a PROTEIN or protein and lipid (FAT) coat. The nucleic acid of the virus interferes with the nucleic acid-synthesizing mechanism of the host cell, organizing it to produce more viral nucleic acid. Viruses cause many diseases (e.g., mosaic diseases of many cultivated plants, myxomatosis, foot and mouth disease, the common cold, influenza, measles, poliomyelitis). Many plant viruses are transmitted by insects (e.g., aphids), some by eelworms. Animal viruses are spread by contact, droplet infection or by insect VECTORS (e.g., yellow fever virus is transmitted by the mosquito *Aedes aegypti*), and some are spread by the exchange of body fluids. *See also* ACQUIRED IMMUNE DEFICIENCY SYNDROME, BACTERIOPHAGE, RETROVIRUS.

Visean. The second oldest STAGE of the CARBONIFEROUS System in Europe.

vitalism. The belief that life originates in a vital principle that is distinct from chemical and physical processes and that is not susceptible to examination by analytical techniques. The idea dates at least from the time of Aristotle, but fell from favour in modern times as explanations were found for the phenomena supposedly caused by the vital principle. *See also* HOLISM, REDUCTIONISM.

vitamin. An organic food substance needed in very small amounts by HETEROTROPHIC organisms. Vitamins play essential roles in METABOLISM (e.g., as constituents of ENZYME systems) and their deficiency causes diseases.

vitamin A (axerophthol). A fat-soluble substance required by vertebrates and present in liver, milk and animal fats. Carotene is converted by animals into vitamin A. Deficiency causes impaired activity of epithelia (e.g., the lining of the respiratory passages and alimentary canal), drying of the cornea and night blindness. The vitamin is a constituent of one of the pigments in the retina.

vitamin B. A complex of water-soluble substances. *See also* BIOTIN, COBALAMINE, FOLIC ACID, NICOTINIC ACID, PANTOTHENIC ACID, PYRIDOXINE, RIBOFLAVIN, THIAMINE.

vitamin C (ascorbic acid). A water-soluble substance needed by humans and some other animals for the formation of COLLAGEN and thus for the maintenance of capillary walls and the healing of wounds. Fresh vegetables and fruit (especially citrus fruits) are important sources of the vitamin. Deficiency causes scurvy, bleeding in the mucous membranes and bleeding under the skin and into the joints.

vitamin D (calciferol). A group of fat-soluble substances (sterols) present in animal fats and required by vertebrates. In humans, vitamin D is synthesized by the action of ULTRAVIOLET LIGHT from a precursor present in the skin. The vitamin is essential for the metabolism of calcium and phosphorus, and deficiency in children causes rickets.

vitamin E (tocopherol). The antisterility vitamin required by vertebrates. It is fat-soluble and present in seeds and green leaves. Deficiency results in abortion in females and sterility in males.

vitamin F. A complex of essential FATTY ACIDS (including linoleic, linolenic and arachidonic acids) needed by various animals and possibly by humans. Deficiency in humans may lead to certain kinds of eczema.

vitamin H. *See* BIOTIN.

vitamin K A substance required by mammals and birds for the clotting of blood. Some of the human requirements may be provided by BACTERIA living in the alimentary canal. It is possible that a further substance, vitamin P (for permeability) affects the permeability of capillary blood vessels, so assisting in the control of bleeding.

vitamin P. *See* VITAMIN K.

Vita-Soy. A high-protein beverage marketed in Hong Kong. *See also* NOVEL PROTEIN FOODS.

vitelline membrane. *See* EGG MEMBRANE.

Vitis (vines). A genus of DICOTYLEDONEAE (family: Vitidaceae) that are climbing plants in which the tendrils coil round supports or force their way into crevices and cement themselves there. *V. vinifera* is the common cultivated species, yielding grapes for winemaking and the production of dried fruits (currants, raisins, sultanas). *V. aestivalis* (summer grape) and *V. labrusca* (fox grape) have been introduced into Europe from North America, as they are more resistant to the plant louse *Phylloxera*, a troublesome pest in vineyards.

vitreous. Glassy; used in descriptions of LUSTRE.

viviparous. (1) Applied to seeds (e.g., MANGROVE) that germinate within the fruit or to plants that reproduce vegetatively (*see* VEGETATIVE PROPAGATION) and do not form seeds although they may bear flowers (e.g., bulbous rush, *Juncus bulbosus*). (2) Applied to animals in which the EMBRYOS develop within the maternal body, deriving nutrient from it (usually by means of a PLACENTA) and not being separated from it by EGG MEMBRANES. *Compare* OVIPAROUS, OVOVIVIPAROUS.

volatile. Having a low boiling or subliming (*see* SUBLIMATION) pressure (i.e. a high vapour pressure). Example of volatile liquids include ether, camphor, CHLOROFORM and BENZENE.

volatiles. Elements and compounds dissolved in MAGMA that would, at atmospheric pressure, have been gases at the temperature of the magma.

volcanic dust. The finest PYROCLASTIC ash ejected by an explosive eruption. Volcanic dust sometimes travels great distances in the upper atmosphere, causing spectacular sunsets. *See also* MOUNT AGUNG.

volcanic rock. (1) Rock formed by volcanic action at the Earth's surface (i.e. includes LAVA and, usually, PYROCLASTIC rocks). This is the most frequent usage. Volcanic is contrasted with HYPABYSSAL and PLUTONIC. (2) Rock formed by volcanic action plus rock formed by the associated intrusive activity. With this meaning, volcanic rocks include some, if not all, rocks classified as hypabyssal by most authors. (3) Loosely, fine-grained and/or glassy.

volcano. A vent or fissure through which MAGMA, gases and solids are ejected from the Earth's CRUST. The shape of the mountain built up by central vent eruptions seems to depend on the viscosity, gas content and rate of extrusion of the magma. Volcanoes often erupt in different manners within a short space of time, but the types of eruptions have been named after volcanoes characteristically erupting in that particular way: (a) Hawaiian: the production of very mobile LAVA with very little explosive activity, apart from lava fountains, which produces very broad, low-angle cones (called shield volcanoes). (b) Strombolian: frequent small eruptions interspersed with the evolution of more ACIDIC magma which produces a steeper-sided composite cone of interbedded lava and TEPHRA. (c) Vulcanian: more infrequent eruptions than Strombolian, coupled with more viscous magma. (d) Vesuvian: very explosive eruptions that occur whenever the plug of viscous lava in the vent is blown out (particularly violent eruptions are called plinian, after Pliny the Elder, who lost his life in one). (e) Peléan: extrusion of RHYOLITE spines and the liberation of NUEÉS ARDENTES.

Volgian. A STAGE of the JURASSIC System.

volt (V). The derived SI unit of electric potential, being the difference of potential between two points on a wire conducting a constant current of 1 ampere when the power dissipated between the points is 1 watt; named after Alessandro Volta (1745–1827).

voluntary muscle. *See* MUSCLE.

Volz photometer. An instrument used in air pollution monitoring to make low-precision measurements of the intensity of direct sunlight at wavelengths defined by narrow-band interference filters.Readings can be taken by comparatively unskilled observers, and the instrument does not require completely cloudless skies. The Voltz photometer is thus suitable for the widespread collection of data. *See also* LIDAR.

vortex. A flow of a fluid in which the STREAMLINES form concentric circles. There are two kinds of vortex, forced and free, depending on whether torque is applied externally. For example, a forced vortex can be made in a liquid contained in a vessel and stirred with a paddle; a free vortex occurs when the liquid is allowed to leave the cylinder through a small hole in the bottom, in which case the force is provided by gravity acting on the fluid. Vortices occur when a solid body moves through a fluid unless the body is designed (e.g., in an aerofoil) to avoid creating them. *See also* BATH PLUG VORTEX, WING-TIP VORTICES.

vugh. A cavity in a rock, usually with a lining of well-formed crystals.

vulcanian volcano. *See* VOLCANO.

vulnerable. *See* RARITY.

vultures. General name for about 20 species of birds of prey that are distributed widely in tropical and subtropical regions, especially the latter, and that are divided into two distinct groups: Old and New World. The Old World species form a subfamily (Aegypiinae) of the hawk family (Accipitridae). The New World species constitute the family Cathartidae and include the largest living land bird capable of flight – the Andean condor (*Vultur gryphus*). Vultures depend on THERMAL SOARING (mostly in ANABATIC WINDS) whereby they remain air-borne in search of carrion. They have a wing form suited to low airspeed and narrow, circling flight, with low span:chord ratio and separated primaries at the wing tips.

W

W. (1) *See* WATT. (2) *See* TUNGSTEN.

wad. An amorphous mixture of several hydrous manganese oxide minerals commonly occurring in the oxidized zone of ore deposits, and also in BOG and shallow marine deposits.

waders. *See* CHARADRIIDAE.

wadi. *See* ARROYO.

wake. A disturbance of the air behind a moving object, or a similar flow behind a fixed object in an airstream. Wakes in the atmosphere are usually characterized by strong TURBULENCE and low mean velocity relative to the object and may exhibit DOWN-WASH.

walking worms. *See* ONYCHOPHORA.

Wallace, Alfred Russell (1823–1913). An English naturalist who, jointly with DARWIN, CHARLES ROBERT, published the first work advancing the theory of evolution by natural selection and who developed the idea of dividing the world into ZOOGEOGRAPHICAL REGIONS. *See also* WALLACE'S LINE.

Wallace's line. A line drawn by WALLACE, ALFRED RUSSELL, to mark the separation between the distinct faunas of the Oriental and Australian ZOOGEOGRAPHICAL REGIONS. In the Australian region the characteristic mammals are marsupials (*see* MARSUPIALIA) and monotremes (*see* MONOTREMATA), in the Oriental region all the mammals are placentals (*see* EUTHERIA). The line follows a deep-water channel running south-east of the Philippine Islands, between Borneo and Celebes, and between Bali and Lombok.

wandering albatross. *See* DIOMEDEIDAE.

warfarin (200 coumarin). A poison used to kill rodents, especially rats and mice. It acts by reducing the ability of the blood to clot, so that internal bleeding occurs. Strains of rodents resistant to warfarin are now widely distributed.

warm-bloodedness. *See* HOMOIOTHERMY.

warm front. The boundary between two AIR MASSES of different temperatures, moving so that the warmer air is advancing into, and rising over, the colder air. The passage of a warm front over a fixed point is preceded by falling pressure and steady rain, followed by a sudden increase in temperature, veering wind (*see* BACKING WIND) and clearing skies.

warm rain. Rain falling from water clouds as opposed to ice clouds.

warm ridge. A ridge of high pressure (e.g., ahead of the advancing front of a DEPRESSION) in which clear skies give warm, sunny weather.

warm-season plant. A plant that grows mainly during the warmer part of the year, usually late spring and summer.

swarm sector. The region between the COLD FRONT and WARM FRONT associated with a DEPRESSION.

Washington Convention on International Trade in Endangered Species of Wild Flora and Fauna. *See* CITES

washout. The removal of AEROSOLS in a layer of the atmosphere by impact with rain-drops falling from above.

wasps. *See* ACULEATA.

waste. Any substance, solid, liquid or gaseous, that remains as a residue or is an incidental by-product of the processing of a substance and for which no use can be found by the organism or system that produces it. Where waste is produced as a consequence of human activities a method of disposing of it safely must be devised. The term includes sewage and household refuse as well as INDUSTRIAL WASTE. *See also* HAZARDOUS WASTE, RADIOACTIVE WASTE.

water balance. The amount of ingoing and outgoing water in a system, which are assumed to be equal in the long term so that the WATER BUDGET will balance.

water bears. *See* TARDIGRADA.

water budget. A measure of the incoming and outgoing water from a region, including rainfall, evaporation, run-off, seepage and perhaps with special attention to the ABLATION of snow, EVAPOTRANSPIRATION from vegetation, DEW or other aspects relevant to particular interests (e.g., agriculture).

water cloud. A cloud in which the particles are water droplets in the liquid phase.

water fleas. *See* BRANCHIOPODA.

waterfowl. *See* ANATIDAE.

water hyacinth (*Eichhornia*). A genus of tropical freshwater plants (family: Pontederiaceae). *E. crasspipes* has become a noxious weed in the USA, Australia and Africa. Highly prolific, and reproducing mainly vegetatively, it can double its numbers in 8–10 days in water at a temperature of 10°C or more, provided nutrients are present. It was introduced to North America in 1884 by visitors to the New Orleans Cotton Exposition, who brought specimens from Venezuela. Water hyacinth has been used experimentally to provide fertilizer and foodstuffs for livestock, and the hippopotamus and SIRENIA (dugongs and manatees) have been suggested as animals that might be farmed in order to clear choked waterways, the animals also providing food for humans.

water injection (WI). A technique for reducing the formation, and so the emission, of pollutants from INTERNAL COMBUSTION ENGINES. Research has shown that WI is the most efficient way of reducing pollutant emissions, but it requires vehicles to carry water tanks as large as fuel tanks, the water may freeze and conventional antifreeze compounds produce exhaust pollutants, and there may be long-term corrosion in the engine.

water meadow. *See* MEADOW.

watershed. (1) In British usage, the divide separating one CATCHMENT from another. (2) In US usage, a collecting area into which water drains (i.e. what in the UK is called a catchment).

water slater. *See* MALACOSTRACA.

water spout. A cyclonic storm similar to a tornado (*see* BATH PLUG VORTEX) that occurs over water and forms a dense funnel-shaped cloud by entraining water droplets from the surface.

water table. The upper surface of GROUND WATER whose level varies according to the quantity of water contained in the ground water and the amount lost (e.g., through abstraction). The water table may reach the surface (e.g., at times of flooding). *See also* AQUIFER, METEORIC WATER, VADOSE WATER.

water vapour. Water in the gaseous phase that enters the atmosphere by evaporation from the sea and lakes, and from damp vegetation and by transpiration. It is responsible for most of the absorption of radiation that occurs in the atmosphere. It is removed by condensation (DEW or CLOUD formation) leading often to PRECIPITATION and is thus a vital element in the determination of weather quality. *See also* HYDROLOGICAL CYCLE.

watt (W). The derived SI unit of power,

equal to 1 joule per second, which is the energy expended per second by an unvarying electric current of 1 ampere flowing through a conductor the ends of which are maintained at a potential difference of 1 volt; named after James Watt (1736–1819).

wattles. *See* ACACIA.

waveband. A segment of the spectrum of wave frequencies (*see* FREQUENCY) (e.g., the long, medium or short wavebands of radio waves). Sound wavebands are measured in OCTAVES.

wave cloud. A cloud situated in the crest of a MOUNTAIN WAVE (lee wave) and as a consequence almost stationary, with condensation of cloud at the upwind edge and evaporation in the descending air at the downwind edge. Wave clouds usually have a characteristically smooth outline (*see* WHALEBACK CLOUD) often showing iridescence.

wave cyclone. A DEPRESSION that forms at a wave or kink in a FRONT. *See also* FRONTAL WAVE.

wave front. A theoretical surface composed of points at which the PHASE of a wave is the same.

wave hole. A hole in a layer of cloud caused by the descent of the air and evaporation of cloud in the trough of a MOUNTAIN WAVE (lee wave).

wavelength. The distance between the crests of a SINE WAVE or, more correctly, the perpendicular distance between two WAVE FRONTS in which the phases differ by one period. The wavelength of a sound wave is equal to the speed of sound divided by the frequency.

wave soaring. The technique used by glider pilots of soaring in the upslope side of a MOUNTAIN WAVE (lee wave).

Wb. *See* WEBER.

wearing course. *See* PAVEMENT.

weathering. The disintegration of rocks or unconsolidated mineral particles near the ground surface by wind, water, ice or chemical action, and including biological activity, but with little or no transport of the materials released except by gravity. *Compare* EROSION.

weathering series. A sequence of common primary SILICATE MINERALS in order of their resistance to WEATHERING; OLIVINE ➟ PYROXENE ➟ HORNBLENDE ➟ BIOTITE ➟ muscovite ➟ FELDSPARS (calcium ➟ sodium ➟ potassium) ➟ QUARTZ (most resistant).

weather map. A chart on which meteorological variables are plotted over an extensive geographical area for a particular time. The most common form is the surface pressure chart on which ISOBARS are shown.

weather modification. The artificial stimulation of rain, and the prevention of hail and tornadoes, which has been attempted by seeking to interfere with the mechanisms that cause freezing and/or agglomeration of cloud particles into FALLOUT. Success has been restricted to occasional minor modifications of clouds and is unlikely on a significant or economic scale because of the efficacy of natural mechanisms. In view of the insuperable task of mounting a definitive statistical test of a weather modification technique, claims of success are likely to continue.

weather radar. A radar (*see* ECHO) carried by ships or aircraft that is designed to detect unfavourable weather on the route by means of the reflection of radar waves by precipitation.

weather ship. A ship equipped with meteorological instruments, providing routine observations at a fixed station at sea. It may also be used for research purposes.

weber (Wb). The derived SI unit of magnetic FLUX, being the flux that, linking a circuit of one turn, produces in it an electromotive force (emf) of 1 volt as it reduces to zero at a uniform rate in 1 second; named after Wilhelm Weber (1804–91).

Weichselian. *See* ICE AGE.

weight of the atmosphere. The cause of atmospheric pressure. It is about 10.3 tonnes per square metre, or 5.3 x 10^{15} tonnes for the whole atmosphere.

weismannism. The theory advanced by the German biologist August Weismann (1834–1914) of the continuity of the germ plasm and the non-inheritance of acquired characters (*see* LAMARCK, JEAN BAPTISTE DE). According to this theory the reproductive cells are not influenced by the body cells because they are set aside early in development. Weismann emphasized that the body (soma) is a product of the germ cells, and not vice versa, and that effective changes could take place only through germ cells.

Wellman–Lord process. A US process for removing SULPHUR DIOXIDE from FLUE GASES, using ELECTROSTATIC PRECIPITATORS to remove FLY ASH and a WET SCRUBBER. The process produces concentrated sulphur dioxide gas which is used in the manufacture of SULPHURIC ACID or from which SULPHUR is extracted. The process is 90 percent efficient and is being fitted at some UK coal-fired power stations.

Welwitschia. See GNETALES.

Wenlockian. The second oldest SERIES of the SILURIAN System in Europe.

West African Floral Region. The part of the PALAEOTROPIC REALM that comprises the southern coastal region of West Africa and Africa east to Lake Tanganyika and south to central Angola.

westerlies. Belts of wind that occur in latitudes between 40° and 60° in which southwest winds predominate in the northern hemisphere and north-westerlies in the southern. *See also* GENERAL CIRCULATION.

westerly type. A type of weather prevalent in temperate latitudes in which the winds at all latitudes in the TROPOSPHERE are from a westerly point. It gives variable weather with passing CYCLONES and moving ANTI-CYCLONES.

Westphalian. The fourth oldest STAGE of the CARBONIFEROUS System in Europe.

wet adiabatic. A process (or on the TEPHIGRAM a line representing a process) in which air saturated with water is expanded or compressed without external heat addition. In the atmosphere this is usually caused by changes in altitude.

wet adiabatic instability. A condition in the atmosphere in which vertical movements of moist air tend to increase. Such a situation will occur if the actual LAPSE RATE of rising (or descending) air is greater than the ADIABATIC LAPSE RATE. The behaviour of moist air is more complex than that of dry air since the cooling of the air may cause the condensation and loss of water, with a consequent warming. Thus the critical lapse rate for descending moist air is the WET ADIABATIC LAPSE RATE only if the air retains sufficient water to remain saturated.

wet adiabatic lapse rate (saturated adiabatic lapse rate). The drop in temperature of a moving parcel of saturated air per unit increase in height in adiabatic ascent (i.e. there is no exchange of heat between the parcel and the surrounding air). The wet adiabatic lapse rate is much lower than the DRY ADIABATIC LAPSE RATE, being about one-third of its value at 300K, about two-thirds at 273K and about 95 percent at 240K.

wet bulb. A thermometer bulb maintained wet with distilled water, usually by means of a muslin wick. The temperature indicated by this bulb (*see* WET-BULB TEMPERATURE) can be used together with the dry bulb to give a measure of humidity. *See also* ABSOLUTE HUMIDITY.

wet-bulb temperature. The lowest temperature to which air can be cooled by evaporating water into it, the air supplying the heat for evaporation.

wetland. An area covered permanently, occasionally or periodically by fresh or salt water up to a depth of 6 metres (e.g., flooded pasture land, marshland, inland lakes, rivers and their estuaries, intertidal mud flats). *See also* LAND DRAINAGE.

Wetlands of International Importance. WETLAND areas that have been designated by the IUCN and UNESCO as of global importance for conservation and study. In 1985 there were 329 such areas in the world as a whole (13 in Africa, 17 in North America, 3 in South America, 32 in Asia, 224 in Europe, 12 in the USSR and 28 in Oceania) with a total area of 193 million hectares.

wet scrubber. An ABSORPTION TOWER that is used to remove polluted gases from a waste gas stream by contact with a liquid (e.g., hydrogen chloride being absorbed in water).

whaleback cloud. A lenticular WAVE CLOUD so named by seamen in high latitudes. The cloud is situated in strong winds over islands and steep coasts, and the name is derived from the smooth shape of the top.

whales. *See* BALAENOIDEA, CETACEA, ODONTOCETI.

wheat. *See* GRAMINEAE.

wheel animalcules. *See* ROTIFERA.

whelks. *See* GASTROPODA.

whin. *See* LEGUMINOSAE.

whirling psychrometer. A psychrometer in which wet and dry bulb thermometers (*see* WET BULB) are mounted on a pivot about which they can be swung by hand to provide an air flow past the wet bulb so that the humidity of the air next to the bulb is always close to the ambient value.

whirlwind. A small, near-vertical VORTEX that forms in conditions of exceptionally strong convection (e.g., in deserts, where it may form a DUST DEVIL, or near large fires).

white bicycles. Bicycles, painted white, that were used experimentally to provide free transport in several cities, including Amsterdam and Oxford, and that were provided officially in Stockholm at the 1972 United Nations Conference on the Human Environment. *See also* WITKAR.

white blood cells. *See* BLOOD CORPUSCLE.

white bryony. *See* CUCURBITACEAE.

white fish. Marine fish with white flesh, especially flatfish and species belonging to the cod family (*see* GADIDAE).

white horizontal arc. A rare arc seen in sunshine on clouds of ice crystals which have both horizontal and vertical faces that act as mirrors. The arc passes through the observer's shadow.

white matter. The part of the CENTRAL NERVOUS SYSTEM of vertebrates that consists mainly of nerve fibres. Generally it lies outside the GREY MATTER, except in the CEREBRAL HEMISPHERES and CEREBELLUM of higher vertebrates.

white noise. Random noise that has equal energy at every FREQUENCY in a particular WAVEBAND.

WHO. *See* WORLD HEALTH ORGANIZATION.

WI. *See* WATER INJECTION.

wilderness. An area of land that has never been permanently occupied by humans, ploughed, mined or developed in any way, and that therefore exists in a natural state. The concept is based on a long history of non-interference by humans and not at all on the particular ECOSYSTEMS the area supports. In the USA, wilderness areas are formally designated. Within them no roads are built, no development permitted, and their physical resources may be exploited only by presidential authorization under very special circumstances. In other countries, wilderness areas may be set aside within NATIONAL PARKS, but wilderness tends to used interchangeably with national park.

Wilderness Society. A US voluntary organization concerned mainly with urging the official designation of suitable areas as WILDERNESS and the conservation of wilderness areas, but that also campaigns actively on related conservation and environmental issues.

Wildfowl Trust. A UK society, founded in 1946, that promotes the study and conservation of wildfowl (ducks, geese and swans). It maintains several collections in England and breeds wildfowl from all over the world, including the Hawaiian Goose (nene), which was saved from extinction and returned to the wild. The Wildfowl Trust has also established extensive WETLAND reserves in the UK. *See also* DECOY DUCK POND.

Wildlife and Countryside Act 1981. A UK Act of Parliament dealing with the protection of wildlife, nature conservation generally, including access to the countryside and national parks (as these are defined in the UK) and with rights of way. It includes lists of the species protected under its provisions and the degree of protection afforded.

wild rice (Indian rice, Tuscacora rice, *Zizania aquatica*). An annual aquatic grass (*see* GRAMINEAE) native to eastern North America, which grows to a height of about 4 metres, with a PANICLE up to 60 centimetres long bearing pendulous male SPIKELETS on its spreading lower branches and APPRESSED female spikelets on the erect upper branches. The seed is used as food for waterfowl and for humans, and is reputed to have a higher protein and vitamin content than true rice.

willows. *See* SALICACEAE.

willy nilly. *See* TROPICAL CYCLONE.

wind classification. A system designed to emphasize the forces mainly responsible for the characteristics of wind. GEOSTROPHIC WINDS have a balance between pressure gradient and the CORIOLIS force and blow along the ISOBARS. GRADIENT WINDS are a modification in which the curvature of the isobars and of the flow is important, as in a CYCLONE. Isallobaric winds are caused by rapidly changing pressure patterns. KATABATIC winds (cold downslope) and ANABATIC WINDS (warm upslope) are shallow and local, as are ANTITRIPTIC WINDS (friction dominated) which are exemplified by katabatic winds and cold outflows from storms. Land and sea BREEZES are antitriptic, but the Coriolis force becomes important with time. The THERMAL WIND is a geostrophic component. The ageostrophic component is due to acceleration or friction. *Compare* BERG WINDS, MOUNTAIN WAVES.

wind drag on the sea. The basic cause of most ocean currents, reacting with temperature gradients and the Earth's rotation. The drag of the sea also significantly reduces the wind speed, especially in the trade-winds where air moving towards the equator is accelerated towards the west. The influence of the sea is important because of its large area and the greater persistence of wind direction than over land.

wind erosion. *See* EROSION.

wind farm. An installation for the generation of electricity that is based on wind turbines and has a rated output comparable to that of a conventional power plant (i.e. in the gigawatt range). A wind farm consists of 1000 or more turbines and feeds its output into the public supply, usually through a grid system. *See also* AEROGENERATOR.

wind measurement. The measurement of wind may be achieved by the direct tracking of BALLOONS, optically or by radar (*see* ECHO). Doppler (*see* DOPPLER SHIFT) radar may be used to determine the velocity of rain or other air-borne objects, their horizontal motion being attributed to wind. The displacement of clouds as seen by satellite photography may be used, provided clouds can be satisfactorily identified as moving with the wind.

windmill. A machine with a rotor that is moved slowly by the wind to produce mechanical power, used originally to mill grain and pump water. Windmills may take various forms including the traditional, large horizontal-axis mill with two, three or four blades, and vertical-axis types. *See also* AEROGENERATOR, DARRIEUS GENERATOR, PANEMONE, SAVONIUS ROTOR, WIND POWER.

wind power. Mechanical or electrical power generated by a WINDMILL using the kinetic energy of the wind as the prime energy source.

wind profile. The variation of wind characteristic (e.g., mean speed, direction, turbulence level) with altitude.

wind rose. A diagram summarizing the frequencies of winds of different strengths and directions as measured at a specified point over an extended period of time. The most common form consists of a circle whose radius is proportional to the frequency of calms, from which a number of bar symbols radiate, one for each wide direction band, with lengths proportional to the frequency of occurrence. The bars are often subdivided to represent the contributions from different wind-strength bands.

wind shadow thermals. The strength of the wind and the terrain determine the intensity of TURBULENCE. Where the wind is reduced (e.g., on the lee side of a hill) the depth of air warmed by the ground heated in sunshine is less, and the maximum temperature reached greater. Intense thermal upcurrents rise from such places from time to time. A special case is a field of ripe wheat: the temperature within the crop is higher than in a green crop, and a gust of wind bending down the stalks releases a body of very warm air.

wind shear. In general, the rate of change of the wind vector with distance in a direction perpendicular to the wind direction. Usually, the term implies the change of wind with height, and near the ground it is often used to designate the vertical gradient of horizontal velocity.

wind variation. The wind varies in time and place. In the wake of a building, tree or cliff it may fluctuate by 50 percent or more in a very few seconds and vary in direction by 180°. With the passage of storms and with time of day and season it varies over minutes, hours, days and months. Even one year may differ markedly from another at the same place. These variations have to be taken into account in defining 'the wind', its average or mean, and in determining the exposure of the measuring instrument.

winged bean (*Psophocarpus tetragonolobus*). A tropical LEGUME, cultivated in Papua New Guinea and parts of south-east Asia, whose leaves, tuberous (*see* TUBER) roots, seeds and pods are all edible by humans or farm livestock. The name refers to four wing-like structures running the length of the pod. The beans contain 34 percent PROTEIN, rich in LYSINE, and 18 percent edible oil; the roots contain 20 percent protein. The plant appears to suffer no serious diseases and has no important pests and has been identified as having considerable potential as a food and fodder crop in the tropics.

wing-tip vortices. A pair of counter-rotating vortices (*see* VORTEX) that are formed in the wakes of aircraft and lie along the direction of motion. In general, they result from the curling up of the vortex sheet generated at the trailing edge of a lifting aerofoil which varies in section along its span, but for constant-section wings they form at the wing tip. The direction of rotation is such as to induce a downward deflection of the air between the two vortices, which is sometimes referred to as downwash.

winkles. *See* GASTROPODA.

winze. In mining, a nearly vertical opening or tunnel sunk downwards from a DRIVE or LEVEL.

witches' broom. *See* GALL.

Witkar. A car, powered by electric batteries, designed to carry two persons and some luggage, that was introduced experimentally in Amsterdam as an alternative to conventional private cars in the city centre. The Witkars worked among a number of kerbside stations, where their batteries were recharged during resting periods. The cars were used by subscribers, each of whom paid an annual subscription plus a charge for the distance travelled in the cars. Subscribers received a magnetic key that unlocked the cars and started their motors. Before starting, the user dialled ahead to reserve a parking place at the desired destination. The scheme was invented by a Dutch engineer, Luud Schimmelpenninck. The cars were withdrawn at the end of the initial period because they were found to be too expensive. *See also* WHITE BICYCLES.

wolframite ((Fe,Mn) WO_4). A mineral that is a tungstate of iron and manganese in varying proportions and is a major ore mineral of TUNGSTEN. It occurs with CASSITERITE and SCHEELITE in contact metamorphic (*see* CONTACT METAMORPHISM) zones and in HYDROTHERMAL deposits adjacent to ACIDIC IGNEOUS rocks, and in PLACER deposits. Tungsten is used mainly in the manufacture of tungsten carbide (WC) used for cutting edges, dies and armour-piercing shells, and in alloy steels.

wood. *See* XYLEM.

wood alcohol. *See* METHANOL.

woodcocks. *See* CHARADRIIDAE.

woodland. Vegetation dominated by trees which form a distinct but sometimes open canopy.

woodland hawthorn. *See* CRATAEGUS.

woodlice. *See* MALACOSTRACA.

woodpeckers. *See* PICIDAE.

wood-wasps. *See* HYMENOPTERA.

woolsorters' disease. *See* ANTHRAX.

work hardening. *See* DUCTILITY.

World Bank. *See* INTERNATIONAL BANK FOR RECONSTRUCTION AND DEVELOPMENT.

World Conservation Strategy. A document, published in 1980 by the INTERNATIONAL UNION FOR CONSERVATION OF NATURE AND NATURAL RESOURCES, that analyzed the issues of resource use, conservation and economic development, and recommended ways in which sustainable development might be based on the most efficient use of resources. It was presented to governments with a recommendation that each country should prepare its own national conservation strategy on similar lines.

World Health Organization (WHO). A United Nations agency, based in Geneva, that was founded in 1948 as a successor to the League of Nations Health Organization and the International Office of Public Health. It is charged with various duties connected with the control of disease epidemics, quarantine and the quality of medicinal drugs, but more widely it aims to promote international cooperation in improving the health of the world population, especially in THIRD WORLD countries.

World Heritage Sites. Areas designated by the IUCN and UNESCO as being of global importance for conservation and study. In 1985 there were 63 such sites in the world as a whole (20 in Africa, 19 in North America, 6 in South America, 6 in Asia, 7 in Europe, and 5 in Oceania). *See also* GORDON AND FRANKLIN.

World Resources Institute (WRI). An organization founded in 1982 to study the relationships between economic development strategies and the environment. It is funded privately and by the United Nations and national governments.

Worldwatch Institute (WWI). A US-based organization that studies global environmental issues, especially those related to economic development and publishes an annual *State of the World* report, covering a wide range of topics.

World Wildlife Fund (WWF). *See* INTER-
NATIONAL UNION FOR CONSERVATION OF NA-
TURE AND NATURAL RESOURCES.

wrench fault. *See* FAULT.

WRI. *See* WORLD RESOURCES INSTITUTE.

Wright, Frank Lloyd (1867–1959). A US
architect and planner, whose concept of
'organic architecture' was radically
different from the high-density planning of
Le Corbusier (*see* JEANNERET, CHARLES
EDOUARD). He used styles and materials that
complemented the surroundings of
buildings. His ideal city was called Usonia
(after Butler's *Erewhon*) in which
inhabitants were largely self-sufficient, each
having an acre or so of land to grow food,
and commuting to nearby factories to
augment their incomes with occasional paid
work. Wright aimed to eliminate
conventional city life and to develop the
quality of rural life.

Wuchereria bancroft. *See* FILIARIA.

WWF. World Wildlife Fund (*see* INTERNA-
TIONAL UNION FOR CONSERVATION OF NATURE
AND NATURAL RESOURCES).

WWI. *See* WORLDWATCH INSTITUTE.

X

xanthism (xanthochroism). A colour variation in which the normal colour of an animal is more or less replaced by yellow pigments.

Xanthophyceae. *See* XANTHOPHYTA.

xanthophylls. *See* CAROTENOIDS.

Xanthophyta (Xanthophyceae, Heterokontae). Yellow–green algae; a group of mainly freshwater and terrestrial ALGAE which contain CAROTENOID pigments as well as CHLOROPHYLL. Most are unicellular and non-motile, some are colonial or filamentous. A few are colourless and are SAPROPHYTES or ingest food particles like PROTOZOA.

X chromosome. *See* SEX CHROMOSOMES.

xenia. Characteristics (e.g., colour) produced in the ENDOSPERM of a seed by the influence of the POLLEN nucleus which fuses with the primary endosperm nucleus. *See also* EMBRYO SAC.

xenogamy. Cross-fertilization in plants.

xenolith. An inclusion of pre-existing rock in an IGNEOUS rock, literally 'stranger stone'.

xenoparasite. *See* PARASITISM.

xerad. *See* XEROPHYTE.

xerarch succession. *See* XEROSERE.

xeromorphy. The possession of features characteristic of a XEROPHYTE

xerophyte (xerad). A plant that lives in a dry habitat and is able to endure prolonged drought. Some xerophytes (e.g., cactus) store water in SUCCULENT tissue for use when none is available from the soil and have STOMATA which close during the day. Other features of xerophytes that tend to check TRANSPIRATION are reduction in leaf size, thickening of the CUTICLE, sunken or protected stomata, and the possession of hairs. In some (e.g., gorse, *Ulex*) both leaves and stems are modified to resist shrinkage, forming stiff spines containing abundant SCLERENCHYMA. Many xerophytes can endure long periods of wilting. Deciduous trees become xeromorphic when they shed their leaves.

xerosere (xerarch succession). The stages in a plant SUCCESSION that begins in a dry site and progresses towards moister conditions.

Xiphosura (king crabs, horseshoe crabs). Aquatic ARACHNIDA with a body enclosed in a carapace. *Limulus*, the sole living genus, has existed since TRIASSIC times and grows to 50 centimetres in length. The four present-day species are shore-living burrowing animals that feed on worms and molluscs.

X-ray analysis. The identification of crystalline solids and the solution of their crystalline structures by examining the way they diffract X-rays.

xylem (wood). VASCULAR tissue through which water containing dissolved mineral salts is transported in PTERIDOPHYTA and seed plants. Xylem also provides mechanical support. It consists of TRACHEIDS and/or vessels, fibres and PARENCHYMA, and forms the bulk of the stems and roots in mature woody plants.

xylophagous. Wood-eating.

Y

yardangs. Ridges formed by wind carrying sand at low level. Sand-blasting of existing grooves in rocks enlarges these into furrows, leaving the sharp ridges between them. *See also* CORRASION.

Y chromosome. *See* SEX CHROMOSOMES.

yeasts. Many species of unicellular FUNGI, most of which belong to the ASCOMYCETES and reproduce by BUDDING. The genus *Saccharomyces* is used in brewing and wine-making because in low oxygen concentrations it produces zymase, an ENZYME system that breaks down sugars to alcohol and carbon dioxide. *Saccharomyces* is also used in bread-making. Some yeasts are used as a source of protein (*see* NOVEL PROTEIN FOODS) and of vitamins of the B group (*see* VITAMIN B).

yellowcake. *See* NUCLEAR FUEL CYCLE.

yellow–green algae. *See* XANTHOPHYTA.

yew. *See* TAXACEAE.

yolk sac. An embryonic membrane that encloses and absorbs the yolk in reptiles, birds and many fish. In mammal EMBRYOS the yolk sac absorbs nutrient from secretions of the uterus before the PLACENTA has formed. In some MARSUPIALIA the yolk sac forms a poorly developed placenta.

Z

Z. *See* ATOMIC NUMBER.

zeatin. One of the CYTOKININ group of plant HORMONES.

Zechstein. The upper division of the PERMIAN System in north-western Europe. The Zechstein contains substantial deposits of EVAPORITES.

Zero Population Growth (ZPG). US voluntary organization, active in the 1960s and the early 1970s, whose aim was to draw attention to problems associated with a rapid increase in the size of human populations and to urge Americans to aim for a stable population in which birth rates and death rates were in balance.

zero sum game. In GAME THEORY, a competition in which one contestant wins while others lose. If the winner is awarded a score of +1 and the loser a score of –1 the sum of the scores is 0, hence the name.

ZETA (Zero Energy Thermonuclear Apparatus). A torus-shaped (i.e. 'doughnut-shaped') apparatus used at the UK Atomic Energy Research Establishment, Harwell, to study controlled thermonuclear reactions. *See also* FUSION REACTOR.

zinc (Zn). A hard, bluish–white metallic element that occurs as calamine, zincite and SPHALERITE (zinc blende). It is used in alloys, especially BRASS, and in the GALVANIZING of iron. It is an essential MICRONUTRIENT. A_r = 65.37; Z = 30; SG 7.14; mp 419°C; bp 907°C.

zinc blende. *See* SPHALERITE.

zineb. A fungicide of the dithiocarbamate (*see* CARBAMATES) group used to control diseases such as potato blight and downy mildews. It can be irritating to the eyes and skin.

zircon (zirconium silicate, $ZrSiO_4$). A mineral that is a common ACCESSORY MINERAL in IGNEOUS rock, particularly sodium-rich PLUTONIC varieties. Zirconium has industrial uses in photoflash bulbs, in ALLOYS, especially for nuclear engineering (zircon alloys are used in the cladding of FUEL ELEMENTS), and the mineral zircon has been used extensively for RADIOMETRIC DATING using uranium–lead. The lead atoms are much larger than the zirconium and would not be expected to substitute for zirconium in the growing crystal lattice in the MAGMA so all lead must be the product of the decay of uranium. Zircon is a very resistant mineral and survives high-grade METAMORPHISM and so can be used to date the original SEDIMENTARY or IGNEOUS rock. Zircon is a common accessory in sediments and is used in correlating sediments by their heavy mineral content. Zircon is worked in PLACER deposits.

Zn. *See* ZINC.

zonal flow. West-to-east airflow.

zonal index. A measure of the strength of the atmospheric circulation in a specified large area of the globe (e.g., the temperate zone), often expressed in the form of a pressure difference.

zone fossil. *See* INDEX FOSSIL.

zone of silence. When anomalous audibility occurs at large distances from a sound source, a zone of silence with relatively low audibility is frequently observed closer to the source, but beyond the distance reached by the propagating SOUND WAVES.

zone time. A local time system in which 24 time zones, each covering 15° of longitude, are designated by letters of the alphabet. Greenwich Mean Time is designated Z time.

zoning. System of land use planning based on boundaries inside which areas can be used only for specified purposes (e.g., agriculture, dwellings, recreations, industry, etc.). Zoning has been used in the USA and Danish planning is based on zoning.

zoobiotic. Applied to an organism that lives parasitically (*see* PARASITISM) on an animal.

zoochore. A plant whose reproductive structures are dispersed by animals (e.g., burdock, *Arctium*; blackberry, *Rubus fruticosus*).

zoocoenose. An animal COMMUNITY.

zoogeographical regions (faunal regions, zoogeographic realms). Regions of the world with distinct natural faunas. (a) Arctogea comprises the Palaearctic Region (Europe, North Africa, Asia as far south as the Himalayas), the Nearctic Region (Greenland, North America as far south as central Mexico), the Ethiopian Region (Africa south of the Sahara) and the Oriental Region (India, Indochina, Malaysia, the Philippines and Indonesian islands west of WALLACE'S LINE). The Palaearctic and Nearctic are sometimes combined as the Holarctic. (b) Neogea (Neotropical Region) comprises South America, Central America, the West Indies and southern Mexico. Llamas, armadillos, arboreal sloths and opossums are native to this region. (c) Notogea (Australian or Australasian Region) comprises Australia, New Zealand, most of the Pacific islands and the Indonesian islands east of Wallace's line. This region contains all the monotremes (*see* MONOTREMATA), most of the marsupials (*see* MARSUPIALIA), and very few native placental mammals (*see* EUTHERIA).

zoography. That branch of zoology concerned with the description of animals.

zooid. An individual member of a colony of animals (e.g., a POLYP in a CORAL colony).

zoonosis. A disease of animals that is transmitted to humans (e.g., brucellosis).

zoophyte. An animal (e.g., BRYOZOA, many CNIDARIA) which has a plant-like form.

zooplankton. *See* PLANKTON.

zoosis. Any disease caused by a parasitic animal (*see* PARASITISM).

zooxanthellae. Parasitic (*see* PARASITISM) or symbiotic (*see* SYMBIOSIS) single-celled organisms (*see* DINOFLAGELLATES) bearing two flagella (*see* FLAGELLUM) that live in association with many marine invertebrate animals. They bear characteristics of both plants (they manufacture nutrients by PHOTOSYNTHESIS) and animals (they are mobile), and some authorities class them as ALGAE (plants), others as PROTOZOA (animals). *See also* CORAL REEF, REEF ECOSYSTEM.

Zosteraceae. An extratropical family of MONOCOTYLEDONEAE, comprising marine herbs that grow submerged along coasts and are water-pollinated. *Zostera* (eel-grass) grows on gently sloping shores (e.g., The Wash in Norfolk, UK) and is a favourite food of the Brent goose and wigeon. The plant is used in some parts of the world as stuffing and packing material and as thermal insulation in hollow cavity walls.

ZPG. *See* ZERO POPULATION GROWTH.

Z time. *See* ZONE TIME.

zygomorphic. *See* BILATERALLY SYMMETRICAL.

zygote. A cell formed by the fusion of two GAMETES.

zymase. *See* YEASTS.

zymogenous. Applied to soil organisms that show a marked increase in metabolic activity, including their rate of reproduction, after the addition of organic material to the soil. *Compare* AUTOCHTHONOUS.

zymoprotein. *See* ENZYME.